Earth Observations for Geohazards

Volume 1

Special Issue Editors

Zhenhong Li
Roberto Tomás

MDPI

Special Issue Editors
Zhenhong Li
Newcastle University
UK

Roberto Tomás
University of Alicante
Spain

Editorial Office
MDPI AG
St. Alban-Anlage 66
Basel, Switzerland

This edition is a reprint of the Special Issue published online in the open access journal *Remote Sensing* (ISSN 2072-4292) from 2015–2017 (available at: http://www.mdpi.com/journal/remotesensing/special_issues/earth_geohazards).

For citation purposes, cite each article independently as indicated on the article page online and as indicated below:

Author 1; Author 2; Author 3 etc. Article title. *Journal Name.* **Year**. Article number/page range.

ISBN 978-3-03842-396-6 (Pbk) Vol. 1-2   ISBN 978-3-03842-398-0 (Pbk) Vol. 1
ISBN 978-3-03842-397-3 (PDF) Vol. 1-2   ISBN 978-3-03842-399-7 (PDF) Vol. 1

Table of Contents

Chapter 2: Landslide Hazards

About the Guest Editors

Zhenhong Li is Professor of Imaging Geodesy in the School of Civil Engineering and Geosciences at Newcastle University. Professor Li has approximately twenty years of research experience in Space Geodesy and Remote Sensing (mainly InSAR and GNSS) and their application to geohazards (e.g., earthquakes, landslides and land subsidence). He specialises in the development of InSAR atmospheric corrections and time-series algorithms for precisely mapping surface movements, and has made several original contributions to the direct estimation and/or mitigation of the effects of atmospheric water vapour on InSAR measurements. His recent major achievements include the generation of the first interferogram from Chinese Gaofen-3 mission together with some collaborators, which is also the first interferogram from Chinese civilian radar missions. He has investigated a series of large earthquakes (e.g., *Sumatra (Indonesia, 2007), Wenchuan (China, 2008), Yushu (China, 2011), Van (Turkey, 2011), Tohoku (Japan, 2011), and Gorkha (Nepal, 2015)*) and active landslides (e.g., *Huangtupo, Shuping, and Daguangbao landslides, China*). He is a Fellow of the International Association of Geodesy (IAG), a Senior Member of the Institute of Electrical and Electronics Engineers (IEEE), an Associate Editor for *Advances in Space Research* and *Remote Sensing*, and a Guest Editor for three Special Issues of *Remote Sensing* and one Special Issue of *Sensors*.

Roberto Tomás is a Civil and Geological engineer. Currently, he is Associate Professor in Geotechnical Engineering at the Department of Civil Engineering of the University of Alicante. He is co-director of the Associate Research Unit IGME-UA of ground movements monitoring using radar interferometry; leader of the Geotechnical and Structural Engineering research group; member of the Geohazard InSAR Laboratory and Modeling Group of the Geological Survey from Spain; member of the Institute of the Water and the Environmental Sciences of the University of Alicante; member of the UNESCO Working Group on Land Subsidence and chair of the Spanish Working Group on land subsidence (SubTer). He has worked on many research projects and has published numerous contributions focused on geohazards (i.e., landslides and land subsidence) monitoring using remote sensing techniques (i.e., LiDAR, InSAR and photogrammetry), modelling, characterization of the effects induced by these phenomena on infrastructures and geotechnical characterization of soils/rocks. For more information, please visit https://personal.ua.es/en/roberto-tomas.

Preface to "Earth Observations for Geohazards"

Earth Observations (EO) encompasses different types of sensors (e.g., SAR, LiDAR, Optical and multispectral) and platforms (e.g., satellites, aircraft, and Unmanned Aerial Vehicles) and enables us to monitor and model geohazards over regions at different scales in which ground observations may not be possible due to physical and/or political constraints. EO can provide high spatial, temporal and spectral resolution, stereo-mapping and all-weather-imaging capabilities, but not by a single satellite at a time. Improved satellite and sensor technologies, increased frequency of satellite measurements, and easier access and interpretation of EO data have all contributed to the increased demand for satellite EO data. EO, combined with complementary terrestrial observations and with physical models, have been widely used to monitor geohazards, revolutionizing our understanding of how the Earth system works.

This book is the first volume of the 2-volume collection of scientific contributions focusing on innovative EO methods and applications for monitoring and modeling geohazards, consisting of two chapters: (1) earthquake hazards, and (2) landslide hazards. The second volume of this book series contains another two chapters: (1) land subsidence hazards, and (2) New EO techniques and services.

Finally, we would like to take this opportunity to thank all authors, editors, reviewers, and supporters for their hard work and dedication that made this Special Issue possible.

Zhenhong Li and Roberto Tomás
Guest Editors

remote sensing

MDPI

Editorial

Earth Observations for Geohazards: Present and Future Challenges

Roberto Tomás [1] and Zhenhong Li [2,*]

[1] Departamento de Ingeniería Civil, Escuela Politécnica Superior, Universidad de Alicante, P.O. Box 99, E-03080 Alicante, Spain; roberto.tomas@ua.es

[2] COMET, School of Civil Engineering and Geosciences, Newcastle University, Newcastle upon Tyne NE1 7RU, UK

* Correspondence: Zhenhong.Li@newcastle.ac.uk; Tel.: +44-191-208-5704

Academic Editor: Prasad S. Thenkabail

Received: 22 February 2017; Accepted: 22 February 2017; Published: 24 February 2017

Abstract: Earth Observations (EO) encompasses different types of sensors (e.g., Synthetic Aperture Radar, Laser Imaging Detection and Ranging, Optical and multispectral) and platforms (e.g., satellites, aircraft, and Unmanned Aerial Vehicles) and enables us to monitor and model geohazards over regions at different scales in which ground observations may not be possible due to physical and/or political constraints. EO can provide high spatial, temporal and spectral resolution, stereo-mapping and all-weather-imaging capabilities, but not by a single satellite at a time. Improved satellite and sensor technologies, increased frequency of satellite measurements, and easier access and interpretation of EO data have all contributed to the increased demand for satellite EO data. EO, combined with complementary terrestrial observations and with physical models, have been widely used to monitor geohazards, revolutionizing our understanding of how the Earth system works. This Special Issue presents a collection of scientific contributions focusing on innovative EO methods and applications for monitoring and modeling geohazards, consisting of four Sections: (1) earthquake hazards; (2) landslide hazards; (3) land subsidence hazards; and (4) new EO techniques and services.

Keywords: earth observation; EO; geohazards; earthquake; landslide; land subsidence; InSAR; LiDAR; optical; images; displacement; deformation; damage assessment; satellite; monitoring

1. Introduction

Geohazards are often defined as the events related to the geological state and processes that pose potential risks to people, properties and/or the environment, which can be classified within two main categories: natural hazards (such as earthquakes, landslides, volcanic eruptions, tsunamis, and floods) and human-induced hazards (such as land subsidence due to groundwater-extraction, water contamination, and atmosphere pollution). Geohazards could cause enormous human and economic losses and disruption, which continue to grow worldwide. In the past decades, the annual cost of natural hazards has increased dramatically [1]. Earthquakes represent one of the most devastating geohazards in terms of human suffering and economic damage, but the major cause of casualties, infrastructural damage, and economic losses, is the secondary hazard of landslides [2]. Volcanic eruptions also represent a significant proportion of geohazards [2], and major eruptions can modulate regional or global atmospheric composition and climate in detrimental ways. Land subsidence due to anthropogenic processes, such as extraction of groundwater, gas, oil, and coal, is another worldwide geohazard that affects wide areas, causing infrastructure damage and increasing flood risk [3,4]. Better decisions require better knowledge to characterize, monitor and model geohazards and then mitigate their impacts on people and the environment. During the past decades, Earth Observation (EO) has been widely applied to disaster risk management (including disaster preparation, response, recovery and mitigation), especially disaster response [5], since it provides extremely useful information for researchers, decision makers and plan makers.

EO is the gathering of information about the Earth using remote sensing technologies, which are often supported by ground surveying techniques. EO has considerably changed our ways of seeing the world, providing a framework to precisely map and monitor large-scale phenomena in a timely way. EO from space and aircraft, combined with complementary terrestrial observations and physical models, have been used to monitor geohazards. An important aspect of space-based (and airborne) EO is that we can investigate areas in which ground observations are not possible due to physical or political constraints. EO techniques can be classified according to sensor types, e.g., passive or active, optical, radar (radio detection and ranging), LiDAR (Laser Imaging Detection and Ranging), or multispectral/hyperspectral. They can be also classified according to the platforms in which the sensors are installed: satellite-based, aircraft-based, Unmanned Aerial Vehicle (UAV) based and ground-based.

This Special Issue contains a collection of articles focusing on the use of EO techniques for the investigation of geo-hazards. We received a total of 79 manuscripts for consideration of publication, which were carefully reviewed by external and independent experts in their respective fields. Forty-three of these manuscripts were accepted for publication. These studies utilize the state-of-the-art EO techniques to map, characterize, monitor and model a range of geohazards, including earthquakes, landslides, land subsidence, and tsunamis. The following sections include a study of present and future trends and challenges on the use of EO and an overview of the 43 contributions in this Special Issue.

2. Use of EO for Geohazards throughout Bibliometric Data

Although EO techniques were firstly used for military and security applications in the Cold War [6], they have become available for a wide range of applications and users during the past few decades. In this section, we illustrate the increasing use of EO techniques throughout a bibliometric study on the occurrence of the term "Earth Observation" in the Web of Science's bibliographical database similar to previous studies (e.g., [7]). Our searches are restricted to the "Remote sensing" and "Geology" segments. Figure 1 shows the increasing number of yearly publications about the above-mentioned terms. It is clear in Figure 1 that the number of EO publications has started to increase since the 1990s with an apparent exponential-like trend, indicating that EO literature has been growing considerably. Although EO techniques were born last century, it is still a topical issue probably due to its inherent high technological components and wide range of potential applications.

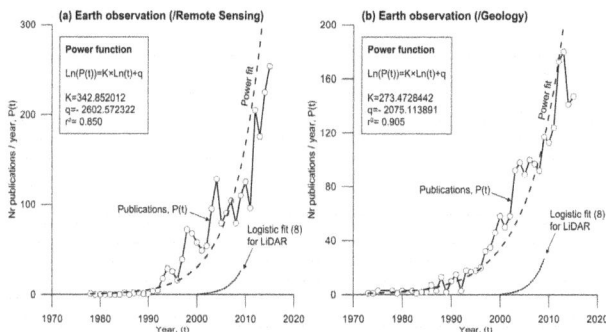

Figure 1. Evolution of the number of publications per year (dots) in the database Web of Science (WOS) containing the term "Earth observation" in the field of TOPICS between 1968 and 2015. Note that the searches are restricted to (**a**) "Remote Sensing"; and (**b**) Geology in the Web of Science. The dashed lines represent the best fit power function of the dots. The dotted lines correspond to the logistic fit performed by [8] for the terms "LiDAR" or "Laser scanning" in the GEOREF database.

3. Overview of Contributions

As mentioned earlier, forty-three papers were published in this Special Issue. These studies cover a range of geohazards including earthquakes, landslides, land subsidence, and tsunamis (Figure 2a). Figure 2b shows all the different types of EO sensors used in the published papers such as Synthetic Aperture Radar (SAR), LiDAR, optical, multispectral, GPS and Altimetry, and Figure 2c shows all the different platforms such as satellites, aircraft, Unmanned Aerial Vehicles (UAV) and ground-based platforms. It is clear in Figure 2 that satellite SAR has been widely employed to investigate geohazards due to its all-weather imaging capabilities. It should also be noted that these 43 studies cover different EO usages including geohazard mapping, monitoring and modelling. An overview of all the contributions are presented in the following sections.

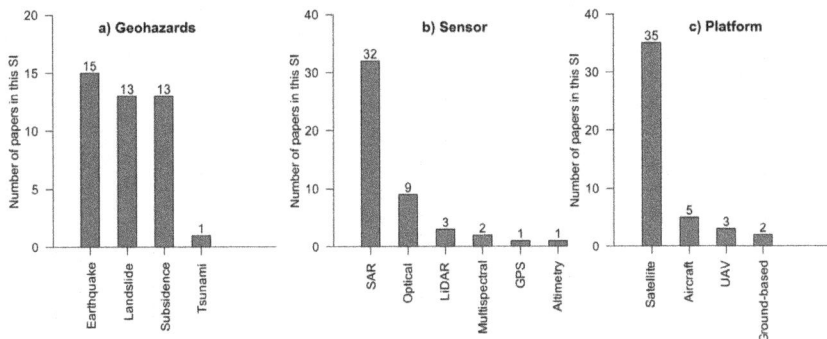

Figure 2. Number of papers published in this Special Issue (SI) classified according to: (**a**) the types of geohazards; (**b**) the types of sensors employed; and (**c**) the types of the sensor platforms. Note that (i) the total number in (**a**) is less than 43 since one article addresses an EO method that could be applied to various geohazards; and (ii) the total numbers in (**b**,**c**) are greater than 43 because multiple sensors and/or platforms were utilised in some papers.

3.1. Earthquake Hazards

Earthquakes represent an increasing risk of human loss and severe economic damage as vulnerable populations grow in areas of seismic hazard. Observations of the seismic cycle not only give insight into the mechanics of a fault, but also play key roles in estimating the likelihood of future earthquakes. Interseismic motions build up stress and lead to earthquakes. Zhu et al. [9] employ two tracks of ENVISAT ASAR images to determine the deformation rate maps of the Altyn Tagh Fault (ATF), and then calculate the regional strain rate field using a multi-scale wavelet method. Their results suggest a left-lateral slip rate of 8.0 ± 0.7 mm/year and a locking depth of 14.5 ± 3 km, which is in agreement with previous GPS and ERS InSAR results.

Postseismic transient deformation is a process that contributes to regional stress evolution, modifying the background tectonic plate motions. Liu et al. [10] present a method to model postseismic deformation time series with the combination model of afterslip and viscoelastic relaxation, and then simultaneously estimate the time-dependent afterslip distribution and the viscosity beneath the earthquake zone. It is reported that (i) the preferred time-dependent afterslip of the 2009 Mw 6.3 Dachaidan, China earthquake mainly occurs in the upper 9.1 km, and increases with time; and (ii) the preferred lower bound of the viscosity beneath the Qaidam Basin's northern side is 1×10^{19} Pa·s, close to that beneath its southern side but different from those in other parts of the Tibetan Plateau, indicating that the viscosity structure beneath the Tibetan Plateau may vary laterally.

The deformation occurring during an earthquake is referred to as coseismic. Observations of coseismic deformation are often used to determine earthquake source parameters, slip distributions, and even the rupture histories. Xu et al. [11] generate continent-wide two-dimensional (2D) (east–west and vertical) coseismic displacement maps for the 2015 Mw 8.3 Illapel, Chile earthquake using Sentinel-1A TOPS imagery, suggesting that the east–west component (up to 2 m) dominates the 2D surface displacement. Using similar Sentinel-1A data, Solaro et al. [12] also produce coseismic displacement maps for the 2015 Mw 8.3 Illapel, Chile earthquake. Their joint Okada inversion with multiple Sentinel-1A interferograms suggests that most of the slip occurs northwest of the epicenter with a maximum located in the shallowest 20 km; their Finite Element Model indicates that (i) its estimated maximum slip is comparable to the Okada model; and (ii) the von Mises stress distribution agrees with the depth distribution of the aftershock hypocenters. InSAR observations are utilised to investigate the 2003–2004 Bange, China earthquake sequence, involving a series of normal faulting events with Mw > 5.0, indicating that InSAR can provide reliable source parameters of shallow, moderate-sized earthquakes in areas that lack dense seismic networks [13]. Li et al. [14] use Sentinel-1A interferograms to model the 2016 Mw 5.9 Menyuan, China earthquake; they find that the 2016 event has a different focal mechanism from a previous Ms 6.5 earthquake although both are at the two ends of a secondary fault, which is believed to reflect the left-lateral strike-slip characteristics of the Lenglongling fault zone. Multi-platform InSAR observations are employed to determine the coseismic and postseismic slip distributions of the 2011 Mw 7.1 Van, Turkey earthquake, indicating that the upper 7–9 km of the fault, unruptured during the coseismic phase, underwent afterslip in the postseismic phase that may have reduced the seismic potential in its whole length from NW to SE [15].

3.2. Landslide Hazards

Landslides can be triggered by many different mechanisms, such as sudden large earthquakes, constant seismic activity in a tectonically active area, monsoonal rainfall or storms. There are a range of factors that affect landslide motion, including topography, geology, vegetation, precipitation and anthropogenic factors (building roads, deforestation or agricultural terraces). In total, 13 papers on landslides are published in this Special Issue. Mapping is the first topic covered by these papers. Landslide inventory maps document the extent of landslide phenomena in an area and contain relevant information that can be exploited in different ways [16]. Therefore, EO techniques can play an important role in landslide mapping. Al-Rawabdeh et al. [17] present an automated approach to use unmanned aerial vehicles (UAVs) and Semi-Global dense Matching techniques to identify and extract landslide scarps. Watanabe et al. [18] use airborne L-band fully polarimetric SAR to detect landsliding areas induced by Typhoon Wipha on 16 October 2013 on Izu Oshima Island (Japan). Plank et al. [19] propose a fast procedure for mapping landslides based on change detection between pre-event optical imagery and the polarimetric entropy derived from post-event very high resolution (VHR) polarimetric SAR data.

Monitoring landslides is a crucial task to understand the mechanisms, adopt preventive measures and reduce casualties and infrastructure damage [20]. This assignment has been revolutionized by EO techniques. Several contributions are focused on landslide monitoring, proposing and applying novel procedures, using different sensors and platforms and evaluating the quality of the results. Qu et al. [21] develop the hybrid-SAR procedure to combine both amplitude-based and phase-based methods to map and monitor large landslides exhibiting different deformation magnitudes, sliding modes and slope geometries. Using the Slumgullion landslide (southwestern Colorado, USA) as an example, Wang et al. [22] propose a fully polarimetric SAR offset tracking method to improve the precision of landslide movement monitoring. Taking the Shuping landslide (Three Gorges, China) as a case study, Sun and Muller [23] demonstrate the capability of sub-Pixel Offset Tracking techniques to monitor relatively fast slope movements in densely vegetated areas with and without the presence of artificial corner reflectors. Bardi et al. [24] integrate ground-based and satellite InSAR data to study the Åknes rockslide (western coast of Norway). Kropáček et al. [25] monitor the displacements of a large landslide on the western escarpment of the Main Ethiopian Rift (Debre Sina) by means of a multisensor and multitechnique approach. Fernández et al. [26] use Unmanned Aerial Vehicles imagery and high resolution photogrammetry to monitor horizontal and vertical displacements of a landslide affecting olive groves in La Guardia de Jaén (Spain). Finally, in Hsieh et al. [27], the errors of digital terrain models (DTMs) derived using different techniques are discussed.

Modelling is also a hot topic in the landslide field. The relationships between triggering factors and the landslide kinematics are a key aspect for the subsequent prediction of future episodes of activities. Jiang et al. [28] propose a sequential data assimilation method (i.e., Ensemble Kalman filter) to couple the surface displacements of the Shuping landslide (Three Gorges, China) derived from the Pixel Offset Tracking technique with hydrological factors. De Novellis et al. [29] develop three-dimensional (3D) Finite Element Models (FEM) of the Ivancich landslide located in Assisi town (Central Italy) through the integration of geological, geotechnical and satellite datasets.

3.3. Land Subsidence Hazards

Land subsidence is an increasing problem worldwide that has strongly attracted the attention of the InSAR community due to the high capability of InSAR techniques for the study of this type of phenomena. This is evidenced by the fact that all the contributions dedicated to this topic in this Special Issue use InSAR techniques for characterizing and monitoring land subsidence.

Beijing and Tianjin are regional economic drivers in Northern China. The aquifers systems in this region have been massively exploited and land subsidence has become evident. Three papers published in this Special Issue are focused on the spatio-temporal distribution pattern and the characterization of land subsidence as well as its triggering and conditioning factors. Zhang et al. [30] study land subsidence in the Beijing-Tianjin-Hebei region from 1992 to 2014 using ERS-1/2, ENVISAT ASAR and RADARSAT-2 images. Similarly, Liu et al. [31] present their results from C-band ENVISAT ASAR and L-band ALOS PALSAR imagery covering the period 2007–2009 , implying Line of Sight (LOS) displacements up to 170 mm/yr. Chen et al. [32] employ Small Baseline InSAR technique to process ENVISAT ASAR and TerraSAR-X stripmap images collected from 2003 to 2011 and observe a maximum subsidence in the eastern part of Beijing with a rate greater than 100 mm/year; they also find some relationships between land subsidence and different conditioning and triggering factors (e.g., groundwater levels, soft soil thickness and active faults). This contribution has attracted attention of a wide range of prestigious international media (e.g., *The Guardian*, *The Telegraph*, *Huffington Post*, *Forbes* and BBC), and is ranked in the top 5% of all research outputs ever tracked by Altmetric, a system that tracks the online attention for a specific piece of research (See: https://mdpi.altmetric.com/details/8441790#score). This contribution is also selected as TOP 10 published articles in 2016 by MDPI (http://blog.mdpi.com/2017/02/20/mdpi-altmetrics-top-10-published-articles-in-2016).

Many coastal areas in the world are experiencing land subsidence due to various factors. The combination of land lowering with rising water levels due to global sea-level rise can make coastal areas especially vulnerable to flooding [33]. Two papers in this Special Issue study land subsidence in coastal areas [34,35]. Cianflore et al. [34] analyze different causes contributing to land subsidence observed in the ancient Greek colony of the Sibari Plain (Southern Italy) using ENVISAT ASAR and Cosmo-SkyMED images. Xu et al. [35] focus on the land subsidence affecting the land reclaimed from the sea in Shenzhen (SE China). These authors use a Point Target based Small Baseline Subset InSAR approach to process ascending and descending ENVISAT ASAR images acquired during the period from 2007 to 2010, and observe subsidence rates up to 25 mm/year.

Yang et al. [36] perform correlation analyses between potential triggering and conditioning factors and land subsidence in the Linfen–Yuncheng basin (China) derived from ENVISAT ASAR data collected in 2009–2010. The authors conclude that the observed land subsidence occurs within the fault-controlled basin and is mainly caused by groundwater withdrawal. Similarly, Bai et al. [37] observe a maximum subsidence rate of up to −67.3 mm/year in Wuhan, China using TerraSAR-X images from 2009 to 2010, which are believed to be mainly caused by anthropogenic activities, natural compaction and karst dissolution. A seasonal pattern of displacements is also noticed near the Yangtze River by these authors. Caló et al. [38] use eight-year ENVISAT ASAR images to investigate land subsidence mainly due to groundwater overextraction in the Konya plain, Turkey, and a joint analysis with GRACE data suggests that the groundwater depletion is not limited to the study site but affects a wider region in the Anatolian Plateau.

Land subsidence information from InSAR observations can be exploited for different purposes. Pacheco-Martínez et al. [39] present a novel approach to combine InSAR and gravimetric surveys for risk management related to land subsidence and surface ground faulting generation. Boni et al. [40] describe a novel methodology for the exploitation of InSAR data to support geological interpretation in areas affected by land subsidence, uplift and seasonal displacements.

Mining-induced subsidence is another important anthropogenic geohazard. Ma et al. [41] study mining subsidence in Shendong Coalfield (China), suggesting that the extent of subsidence exhibits a progressive increase of 13.09 km^2 per month during the period from 2012 to 2013.

Surface displacements associated with the geothermal field of Yangbajing (China) are studied and modelled by Hu et al. [42] using ENVISAT ASAR images collected from 2006 to 2010, allowing for the interpretation of the volume changes produced in the geothermal field.

Zhou et al. [43] focus on the combination of InSAR observation and numerical modelling to obtain the physical parameters of the Earth-dam displacements of the Shuibuya Dam (China), and these parameters are then used to predict future behaviors of the dam.

3.4. New EO Techniques and Services

Several papers in this Special Issue are focused on new EO techniques and services for geohazard management. Ding et al. [44] develop new methods to reduce stripe artifacts (SA) and Topographic Shadowing Artifacts (TSA) in surface displacement maps retrieved from Landsat 8 optical images. Their experiments indicate that their algorithms could improve the precision of surface displacement maps (near 15%).

Since parameters of satellite orbits of historical declassified intelligence satellite photography (DISP) imagery are not available and ground control points (GCPs) are lacking, Zhou et al. [45] develop a second order polynomial equation-based block adjustment model for orthorectification that provides accuracy in the level of 2.0 pixels (i.e., approximately 2.0–4.0 m) in the assembling of the imagery for geohazard mapping. Chen et al. [46] illustrate the potential of the BeiDou Navigation Satellite System (BDS) to serve as a fast and reliable early warning system of tsunamis. Cignetti et al. [47] propose an iterative procedure which is applied through the Parallel-SBAS web-tool within the Grid Processing-on-Demand (G-POD) environment to improve SAR data selection and processing by minimizing the temporal decorrelation effects over high mountain regions to obtain mean deformation velocity maps and displacement time series.

Following large geohazards, especially those associated with widespread destruction and high mortality, rapid, accurate and reliable damage assessment is essential to obtain information to guide response activities in the critical post-event hours. Using airborne LiDAR data, He et al. [48] develop a 3D shape descriptor to detect surface- and structure-damaged roofs. Using airborne oblique images, Vetrivel et al. [49] develop a Visual-Bag-of-Words (BoW) based damage classification to detect structure-damaged areas. Ma et al. [50] present an automatic procedure to generate cloudless backdrop and disaster change-detection maps from optical imagery, whilst Xie et al. [51] demonstrate a framework to combine aerial remote sensing imagery with crowdsourcing to support wide-area assessments of building collapses. Using high resolution multispectral and panchromatic remote sensing data, Cooner et al. [52] demonstrate the effectiveness of multilayer feedforward neural networks, radial basis neural networks, and Random Forests in detecting earthquake damage. It is demonstrated in [53] that a combination of a post-event very high resolution SAR image with a pre-event building footprint map can be effectively used to detect damaged buildings, and Zhai et al. [54] demonstrated the feasibility to use a single post-event PolSAR image to assess building damage.

4. Current Challenges and Future Trends of EO for Geohazards

In this section, we analyze the evolution of the annual production of EO publications recorded in the Web of Science's bibliographical database in the fields of Remote sensing and Geology. Several previous studies [55–57] conclude that a power model (q > 0, K > 1) can properly explain the growth of publications and new authors in different fields of science. Therefore, to explore the evolution of EO publications, we fit power functions to the available data (Figure 1). It is clear in Figure 1 that the first work in this discipline was published in the middle of the 1970s, but the take-off of this discipline took place in the 1990s. Since then, the number of EO papers has exhibited an exponential-like growth. Furthermore, the fittings show that the growth rate (K) in the number of publications is higher in the remote sensing segment than in the Geology one, because remote sensing includes all the applications of EO not restricted to geology. In order to provide a reference value to be compared with our data, we also plot the best-fit logistic functions of the publications focused on LiDAR or Laser scanning proposed by Derron and Jaboyedoff [8]. Although the data are not directly comparable since their publication data were extracted from the GEOREF database rather than from the Web of Science (WOS), which is usually more restrictive, and the LiDAR discipline is much younger that EO, the initial trends of both EO and LiDAR publications are quite similar.

The high number of manuscripts (i.e., 79) received to be considered for this Special Issue is also a good indicator of the increasing prosperity of the EO community. Most of the papers published in this Special Issue are focused on the post-event exploitation of EO products, i.e., they use EO data to investigate disasters after the event occurs. There is a delay between the event occurrence and the delivery of the useful EO-derived information for decision-makers. Therefore, a great effort is needed in the coming years to reduce the response time after disasters. It should be noted that EO can also be widely applied to disaster preparation, recovery and mitigation. On 1 January 2016, the 17 Sustainable Development Goals (SDGs) of Transforming Our World: the 2030 Agenda for Sustainable Development adopted by world leaders at the 2015 UN Sustainable Development Summit officially came into force, and will run through 2030 and applies to every country. EO and its derived information are specifically demanded to serve the 2030 Agenda by monitoring the 17 SDGs and associated 169 targets, planning and tracking progress, and helping nations and other stakeholders make informed decisions.

5. Conclusions

To conclude, EO has reached some degree of maturity although it is still growing fast following a power trend. The recent success of EO for the investigation of geohazards is mainly due to the development of new EO sensors and techniques, the improved capabilities of EO data acquisition and analysis, and the advance in the production of standardized products to be used by planners and decision makers. It is expected that more and more real-time EO products and services will be emerging in the near future, which will enable us to better manage geohazards.

Acknowledgments: Guest editors would like to take this opportunity to thank all authors, editors, reviewers, and supporters for the hard work and dedication that made this Special Issue possible. Part of this work was supported by the UK Natural Environmental Research Council (NERC) through the Centre for the Observation and Modelling of Earthquakes, Volcanoes and Tectonics (COMET, ref.: come30001) and the LICS and CEDRRIC projects (refs. NE/K010794/1 and NE/N012151/1, respectively), European Space Agency through the ESA-MOST DRAGON-4 projects (ref. 32244) and the Spanish Ministry of Economy and Competitiveness and EU FEDER funds under projects TIN2014-55413- C2-2-P and ESP2013-47780-C2-2-R.

Author Contributions: R.T. and Z.L. are guest editors of this Special Issue, and both wrote and edited the editorial.

Conflicts of Interest: The authors declare no conflict of interest.

References

1. Hyndman, D.; Hyndman, D. *Natural Hazards and Disasters*, 5th ed.; Cebgage Learning: Boston, MA, USA, 2017.
2. Aleotti, P.; Chowdhury, R. Landslide hazard assessment: Summary review and new perspectives. *Bull. Eng. Geol. Environ.* **1999**, *58*, 21–44. [CrossRef]

3. Poland, J.F. *Guidebook to Studies of Land Subsidence Due to Ground-Water Withdrawal*; United Nations Educational, Scientific and Cultural Organization: Chelsea, UK, 1984; p. 340.

4. Galloway, D.L.; Jones, D.R.; Ingebritsen, S.E. *Land Subsidence in the United States*; U.S. Geological Survey: Reston, VA, USA, 1999; p. 177.

5. Denis, G.; de Boissezon, H.; Hosford, S.; Pasco, X.; Montfort, B.; Ranera, F. The evolution of earth observation satellites in europe and its impact on the performance of emergency response services. *Acta Astron.* **2016**, *127*, 619–633. [CrossRef]

6. Tatem, A.J.; Goetz, S.J.; Hay, S.I. Fifty years of earth observation satellites: Views from above have lead to countless advances on the ground in both scientific knowledge and daily life. *Am. Sci.* **2008**, *96*, 390–398. [CrossRef] [PubMed]

7. Abellan, A.; Derron, M.-H.; Jaboyedoff, M. "Use of 3D point clouds in geohazards" special issue: Current challenges and future trends. *Remote Sens.* **2016**, *8*, 130. [CrossRef]

8. Derron, M.H.; Jaboyedoff, M. Preface "LiDAR and DEM techniques for landslides monitoring and characterization". *Nat. Hazards Earth Syst. Sci.* **2010**, *10*, 1877–1879. [CrossRef]

9. Zhu, S.; Xu, C.; Wen, Y.; Liu, Y. Interseismic deformation of the altyn tagh fault determined by interferometric synthetic aperture radar (INSAR) measurements. *Remote Sens.* **2016**, *8*, 233. [CrossRef]

10. Liu, Y.; Xu, C.; Li, Z.; Wen, Y.; Chen, J.; Li, Z. Time-dependent afterslip of the 2009 mw 6.3 dachaidan earthquake (China) and viscosity beneath the qaidam basin inferred from postseismic deformation observations. *Remote Sens.* **2016**, *8*, 784. [CrossRef]

11. Xu, B.; Li, Z.; Feng, G.; Zhang, Z.; Wang, Q.; Hu, J.; Chen, X. Continent-wide 2-d co-seismic deformation of the 2015 mw 8.3 illapel, chile earthquake derived from sentinel-1a data: Correction of azimuth co-registration error. *Remote Sens.* **2016**, *8*, 376. [CrossRef]

12. Solaro, G.; De Novellis, V.; Castaldo, R.; De Luca, C.; Lanari, R.; Manunta, M.; Casu, F. Coseismic fault model of mw 8.3 2015 illapel earthquake (CHILE) retrieved from multi-orbit sentinel1-A dinsar measurements. *Remote Sens.* **2016**, *8*, 323. [CrossRef]

13. Ji, L.; Xu, J.; Zhao, Q.; Yang, C. Source parameters of the 2003–2004 bange earthquake sequence, central tibet, china, estimated from insar data. *Remote Sens.* **2016**, *8*, 516. [CrossRef]

14. Li, Y.; Jiang, W.; Zhang, J.; Luo, Y. Space geodetic observations and modeling of 2016 mw 5.9 menyuan earthquake: Implications on seismogenic tectonic motion. *Remote Sens.* **2016**, *8*, 519. [CrossRef]

15. Trasatti, E.; Tolomei, C.; Pezzo, G.; Atzori, S.; Salvi, S. Deformation and related slip due to the 2011 van earthquake (turkey) sequence imaged by sar data and numerical modeling. *Remote Sens.* **2016**, *8*, 532. [CrossRef]

16. Guzzetti, F.; Mondini, A.C.; Cardinali, M.; Fiorucci, F.; Santangelo, M.; Chang, K.-T. Landslide inventory maps: New tools for an old problem. *Earth-Sci. Rev.* **2012**, *112*, 42–66. [CrossRef]

17. Al-Rawabdeh, A.; He, F.; Moussa, A.; El-Sheimy, N.; Habib, A. Using an unmanned aerial vehicle-based digital imaging system to derive a 3D point cloud for landslide scarp recognition. *Remote Sens.* **2016**, *8*, 95. [CrossRef]

18. Watanabe, M.; Thapa, R.; Shimada, M. Pi-sar-l2 observation of the landslide caused by typhoon wipha on izu oshima island. *Remote Sens.* **2016**, *8*, 282. [CrossRef]

19. Plank, S.; Twele, A.; Martinis, S. Landslide mapping in vegetated areas using change detection based on optical and polarimetric sar data. *Remote Sens.* **2016**, *8*, 307. [CrossRef]

20. Angeli, M.-G.; Pasuto, A.; Silvano, S. A critical review of landslide monitoring experiences. *Eng. Geol.* **2000**, *55*, 133–147. [CrossRef]

21. Qu, T.; Lu, P.; Liu, C.; Wu, H.; Shao, X.; Wan, H.; Li, N.; Li, R. Hybrid-Sar technique: Joint analysis using phase-based and amplitude-based methods for the xishancun giant landslide monitoring. *Remote Sens.* **2016**, *8*, 874. [CrossRef]

22. Wang, C.; Mao, X.; Wang, Q. Landslide displacement monitoring by a fully polarimetric sar offset tracking method. *Remote Sens.* **2016**, *8*, 624. [CrossRef]

23. Sun, L.; Muller, J.-P. Evaluation of the use of sub-pixel offset tracking techniques to monitor landslides in densely vegetated steeply sloped areas. *Remote Sens.* **2016**, *8*, 659. [CrossRef]

24. Bardi, F.; Raspini, F.; Ciampalini, A.; Kristensen, L.; Rouyet, L.; Lauknes, T.; Frauenfelder, R.; Casagli, N. Space-borne and ground-based insar data integration: The knes test site. *Remote Sens.* **2016**, *8*, 237. [CrossRef]

25. Kropáček, J.; Vařilová, Z.; Baroň, I.; Bhattacharya, A.; Eberle, J.; Hochschild, V. Remote sensing for characterisation and kinematic analysis of large slope failures: Debre sina landslide, main ethiopian rift escarpment. *Remote Sens.* **2015**, *7*, 16183–16203. [CrossRef]

26. Fernández, T.; Pérez, J.; Cardenal, J.; Gómez, J.; Colomo, C.; Delgado, J. Analysis of landslide evolution affecting olive groves using uav and photogrammetric techniques. *Remote Sens.* **2016**, *8*, 837. [CrossRef]

27. Hsieh, Y.-C.; Chan, Y.-C.; Hu, J.-C. Digital elevation model differencing and error estimation from multiple sources: A case study from the meiyuan shan landslide in taiwan. *Remote Sens.* **2016**, *8*, 199. [CrossRef]

28. Jiang, Y.; Liao, M.; Zhou, Z.; Shi, X.; Zhang, L.; Balz, T. Landslide deformation analysis by coupling deformation time series from sar data with hydrological factors through data assimilation. *Remote Sens.* **2016**, *8*, 179. [CrossRef]

29. De Novellis, V.; Castaldo, R.; Lollino, P.; Manunta, M.; Tizzani, P. Advanced three-dimensional finite element modeling of a slow landslide through the exploitation of dinsar measurements and in situ surveys. *Remote Sens.* **2016**, *8*, 670. [CrossRef]

30. Zhang, Y.; Wu, H.a.; Kang, Y.; Zhu, C. Ground subsidence in the Beijing-Tianjin-Hebei region from 1992 to 2014 revealed by multiple sar stacks. *Remote Sens.* **2016**, *8*, 675. [CrossRef]

31. Liu, P.; Li, Q.; Li, Z.; Hoey, T.; Liu, G.; Wang, C.; Hu, Z.; Zhou, Z.; Singleton, A. Anatomy of subsidence in tianjin from time series insar. *Remote Sens.* **2016**, *8*, 266. [CrossRef]

32. Chen, M.; Tomás, R.; Li, Z.; Motagh, M.; Li, T.; Hu, L.; Gong, H.; Li, X.; Yu, J.; Gong, X. Imaging land subsidence induced by groundwater extraction in Beijing (China) using satellite radar interferometry. *Remote Sens.* **2016**, *8*, 468. [CrossRef]

33. Holzer, T.L.; Galloway, D.L. Impacts of land subsidence caused by withdrawal of underground fl uids in the united states. *Rev. Eng. Geol.* **2005**, *XVI*, 87–99.

34. Cianflone, G.; Tolomei, C.; Brunori, C.; Dominici, R. Insar time series analysis of natural and anthropogenic coastal plain subsidence: The case of sibari (southern Italy). *Remote Sens.* **2015**, *7*, 15812. [CrossRef]

35. Xu, B.; Feng, G.; Li, Z.; Wang, Q.; Wang, C.; Xie, R. Coastal subsidence monitoring associated with land reclamation using the point target based SBAS-INSAR method: A case study of shenzhen, China. *Remote Sens.* **2016**, *8*, 652. [CrossRef]

36. Yang, C.-S.; Zhang, Q.; Xu, Q.; Zhao, C.-Y.; Peng, J.-B.; Ji, L.-Y. Complex deformation monitoring over the Linfen–Yuncheng basin (China) with time series insar technology. *Remote Sens.* **2016**, *8*, 284. [CrossRef]

37. Bai, L.; Jiang, L.; Wang, H.; Sun, Q. Spatiotemporal characterization of land subsidence and uplift (2009–2010) over wuhan in central china revealed by terrasar-X insar analysis. *Remote Sens.* **2016**, *8*, 350. [CrossRef]

38. Caló, F.; Notti, D.; Galve, J.; Abdikan, S.; Görüm, T.; Pepe, A.; Balik Şanli, F. Dinsar-based detection of land subsidence and correlation with groundwater depletion in konya plain, turkey. *Remote Sens.* **2017**, *9*, 83. [CrossRef]

39. Pacheco-Martínez, J.; Cabral-Cano, E.; Wdowinski, S.; Hernández-Marín, M.; Ortiz-Lozano, J.; Zermeño-de-León, M. Application of insar and gravimetry for land subsidence hazard zoning in aguascalientes, mexico. *Remote Sens.* **2015**, *7*, 15868. [CrossRef]

40. Bonì, R.; Pilla, G.; Meisina, C. Methodology for detection and interpretation of ground motion areas with the A-dinsar time series analysis. *Remote Sens.* **2016**, *8*, 686. [CrossRef]

41. Ma, C.; Cheng, X.; Yang, Y.; Zhang, X.; Guo, Z.; Zou, Y. Investigation on mining subsidence based on multi-temporal insar and time-series analysis of the small baseline subset—case study of working faces 22201–1/2 in bu'ertai mine, shendong coalfield, China. *Remote Sens.* **2016**, *8*, 951. [CrossRef]

42. Hu, J.; Wang, Q.; Li, Z.; Zhao, R.; Sun, Q. Investigating the ground deformation and source model of the yangbajing geothermal field in tibet, china with the wls insar technique. *Remote Sens.* **2016**, *8*, 191. [CrossRef]

43. Zhou, W.; Li, S.; Zhou, Z.; Chang, X. Insar observation and numerical modeling of the earth-dam displacement of shuibuya dam (China). *Remote Sens.* **2016**, *8*, 877. [CrossRef]

44. Ding, C.; Feng, G.; Li, Z.; Shan, X.; Du, Y.; Wang, H. Spatio-temporal error sources analysis and accuracy improvement in landsat 8 image ground displacement measurements. *Remote Sens.* **2016**, *8*, 937. [CrossRef]

45. Zhou, G.; Yue, T.; Shi, Y.; Zhang, R.; Huang, J. Second-order polynomial equation-based block adjustment for orthorectification of disp imagery. *Remote Sens.* **2016**, *8*, 680. [CrossRef]

46. Chen, K.; Zamora, N.; Babeyko, A.; Li, X.; Ge, M. Precise positioning of bds, BDS/GPS: Implications for tsunami early warning in South China sea. *Remote Sens.* **2015**, *7*, 15814. [CrossRef]

47. Cignetti, M.; Manconi, A.; Manunta, M.; Giordan, D.; De Luca, C.; Allasia, P.; Ardizzone, F. Taking advantage of the esa G-pod service to study ground deformation processes in high mountain areas: A valle d'aosta case study, northern italy. *Remote Sens.* **2016**, *8*, 852. [CrossRef]

48. He, M.; Zhu, Q.; Du, Z.; Hu, H.; Ding, Y.; Chen, M. A 3D shape descriptor based on contour clusters for damaged roof detection using airborne LiDAR point clouds. *Remote Sens.* **2016**, *8*, 189. [CrossRef]

49. Vetrivel, A.; Gerke, M.; Kerle, N.; Vosselman, G. Identification of structurally damaged areas in airborne oblique images using a visual-bag-of-words approach. *Remote Sens.* **2016**, *8*, 231. [CrossRef]

50. Ma, Y.; Chen, F.; Liu, J.; He, Y.; Duan, J.; Li, X. An automatic procedure for early disaster change mapping based on optical remote sensing. *Remote Sens.* **2016**, *8*, 272. [CrossRef]

51. Xie, S.; Duan, J.; Liu, S.; Dai, Q.; Liu, W.; Ma, Y.; Guo, R.; Ma, C. Crowdsourcing rapid assessment of collapsed buildings early after the earthquake based on aerial remote sensing image: A case study of yushu earthquake. *Remote Sens.* **2016**, *8*, 759. [CrossRef]

52. Cooner, J.A.; Shao, Y.; Campbell, B.J. Detection of urban damage using remote sensing and machine learning algorithms: Revisiting the 2010 Haiti earthquake. *Remote Sens.* **2016**, *8*, 868. [CrossRef]

53. Gong, L.; Wang, C.; Wu, F.; Zhang, J.; Zhang, H.; Li, Q. Earthquake-induced building damage detection with post-event sub-meter vhr terrasar-X staring spotlight imagery. *Remote Sens.* **2016**, *8*, 887. [CrossRef]

54. Zhai, W.; Shen, H.; Huang, C.; Pei, W. Building earthquake damage information extraction from a single post-earthquake polsar image. *Remote Sens.* **2016**, *8*, 171. [CrossRef]

55. Egghe, L.; Ravichandra Rao, I.K. Classification of growth models based on growth rates and its applications. *Scientometrics* **1992**, *25*, 5–46. [CrossRef]

56. Gupta, B.M.; Karisiddappa, C.R. Modelling the Growth of Literature in the Area of Theoretical Population Genetics. *Scientometrics,* **2000**, *49*, 321–355. [CrossRef]

57. Gupta, B.M.; Sharma, P.; Kumar, S. Growth of world and Indian physics literature. *Scientometrics,* **1999**, *44*, 5–16. [CrossRef]

Chapter 1:
Earthquake Hazards

remote sensing

MDPI

Article

Interseismic Deformation of the Altyn Tagh Fault Determined by Interferometric Synthetic Aperture Radar (InSAR) Measurements

Sen Zhu [1], Caijun Xu [1,2,3,*], Yangmao Wen [1,2,3] and Yang Liu [1,2,3]

[1] School of Geodesy and Geomatics, Wuhan University, Wuhan 430079, China; zs0255@163.com (S.Z.);
    ymwen@sgg.whu.edu.cn (Y.W.); yang.liu@sgg.whu.edu.cn (Y.L.)
[2] Key Laboratory of Geospace Environment and Geodesy, Ministry of Education, Wuhan 430079, China
[3] Collaborative Innovation Center of Geospatial Technology, Wuhan 430079, China
*   Correspondence: cjxu@sgg.whu.edu.cn; Tel.: +86-27-6877-8805; Fax: +86-27-6877-8371

Academic Editors: Zhenhong Li, Roberto Tomas, Zhong Lu and Prasad S. Thenkabail
Received: 27 November 2015; Accepted: 4 March 2016; Published: 11 March 2016

Abstract: The Altyn Tagh Fault (ATF) is one of the major left-lateral strike-slip faults in the northeastern area of the Tibetan Plateau. In this study, the interseismic deformation across the ATF at $85°E$ was measured using 216 interferograms from 33 ENVISAT advanced synthetic aperture radar images on a descending track acquired from 2003 to 2010, and 66 interferograms from 15 advanced synthetic aperture radar images on an ascending track acquired from 2005 to 2010. To retrieve the pattern of interseismic strain accumulation, a global atmospheric model (ERA-Interim) provided by the European Center for Medium Range Weather Forecast and a global network orbital correction approach were applied to remove atmospheric effects and the long-wavelength orbital errors in the interferograms. Then, the interferometric synthetic aperture radar (InSAR) time series with atmospheric estimation model was used to obtain a deformation rate map for the ATF. Based on the InSAR velocity map, the regional strain rates field was calculated for the first time using the multi-scale wavelet method. The strain accumulation is strongly focused on the ATF with the maximum strain rate of 12.4×10^{-8}/year. We also show that high-resolution 2-D strain rates field can be calculated from InSAR alone, even without GPS data. Using a simple half-space elastic screw dislocation model, the slip-rate and locking depth were estimated with both ascending and descending surface velocity measurements. The joint inversion results are consistent with a left-lateral slip rate of 8.0 ± 0.7 mm/year on the ATF and a locking depth of 14.5 ± 3 km, which is in agreement with previous results from GPS surveys and ERS InSAR results. Our results support the dynamic models of Asian deformation requiring low fault slip rate.

Keywords: InSAR; AltynTagh Fault; interseismic deformation; geodetic inversion; slip rate

1. Introduction

The ongoing active continental collision between the Indian and Eurasian plates has created the massive topography of the Tibetan Plateau over the last 50 million years. The tectonic processes underlying this region are still not completely understood and a number of models have been proposed to explain the dynamics of the area [1–4]. The active Altyn Tagh Fault (ATF) is a distinctive feature of Tibetan geography, inscribing a fairly linear trace over 1500 km across northern Tibet (Figure 1), separating the high plateau in the south from the low Tarim Basin in the north. The Tibetan Plateau south of the ATF has an average elevation of ~4000 m, whereas the northern Tarim Basin has an average elevation of only ~1000 m. The ATF begins in the west at the Pamir Mountains and ends in the east near the Qilian Mountains. It is generally accepted that the ATF is divided into three segments, the west, middle, and east, between $84°E$ and $94°E$ [5]. It is a major tectonic element in the Cenozoic Indo–Asian collision zone [6] that defines the northern margin of the Tibetan Plateau. It plays a key role in identifying the Tibet deformation patterns and its slip rates have constituted an important constraint for deriving the velocity field of East Asia [7]. Some studies argued that the ATF might accommodate as much as one-third of the overall convergence between India and Siberia [7]. Zhang *et al.* [8] derived the velocity field of the present day tectonic deformation in the Tibetan Plateau using 553 GPS observations. The velocity distribution suggested that the relative movement between the Indian and Eurasian plates was adjusted and absorbed by crustal shortening and internal strike-slip shearing in the Tibetan Plateau. In particular, the Himalaya Mountains absorbed 44%–53% of the total shortening, while the Qaidam Basin, Qilian Mountains, and Altyn Mountains in the north absorbed 15%–17%, and the interior of the plateau absorbed the remaining 32%–41%. It has been regarded as either a lithospheric-scale strike-slip fault promoting the eastward extrusion of Tibet [9] or a crustal-scale transfer fault linking thrust belts [10].

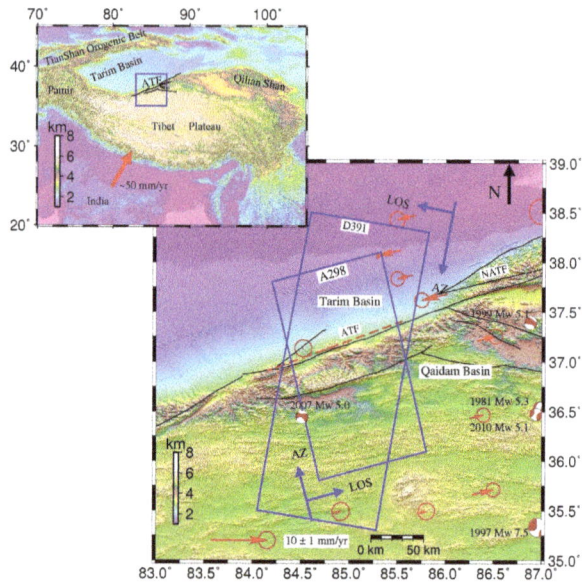

Figure 1. Location and velocity map of the Altyn Tagh Fault (ATF). The top left figure shows the whole tectonic background of the ATF, the low right figure is the zoom-in of the blue rectangular in it and shows the velocity map. In the low right part, the black solid line indicates the ATF. The blue rectangle is the InSAR frame of D391 and A298. The red dashed line is the estimated fault location, and the blue arrows represent the azimuth and LOS direction. The focal mechanism symbols represent the events after 1976 from the Harvard CMT catalogue. The black arrow shows the north direction. The red arrows with ellipses show the GPS velocity projected to the rough fault line. The GPS velocity is from Ge *et al.* [11].

There are two end-member models for the continental deformation mechanism in Tibet accommodating the Indian sub-continent collision with Eurasia: block and microplate models and continuum models. To discriminate between these two models, we need to determine the slip rate of the ATF accurately. Late Quaternary slip rates have been reported along most of the ATF at both decadal and millennial time scales. Geodetic measurements at the decadal time scale indicating that the ATF slips at ~10 mm/year [12–15] have been used to support continuum deformation of the Tibetan Plateau [1,2,8,15] or block-like deformation [3,4]. Elliott *et al.* [16] investigated the same location as this study using ERS data and the stacking method and concluded that the slip rate was 11 ± 5 mm/year, supporting continuum deformation. In contrast, He *et al.* [17] tried to determine the slip rate using a GPS profile, and their conclusion was 9.0 ± 4.0 mm/year, consistent with block-like deformation. At a millennial time scale, the geological slip rate is given as 27 ± 7 mm/year from detailed measurements of riser offsets of different geomorphic terraces using radiocarbon and cosmic ray exposure dating of the offset terraces [18].

In the past two decades, InSAR [19] has become a widely used deformation mapping tool in studying geophysical process. Since the first coseismic interferogram was published [20] in the 1992 Mw 7.4 Landers earthquake, InSAR has made significant contributions to earthquake cycle research (e.g., [20–24]). Wright *et al.* [21] first retrieved the interseismic slip rate of strike-slip fault using stacking method. Walters *et al.* [22] improved Wright and coworkers' results using SAR data from two look directions, reducing the range of uncertainties of slip rate and locking depth by 60%. At present, InSAR is regarded as a significant tool for tectonic geomorphology and seismic hazard assessment [25]. In this paper, InSAR observations are the major source of data for the determination of slip rate and locking depth of the ATF.

Although geodetic measurements are consistent with others that estimated ~10 mm/year, they differ significantly from rates determined by geological methods (27 ± 7 mm/year; [18]) and have significant uncertainties. In this study, we attempted to determine a highly accurate slip rate of the left-strike ATF using the interferometric synthetic aperture radar (InSAR) time series with atmospheric estimation model (TS + AEM) package with the same descending track (D391) as Elliott *et al.* [16], as well as an ascending track (A298) of ENVISAT data after the atmospheric and orbital errors were removed. In order to determine whether there is significant strain accumulation away from the major faults, the strain rate field was calculated based on horizontal velocity field using the multi-scale waveform method proposed by Tape *et al.* [26]. The fault slip rates are inverted based on a half-space elastic screw dislocation model. Finally, we interpreted the geodetic velocity field and investigated the essence of crustal deformation in this region.

2. Interseismic Rate Map from InSAR Time Series

InSAR has the potential to measure interseismic strain accumulation at a dense spatial scale. It is widely used to estimate the slip rates of major faults [16,21,22,27,28] since Massonnet *et al.* successfully extracted the coseismic deformation of the 1992 Landers earthquake [20]. Wright *et al.* [21] werethe first to use InSAR for studying the interseismic deformation. InSAR has been proven to be a important tool in researching the seismic cycle deformation after many successful applications in coseismic [20], postseismic [29] and interseismic [16,21,22,27,28] deformation. Walters *et al.* [22] improved Wright and coworkers' results [21] by using SAR data from two different look directions, in their paper, they stated that the uncertainties in slip rate and locking depth are reduced up to 60% which is a huge improvement. In our study, we used two directions SAR data in order to obtain the deformation of ATF following Waltersand coworkers' method [22]. Crustal deformation signals measured over short time scales are dominated by orbital and atmospheric errors. Interseismic deformation can be extracted from several interferograms when the atmospheric and orbital errors are removed.

2.1. InSAR Data and Processing

Synthetic aperture radar images from the advanced SAR (ASAR) instrument on board the ENVISAT satellite were used in this study to investigate interseismic deformation across the ATF at 85°E. The data set was acquired between June 2003 and June 2010, and includes 33 descending images and 15 ascending images. The raw data were processed with repeat orbit interferometry package (ROI_PAC) Version 3.1 beta [30]. Precise DORIS orbital data for ENVISAT satellite provided by ESA has been used for interferometric processing. All of the interferograms were produced using the NSBAS package [31,32]. Image pairs were selected by spatial and temporal baselines: (i) B_t < 1 year and B_s < 500 m; (ii) B_t < 3 year and B_s < 300 m; and (iii) B_t < 5 year and B_s < 100 m (Figure 2). In ascending track 298, the 10 October 2005data extends the time baseline from 2 to 5 years, which is significant for the time series deformation analysis. Unlike ROI_PAC, the NSBAS takes into account the DEM when co-registering images [33], which improved the interferometric processing and increased the coherence of interferograms.

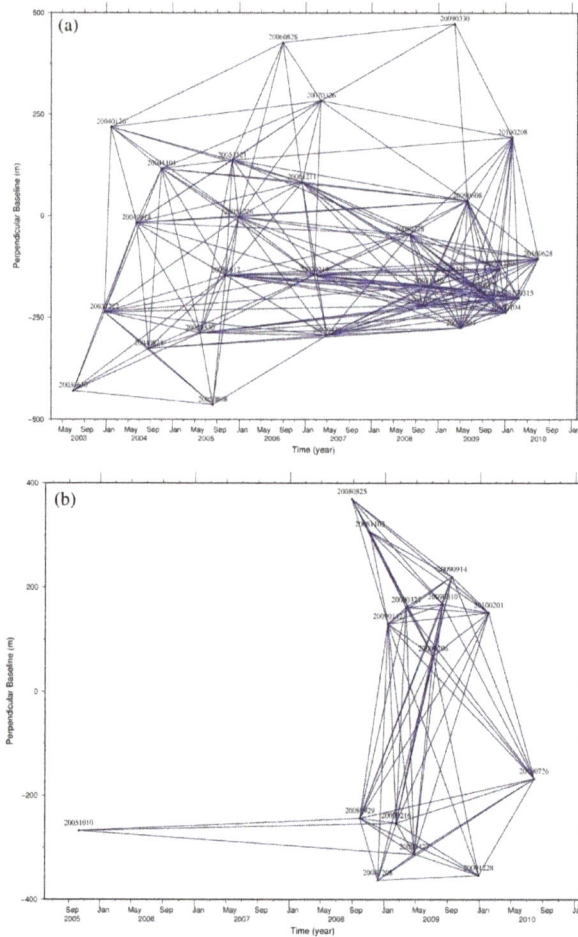

Figure 2. Temporal and spatial baseline of InSAR interferograms: (**a**) D391 and (**b**) A298.

During SAR data processing, we multilooked the SAR images 4 and 20 times in the azimuth and range directions, respectively, during processing to improve the signal-to-noise ratio. The phase component of the topography was removed using the three-arc second shuttle radar topography mission (SRTM) DEM [34] with the ESA DORIS precision orbits. A power spectrum filter [35] was applied to further reduce the phase noise to improve the interferogram coherence. Most interferograms were unwrapped using the branch-cut method [36], but some of them needed manual unwrapping (bridge) across the fault due to the loss of coherence. The unwrapping errors were detected and removed by a phase-closure technique [37,38]. After phase closure check, we correct the atmospheric errors and orbital ramps (Figure 3) respectively. The whole process flow is summarized in Figure 4. The coherence of north areas of the ATF was commonly higher than that of the southern part because of snow cover on the southern mountains (Figure 5). Areas near the ATF have poor coherence even compared with the southern part as a result of the topography.

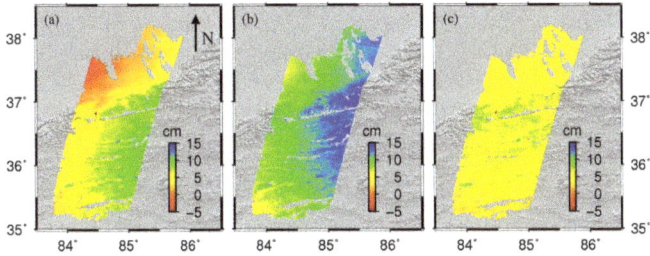

Figure 3. Example interferogram (17 October 2005–19 February 2007) showing the effects of APS and orbit correction. (**a**) observation interferogram; (**b**) APS corrected interferogram (using TRAIN software); (**c**) orbit corrected interferogram (using Biggs's network correction method).

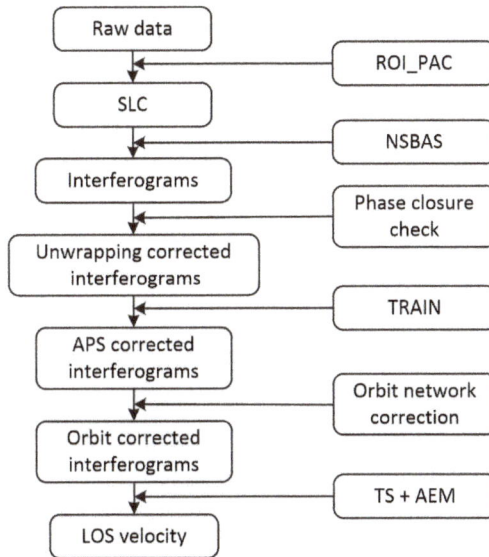

Figure 4. Flow chart of data processing, from raw data to final LOS velocity.

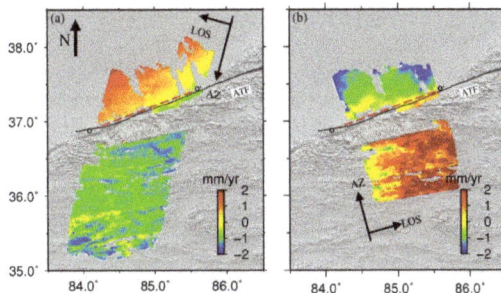

Figure 5. (**a**) Ground surface mean LOS velocity maps from InSAR time series analysis for D391. The fat black arrow shows the north direction. The dashed line indicates the fault location used for the 2-D modeling. The OO′ line stands for the fault location. (**b**) Ground surface mean LOS velocity maps from InSAR time series analysis for A298.

2.2. Atmospheric Correction

The greatest factor limiting InSAR measurement accuracy is atmospheric delay, particularly from tropospheric water vapor [39].This is exacerbated for the ATF due to its relief of ~4 km [16]. Crustal deformation signals measured over short time scales are dominated by orbital and atmospheric errors. The atmospheric phase delay, mainly derived from the temporal variation of stratified troposphere, may reach tens of centimeters [40,41].

There are two groups of correction methods to estimate atmospheric phase delay, the empirical and the predictive methods. Assuming that the relationship between the atmospheric delays and elevation is approximately linear, a first order function [16,41,42] could be used to estimate the conversion coefficient. Unfortunately, empirical methods cannot be easily used when the expected deformation signal correlates with topography, which is the case for such major topographic steps like ATF [16]. Predictive methods are based on inputs from external data sets to compute synthetic delay maps and directly correct for tropospheric delays in interferograms. The external data sets include local meteorological data [43], GPS zenith delay measurements [44–47], satellite multi-spectral imagery [48,49], and outputs from local meteorological models constrained by local data collection [50–52]. These external data have proven successful and accurate for atmospheric correction; however, rarely available local data limit the popularity of this method.

Here, we use the method from Bekaert *et al.* [33,53,54] to estimate the atmospheric phase delay maps using the ERA-I global atmospheric model reanalysis product [55] obtained from the European Center for Medium-Range Weather Forecasts. The ERA-I provides estimates of temperature, water vapor partial pressure, and geopotential height along 37 pressure levels, on a global $0.7° \times 0.7°$ grid at regular 6-hour intervals from 1989 to present. Both hydrostatic and wet contributions to the phase delay are taken into account in this approach (Figure 3). This method accounts for spatial variability of the tropospheric properties and successfully captures tropospheric signals over large regions. In Figure 3, for example, we removed the elevation-correlated atmospheric path delay from original observations to correct for atmospheric errors; in Figure 3b, the clear difference across the fault in Figure 3a is removed denoting the reduction of the elevation-correlated atmospheric delay. We have calculated the standard deviation(σ) reduction of the APS correction and find that σ is reduced in 38% by ECMWF-based correction. The results confirm that the ERA-I model can overcome limitations of empirical models [56].

2.3. Orbital Correction

Because of our imperfect knowledge of the state vector of the ENVISAT satellite, even after the effects of baseline separation have been removed, an orbital error remains in the interferogram [57]. A clear linear ramp is shown in atmospheric corrected interferogram (Figure 3b). It is caused by long wavelength orbital errors and the orbital correction is needed before further processing. A linear approximation is usually sufficiently accurate for short strips (100–200 km along track) and a quadratic approximation is required for longer strips. This approach is sufficient because the deformation signal represents a small portion of interferometric phase, such as volcanology and earthquakes [58]. Hence, it is possible to define a "far-field" area of the interferogram and to re-estimate the baseline parameters [37]. However, there is no true "far-field" in such a large area for studies of interseismic deformation, so the baseline for individual interferograms to deal with the orbital contributions cannot be precisely re-estimated.

For this reason, Biggs *et al.* [37] presented a global network orbital correction method to deal with the issue of orbital contributions at a later stage of re-estimating the baseline. Here, we use this method to remove orbital errors remaining in the interferograms (Figure 3c). It can be seen from Figure 3c that a phase ramp is removed from the atmospheric corrected interferogram (Figure 3b) after the network orbital correction. The σ reduction of this step is 13% of the interferogram. This correction technique also accounts for other long wavelength signals, such as ionospheric and long wavelength tropospheric contributions as well as errors in orbit [37].

2.4. Rate Map

A LOS rate map is calculated from the interferograms corrected for both atmospheric and orbital errors using the InSAR TS + AEM package developed at the University of Glasgow based on the SBAS algorithm [59–61] as the flow chart in Figure 4. The RMS between LOS deformation derived from the InSAR TS + AEM package and GPS measurements is smaller than 0.5 mm/year, which was validated independently by Li *et al.* (*Abstract G13B-07 presented at Fall Meeting.* 2010, AGU, San Francisco, CA, USA) and Hammond *et al.* [62].

We obtained the deformation rate map for the ATF from corrected interferograms using the InSAR TS + AEM approach (Figure5). It indicates that the deformation rate of ATF ranges from –2 mm/year to 2 mm/year in the LOS direction. One clear point is that the LOS velocity changes across the ATF in both tracks. The opposite gradient trends on ascending and descending tracks were consistent with left-lateral motion of the ATF system. There is no clear correlation between the topography and the mean-velocity map, indicating that the topography-related atmospheric errors have been successfully removed and that no significant residual atmospheric effect related to the topography is affecting the results.

To test the effect of the isolated 10 October 2005 acquisition to the final velocity results, we performed the inversion without the 10 October 2005 data. The result is shown in Figure 6: the left one (Figure 6a) is the velocity of data set including 10 October 2005 data, whereas the right one (Figure 6b) is velocity result from data set without 10 October 2005 data. From the velocity map we can know that the 10 October 2005 data actually affects the result in some degree (the range of LOS velocity changes from –2.00~1.72 mm/year of data set including 10 October 2005 data to –2.20~1.92 mm/year without 10 October 2005 data), however the first-order deformation characteristics is consistent with the results obtained from all the dataset.

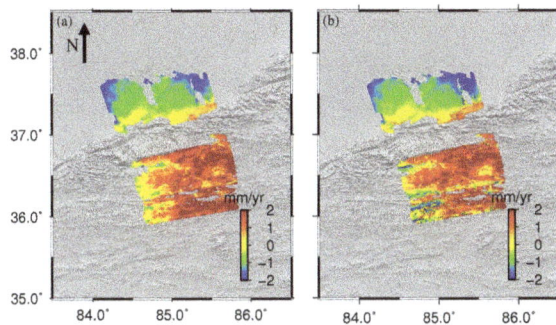

Figure 6. LOS velocity map of Ascending track 298 with (**a**) and without (**b**) the 10 October 2005 acquisition.

2.5. Strain Rate Map

Strain fields of a region describe the geometric changes caused by deformation and, therefore, its characteristic pattern and amplitude are very important to understanding the present-day tectonic processes and underlying driving forces in this area. The strain rate tensor can be used to study crustal deformation and characterize geodynamic processes, such as strain accumulation, independent of a reference frame. Therefore, it is common to derive the distribution of strain rate fields to capture the tectonic mechanism of a region.

To better constrain the deformation characteristics of the ATF, we calculated the strain rate field in the spherical coordinates from velocity fields using the wavelet-based multi-scale estimation method proposed by Tape *et al.* [26]. This method can localize a given deformation field in space and scale as well as detect outliers in the observation set. Based on the relative less normal deformation from a GPS study [11], we resolved the along-fault velocity using ascending and descending track LOS velocities and used it as the total horizontal velocity. Then we extrapolated the velocity field of this area using the Kriging interpolation method (Figure 7a). Finally, we calculated the three invariant quantities, related dilatation rate, strain rate and rotation rate, which do not depend on the coordinate system.

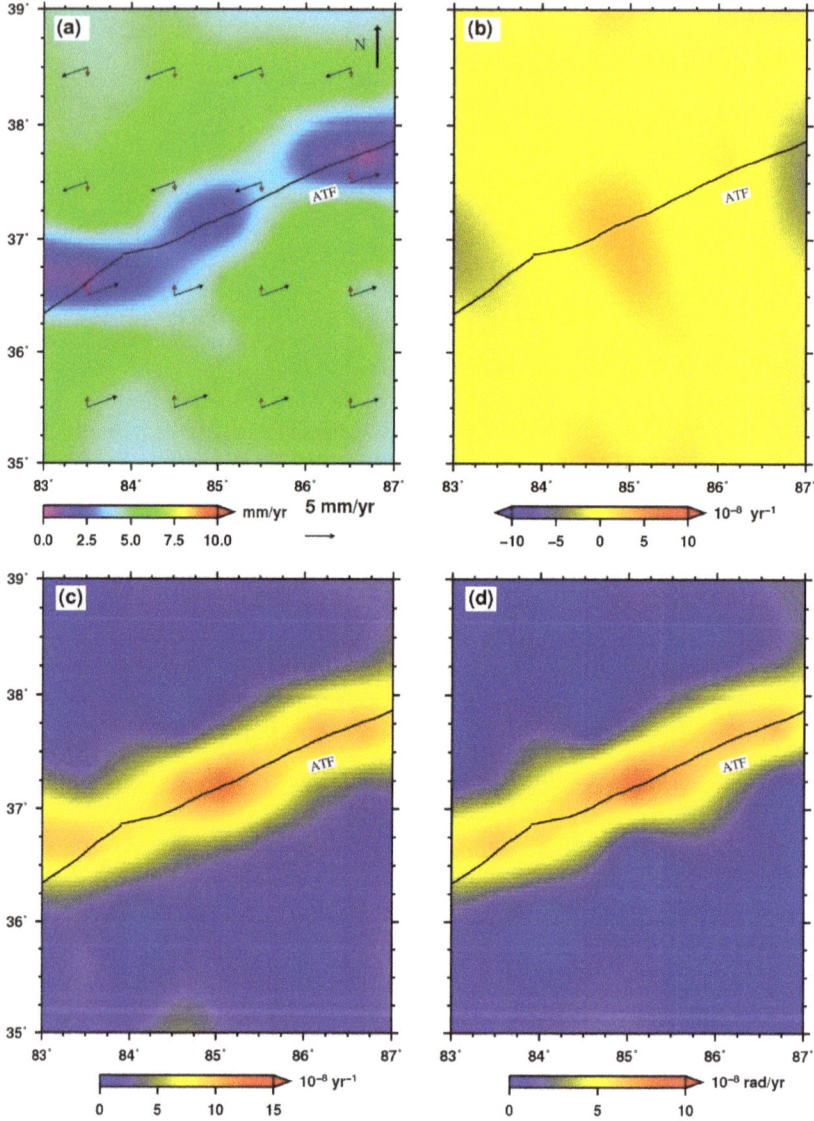

Figure 7. ATF strain map: (**a**) velocity map, the red arrows show the vertical velocity (directs north means upmotions and south means down motions) and the fat black arrow shows the north direction; (**b**) dilatation rate; (**c**) strain rate and (**d**) rotation rate.

The strain rate field (Figure 7c) exhibits the expected interseismic pattern for a locked fault with a maximum along the fault and symmetric on either side. Strain accumulation is strongly localized on the ATF. The northern Tarim Basin and southern high plateau regions are straining at very low rates in comparison to the major strike-slip faults. The maximum strain rate along the fault is approximately 12.4×10^{-8}/year. The dilatation rate was nearly zero. The rotation rate has a maximum of approximately 10×10^{-8}/year, indicating that rotation can partly account for the deformation in the vicinity of the fault. In comparison to previous studies, our results are consistent in one-order features, although we only calculate the strain rate field across the ATF rather than the whole Tibetan Plateau. The strain rate fields in Figure 7c reproduce the same first-order features as the velocity field, and show that strain is strongly focused on the main left-lateral strike-slip fault. In this study, we have shown that it is possible to create 2-D strain maps using InSAR data from two observations, even in the absence of GPS observations.

3. Modeling

As seen from the strain field in Figure 7c, the strain strongly focused on the main left-lateral strike-slip fault, so we can assume that the main contributions to the deformation are from the ATF. To aid in the interpretation of the first-order characteristics of the LOS velocity map, a simple elastic model with single fault geometry was used to invert the slip-rate and locking depth of the ATF. Although one more complex model may be more appropriate for this fault, the lack of data near the fault makes it hard to constrain the deformation using a complex model. Consequently, we assumed a strike-slip fault with $90°$ dip angle. We assumed that LOS deformation is a combination of fault parallel deformation and vertical deformation, ignoring normal fault deformation, which is justified by GPS measurements [11]. We modeled the fault as a buried infinite screw dislocation in a homogeneous, isotropic elastic half-space, where aseismic slip occurs at a rate (s) below a locking depth (d) during the interseismic period.

Under this assumption, we converted displacement of two profiles in LOS direction into horizontal displacements parallel to the fault and vertical displacements. We accounted for both horizontal and vertical displacement with two data tracks in our inversion rather than accounting for only horizontal displacement [16] with only one track, in our inversion process the contribution from vertical components is subtracted from the InSAR observations. The relationship between LOS displacements and horizontal and vertical displacement is given by:

$$d_{los} = u_n sin\varphi sin\theta - u_e cos\varphi sin\theta + u_u cos\theta \tag{1}$$

where φ is the azimuth angle for the satellite orbit, θ is the incidence angle, and $u_{(n,e,u)}$ are the north, east, and up components of displacement, respectively. We decomposed the horizontal component of the fault to parallel and normal components and ignored the normal component. As it is one-third smaller with respect to the parallel component and nearly vertical to the LOS direction, its projection to the LOS is negligible. Thus, Equation (1) can be rewritten as:

$$d_{los} = u_h cos\alpha sin\theta + u_u cos\theta \tag{2}$$

where α is the angle between the fault line and horizontal projection of LOS. We thus have two unknowns (u_h, u_u) and two equations, so it is possible to solve the equations.

The slip rate (s) and locking depth (d) of the fault were determined based on the Savage and Burford [63] analytical solution, which explains the surface velocity (y) parallel to the fault at a given distance (x) from the fault as:

$$y = \frac{s}{\pi} tan^{-1}\frac{x}{d} \tag{3}$$

We performed an initial parameter search over the 0.5–10.0 mm/year for slip rate and 5–35 km locking depth with the simulated annealing algorithm [64,65]. Our best-fit model for joint inversion, corresponding to the minimum of RMS misfit, yields a slip rate of 8.0 mm/year and a locking depth of 14.5 km. The simulation and residuals between observations and simulating results of D391 and A298 are presented in Figures 8 and 9 respectively. The residuals of A298 are not as smooth as D391. This probably results from fewer image dates in the inversion, which again confirms that additional observations can reduce errors.

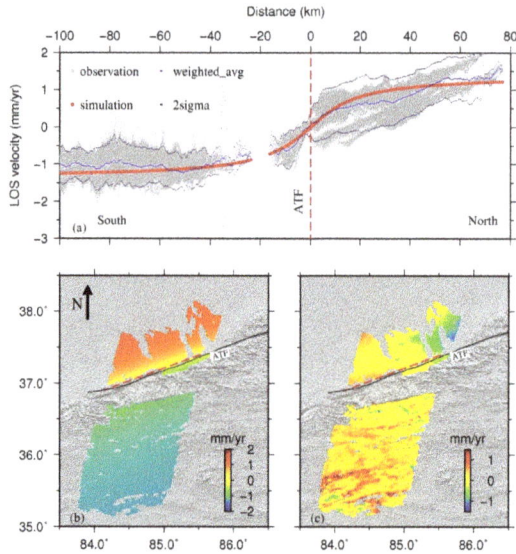

Figure 8. (a) D391 mean LOS velocity profiles (gray dots) and the weighted average profile (dark blue) with 2σ deviation (light blue). All of the points shown on the left are projected onto the profile as gray dots. The red line is the best-fit model with a slip rate of 8.0 mm/year and 14.5 km locking depth; (b) Model simulation map from the best-fit model values. The fat black arrow shows the north direction; (c) Residuals between model and observation results. The near zero residuals demonstrate the inversion goodness.

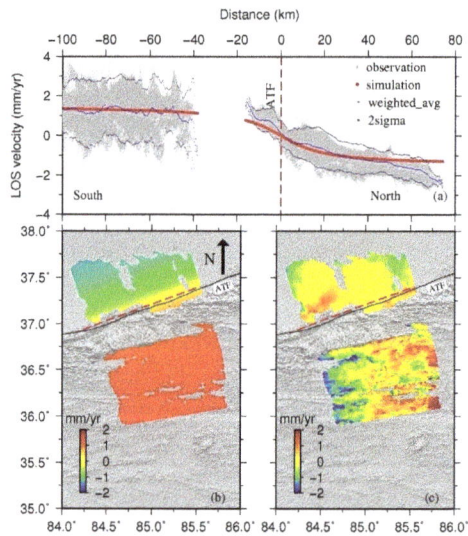

Figure 9. (a) A298 mean LOS velocity profiles (red dots) and the weighted average profile (dark blue) with 2σ deviation (light blue). All the points shown in the left are projected onto the profile as red dots. The red line is the best-fit model with a slip rate of 8.0 mm/year and 14.5 km locking depth; (b) Model simulation map from the best-fit model values. The fat black arrow shows the north direction; (c) Residuals between model and observation (Figure 5b) results.

Regarding error associated with these fault parameters, the main source remaining in the mean velocity map is related to residual turbulent atmospheric perturbations and orbital error in the interferograms. Therefore, we attempted to estimate the impact of such perturbations in the fault parameters by generating 100 perturbed mean velocity maps and then using the same inversion scheme as for the non-perturbed dataset. The parameter uncertainties were determined using the standard deviation of the slip and locking depth estimate from the 100 runs, which resulted in a slip rate of 8.0 ± 0.7 mm/year and 14.5 ± 3 km for locking depth (Figure 10).

We also performed the parameter search for each track data and assumed no vertical displacements. The best-fit parameters for D391 are 8.2 mm/year of slip and a locking depth of 16 km (blue dots in Figure 10), whereas the best-fit parameters for A298 are a slip rate of 8.3 mm/year and 14.8 km of locking depth (green dots in Figure 10). The results from the two tracks are in good agreement with each other. Figures 8 and 9 show profiles of the LOS velocity perpendicular to the strike direction of the ATF for the descending and ascending tracks, respectively, which are consistent overall with a classic arctangent shape predicted by screw dislocation models across strike-slip faults [63]. The residual maps show that the LOS velocity of the descending track (D391) fits well with the model, whereas the ascending track (A298) has some significant misfits in the southern part. This may be caused by unwrapping errors or less coherence resulting from a smaller number of SAR images.

The joint inversion results of both tracks are slightly slower because we ignored the vertical fault motion contribution in the single track inversion, as it is small relative to horizontal motion along the fault. We note that the joint inversion results are more concentrated near the best-fit parameters than the single track inversion, which demonstrates the importance of including the vertical displacements to better constrain the ATF deformation. The uncertainty for both slip rate and locking depth was reduced using SAR data from two directions as opposed to one single direction.

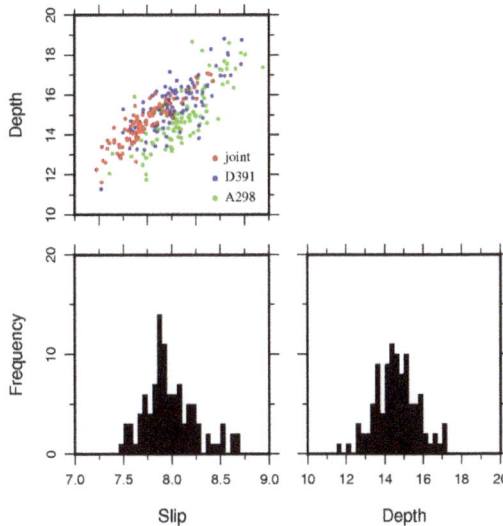

Figure 10. Results from the Monte Carlo error analysis. The red dots represent joint inversion whereas blue (D391) and green (A298) dots represent single inversions. The higher frequencies of slip rate and depth concentrate near 8.0 mm/year and 14.5 km, respectively.

4. Discussion

After planar orbital and linear atmospheric corrections, we computed the strain rate fields using a multi-scale wavelet method. The strain fields show that deformation across the ATF is concentrated on the fault. England and Molnar [2] calculated crustal velocity fields and strain rates within Asia from estimates of Quaternary slip rates on faults and baseline length rates between GPS sites. They suggested that the strain rate fields determined by geodetic measurements taken on a decadal timescale are consistent with geological observations on a 10^4 years timescale. Allmendinger *et al.* [66] derived 2-D strain and rotation rate fields from a GPS velocity field using both the nearest neighbor and distance weighted approach. Their results showed that the principal infinitesimal strain rate axes in Tibet are consistent with large, long term geological structures similar to England and Molnar [2]. Gan *et al.* [67] deduced the strain rate fields using ~726 GPS velocities around the plateau after removing the rotational component of the entire plateau. Although the method and data set are different from England and Molnar [2], their results were generally consistent in all of the relevant areas within the plateau. Our strain fields are consistent with the first-order characteristics across the major faults from previous studies [2]. This study confirms the possibility of creating 2-D strain rate maps from InSAR data and also shows that InSAR observations can constrain velocity and strain maps with respect to GPS data.

We inverted the slip rate and fault depth using a screw dislocation model. Our preferred model implies a left-lateral displacement rate across the ATF of 8.0 $\pm$ 0.7 mm/year and 14.5 $\pm$ 3 km locking depth. The locking depth was not as well constrained as the other geodetic inversion results [14]. We note that the slip rate is in good agreement with GPS measurements of 9 $\pm$ 4 mm/year [17] almost 1° east of our study area. It is also consistent with InSAR results of 11 $\pm$ 5 mm/year estimated from ERS data [16]. Similar to other geodetic measurements [15–17,68], we also identified the asymmetric deformation pattern between the two sides of the fault. The possible mechanisms to explain this interseismic velocity asymmetry maybe a variation of the elastic crustal thickness from rheological contrast on two sides of the ATF [68] or variation in fault dip angle.

Our results differ significantly from slip rates determined by geological methods (27 $\pm$ 7 mm/year; [18]) based on detailed measurements of different geomorphic terrace riser offsets using radiocarbon and cosmic ray dating. The discrepancy between these two methods could be due to the different time scales if both methods have no errors or slip rates vary with time like faults in Southern California [69]. Cowie *et al.* reported earlier that fault slip rates can vary in time or along strike the fault, both due to clustering or quiescence periods of the fault activity or due to long term effects [70,71]. Roberts *et al.* showed the first example of such fault slip rate variability from active faults [72]. Burchfiel *et al.* [73] suggested that whether the geodetically observed slip rates recorded over the last 5–10 years agree or are representative to the longer-term geologically derived slip rates is debatable. Indeed there are several examples where such a discrepancy occurs between geological and geodetic rates in several settings worldwide (e.g., Papanikolaou *et al.* [74]; Kenner and Simons, [75]) However, some geological authors suggest a low slip rate of the ATF that is consistent with our results. Yin *et al.* [76] derived an average rate of 9 $\pm$ 2 mm/year since 49 Ma. This significant difference was reconciled by geologic [77] and recent geodetic [17] data, which both agree on a slip rate of 9.0 $\pm$ 4.0 mm/year. As Cog will *et al.* [77,78] argued that the apparent discrepancy in slip rate along this fault might result more from systematic biases in geomorphic reconstructions rather than from true secular variation in slip rate.

Our consistent low slip rates along the fault are in consistent with the historical seismicity in this region. Only two great historical earthquakes have occurred along the ATF, both in 1924 west of 85°E [79,80]. In fact, contemporary instrumental recordings also show a low level of seismicity along the ATF. Paleoseismo logical studies suggest recurrence intervals of major earthquakes in the range of 700 $\pm$ 400 and 1100 $\pm$ 300 years for offsets of 4 and 7 m, respectively, consistent with a slip rate of ~10 mm/year [79,81,82].

Elliott *et al.* [16] measured the interseismic slip rate of the ATF using a stacking method with one single descending track of ERS data (1993–2000); their slip rate estimate was 11 $\pm$ 5 mm/year with significant uncertainty. Here, we used the InSAR TS + AEM method with 7 years (2003–2010) of ENVISAT ASAR data and both ascending and descending tracks; our slip rate estimate was

8.0 ± 0.7 mm/year. Our results are in agreement with results of Elliott *et al.* from ERS data [16] and GPS measurements (9 ± 4 mm/year; [17]). We used the same track ENVISAT ASAR data as [16] with an additional two years of descending track data and five years of ascending track data. The difference in slip rates between our study and Elliott *et al.* [16] is probably a result of time-varying slip rate or stacking method assumptions that the atmospheric effect was normally distributed. However, because the topographic variation across the ATF is large, maybe it is not appropriate to assume normal distribution atmospheric errors. We had to remove any topography-correlated atmospheric errors. This study again demonstrates the importance of atmospheric and orbital corrections in InSAR time series processing. We determined our slip rate from 216 interferograms, while they used only 66 interferograms, which possibly cannot constrain the InSAR slip rate as well. Additionally, because we used ascending and descending track data, we were able to calculate the velocity parallel to the ATF as well as vertical velocity from two observing geometries and thus remove the projection of vertical deformation velocity from the LOS velocity. Elliott *et al.* [16] simply ignored the vertical velocity whose contributions are subtracted from the InSAR observations in our study, so their result was a little higher than ours. It is likely our calculated velocity is closer to the real velocity. We were also able to reduce the range of uncertainty for both slip rate and locking depth using two look directions of data because we used both descending and ascending track data, in comparison to the results of [16] that are based on descending track data only. This improvement demonstrates the importance of using SAR data from two look directions for interseismic studies.

It is worth noting that we simplify the model by disregarding deformation produced by the other possible surrounding secondary faults in the study area. The real tectonic setting is definitely more complex than the simple model we have used in this study. The reasons we used this simple model are that reliable SAR data near the fault were not available, which may be insufficient for a distributed slip inversion in this case; and the fault-normal and vertical deformations across the ATF are relative less with respect the strike slip [1,2,14,15]. Our results may include contributions of surrounding secondary faults or other mechanisms like the shrinkage or expansion of clay minerals, or even excessive water pumping or significant seasonal variations of the water table. In addition, our SAR data covered only seven years, the low temporal resolution results prevent us from extracting the varying history of fault slip rate and locking depth. Our SAR data covered only a segment of the whole ATF, we assume that our results from SAR data can represent the first-order deformation characteristics of the ATF. The detailed deformation needs further geodetic or geological study.

In Figure 7a, we have shown the vertical component of the interpolated LOS velocity obtained from two tracks of SAR data. If we assumed that the Tarim basin is static, then the south part across the ATF is uplifting with respect to the Tarim basin with nearly 3 mm/year velocity in agreements with Bendickand coworkers' [14] and England and Molnar's [1] results. England and Molnar [1] estimated a shortening rate of 6 ± 4 mm/year across the central section of the ATF. Bendick *et al.* [14] reported a contraction rate of 3 ± 1 mm/year across the same section. Chen *et al.* [83] estimated 5 ± 2 mm/year across the east section at ~95°E. Shen *et al.* [15] estimate 0 ± 2 mm/year convergence across the central Altyn Tagh system, which indicates strongly that north-south shortening across the ATF is negligible. Also, geologic mapping along the NATF by Cog will *et al.* [84] suggested that it is mostly a left slip fault with only minor compression.

Previous geodetic results based on GPS [14,17,79] and InSAR [16,68] data suggested a slip rate of~10 mm/year at different points across the ATF. Our interseismic strain accumulation profiles across the ATF are in good agreement with these previous GPS measurements and InSAR results. The slip rates and locking depth for the ATF estimated from both GPS and InSAR studies and from geological studies [77,78] are also consistent with our results. This consistency indicates that the ATF played only a secondary role in the northeastward extrusion of material in the Tibetan Plateau [79], with most of the extrusion occurring in the interior of the plateau. Therefore, our results are close to the continuum models of Asian deformation requiring low fault slip rates [2,85].

5. Conclusions

Using a multi-scale wavelet method, we calculated the strain rate fields of the ATF region and determined that the strain was mainly localized along the fault, which is generally consistent with previous studies. We also show that high-resolution 2-D strain rate fields can be determined from InSAR data alone, especially in regions with spatially sparse GPS data. Based on the orbital ramp and atmospheric effect correction, we inverted the slip rate and locking depth using a screw dislocation model with ascending and descending track ENVISAT data. We performed the inversions for single track data and jointly. Our joint inversion result indicates that the ATF has a slip rate of 8.0 ± 0.7 mm/year and locking depth of 14.5 ± 3 km. The joint inversion slip rate is a little slower than the single track inversion because of vertical deformation in the inversion. Our results are more accurate than those of Elliott *et al.* [16] owing to a greater number of observations in our study and the fact that we accounted for the vertical deformation in the inversion process. This study shows the importance of using both descending and ascending data in interseismic InSAR studies. As a result of the more frequent data acquisition from satellite missions, such as Sentinel-1A, ALOS-2 and the upcoming Sentinel-1B, similar studies will become increasingly common. The relative low slip rate indicates that the ATF does not transfer a significant part of the convergence between India and Asia into the northeastward extrusion of the Tibetan Plateau [76], and the deformation pattern is close to the continuum deformation models [2].

Acknowledgments: This work is co-supported by the National Natural Science Foundation of China (key program) under grants No.41431069, the National Key Basic Research Development Program (973 program) under grants No. 2013CB733304 and 2013CB733303, the China National Special Fund for Earthquake Scientific Research in Public Interest under grants No.201308009, the National Natural Science Foundation of China under grants No. 41204010 and 41404007and the Special Project of Basic Work of Science and Technology under grants No. 2015FY210400). The Envis at SAR data are provided by ESA through C1F project (ID: 29558).

Author Contributions: Caijun Xu led the research work, proposed the crucial suggestions and structure of this manuscript. Sen Zhu and Yangmao Wen designed the research, and organized the contents of the manuscript. Sen Zhu processed data and wrote the first draft of the manuscript. CaijunXu, Yangmao Wen and Yang Liu reviewed and edited the manuscript.

Conflicts of Interest: The authors declare no conflict of interest.

References

1. England, P.; Molnar, P. Active deformation of Asia: From kinematics to dynamics. *Science* **1997**, *278*, 647–650. [CrossRef]
2. England, P.; Molnar, P. Late Quaternary to decadal velocity fields in Asia. *J. Geophys. Res.* **2005**, *110*, B12401. [CrossRef]
3. Meade, J. Present-day kinematics at the India-Asia collision zone. *Geology* **2007**, *35*, 81–84. [CrossRef]
4. Thatcher, W. Microplate model for the present-day deformation of Tibet. *J. Geophys. Res.* **2007**, *112*, B01401. [CrossRef]
5. Zhang, L.; Ye, G.; Jin, S.; Wei, W.; Unsworth, M.; Jones, A.; Jing, J.; Dong, H.; Xie, C.; Le Pape, F.; *et al.* Lithospheric electrical structure across the Eastern Segment of the Altyn Tagh Fault on the Northern Margin of the Tibetan Plateau. *Acta Geol. Sin.* **2015**, *89*, 90–104.
6. Molnar, P.; Tapponnier, P. Cenozoic tectonics of Asia: Effects of a continental collision. *Science* **1975**, *189*, 419–426. [CrossRef] [PubMed]
7. Avouac, J.-P.; Tapponnier, P. Kinematic model of active deformation in central Asia. *Geophys. Res. Lett.* **1993**, *20*, 895–898. [CrossRef]
8. Zhang, P.Z.; Shen, Z.K.; Wang, M.; Gan, W.; Bürgmann, R.; Molnar, P.; Wang, Q.; Niu, Z.; Sun, J.; Wu, J. Continuous deformation of the Tibetan Plateau from Global Positioning System data. *Geology* **2004**, *32*, 809–812. [CrossRef]
9. Wittlinger, G.; Tapponnier, P.; Poupinet, G.; Jiang, M.; Shi, D.; Herquel, G.; Masson, F. Tomographic evidence for localized lithospheric shear along the AltynTaghFault. *Science* **1998**, *282*, 74–76. [CrossRef] [PubMed]

10. Burchfiel, B.C.; Deng, Q.; Molnar, P.; Royden, L.; Wang, Y.; Zhang, P.; Zhang, W. Intracrustal detachment within zones of continental deformation. *Geology* **1989**, *17*, 748–752. [CrossRef]

11. Ge, W.-P.; Molnar, P.; Shen, Z.-K.; Li, Q. Present-day crustal thinning in the southern and northern Tibetan Plateau revealed by GPS measurements. *Geophys. Res. Lett.* **2015**, *42*, 5227–5235. [CrossRef]

12. Wallace, K.; Yin, G.; Bilham, R. Inescapable slow slip on the Altyn Tagh Fault. *Geophys. Res. Lett.* **2004**, *31*, L09613. [CrossRef]

13. Houseman, G.; England, P. Crustal thickening *vs.* lateral expulsion in the Indian-Asian continental collision. *J. Geophys. Res.* **1993**, *98*, 12233–12249. [CrossRef]

14. Bendick, R.; Bilham, P.; Freymueller, J.T.; Larson, K.M.; Yin, G. Geodetic evidence for a low slip rate in the Altyn Tagh fault system. *Nature* **2000**, *386*, 61–64.

15. Shen, Z.-K.; Wang, M.; Li, Y.; Jackson, D.D.; Yin, A.; Dong, D.; Fang, P. Crustal deformation along the Altyn Tagh fault system, western China, from GPS. *J. Geophys. Res.* **2001**, *106*, 30607–30622. [CrossRef]

16. Elliott, J.R.; Biggs, J.; Parsons, B.; Wright, T.J. InSAR slip rate determination on the Altyn Tagh Fault, northern Tibet, in the presence of topographically correlated atmospheric delays. *Geophys. Res. Lett.* **2008**, *35*, L12309. [CrossRef]

17. He, J.; Vernant, P.; Chéry, J.; Wang, W.; Lu, S.; Ku, W.; Xia, W.; Bilham, R. Nailing down the slip rate of the AltynTaghFault. *Geophys. Res. Lett.* **2013**, *40*, 5382–5386. [CrossRef]

18. Mériaux, A.-S.; Ryerson, F.J.; Tapponnier, P.; Van der Woerd, J.; Finkel, R.C.; Xu, X.; Xu, Z.; Caffee, M.W. Rapid slip along the central Altyn Tagh Fault: Morphochronologic evidence from Cherchen He and Sulamu Tagh. *J. Geophys. Res.* **2004**, *109*, B06401. [CrossRef]

19. Rosen, P.A.; Hensley, S.; Zebker, H.A.; Webb, F.H.; Fielding, E.J. Surface deformation and coherence measurements of Kilauea Volcano, Hawaii, from SIR-C radar interferometry. *J. Geophys. Res. Planets* **1996**, *101*, 23109–23125. [CrossRef]

20. Massonnet, D.; Rossi, M.; Carmona, C.; Adragna, F.; Peltzer, G.; Feigl, K.; Rabaute, T. The displacement field of the Landers earthquake mapped by radar interferometry. *Nature* **1993**, *364*, 138–142. [CrossRef]

21. Wright, T.; Parsons, B.; Fielding, E.J. Measurement of interseismic strain accumulation across the North Anatolian Fault by satellite radar interferometry. *Geophys. Res. Lett.* **2001**, *28*, 2117–2120. [CrossRef]

22. Walters, R.J.; Holley, R.J.; Parsons, B.; Wright, T.J. Interseismic strain accumulation across the North Anatolian Fault from Envisat InSAR measurements. *Geophys. Res. Lett.* **2011**, *38*, L05303. [CrossRef]

23. Xu, C.J.; Wang, H.; Jiang, G.Y. Study on crustal deformation of Wenchuan Ms8.0 earthquake using wide-swath scan SAR and MODIS. *Geod. Geodyn.* **2011**, *2*, 1–6.

24. Liu, Y.; Xu, C.J.; Wen, Y.M. InSAR measurement of surface deformation between two Da-Qaidam Mw6.3 earthquakes and joint analysis with coseismic rupture. *Geod. Geodyn.* **2016**, *36*, 110–114.

25. Papanikolaou, I.D.; Balen, R.V.; Silva, P.G.; Reicherter, K. Geomorphology of active faulting and seismic hazard assessment: New tools and future challenges. *Geomorphology* **2015**, *237*, 1–13. [CrossRef]

26. Tape, C.; Musé, P.; Simons, M.; Dong, D.; Webb, F. Multiscale estimation of GPS velocity fields. *Geophys. J. Int.* **2009**, *179*, 945–971. [CrossRef]

27. Walters, R.J.; Parsons, B.; Wright, T.J. Constraining crustal velocity fields with InSAR for Eastern Turkey: Limits to the block-like behavior of Eastern Anatolia. *J. Geophys. Res.* **2014**, *119*, 5215–5234. [CrossRef]

28. Garthwaite, M.C.; Wang, H.; Wright, T.J. Broadscale interseismic deformation and fault slip rates in the central Tibetan Plateau observed using InSAR. *J. Geophys. Res. Solid Earth* **2013**, *118*, 5071–5083. [CrossRef]

29. Ryder, I.; Parsons, B.; Wright, T.J.; Funning, G. Post-seismicmotion following the 1997 Manyi, Tibet earthquake, InSAR observations and modelling. *Geophys. J. Int.* **2007**, *169*, 1009–1027. [CrossRef]

30. Rosen, P.A.; Henley, S.; Peltzer, G.; Simons, M. Update repeat orbit interferometry package released. *Eos Trans. AGU* **2004**, *85*, 47. [CrossRef]

31. Doin, M.P.; Guillaso, S.; Jolivet, R.; Lasserre, C.; Lodge, F.; Ducret, G.; Grandin, R. Presentation of the small baseline NSBAS processing chain on a case example: The Etna deformation monitoring from 2003 to 2010 using Envisat data. In Proceedings of the European Space Agency Symposium "Fringe", ESA SP-697. Frascati, Italy, 19–23 September 2011.

32. NSBAS. Available online: http://efidir.poleterresolide.fr/index.php/effidir-tools/nsbas (accessed on 8 March 2016).

33. Doin, M.P.; Lasserre, C.; Peltzer, G.; Cavalie, O.; Doubre, C. Corrections of stratified tropospheric delays in SAR interferometry: Validation with global atmospheric models. *J. Appl. Geophys.* **2009**, *69*, 35–50. [CrossRef]

34. Farr, T.G.; Rosen, P.A.; Caro, E.; Crippen, R.; Duren, R.; Hensley, S.; Alsdorf, D. The shuttle radar topography mission. *Rev. Geophys.* **2007**, *45*. [CrossRef]
35. Goldstein, R.M.; Werner, C.L. Radar interferogram filtering for geophysical applications. *Geophys. Res. Lett.* **1998**, *25*, 4035–4038. [CrossRef]
36. Goldstein, R.M.; Zebker, H.A.; Werner, C.L. Satellite radar interferometry—Two-dimensional phase unwrapping. *Radio Sci.* **1988**, *23*, 713–720. [CrossRef]
37. Biggs, J.; Wright, T.J.; Lu, Z.; Parsons, B. Multi-interferogram method for measuring interseismic deformation: Denali fault, Alaska. *Geophys. J. Int.* **2007**, *170*, 1165–1179. [CrossRef]
38. Wen, Y.; Li, Z.; Xu, C.; Ryder, I.; Bürgmann, R. Postseismic motion after the 2001 MW7.8 Kokoxili earthquake in Tibet observed by InSAR time series. *J. Geophys. Res.* **2012**, *117*, B08405. [CrossRef]
39. Puysségur, B.; Michel, R.; Avouac, J.-P. Tropospheric phase delay in interferometric synthetic aperture radar estimated from meteorological model and multispectral imagery. *J. Geophys. Res.* **2007**, *112*, B05419. [CrossRef]
40. Hanssen, R. *Radar Interferometry: Data Interpretation and Error Analysis*; Kluwer Academic Publishers: Dordrecht, The Netherlands, 2001.
41. Cavalié, O.; Doin, M.-P.; Lasserre, C.; Briole, P. Ground motion measurement in the Lake Mead area, Nevada, by differential synthetic aperture radar interferometry time series analysis: Probing the lithosphere rheological structure. *J. Geophys. Res.* **2007**, *112*, B03403. [CrossRef]
42. Wicks, C.W.; Dzurisin, D.; Ingebritsen, S.; Thatcher, W.; Lu, Z.; Iverson, J. Magmatic activity beneath the quiescent three sisters volcanic center, central Oregon Cascade Range, USA. *Geophys. Res. Lett.* **2002**, *29*. [CrossRef]
43. Delacourt, C.; Briole, P.; Achache, J.A. Tropospheric corrections of SAR interferograms with strong topography: Application to Etna. *Geophys. Res. Lett.* **1998**, *25*, 2849–2852. [CrossRef]
44. Williams, S.; Bock, Y.; Fang, P. Integrated satellite interferometry: Tropospheric noise, GPS estimates and implications for interferometric synthetic aperture radar products. *J. Geophys. Res.* **1998**, *103*, 27051–27067. [CrossRef]
45. Li, Z.; Fielding, E.J.; Cross, P.; Muller, J.-P. Interferometric synthetic aperture radar atmospheric correction: GPS topography-dependent turbulence model. *J. Geophys. Res.* **2006**, *111*, B02404. [CrossRef]
46. Onn, F.; Zebker, H.A. Correction for interferometric synthetic aperture radar atmospheric phase artifacts using time series of zenith wet delay observations from a GPS network. *J. Geophys. Res.* **2006**, *111*, B09102. [CrossRef]
47. Li, Z.; Fielding, E.J.; Cross, P.; Muller, J.-P. Interferometric synthetic aperture radar atmospheric correction: medium-resolution imaging spectrometer and advanced synthetic aperture radar integration. *Geophys. Res. Lett.* **2006**, *33*, L06816. [CrossRef]
48. Li, Z.W.; Xu, W.B.; Feng, G.C.; Hu, J.; Wang, C.C.; Ding, X.L.; Zhu, J.J. Correcting atmospheric effects on InSAR with MERIS water vapour data and elevation-dependent interpolation model. *Geophys. J. Int.* **2012**, *189*, 898–910. [CrossRef]
49. Wadge, G.; Dodson, A.; Waugh, S.; Veneboer, T.; Puglisi, G.; Mattia, M.; Baker, D.; Edwards, S.C.; Edwards, S.J.; Clarke, P.J.; *et al.* Atmospheric models, GPS and InSAR measurements of the tropospheric water vapour field over Mount Etna. *Geophys. Res. Lett.* **2002**, *29*, 1905. [CrossRef]
50. Foster, J.; Brooks, B.; Cherubini, T.; Shacat, C.; Businger, S.; Werner, C.L. Mitigating atmospheric noise for InSAR using a high resolution weather model. *Geophys. Res. Lett.* **2006**, *33*, L16304. [CrossRef]
51. Foster, J.; Kealy, J.; Cherubini, T.; Businger, S.; Lu, Z.; Murphy, M. The utility of atmospheric analyses for the mitigation of artifacts in InSAR. *J. Geophys. Res.* **2013**, *118*, 748–758. [CrossRef]
52. Li, Z.; Fielding, E.J.; Cross, P.; Preusker, R. Advanced InSAR atmospheric correction: MERIS/MODIS combination and stacked water vapour models. *Int. J. Remote Sens.* **2009**, *30*, 3343–3363. [CrossRef]
53. Bekaert, D.; Hooper, A.; Wright, T. A spatially variable power law tropospheric correction technique for InSAR data. *J. Geophys. Res.* **2015**, *120*, 1345–1356. [CrossRef]
54. Bekaert, D.; Walters, R.; Wright, T.; Hooper, A.; Parker, D. Statistical comparison of InSAR tropospheric correction techniques. *Remote Sens. Environ.* **2015**, *170*, 40–47. [CrossRef]
55. Dee, D.P.; Uppala, S.; Simmons, A.; Berrisford, P.; Poli, P.; Kobayashi, S.; Andrae, U.M.; Balmaseda, A.; Balsamo, G.; Bauer, P.; *et al.* The ERA-Interim reanalysis: Configuration and performance of the data assimilation system. *Q. J. R. Meteorol. Soc.* **2011**, *137*, 553–597. [CrossRef]

56. Walters, R.J.; Elliott, J.R.; Li, Z.; Parsons, B. Rapid strain accumulation on the Ashkabad fault (Turkmenistan) from atmosphere-corrected InSAR. *J. Geophys. Res.* **2013**, *118*, 1–17. [CrossRef]

57. Zebker, H.; Rosen, P.; Goldstein, R.M. On the derivation of coseismic displacement fields using differential radar interferometry: The Landers earthquake. *J. Geophys. Res.* **1994**, *99*, 19617–19634. [CrossRef]

58. Pritchard, M.E.; Simons, M. A satellite geodetic survey of large-scale deformation of volcanic centres in the central Andes. *Nature* **2002**, *418*, 167–171. [CrossRef] [PubMed]

59. Berardino, P.; Fornaro, G.; Lanari, R.; Sansosti, E. A new algorithm for surface deformation monitoring based on small baseline differential SAR interferograms. *IEEE Trans. Geosci. Remote Sens.* **2002**, *40*, 2375–2383. [CrossRef]

60. Mora, O.; Mallorqui, J.J.; Broquetas, A. Linear and nonlinear terrain deformation maps Fromareduced set of interferometric SAR images. *IEEE Trans. Geosci. Remote Sens.* **2003**, *41*, 2243–2253. [CrossRef]

61. Lundgren, P.; Hetland, E.A.; Liu, Z.; Fielding, E.J. Southern San Andreas-San Jacinto fault system slip rates estimated from earthquake cycle models constrained by GPS and interferometric synthetic aperture radar observations. *J. Geophys. Res.* **2009**, *114*. [CrossRef]

62. Hammond, W.C.; Blewitt, G.; Li, Z.; Plag, H.P.; Kreemer, C. Contemporary uplift of the Sierra Nevada, western United States, from GPS and InSAR measurements. *Geology* **2012**, *40*, 667–670. [CrossRef]

63. Savage, G.D.; Burford, R.O. Geodetic determination of relative plate motion in central California. *J. Geophys. Res.* **1973**, *5*, 832–845. [CrossRef]

64. Shirzaei, M.; Walter, T.R. Randomly iterated search and statistical competency as powerful inversion tools for deformation source modeling: Application to volcano interferometric synthetic aperture radar data. *J. Geophys. Res.* **2009**, *114*, B10401. [CrossRef]

65. He, P.; Wen, Y.; Xu, C.; Liu, Y.; Fok, H.S. New evidence for active tectonics at the boundary of the Kashi depression, China, from time series InSARobservations. *Tectonophysics* **2015**, *653*, 140–148. [CrossRef]

66. Allmendinger, R.W.; Reilinger, R.; Loveless, J. Strain and rotation rate from GPS in Tibet, Anatolia, and the Altiplano. *Tectonics* **2007**, *26*, TC3013. [CrossRef]

67. Gan, W.J.; Zhang, P.Z.; Shen, Z.K.; Niu, Z.J.; Wang, M.; Wan, Y.G.; Zhou, D.M.; Cheng, J. Present-day crustal motion within the Tibetan Plateau inferredfrom GPS measurements. *J. Geophys. Res.* **2007**, *112*, B08416. [CrossRef]

68. Jolivet, R.; Cattin, R.; Chamot-Rooke, N.; Lasserre, C.; Peltzer, G. Thin-plate modeling of interseismic deformation and asymmetry across the Altyn Tagh Fault zone. *Geophys. Res. Lett.* **2008**, *35*, L02309. [CrossRef]

69. Dolan, J.F.; Bowman, D.D.; Briole, C.G. Long-range and long term fault interactions in southern California. *Geology* **2007**, *35*, 855–858. [CrossRef]

70. Cowie, P.A.; Gupta, S.; Dawers, N.H. Implications of fault array evolution for synriftdepocentre development: Insights from a numerical fault growth model. *Basin Res.* **2000**, *12*, 241–261. [CrossRef]

71. Cowie, P.A. A healing-reloading feedback control on the growth rate of seismogenic faults. *J. Struct. Geol.* **1998**, *20*, 1075–1087. [CrossRef]

72. Roberts, G.P.; Michetti, A.M.; Cowie, P.; Morewood, N.C.; Papanikolaou, I. Fault slip-rate variations during crustal-scale strain localisation, Central Italy. *Geophys. Res. Lett.* **2002**, *29*, 91–94. [CrossRef]

73. Burchfiel, B.C.; Wang, E. Northwest-trending, Middle Cenozoic, left-lateral faults in southern Yunnan, China, and their tectonic significance. *J. Struct. Geol.* **2003**, *25*, 781–792. [CrossRef]

74. Papanikolaou, I.D.; Roberts, G.P.; Michetti, A.M. Fault scarps and deformation rates in Lazio-Abruzzo, Central Italy: Comparison between geological fault slip-rate and GPS data. *Tectonophysics* **2005**, *408*, 147–176. [CrossRef]

75. Kenner, S.J.; Simons, M. Temporal clustering of major earthquakes along individual faults due to post-seismic reloading. *Geophys. J. Int.* **2005**, *160*, 179–194. [CrossRef]

76. Yin, A.; Rumelhart, P.E.; Butler, R.; Cowgill, E.; Harrison, T.M.; Foster, D.A.; Ingersoll, R.V.; Zhang, Q.; Zhou, X.Q.; Wang, X.F.; *et al.* Tectonic history of the Altyn Tagh Fault system in northern Tibet inferred from Cenozoic sedimentation. *Geol. Soc. Am. Bull.* **2002**, *114*, 1257–1295. [CrossRef]

77. Cowgill, E. Impact of riser reconstructions on estimation of secular variation in rates of strike slip faulting: Revisiting the Cherchen River site along the Altyn Tagh Fault, NW China. *Earth Planet. Sci. Lett.* **2007**, *254*, 239–255. [CrossRef]

78. Cowgill, E.; Gold, R.D.; Chen, X.; Wang, X.-F.; Arrowsmith, J.R.; Southon, J.R. Low quaternary slip rate reconciles geodetic and geologic rates along the Altyn Tagh Fault, northwestern Tibet. *Geology* **2009**, *37*, 647–650. [CrossRef]

79. Zhang, P.Z.; Molnar, P.; Xu, X. Late Quaternary and present-day rates of slip along the Altyn Tagh Fault, northern margin of the Tibetan Plateau. *Tectonics* **2007**, *26*, TC5010. [CrossRef]

80. Gu, G.; Lin, T.; Shi, Z. *Catalogue of Chinese Earthquakes (1831 BC-1969 AD)*; Science Press: Beijing, China, 1989; pp. 1373–1388.

81. Washburn, Z.; Arrowsmith, J.R.; Forman, S.L.; Cowgill, E.; Wang, X.-F.; Zhang, Y.-Q.; Chen, Z.-L. Late Holocene earthquake history of the central Altyn Tagh Fault, China. *Geology* **2001**, *29*, 1051–1054. [CrossRef]

82. Washburn, Z.; Arrowsmith, J.R.; Dupont-Nivet, G.; Wang, X.-F.; Zhang, Y.-Q.; Chen, Z.-L. Paleoseismology of the Xorxol segment of the central Altyn Tagh Fault, Xinjiang, China. *Ann. Geophys.* **2001**, *46*, 1015–1034.

83. Chen, Z.; Burchfiel, B.C.; Liu, Y.; King, R.W.; Royden, L.H.; Tang, W.; Wang, E.; Zhao, J.; Zhang, X. Global Positioning system measurements from eastern Tibet and their implications for India/Eurasiaintercontinental deformation. *J. Geophys. Res.* **2000**, *105*, 215–227.

84. Cowgill, E.; Yin, A.; Wang, X.F.; Zhang, Q. Is the North Altyn Fault part of a strike-slip duplex along the Altyn Tagh fault system. *Geology* **2000**, *28*, 255–258. [CrossRef]

85. Vergnolle, M.; Calais, E.; Dong, L. Dynamics of continental deformation in Asia. *J. Geophys. Res.* **2007**, *112*, B11403. [CrossRef]

remote sensing

MDPI

Article

Time-Dependent Afterslip of the 2009 Mw 6.3 Dachaidan Earthquake (China) and Viscosity beneath the Qaidam Basin Inferred from Postseismic Deformation Observations

Yang Liu [1,2,3,4,*], Caijun Xu [1,2,3], Zhenhong Li [4], Yangmao Wen [1,2,3], Jiajun Chen [4] and Zhicai Li [5]

[1] School of Geodesy and Geomatics, Wuhan University, Wuhan 430079, China; cjxu@sgg.whu.edu.cn (C.X.); ymwen@sgg.whu.edu.cn (Y.W.)
[2] Key Laboratory of Geospace Environment and Geodesy, Ministry of Education, Wuhan University, Wuhan 430079, China
[3] Collaborative Innovation Center for Geospatial Technology, Wuhan 430079, China
[4] COMET, School of Civil Engineering and Geosciences, Newcastle University, Newcastle upon Tyne NE1 7RU, UK; Zhenhong.Li@newcastle.ac.uk (Z.L.); J.Chen26@newcastle.ac.uk (J.C.)
[5] Department of Geodesy, National Geomatics Center of China, Beijing 100048, China; zcli@nsdi.gov.cn
*   Correspondence: Yang.Liu@sgg.whu.edu.cn; Tel.: +86-27-6877-8404

Academic Editors: Roberto Tomas, Salvatore Stramondo and Prasad S. Thenkabail
Received: 30 June 2016; Accepted: 3 August 2016; Published: 10 August 2016

Abstract: The 28 August 2009 Mw 6.3 Dachaidan (DCD) earthquake occurred at the Qaidam Basin's northern side. To explain its postseismic deformation time series, the method of modeling them with a combination model of afterslip and viscoelastic relaxation is improved to simultaneously assess the time-dependent afterslip and the viscosity. The coseismic slip model in the layered model is first inverted, showing a slip pattern close to that in the elastic half-space. The postseismic deformation time series can be explained by the combination model, with a total root mean square (RMS) misfit of 0.37 cm. The preferred time-dependent afterslip mainly occurs at a depth from the surface to about 9.1 km underground and increases with time, indicating that afterslip will continue after 28 July 2010. By 334 days after the main shock, the moment released by the afterslip is 0.91×10^{18} N·m (Mw 5.94), approximately 24.3% of that released by the coseismic slip. The preferred lower bound of the viscosity beneath the Qaidam Basin's northern side is 1×10^{19} Pa·s, close to that beneath its southern side. This result also indicates that the viscosity structure beneath the Tibet Plateau may vary laterally.

Keywords: afterslip; the 2009 Dachaidan earthquake; viscosity; the Qaidam Basin; postseismic deformation; InSAR

1. Introduction

On 28 August 2009, an Mw 6.3 earthquake occurred at the Dachaidan (DCD) district, Qinghai province in China [1–4]. About ten months before this event, another Mw 6.3 event occurred at almost the same location (Figure 1). The two events threaten human beings moderately. The related rupturing faults are part of the northern fold-and-thrust belts of the Qaidam Basin, which has a strike of northwestern-west (NWW) to southeastern-east (SEE) and divides the northeastern margin of the Tibet Plateau into the Qaidam block and the Qilianshan block [5,6]. This zone has the tectonic characteristics of significant uplift in the geologic history [5,6]. However, because of the complex topography and the arduous working environment, the related observations and explanations are not yet sufficient, in particular for the zone around this event. Previous studies suggested that the occurrence of earthquake can present an opportunity to explore the earthquake behaviors and rheological properties of the regional lithosphere materials [7–18].

Geodetic observations of surface deformation related to the active fault can be used to characterize the behaviors of the earthquake cycle [19]. During the interseismic phase, Global Positioning System (GPS) observations of crustal strain within the Tibet Plateau suggested that the fault belts around the 2009 event are characterized by the thrust movement along vertical direction and the left-lateral movement along horizontal direction [20]. During the coseismic phase, Interferometric Synthetic Aperture Radar (InSAR) observations of surface deformation and the elastic dislocation model inversion using both uniform and distributed slip models indicated that the 2008 event ruptured the lower half of the brittle seismogenic layer, while the 2009 event ruptured the upper half [1,3,21]. During the postseismic phase, InSAR deformation time series for the first 334 days after the 2009 event displayed that the postseismic deformation changes from fast to slow with time, and has a similar spatial pattern with the coseismic deformation [22,23]. On the geophysical interpretation of the postseismic deformation, Feng [22] modeled it using the postseismic afterslip model only.

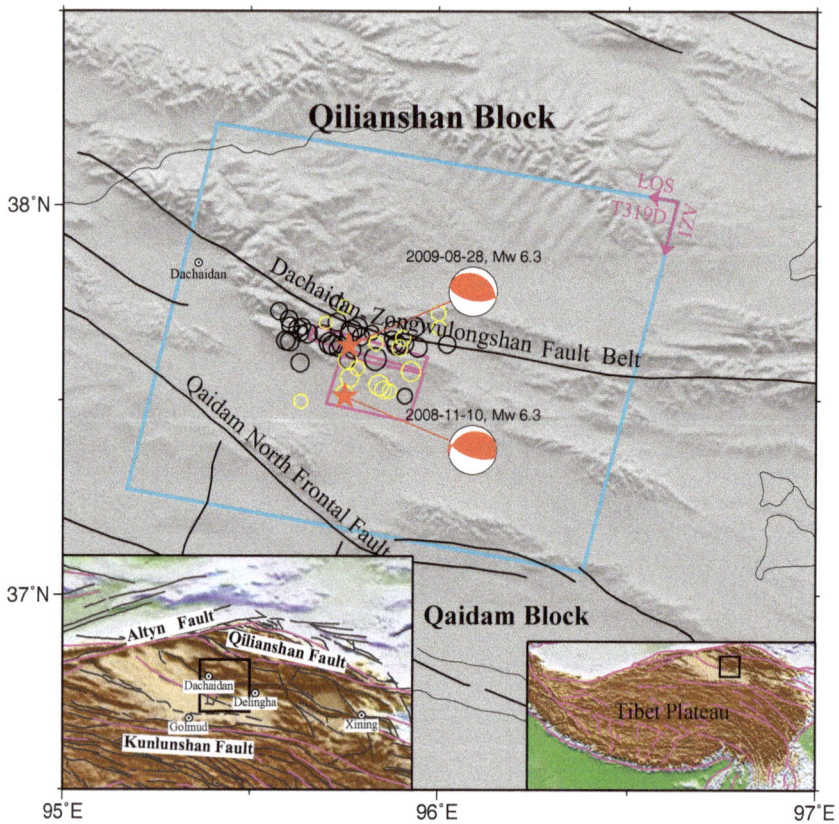

Figure 1. Tectonics associated with the 2009 Mw 6.3 DCD earthquake. The bottom-left and bottom-right insets show the location of the main figure, respectively. The light blue rectangle is the spatial extent of the Envisat Advanced Synthetic Aperture Radar (ASAR) descending Track 319 images, with AZI and LOS referring to satellite azimuth and look directions. The focal mechanisms of the 2008 and 2009 events are from United States Geological Survey (USGS) [4]. The two purple rectangles are the surface projections of the main fault rupturing zones during the 2008 and 2009 events [3,21]. The black and yellow hollow circles are the aftershocks of the 2008 and 2009 events, respectively [4]. The purple and black lines are the active faults from Peltzer and Saucier [24] and Deng et al. [5], respectively.

Three postseismic deformation mechanisms, alone or mixed, have been proposed to interpret the observed movements after an earthquake event, including poroelastic rebound, afterslip, and viscoelastic relaxation. InSAR observations have been successfully used to constrain the possible geophysical mechanisms of some earthquake cases [15,16,25–32]. The mechanism of poroelastic rebound usually influences the postseismic deformation within a few kilometers to fault ruptures in a short period of time, whereas the other two mechanisms can generate postseismic deformation with a similar spatial pattern in the first few years [11,33]. However, the relative significance of the viscoelastic relaxation always increases with time, and the afterslip is usually contrary with it [11,16]. With these results, in this study we will interpret the observed postseismic deformation time series of the 2009 Mw 6.3 DCD earthquake with the combination model of afterslip and viscoelastic relaxation, along with comparisons of afterslips derived from the pure afterslip and the combination models.

Slip models are essential for interpreting postseismic deformation with the model of viscoelastic relaxation and the combination model of afterslip and viscoelastic relaxation [11,28,32]. When modeling with the combination model, previous studies usually considered the viscoelastic relaxation due to coseismic slip only, and neglected that due to the accumulated afterslip [16,28–30,32]. Recent study by Diao et al. [33] adopted a method which can consider the viscoelastic relaxation due to coseismic slip and the accumulated afterslip. In Diao et al. [33], coseismic slip and the accumulated afterslip estimated by pure afterslip model were directly employed as the driving force sources of viscoelastic relaxation. This means that Diao et al. [33] assessed afterslip and viscoelastic relaxation separately, first the afterslip and then the viscoelastic relaxation. With such a processing method, the input accumulated afterslip model may be not very reasonable because the afterslip and the viscoelastic relaxation processes are interactional during the postseismic phase. Therefore, it is necessary to carry out researches on simultaneously estimating the afterslip and viscoelastic relaxation parameter (viscosity).

In this study, we mainly focus on four items, including (1) improving the method in Diao et al. [33] to estimate the time-dependent afterslip distribution and the viscosity simultaneously; (2) modeling the InSAR deformation time series following the 2009 Mw 6.3 DCD earthquake in Liu et al. [23]; (3) seeking the dominant mechanisms responsible for postseismic deformation; and (4) finally investigating the viscosity beneath the northern side of the Qaidam Basin.

2. Data and Layered Model

2.1. Data

Investigating the postseismic deformation process requires the use of data from geodetic observations [7,34]. GPS data has been proven as an effective observation to accurately constrain the postseismic process due to its high horizontal precision and time resolution [35–37]. Unfortunately, for this event, according to Chen et al. [38], there are not enough GPS observations around the earthquake zone. The nearest station is approximately 40 km far away from the epicenter, making it unavailable to use GPS observation to do the following modeling. Meanwhile, InSAR data has also been confirmed as an alternative observation to investigate the postseismic process due to its high vertical precision and spatial resolution [15,16,25–32]. Feng [22] and Liu et al. [23] have derived the postseismic deformation time series. In this study, we prefer to use the observations from Liu et al. [23] because it includes one more SAR image than Feng [22], and has more observations in the near-field region.

The deformation time series in Liu et al. [23] are derived by processing the C-band Envisat ASAR descending Track 319 images with a small baseline subset InSAR technique, and include eight time epochs (Table 1). These epochs correspond to 19 days, 54 days, 124 days, 194 days, 229 days, 264 days, 299 days, and 334 days after the main shock, respectively. The derived InSAR deformation time series during the first 334 days after the event change from fast to slow with time, and display a smaller displacement for the footwall than that for the hanging wall, where the deformation decreases from the middle to both sides (Figure 2).

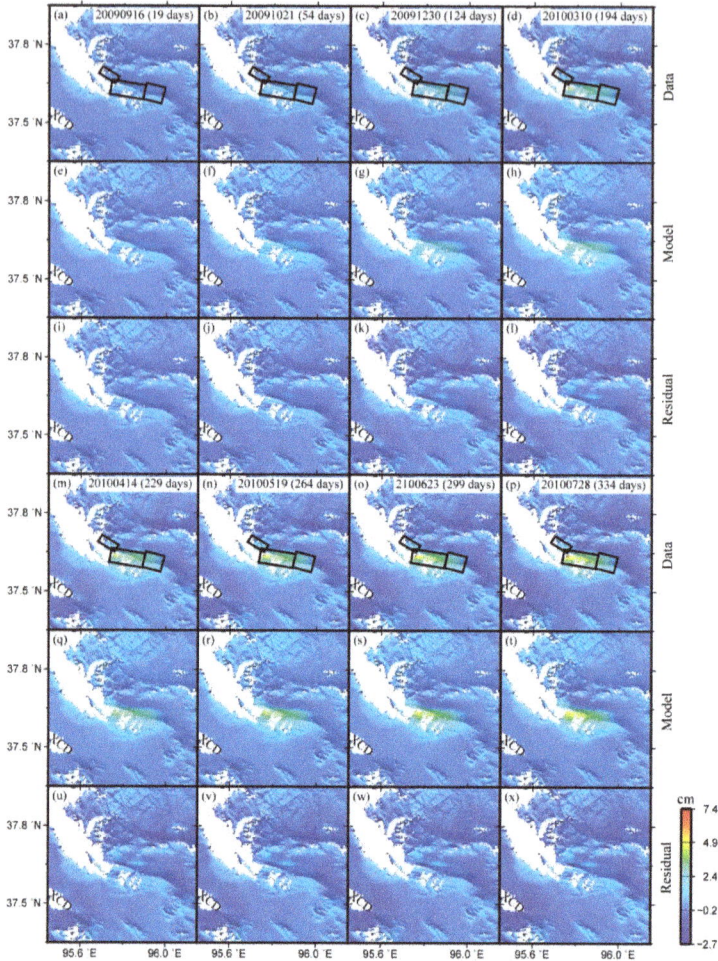

Figure 2. (**a–d**) Observed postseismic deformation time series at the first four time epochs of InSAR observations [23]; (**e–h**) modeled postseismic deformation time series from the combination model of afterslip and viscoelastic relaxation; and (**i–l**) residual deformation time series by subtracting (**e–h**) from (**a–d**); (**m–x**) are for the last four time epochs of InSAR observations [23]. The dates and days after the 2009 Mw 6.3 DCD earthquake are labeled at the top-right of subfigures (**a–d**) and (**m–p**). XCD is the abbreviated form of Xiaochaidan. Positive and negative values indicate motions toward and away from the satellite in the line-of-sight (LOS) direction, respectively.

According to Liu et al. [23], the spatial pattern of the postseismic deformation is similar with that of the coseismic deformation. This may indicate that an underground fault is processing during the postseismic phase similar to that during the coseismic phase. In other words, postseismic afterslip may occur with similar fault geometry and slip distribution as coseismic slip. Meanwhile, the correlation coefficient of 0.73 indicates that the postseismic afterslip may have a different slip distribution, and/or that the other types of postseismic deformation mechanisms may occur. After analyzing the three mechanisms described above, in this study mechanisms of afterslip and viscoelastic relaxation are considered, and their behaviors will be investigated by the modeling experiments in the following content.

Table 1. Observation data used in this study.

Date No.	Date	T [a] (Days)	Alpha [b] (km)	Sigma [c] (cm)	RMS [d] (cm)	Statistic Value [e] (cm)
1	16 September 2009	19	4.72	0.07	0.07	
2	21 October 2009	54	4.72	0.20	0.19	
3	30 December 2009	124	4.72	0.36	0.36	
4	10 March 2010	194	4.24	0.36	0.40	0.31/0.37
5	14 April 2010	229	4.40	0.35	0.41	
6	19 May 2010	264	4.84	0.34	0.42	
7	23 June 2010	299	5.56	0.37	0.44	
8	28 July 2010	334	6.12	0.46	0.47	

[a] Days after the 2009 DCD earthquake; [b] The spatial length of observation errors calculated with 1-D covariance function [39]; [c] The standard deviation of observation errors calculated with 1-D covariance function [39]; [d] The root mean square (RMS) misfit calculated by the combination model of afterslip and viscoelastic relaxation; [e] The values before and after slash are the average standard deviation and the total RMS misfit, respectively.

2.2. Layered Model

It is well known that a layered model is needed to interpret the postseismic deformation with mechanisms related to viscoelastic relaxation [28–32]. In addition, the input coseismic slip, which is considered as a driving force source of viscoelastic relaxation, should also be obtained in the layered model. Previous studies using seismic wave data have constructed some layered models.

An et al. [40] adopted records from 27 fundamental stations located in China and three stations located in Islamabad (Pakistan), Kabul (Afghanistan), and New Delhi (India) to image the 3-D shear velocity structure in Northwestern China. According to An et al. [40], the Qaidam Basin zone has an average S wave velocity of about 3.55 km/s in the 56-kilometer-thick crust and can be divided into four structural layers: the first one has an S wave velocity of 3.00 km/s at the upper 8 km depth, the second one 3.65 km/s at the depth from 8 km down to 18 km, the third one a low-velocity zone at a thick layer, and the fourth one 3.85–3.90 km/s at a thin layer. By referring to this layered model and a series of repeated tests, Liu et al. [41,42] constructed a slightly different structure model including four layers: from top to bottom, a 2.5-kilometer-thick layer with an S wave velocity of 3.15 km/s, a 5-kilometer-thick layer with an S wave velocity of 3.51 km/s, a 15-kilometer-thick layer with an S wave velocity of 3.63 km/s, and a 10-kilometer-thick layer with an S wave velocity of 3.69 km/s. With this layered model, Liu et al. [41,42] successfully derived the focal mechanism solutions of 98 aftershocks of the 2008 Mw 6.3 DCD earthquake, and relocated the aftershock sequences of the 2009 Mw 6.3 DCD earthquake.

Based on these previous studies, a three-layered model is constructed in this study (Table 2). The thickness of the top layer is 8 km with an S wave velocity of 3.00 km/s, the middle one 15 km with an S wave velocity of 3.65 km/s, and the bottom one infinite with an S wave velocity of 3.50 km/s. The corresponding P wave velocities are calculated by assuming the Poisson's ratio equal to 0.25 [43], and the related densities are determined by using the empirical relationship between P wave velocity and density [44]. It should be noted that the parameter of viscosity can be meaningful only in the modeling of the postseismic deformation time series.

Table 2. Layered model used in this study.

Layer No.	Thickness (km)	Vp (km/s)	Vs (km/s)	Density (kg/m^3)	Viscosity (Pa·s)
1	8	5.19	3.00	2430.8	N/A
2	15	6.31	3.65	2790.6	N/A
3	Infinite	6.06	3.50	2707.6	Variable

3. Modeling Method

In the modeling, the observed postseismic deformation time series are considered to relate to two mechanisms, postseismic afterslip and viscoelastic relaxation, where the viscoelastic relaxation is driven by both coseismic slip and the accumulated afterslip. This modeling method has been adopted to interpret the postseismic deformation observation following the earthquake, such as the 2011 Mw 9.0 Tohoku earthquake [33]. Meanwhile, Diao et al. [33] indicated that when investigating the postseismic process, it is necessary to consider the viscoelastic relaxation due to the accumulated afterslip. However, Diao et al. [33] assessed afterslip and viscoelastic relaxation separately, which will be improved to simultaneously estimate these two types of parameters.

The optimization problem for the modeling method can be expressed as:

$$\| Gs^i_{post} - \left(l^i - F^i\left(s_{co}, s^1_{post}, \cdots, s^{i-1}_{post}, \eta\right)\right) \|^2 + \beta^2 H\left(s^i_{post}\right) = min \tag{1}$$

where G is Green's function, s^i_{post} is the postseismic afterslip at the *i*-th time epoch, l^i is the postseismic deformation observation at the *i*-th time epoch, F^i is the surface deformation produced by viscoelastic relaxation at the *i*-th time epoch, which is driven by the coseismic slip and the accumulated afterslip, s_{co} is the coseismic slip, $s^1_{post}, \cdots, s^{i-1}_{post}$ are the postseismic afterslips at the 1-th, $\cdots$, *i*–1-th epoch times, η is the regional viscosity, β^2 is the smoothing factor, and H is the second-order Laplacian operator across the fault plane, which is used to avoid the unreasonable slip oscillation [39].

In Equation (1), coseismic slip, afterslip and viscosity are unknown parameters to be optimized. The Green's function related to afterslip is calculated with the EDGRN/EDCMP software (GFZ, Potsdam, Germany) [45], and the surface deformation produced by viscoelastic relaxation is calculated with the PSGRN/PSCMP software (GFZ, Potsdam, Germany) [46]. The viscosity is assumed to be constant during the observed time period. This assumption is accordant with the practices of modeling postseismic deformation of the 2008 M 6.4 and M 5.9 Nima-Gaize earthquakes, the 2008 Mw 6.3 Dangxiong earthquake, and the 2011 Mw 9.0 Tohoku earthquake [29,32,33], and is also proved to be reasonable by the following modeling experiments (Section 5).

The practical calculations mainly consist of four steps (Figure 3). In step 1, the postseismic viscoelastic relaxation deformation with different viscosities are simulated with the given coseismic slip distribution, the postseismic afterslip time series, the viscosity models, and so on. The input postseismic afterslip time series at the *i*-th time epoch are those estimated from the modeling with the same viscosity at the 1-th, . . . , *i* – 1-th time epochs. In step 2, the differential postseismic deformation is calculated by subtracting the postseismic viscoelastic relaxation deformation in step 1 from the observed postseismic deformation. In step 3, with the differential postseismic deformation in step 2, the postseismic afterslip distribution is estimated using the steepest descent method [47], which has been proved as an effective technique to estimate the coseismic slip and postseismic afterslip [33,48]. In step 4, the trade-off curve between the viscosity and RMS misfit error for the combined observations at all-time epochs is plotted to choose the preferred viscosity, and then the optimal postseismic afterslip time series are determined as those derived with the preferred viscosity.

The advantage of this method is that the afterslip and the viscosity can be estimated simultaneously, compared to most of the existing studies [16,28–32]. This method is also slightly different from that in Diao et al. [33], in which these two types of unknown parameters are estimated separately, first the afterslip distribution and then the viscosity. The separate estimation in Diao et al. [33] might ignore the effect of viscoelastic relaxation on the afterslip distribution, because the input postseismic afterslip time series are directly estimated from modeling postseismic deformation with afterslip only.

During the calculation, the layered model (Table 2) is used to calculate all the postseismic deformation time series of the 2009 Mw 6.3 DCD earthquake. For the viscoelastic relaxation, the part beneath the seismogenic layer is globally considered as a viscoelastic half-space of a Maxwell-type body, and the viscosities are varied from 1×10^{17} Pa·s to 1×10^{22} Pa·s with an index increased by one each time to simulate the viscoelastic relaxation deformation. For the afterslip, the fault geometry is fixed as that derived by Liu et al. [3], and the fault plane is discretized into patches with a size of 1-km length by 1-km width to estimate the afterslip distribution. After modeling for the eight time epochs with all possible viscosities and plotting the trade-off curve between the viscosity and RMS misfit error, the preferred viscosity is chosen when the RMS misfit begins to stabilize. It should be noted that in order to better display the trade-off curve, the RMS misfit is plotted against the log value of the viscosity instead of the viscosity.

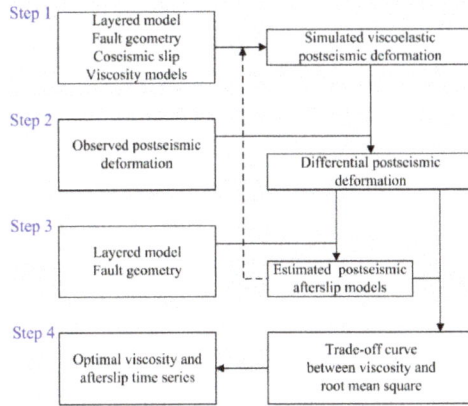

Figure 3. Flowchart of modeling postseismic deformation time series with a combination model of afterslip and viscoelastic relaxation. Both coseismic slip and the accumulated afterslip models occurring before the current time epoch are used to drive the viscoelastic relaxation.

4. Coseismic Slip in the Layered Model

When modeling the postseismic deformation time series with mechanisms related to the viscoelastic relaxation, coseismic slip is provided as the driving force source and then be thought as one key input parameter [28–31]. Therefore, a coseismic slip model that matches the postseismic process should be acquired. To achieve this, the dislocation Green's function should be calculated with the layered model instead of the elastic half-space [32], and the adopted observation dataset should be obtained in the shortest possible time after the earthquake.

Figure 4a shows the coseismic surface displacements for the 2009 Mw 6.3 DCD earthquake derived from two Envisat ASAR descending Track 319 images, which were observed on 14 January 2009 and 16 September 2009, respectively. The interferometric processing was done by using the ROI_PAC software (Caltech/JPL, Pasadena, CA, USA) [49], and the detailed data processing procedures can be found in Liu et al. [23]. The estimation method of coseismic slip distribution is close to that of modeling postseismic deformation with afterslip, with the difference that the postseismic deformation observation is replaced with the coseismic deformation.

Figure 5a,b show the derived coseismic slip distribution and uncertainties in the layered model. The uncertainties refer to the standard deviations in slip estimated from the Monte Carlo calculations with 100 perturbed datasets [3,50]. The fault rupturing model is characterized by three slip asperities, which are located at three fault segments, separately. The maximum slip is about 2.41 m, which is located at the central segment. The slips of three fault segments are mainly located at a depth from 2.5 km to 8.2 km, and are dominated by thrust motion. The estimated geodetic moment is 3.75×10^{18} N·m, equaling to a magnitude of Mw 6.35. The slip uncertainties range from 0 m to 0.12 m, and most of them are less than 0.08 m. For the main slipping fault patches, the uncertainties are obviously less than the corresponding slips, indicating that the slip model in Figure 5a is reliable.

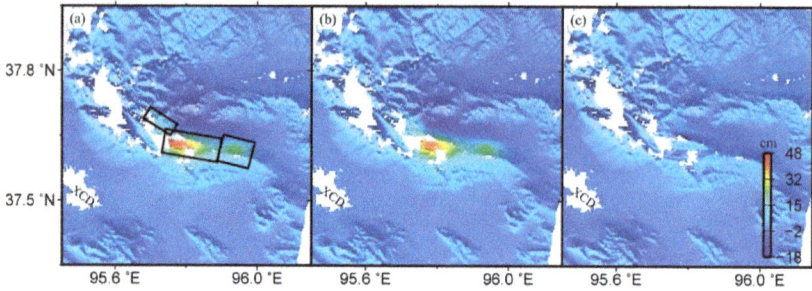

Figure 4. (**a**) Observed coseismic displacements for the 2009 Mw 6.3 DCD earthquake from two Envisat ASAR descending Track 319 images; (**b**) modeled displacements from the coseismic slip distribution in the layered model; and (**c**) residual displacements. Three rectangles in subfigure (**a**) indicate the surface projections of the fault geometry model. XCD is the abbreviated form of Xiaochaidan. Positive and negative values indicate motions toward and away from the satellite in the LOS direction, respectively.

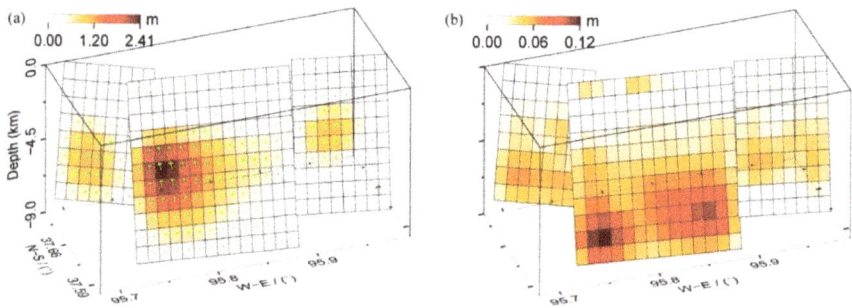

Figure 5. (**a**) Coseismic slip distribution of the 2009 DCD earthquake in the layered model; and (**b**) slip uncertainties estimated from the Monte Carlo calculation with 100 perturbed datasets. Green arrows in (**a**) indicate the slip direction on the corresponding fault patch.

The slip pattern in Figure 5a is generally consistent with that from the homogeneous half-space model [3]. Both models are dominated by thrust motion, have three slip asperities, and show no distinct slip for the upper three rows of fault patches. The maximum slips for the layered and homogeneous models are 2.41 m and 2.44 m, and are both located at a depth between 4.1 km and 4.9 km. Although the pattern and magnitude of these two slip models are very close, we still argue that the model derived from the layered model is a more realistic slip distribution, and adopt it as the driving force source when modeling the observed postseismic deformation time series.

Figure 4b,c show the modeled and residual displacements from the coseismic slip distribution in the layered model. The observed and modeled displacements have a similar deformation pattern, with a decreasing trend from the main deformation zone to both eastern and western sides. The residual displacement shows no distinct deformation zone. These facts indicate that the slip model in Figure 5a can better interpret the observed coseismic surface deformation.

5. Time-Dependent Afterslip and Viscosity

5.1. Time-Dependent Afterslip

Using the modeling method (Section 3) and the observed postseismic deformation time series (Figure 2), the preferred time-dependent afterslip distribution and the viscosity can be estimated simultaneously. Figure 6 shows the temporal and spatial distribution of postseismic afterslip at the eight time epochs of InSAR observations, ranging from 19 days to 334 days after the main shock.

The temporal and spatial features of the postseismic afterslip in Figure 6 are analyzed as follows. On the whole, the afterslip shows continuity and its magnitude increases with time. By 21 October 2009, 54 days after the main shock, the afterslip mainly occurs on the central and eastern segments, and two separate slip zones exist on the central segment. By 30 December 2009, 124 days after the main shock, for the western segments, afterslip begins to occur on the deep patches; for the central segment, afterslip has a trend of extending to the upper zone, the two separate slip zones are almost connected together, and some afterslip begins to occur on the upper-right zone; for the eastern segment, afterslip also has a trend of extending to the upper zone. From 10 March 2010 to 19 May 2010, afterslip for each segment increases with time, and the upper-right slip zone gradually merges into the main bottom-left slip zone for the central segment. From that time until 28 July 2010, 334 days after the main shock, the slip pattern across all fault segments is stable, and its magnitude still increases with time.

The postseismic afterslip distribution 334 days after the main shock (Figure 6h) is then analyzed. It has a complex pattern, with a maximum of 0.302 m slip located at the central segment. For the western segment, the afterslip is dominated by thrust motion with a slight left-lateral strike-slip component; For the central segment, the afterslip is dominated by thrust motion with some right-lateral strike-slip component, except that the main deep slipping zone has a slight left-lateral strike-slip component; For the eastern segment, the afterslip is dominated by thrust motion with some right-lateral strike-slip component. The moment released by the afterslip is 0.91×10^{18} N·m, which is about 24.3% of the main shock and equals to a magnitude of Mw 5.94. This ratio is consistent with previous findings [28,32,51].

The afterslip in Figure 6 is then compared with that in Feng [22]. The significant afterslip zone at a depth of about 3–9 km can be both resolved by the two results, although there exist some differences. The afterslip in this study indicates an upper-right slip zone on the central segment, which was not detected by Feng [22]. However, Feng [22] indicated a slip zone at depths larger than 10 km, which was not found in this study. The maximum afterslip detected by this study is 0.302 m, which differs from that from Feng [22] by about 0.5 m. The released moment is 0.91×10^{18} N·m, which is about one times less than that from Feng [22]. The reason for these differences may be that the two studies have adopted different datasets and inversion methods to do the modeling of postseismic deformation observations.

The uncertainties for the afterslip models on each date are estimated from the Monte Carlo calculations with 100 perturbed datasets, respectively (Figure 7) [3,50]. The spatial features of afterslip uncertainties are also complex, with larger values mostly located at the deeper zone. This is consistent with that in Bie et al. [32], which investigated the errors of the postseismic afterslip time series of the 2008 Mw 6.3 Dangxiong earthquake using the same method. The uncertainty time series for each time epochs are obviously less than the corresponding afterslip time series, and the maximum uncertainty of 0.072 m is about three times less than that of the afterslip, indicating that the afterslip distribution time series are reliable in both temporal and spatial features. It is noted that the afterslip time series uncertainties do not show a trend of increasing with time.

The resolution test demonstrates that the input slip model, which has a slip pattern similar to the main slip (larger than 40% of its maximum afterslip) zone in Figure 6h, can be generally well resolved (Figure 8), indicating the afterslip in Figure 6 are spatially reliable. The slip patches in the fault plane at shallower depth are recovered better than those at deeper depth. The reason for this may be the lower spatial resolution in the fault plane at deeper depth for geodetic inversions, as has been demonstrated by other studies [32,50].

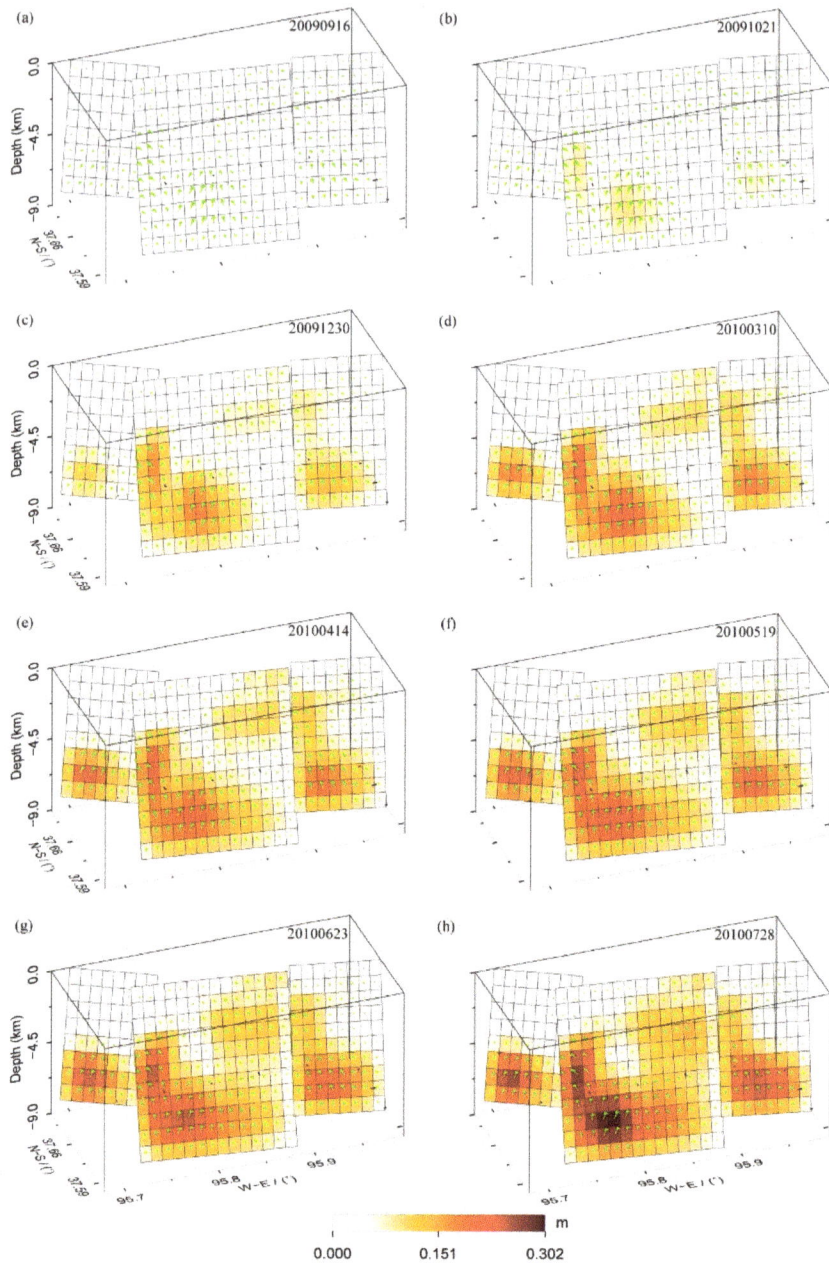

Figure 6. (**a–h**) Temporal and spatial distribution of postseismic afterslip, corresponding successively to the eight time epochs of InSAR observations. The characters labeled in all subfigures are the corresponding observation dates. Green arrows in all subfigures indicate the slip direction on the corresponding fault patch.

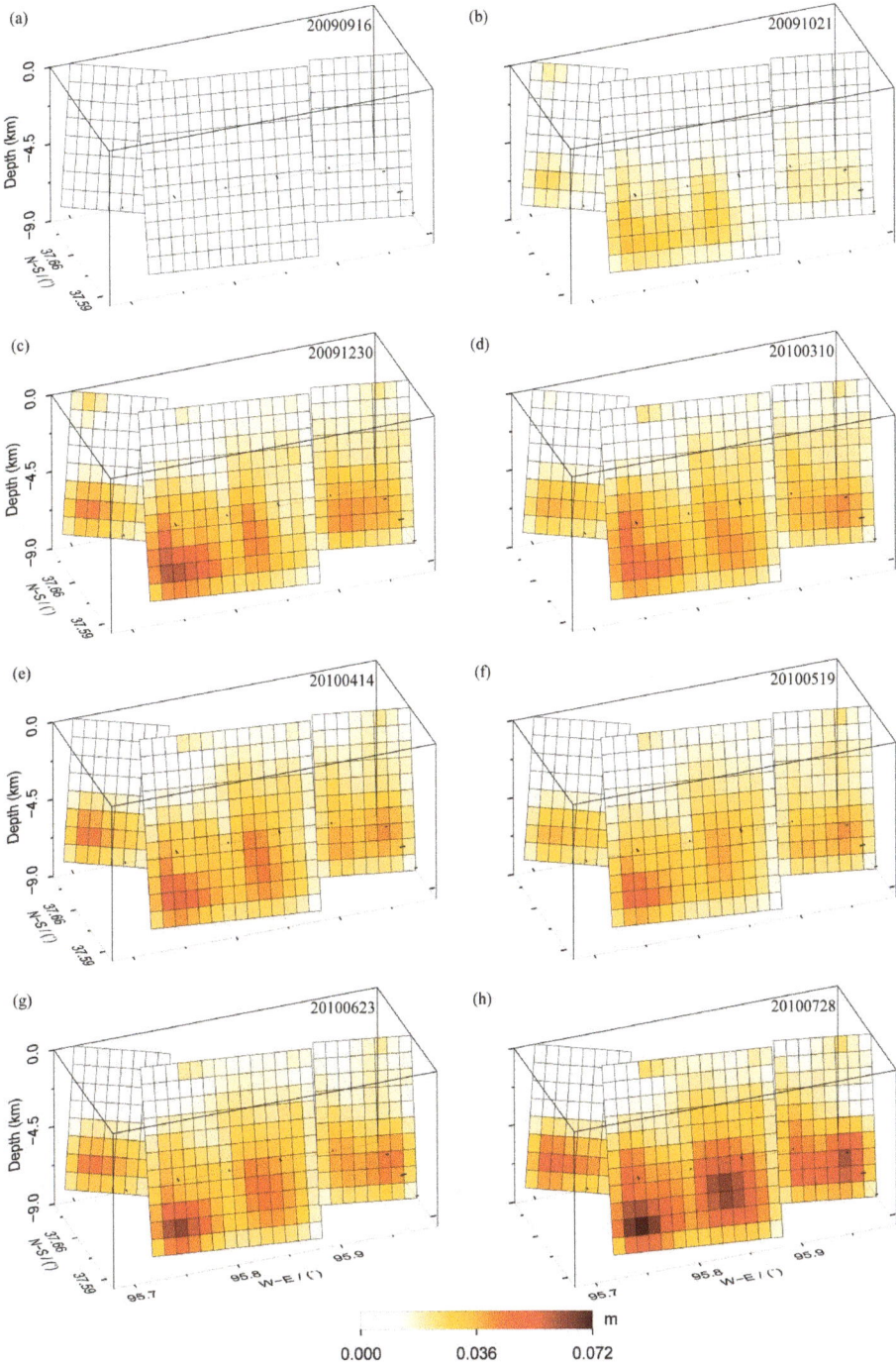

Figure 7. (a–h) Corresponding uncertainties for the afterslip models in Figure 6. These uncertainties for each date are estimated from the Monte Carlo calculation with 100 perturbed datasets, respectively.

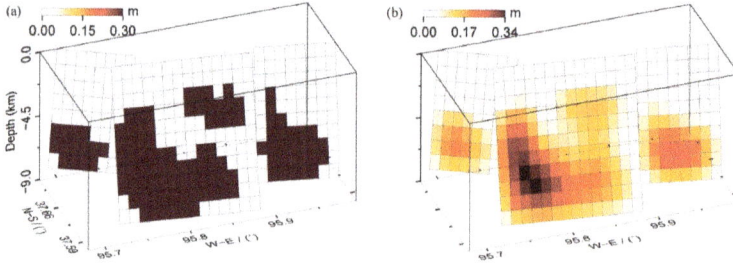

Figure 8. Resolution test. (**a**) Input slip model; and (**b**) recovered slip model. The input model in (**a**) has a slip pattern similar to the main slip (larger than 40% of its maximum afterslip) zone in Figure 6h.

5.2. Viscosity

Figure 9 shows the relationship curve between the log viscosity and the RMS misfit for the combined observations at the eight time epochs. It is clear that the RMS misfit decreases rapidly from 0.72 cm to 0.40 cm and then to 0.37 cm when the viscosity increases from 1×10^{17} Pa·s to 1×10^{18} Pa·s and then to 1×10^{19} Pa·s, and that the RMS misfit almost converges to a constant (0.37 cm) after the viscosity increasing to 1×10^{19} Pa·s. This RMS misfit is slightly larger than the average standard deviation (0.31 cm) of the eight postseismic deformation time series (Table 1).

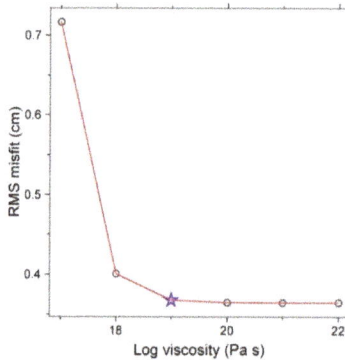

Figure 9. Relationship curve between the log viscosity and the RMS misfit for the combined observations at the eight time epochs. The blue hollow star denotes the preferred lower bound of the viscosity, 1×10^{19} Pa·s. Beyond this value the RMS misfit nearly no longer changes with viscosity.

The changing characteristic between RMS misfit and viscosity in Figure 9 is similar to those obtained by Ryder et al. [29] and Bie et al. [32], which modeled the postseismic InSAR deformation time series of the 2008 M 6.4 and M 5.9 Nima-Gaize earthquakes and the 2008 Mw 6.3 Dangxiong earthquake with a combined model of afterslip and viscoelastic relaxation in a Maxwell half-space. In Ryder et al. [29], the viscosity less than which the RMS misfit residual will increase rapidly is viewed as the lower bound on the Maxwell viscosity. Therefore, for this study the lower bound on the Maxwell viscosity is 1×10^{19} Pa·s.

Figure 10 shows the relationship curves between the log viscosity and the RMS misfit for the eight time epochs of postseismic deformation observations, separately. It is clear that for all of the time epochs, beyond the preferred lower bound of the viscosity (1×10^{19} Pa·s) in Figure 9, the RMS misfits almost no longer change with viscosity. This fact potentially indicates that the viscosity does not change significantly during the first 334 days after the main shock. This is inconsistent with the result from Ryder et al. [28], which modeled the postseismic displacement time series following the 1997 Mw 7.6 Manyi earthquake, and suggested that the viscosity may change with time during the first about three years. The possible reason is that the observing time period in this study is about two times shorter than that in Ryder et al. [28]. If observations with longer periods of time are available, whether the viscosity beneath the 2009 Mw 6.3 DCD earthquake zone changes with time, may be confirmed.

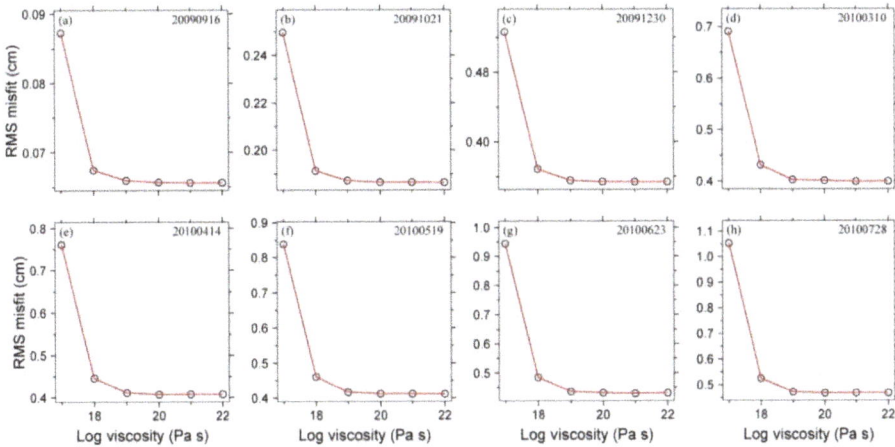

Figure 10. (a–h) Relationship curves between the log viscosity and the RMS misfit for the eight time epochs of postseismic deformation observations. For all of the time epochs, beyond the preferred lower bound of the viscosity in Figure 9, the RMS misfits almost no longer change with viscosity.

The RMS misfits for the eight postseismic deformation observations are 0.07 cm, 0.19 cm, 0.36 cm, 0.40 cm, 0.41 cm, 0.42 cm, 0.44 cm, and 0.47 cm, respectively. These values are in close agreement with the corresponding standard deviations of the observed deformation (Table 1). Figure 2 shows the modeled and residual displacements time series derived from the preferred combination model of afterslip (Figure 6) and viscoelastic relaxation (a viscosity of 1×10^{19} Pa·s). It is clear that the combination model can well explain the observed postseismic deformation time series. For all eight time epochs, the surface uplift along LOS direction can be clearly seen across the hanging wall and has an increasing trend with time, and no significant deformation can be found across the footwall. In addition, the predicted deformation time series across the hanging wall display the spatial features of decreasing from the central to both sides. These temporal and spatial features are consistent with those in the observed deformation time series.

6. Discussion

6.1. Comparison of Inversion Results from Different Methods

To investigate the effect of neglecting viscoelastic relaxation due to the accumulated afterslip on the estimations of the postseismic moment release and the viscosity, the observed postseismic deformation time series were also modeled with afterslip and viscoelastic relaxation driven by coseismic slip only, named as Method 2 in Table 3. Here, the method of modeling with afterslip and viscoelastic relaxation due to both coseismic slip and accumulated afterslip, which is adopted and analyzed above, is named as Method 1. During the modeling, all of the parameters are configured as those in Method 1, with the exception that the viscoelastic relaxation deformation is only driven by the coseismic slip, but not the accumulated postseismic afterslip. With this method, the RMS misfit for all eight observations is consistent with that from Method 1 at the sub-millimeter level. The temporal and spatial distributions of the derived postseismic afterslip are very similar to those derived from Method 1. By 28 July 2010,

334 days after the main shock, the maximum slip, also located at the central segment, is 0.298 m, which is slightly smaller than that (0.302 m) from Method 1; the moment released by postseismic afterslip is 9.02×10^{17} N·m, which is equal to an Mw of 5.937, compared to a moment of 9.07×10^{17} N·m and an Mw of 5.938 from Method 1. The preferred lower bound of the viscosity is 1×10^{19} Pa·s, which is consistent with that from Method 1. In addition, the RMS misfit almost becomes stable after the viscosity increasing to 1×10^{19}. These similarities of the results between Methods 1 and 2 may be mainly related to the magnitudes of the accumulated postseismic afterslip. If the afterslip of the 2009 Mw 6.3 DCD earthquake can have a comparable magnitude to that (a maximum afterslip of 3.8 m 564 days after the main shock) of the 2011 Mw 9.0 Tohoku earthquake, then the effect of visocoelastic relaxation driven by the accumulated afterslip can be significant [33].

Table 3. Statistics of inversion results from different methods.

Method	Maximum Slip [a] (m)	Moment [a] (10^{16} N·m)	Mw [a]	Viscosity (Pa·s)
1	0.302	90.68	5.938	1×10^{19}
2	0.298	90.22	5.937	1×10^{19}
3	0.274	87.26	5.927	N/A

[a] Values for 28 July 2010, 334 days after the main shock.

To investigate the effect of neglecting the viscoelastic relaxation due to both coseismic slip and accumulated afterslip on the estimations of the moment released by postseismic afterslip, the observed postseismic deformation time series were also modeled with afterslip only, named as Method 3 in Table 3. During the modeling, all the parameters are configured as those in Method 1, with the exception that the viscoelastic relaxtion deformation due to the coseismic slip and the accumulated postseismic afterslip is ignored. With this method, the RMS misfit for all eight observations is consistent with those from Methods 1 and 2 at the sub-millimeter level. The temporal and spatial distributions of the derived postseismic afterslip resemble those from Method 1 (Figure 6), but some differences still exist. By 28 July 2010, 334 days after the main shock, the maximum slip, also located at the central segment, is 0.274 m, which is 0.028 m smaller than that from Method 1; the moment released by postseismic afterslip is 8.73×10^{17} N·m (Mw 5.927), which is 0.34×10^{17} N·m smaller than that from Method 1. Likewise, Wen et al. [11] investigated the postseismic deformation of the 2001 Mw 7.8 Kokoxili earthquake, and found that the maximum slip for the preferred combination model is 0.57 m, close to that from the afterslip model. These similarities of the results between Methods 1 and 3 may be mainly related to the length of postseismic observation time period. If postseismic deformation observations of tens, or even hundreds, of years are available, then the differences of results between Methods 1 and 3 would be obviously identified. In which case, the differences of results between Methods 1 and 2 would also be recognized.

A comparison of inversion results among three methods indicates that the effect of coseismic slip on the afterslip estimations is larger than that of the accumulated afterslip. 334 days after the main shock, the maximum afterslip and moment release for Method 2 are only 0.004 m and 0.05×10^{17} N·m less than those from Method 1, compared to 0.028 m and 0.34×10^{17} N·m for Method 3. This is to be expected because the maximum afterslip 334 days after the main shock is only about 12.5% of the maximum coseismic slip.

6.2. Depth of Afterslip for the 2009 DCD Earthquake

Previous postseismic afterslip studies showed that the afterslip can occur at a depth comparable to the coseismic slip, and/or at a shallower depth, and/or below the coseismic rupture [11,52,53]. The different depths may reveal the frictional heterogeneities in and around the coseismic rupturing zone [33,51].

The analysis of postseismic deformation time series for the 2009 Mw 6.3 DCD earthquake indicates that the afterslip mainly occurs at a depth of about 0–9.1 km, rather than being limited to the very shallow upper crust. In comparison with the depth (about 2.5–8.2 km) of coseismic slip derived with the layered model (Section 4), the afterslip may extend from the coseismic slip zone to the shallower and deeper depths, respectively. In addition, during the observed period, the significant afterslip on a more deeper down-dip extension (larger than 9.1 km) of the coseismic slip can be ruled out, due to that the 2008 Mw 6.3 DCD earthquake mainly occurred at a depth of about 9.5–24.5 km [1,3]. This is consistent with the results of Pollitz et al. [54], which argued that the significant deep afterslip was not the dominant modes of postseismic deformation at the time scales of about one year.

Depths of afterslip for some earthquakes in the Tibet Plateau have been investigated [11,28,29,32]. For the 2008 M 6.4 and M 5.9 Nima-Gaize earthquakes and the 2008 Mw 6.3 Dangxiong earthquake, the estimated afterslips mainly occurred at the comparable depth and the up-dip extension depth of coseismic slip [29,32]. For the 1997 Mw 7.6 Manyi earthquake and the 2001 Mw 7.8 Kokoxili earthquake, the estimated afterslips mainly occurred at the comparable depth and the down-dip extension depth of the coseismic slip [11,28]. The depths of afterslip for these earthquakes in the Tibet Plateau are inconsistent with each other. The reason for this may be related to the different frictional properties of materials beneath each earthquake, or the different time scales of postseismic observations.

6.3. Viscosity Structure Beneath the Qaidam Basin

The viscosity structure beneath the Tibet Plateau has been investigated by modeling other types of observations [55–60]. When interpreting the role of the viscosity structure in shaping the present-day topography of the Tibet Plateau, the results suggested that the viscosities range from 10^{16} to 10^{20} Pa·s [55–57]. When interpreting the role of the viscosity structure in producing the surface GPS velocity field in the India-Asia collision zone and the Northern Tibet Plateau, the results indicated that the viscosities range from 10^{18} to 10^{20} Pa·s [58,59]. When explaining the horizontality of palaeolake shorelines in the central Tibet Plateau, the results suggested that the viscosity of the middle to lower crust is at least 10^{19}–10^{20} Pa·s [60]. These studies demonstrated that the viscosity structure beneath the Tibet Plateau may vary.

Postseismic deformation observations can provide another way of quantitating the regional viscosity structure with better accuracy [16]. In the Tibet Plateau, several studies have investigated the regional viscosity with postseismic deformation. Ryder et al. [28,29] analyzed the postseismic InSAR deformation of the 1997 Mw 7.6 Manyi earthquake and the 2008 M 6.4 and M 5.9 Nima-Gaize earthquakes, and inferred the effective viscosities of about 3–10 × 10^{18} Pa·s and 3 × 10^{17} Pa·s beneath the earthquake zones, respectively. Zhang et al. [61] modeled the postseismic leveling observations of the 1973 Luhuo Mw 7.9 earthquake and inferred the viscosities between 10^{19} Pa·s and 10^{21} Pa·s beneath the earthquake zone. Ryder et al. [30] and Wen et al. [11] modeled the postseismic InSAR and/or GPS time series of the 2001 Mw 7.8 Kokoxili earthquake and obtained the steady-state viscosities between 1 × 10^{19} Pa·s and 2 × 10^{19} Pa·s below the earthquake zone. The differences between them might be related to the different time periods and observations used for modeling. With the postseismic GPS observations of the 2008 Mw 7.9 Wenchuan earthquake, Xu et al. [12] found that the viscosity beneath the earthquake zone is larger than 3 × 10^{18} Pa·s. By fitting the postseismic InSAR observations of the 2008 Mw 6.3 Dangxiong earthquake, Bie et al. [32] argued a viscosity of 1 × 10^{18} Pa·s below the earthquake zone.

The 2009 Mw 6.3 DCD earthquake is located at the northern side of the Qaidam Basin (Figure 1), and in this study the preferred lower bound of the viscosity beneath this event is 1 × 10^{19} Pa·s. This value is consistent with those derived from the postseismic deformation of the 2001 Mw 7.8 Kokoxili earthquake [11,30]. It is known that these two earthquakes are located on the northern and southern sides of the Qaidam Basin, respectively. The consistency between the viscosities derived from the postseismic deformation observations of the two earthquakes potentially proves their reliabilities.

The derived viscosity in this study differs one or two orders of magnitude from those estimated from the postseismic deformation of the 2008 M 6.4 and M 5.9 Nima-Gaize earthquakes, the 2008 Mw 7.9 Wenchuan earthquake, and the 2008 Mw 6.3 Dangxiong earthquake [12,29,32]. These events are located on the different parts of the Tibet Plateau, compared to the 2009 Mw 6.3 DCD event. This inconsistency may be related to their relative positions in the Tibet Plateau. In addition, from another side, this fact potentially validates that the viscosity structure beneath the Tibet Plateau may vary laterally.

7. Conclusions

On 28 August 2009, an Mw 6.3 DCD earthquake occurred at the northern side of the Qaidam Basin. We modified a method of modeling the postseismic deformation time series with the combination model of afterslip and viscoelastic relaxation, and then can simultaneously estimate the time-dependent afterslip distribution and the viscosity beneath the earthquake zone. To obtain a more rational driving force source of viscoelastic relaxation, we invert for a coseismic slip model in the layered model, which gives a slip pattern with a maximum slip of 2.41 m comparable to that in the elastic half-space model. With the postseismic deformation observations, we investigate the time-dependent afterslip of the 2009 Mw 6.3 DCD earthquake and the viscosity beneath the northern side of the Qaidam Basin.

The combination model of afterslip and viscoelastic relaxation can interpret the observed postseismic InSAR time series, with a total RMS misfit of 0.37 cm comparable to the average uncertainty of all of the observations. The preferred time-dependent afterslip shows continuity and increases with time, which is mainly located at a depth from the surface to about 9.1 km underground. The changing trend of the moment released by afterslip indicates that the postseismic afterslip is likely to continue 334 days after the main shock. By 28 July 2010, the moment released by the afterslip was 0.91×10^{18} N·m, about 24.3% of the main shock, and equaled a magnitude of Mw 5.94. The simultaneously estimated lower bound of the viscosity beneath the northern side of the Qaidam Basin is 1×10^{19} Pa·s, close to that beneath the southern side of the Qaidam Basin. This viscosity differs from those beneath other parts of the Tibet Plateau potentially indicates that the viscosity structure beneath the Tibet Plateau may vary laterally.

Acknowledgments: This research is supported by the National Natural Science Foundation of China (No. 41404007, 41431069, and 41274030), the State Key Development Program for Basic Research of China (No. 2013CB733304), the Special Project of Basic Work of Science and Technology (No. 2015FY210400), and the China Scholarship Council (No. 201506275015). Part of this work is also supported by the UK Natural Environmental Research Council (NERC) through the LICS and CEDRRiC projects (ref. NE/K010794/1 and NE/N012151/1, respectively), and the ESA-MOST DRAGON-3 projects (ref. 10607 and 10665). We are grateful to editors and three anonymous reviewers for helping us improve the manuscript.

Author Contributions: Yang Liu and Caijun Xu conceived and designed the experiments; Yang Liu, Zhenhong Li, Yangmao Wen, Jiajun Chen, and Zhicai Li performed and analyzed the experiments; all authors wrote the paper.

Conflicts of Interest: The authors declare no conflict of interest.

Abbreviations

The following abbreviations are used in this manuscript:

DCD	Dachaidan
NWW	northwestern-west
SEE	southeastern-east
GPS	Global Positioning System
InSAR	Interferometric Synthetic Aperture Radar
ASAR	Advanced Synthetic Aperture Radar
USGS	United States Geological Survey
XCD	Xiaochaidan
LOS	line-of-sight
RMS	root mean square

References

1. Elliott, J.; Parsons, B.; Jackson, J.; Shan, X.; Sloan, R.; Walker, R. Depth segmentation of the seismogenic continental crust: The 2008 and 2009 Qaidam earthquakes. *Geophys. Res. Lett.* **2011**, *38*. [CrossRef]
2. Liu, W.; Wang, P.; Ma, Y.; Chen, Y. Relocation of Dachaidan Ms 6.4 earthquake sequence in Qinghai province in 2009 using the double difference location method. *Plateau Earthq. Res.* **2011**, *23*, 24–26.
3. Liu, Y.; Xu, C.; Wen, Y.; Fok, H.S. A new perspective on fault geometry and slip distribution of the 2009 Dachaidan Mw 6.3 earthquake from InSAR observations. *Sensors* **2015**, *15*, 16786–16803. [CrossRef] [PubMed]
4. USGS. Magnitude 6.2—Northern Qinghai, China. Available online: http://earthquake.usgs.gov/earthquakes/eqinthenews/2009/us2009kwaf/ (accessed on 23 May 2016).
5. Deng, Q.; Zhang, P.; Ran, Y.; Yang, X.; Min, W.; Chu, Q. Basic characteristics of active tectonics of China. *Sci. China Ser. D* **2003**, *46*, 356–372.
6. Zhang, P.; Deng, Q.; Zhang, G.; Ma, J.; Gan, W.; Min, W.; Mao, F.; Wang, Q. Active tectonic blocks and strong earthquakes in the continent of China. *Sci. China Ser. D* **2003**, *46*, 13–24.
7. Reilinger, R. Evidence for postseismic viscoelastic relaxation following the 1959 M = 7.5 Hebgen Lake, Montana, earthquake. *J. Geophys. Res.* **1986**, *91*, 9488–9494. [CrossRef]
8. Pollitz, F.F.; Peltzer, G.; Bürgmann, R. Mobility of continental mantle: Evidence from postseismic geodetic observations following the 1992 Landers earthquake. *J. Geophys. Res.* **2000**, *105*, 8035–8054. [CrossRef]
9. Pollitz, F.F.; Wicks, C.; Thatcher, W. Mantle flow beneath a continental strike-slip fault: Postseismic deformation after the 1999 Hector Mine earthquake. *Science* **2001**, *293*, 1814–1818. [CrossRef] [PubMed]
10. Tronin, A.A. Satellite remote sensing in seismology. *Remote Sens.* **2009**, *2*, 124–150. [CrossRef]
11. Wen, Y.; Li, Z.; Xu, C.; Ryder, I.; Bürgmann, R. Postseismic motion after the 2001 Mw 7.8 Kokoxili earthquake in Tibet observed by InSAR time series. *J. Geophys. Res.* **2012**, *117*. [CrossRef]
12. Xu, C.; Fan, Q.; Wang, Q.; Yang, S.; Jiang, G. Postseismic deformation after 2008 Wenchuan Earthquake. *Surv. Rev.* **2014**, *46*, 432–436. [CrossRef]
13. Spinler, J.C.; Bennett, R.A.; Walls, C.; Lawrence, S.; González García, J.J. Assessing long-term postseismic deformation following the M7.2 4 April 2010, El Mayor-Cucapah earthquake with implications for lithospheric rheology in the Salton Trough. *J. Geophys. Res.* **2015**, *120*, 3664–3679. [CrossRef]
14. Xu, B.; Xu, C. Numerical simulation of influences of the earth medium's lateral heterogeneity on co-and post-seismic deformation. *Geod. Geodyn.* **2015**, *6*, 46–54. [CrossRef]
15. Hamling, I.J.; Hreinsdóttir, S. Reactivated afterslip induced by a large regional earthquake, Fiordland, New Zealand. *Geophys. Res. Lett.* **2016**, *43*, 2526–2533. [CrossRef]
16. Huang, M.-H.; Bürgmann, R.; Pollitz, F. Lithospheric rheology constrained from twenty-five years of postseismic deformation following the 1989 M w 6.9 Loma Prieta earthquake. *Earth Planet. Sci. Lett.* **2016**, *435*, 147–158. [CrossRef]
17. Wen, Y.; Xu, C.; Liu, Y.; Jiang, G. Deformation and source parameters of the 2015 Mw 6.5 earthquake in Pishan, western China, from Sentinel-1A and ALOS-2 data. *Remote Sens.* **2016**, *8*. [CrossRef]
18. Xu, C.; Xu, B.; Wen, Y.; Liu, Y. Heterogeneous fault mechanisms of the 6 October 2008 Mw 6.3 Dangxiong (Tibet) earthquake using Interferometric Synthetic Aperture Radar observations. *Remote Sens.* **2016**, *8*. [CrossRef]
19. Evans, E.L.; Meade, B.J. Geodetic imaging of coseismic slip and postseismic afterslip: Sparsity promoting methods applied to the great Tohoku earthquake. *Geophys. Res. Lett.* **2012**, *39*. [CrossRef]
20. Gan, W.; Zhang, P.; Shen, Z.-K.; Niu, Z.; Wang, M.; Wan, Y.; Zhou, D.; Cheng, J. Present-day crustal motion within the Tibetan Plateau inferred from GPS measurements. *J. Geophys. Res.* **2007**, *112*. [CrossRef]
21. Liu, Y.; Xu, C.; Wen, Y.; He, P. The InSAR coseismic deformation observation and fault parameter inversion of the 2008 Dachaidan Mw 6.3 earthquake. *Acta Geod. Cart. Sin.* **2015**, *44*, 1202–1209.
22. Feng, W. Modelling Co- and Post-Seismic Displacements Revealed by InSAR, and their Implications for Fault Behaviour. Ph.D. Thesis, University of Glasgow, Glasgow, UK, 2015.
23. Liu, Y.; Xu, C.; Wen, Y.; Li, Z. Post-seismic deformation from the 2009 Mw 6.3 Dachaidan earthquake in the northern Qaidam Basin detected by small baseline subset InSAR technique. *Sensors* **2016**, *16*. [CrossRef] [PubMed]
24. Peltzer, G.; Saucier, F. Present-day kinematics of Asia derived from geologic fault rates. *J. Geophys. Res.* **1996**, *101*, 27943–27956. [CrossRef]

25. Massonnet, D.; Feigl, K.; Rossi, M.; Adragna, F. Radar interferometric mapping of deformation in the year after the Landers earthquake. *Nature* **1994**, *369*, 227–230. [CrossRef]
26. Fialko, Y. Evidence of fluid-filled upper crust from observations of postseismic deformation due to the 1992 Mw 7. 3 Landers earthquake. *J. Geophys. Res.* **2004**, *109*. [CrossRef]
27. Johanson, I.A.; Fielding, E.J.; Rolandone, F.; Burgmann, R. Coseismic and postseismic slip of the 2004 Parkfield earthquake from space-geodetic data. *Bull. Seismol. Soc. Am.* **2006**, *96*, 269–282. [CrossRef]
28. Ryder, I.; Parsons, B.; Wright, T.J.; Funning, G.J. Post-seismic motion following the 1997 Manyi (Tibet) earthquake: InSAR observations and modelling. *Geophys. J. Int.* **2007**, *169*, 1009–1027. [CrossRef]
29. Ryder, I.; Bürgmann, R.; Sun, J. Tandem afterslip on connected fault planes following the 2008 Nima-Gaize (Tibet) earthquake. *J. Geophys. Res.* **2010**, *115*, B03404. [CrossRef]
30. Ryder, I.; Bürgmann, R.; Pollitz, F. Lower crustal relaxation beneath the Tibetan Plateau and Qaidam Basin following the 2001 Kokoxili earthquake. *Geophys. J. Int.* **2011**, *187*, 613–630. [CrossRef]
31. Biggs, J.; Burgmann, R.; Freymueller, J.T.; Lu, Z.; Parsons, B.; Ryder, I.; Schmalzle, G.; Wright, T. The postseismic response to the 2002 M 7.9 Denali Fault earthquake: Constraints from InSAR 2003–2005. *Geophys. J. Int.* **2009**, *176*, 353–367. [CrossRef]
32. Bie, L.; Ryder, I.; Nippress, S.E.J.; Bürgmann, R. Coseismic and post-seismic activity associated with the 2008 Mw 6.3 Damxung earthquake, Tibet, constrained by InSAR. *Geophys. J. Int.* **2014**, *196*, 788–803. [CrossRef]
33. Diao, F.; Xiong, X.; Wang, R.; Zheng, Y.; Walter, T.R.; Weng, H.; Li, J. Overlapping post-seismic deformation processes: Afterslip and viscoelastic relaxation following the 2011 Mw 9.0 Tohoku (Japan) earthquake. *Geophys. J. Int.* **2014**, *196*, 218–229. [CrossRef]
34. Brown, L.D.; Reilinger, R.E.; Holdahl, S.R.; Balazs, E.I. Postseismic crustal uplift near Anchorage, Alaska. *J. Geophys. Res.* **1977**, *82*, 3369–3378. [CrossRef]
35. Shen, Z.K.; Jackson, D.D.; Feng, Y.; Cline, M.; Kim, M.; Fang, P.; Bock, Y. Postseismic deformation following the Landers earthquake, California, 28 June 1992. *Bull. Seismol. Soc. Am.* **1994**, *84*, 780–791.
36. Savage, J.; Svarc, J. Postseismic deformation associated with the 1992 Mw = 7.3 Landers earthquake, southern California. *J. Geophys. Res.* **1997**, *102*, 7565–7577. [CrossRef]
37. Hsu, Y.J.; Bechor, N.; Segall, P.; Yu, S.B.; Kuo, L.C.; Ma, K.F. Rapid afterslip following the 1999 Chi-Chi, Taiwan earthquake. *Geophys. Res. Lett.* **2002**, *29*, 1–4. [CrossRef]
38. Chen, G.; Xu, X.; Zhu, A.; Zhang, X.; Yuan, R.; Klinger, Y.; Nocquet, J.-M. Seismotectonics of the 2008 and 2009 Qaidam earthquakes and its implication for regional tectonics. *Acta Geol. Sin.* **2013**, *87*, 618–628.
39. Parsons, B.; Wright, T.; Rowe, P.; Andrews, J.; Jackson, J.; Walker, R.; Khatib, M.; Talebian, M.; Bergman, E.; Engdahl, E. The 1994 Sefidabeh (eastern Iran) earthquakes revisited: New evidence from satellite radar interferometry and carbonate dating about the growth of an active fold above a blind thrust fault. *Geophys. J. Int.* **2006**, *164*, 202–217. [CrossRef]
40. An, C.; Song, Z.; Chen, G.; Chen, L.; Zhuang, Z.; Fu, Z.; Lu, J.; Hu, J. 3-D shear velocity structure in north-west China. *Chin. J. Geophys.* **1993**, *36*, 317–325.
41. Liu, W.; Zhang, X.; Hu, Y. Accurate location of Dachaidan Ms 6.3 earthquake sequence and earthquake tectonic using the double-difference earthquake location method. *Plateau Earthq. Res.* **2012**, *24*, 20–24.
42. Liu, W.; Zhang, X.; Shi, Y.; Wen, Y.; Zhao, Y. Using green function database and quick moment tensor inversion calculating the focal mechanism solution of aftershocks of Dachaidan Ms 6.4 earthquake in 2008 in Qinghai province. *Northwest. Seismol. J.* **2012**, *34*, 154–160.
43. Owens, T.J.; Zandt, G. Implications of crustal property variations for models of Tibetan plateau evolution. *Nature* **1997**, *387*, 37–43. [CrossRef]
44. Berteussen, K.A. Moho depth determinations based on spectral ratio analysis of NORSAR long-period P waves. *Phys. Earth Planet. Inter.* **1977**, *31*, 313–326. [CrossRef]
45. Wang, R.; Martin, F.L.; Roth, F. Computation of deformation induced by earthquakes in a multi-layered elastic crust-FORTRAN programs EDGRN/EDCMP. *Comput. Geosci.* **2003**, *29*, 195–207. [CrossRef]
46. Wang, R.; Lorenzo-Martin, F.; Roth, F. PSGRN/PSCMP—A new code for calculating co-and post-seismic deformation, geoid and gravity changes based on the viscoelastic-gravitational dislocation theory. *Comput. Geosci.* **2006**, *32*, 527–541. [CrossRef]
47. Wang, R.; Diao, F.; Hoechner, A. SDM—A geodetic inversion code incorporating with layered crust structure and curved fault geometry. In Proceedings of the EGU General Assembly, Vienna, Austria, 7–12 April 2013.

48. Wen, Y.; Xu, C.; Liu, Y.; Jiang, G.; He, P. Coseismic slip in the 2010 Yushu earthquake (China), constrained by wide-swath and strip-map InSAR. *Nat. Hazards Earth Syst. Sci.* **2013**, *13*, 35–44. [CrossRef]

49. Rosen, P.A.; Hensley, S.; Peltzer, G.; Simons, M. Updated repeat orbit interferometry package released. *Eos Trans. AGU* **2004**, *85*. [CrossRef]

50. Wright, T.J.; Lu, Z.; Wicks, C. Source model for the Mw 6.7, 23 October 2002, Nenana Mountain earthquake (Alaska) from InSAR. *Geophys. Res. Lett.* **2003**, *30*. [CrossRef]

51. Lin, Y.-n.N.; Sladen, A.; Ortega-Culaciati, F.; Simons, M.; Avouac, J.-P.; Fielding, E.J.; Brooks, B.A.; Bevis, M.; Genrich, J.; Rietbrock, A. Coseismic and postseismic slip associated with the 2010 Maule Earthquake, Chile: Characterizing the Arauco Peninsula barrier effect. *J. Geophys. Res.* **2013**, *118*, 3142–3159. [CrossRef]

52. Marone, C.J.; Scholtz, C.; Bilham, R. On the mechanics of earthquake afterslip. *J. Geophys. Res.* **1991**, *96*, 8441–8452. [CrossRef]

53. Bürgmann, R.; Ergintav, S.; Segall, P.; Hearn, E.H.; McClusky, S.; Reilinger, R.E.; Woith, H.; Zschau, J. Time-dependent distributed afterslip on and deep below the Izmit earthquake rupture. *Bull. Seismol. Soc. Am.* **2002**, *92*, 126–137. [CrossRef]

54. Pollitz, F.F.; Bürgmann, R.; Thatcher, W. Illumination of rheological mantle heterogeneity by the M7.2 2010 El Mayor-Cucapah earthquake. *Geochem. Geophys. Geosyst.* **2012**, *13*. [CrossRef]

55. Clark, M.K.; Royden, L.H. Topographic ooze: Building the eastern margin of Tibet by lower crustal flow. *Geology* **2000**, *28*, 703–706. [CrossRef]

56. Cook, K.L.; Royden, L.H. The role of crustal strength variations in shaping orogenic plateaus, with application to Tibet. *J. Geophys. Res.* **2008**, *113*. [CrossRef]

57. Bendick, R.; McKenzie, D.; Etienne, J. Topography associated with crustal flow in continental collisions, with application to Tibet. *Geophys. J. Int.* **2008**, *175*, 375–385. [CrossRef]

58. Copley, A.; McKenzie, D. Models of crustal flow in the India-Asia collision zone. *Geophys. J. Int.* **2007**, *169*, 683–698. [CrossRef]

59. Hilley, G.E.; Johnson, K.M.; Wang, M.; Shen, Z.K.; Burgmann, R. Earthquake-cycle deformation and fault slip rates in northern Tibet. *Geology* **2009**, *37*, 31–34. [CrossRef]

60. England, P.C.; Walker, R.T.; Fu, B.; Floyd, M.A. A bound on the viscosity of the Tibetan crust from the horizontality of palaeolake shorelines. *Earth Planet. Sci. Lett.* **2013**, *375*, 44–56. [CrossRef]

61. Zhang, C.; Cao, J.; Shi, Y. Studying the viscosity of lower crust of Qinghai-Tibet Plateau according to post-sesimic deformation. *Sci. China Ser. D* **2008**, *38*, 1250–1257.

remote sensing

MDPI

Article

Continent-Wide 2-D Co-Seismic Deformation of the 2015 Mw 8.3 Illapel, Chile Earthquake Derived from Sentinel-1A Data: Correction of Azimuth Co-Registration Error

Bing Xu [1], Zhiwei Li [1,*], Guangcai Feng [1], Zeyu Zhang [1], Qijie Wang [1], Jun Hu [1] and Xingguo Chen [2]

[1]   School of Geosciences and Info-Physics, Central South University, Changsha 410083, Hunan, China; xubing@csu.edu.cn (B.X.); fredgps@csu.edu.cn (G.F.); zzy1416@163.com (Z.Z.); qjwang@csu.edu.cn (Q.W.); csuhujun@csu.edu.cn (J.H.)

[2]   Beijing Di Kong Software Technology Co., Ltd., Beijing 100083, China; chenxg@diksw.com

*   Correspondence: zwli@csu.edu.cn; Tel.: +86-731-888-30573

Academic Editors: Zhenhong Li, Roberto Tomas, Richard Müller and Prasad S. Thenkabail
Received: 29 February 2016; Accepted: 28 April 2016; Published: 4 May 2016

Abstract: In this study, we mapped the co-seismic deformation of the 2015 Mw 8.3 Illapel, Chile earthquake with multiple Sentinel-1A TOPS data frames from both ascending and descending geometries. To meet the requirement of very high co-registration precision, an improved spectral diversity method was proposed to correct the co-registration slope error in the azimuth direction induced by multiple Sentinel-1A TOPS data frames. All phase jumps that appear in the conventional processing method have been corrected after applying the proposed method. The 2D deformation fields in the east-west and vertical directions are also resolved by combing D-InSAR and Offset Tracking measurements. The results reveal that the east-west component dominated the 2D displacement, where up to 2 m displacement towards the west was measured in the coastal area. Vertical deformations ranging between −0.25 and 0.25 m were found. The 2D displacements imply the collision of the Nazca plate squeezed the coast, which shows good accordance with the geological background of the region.

Keywords: Illapel Earthquake; Sentinel-1A; spectral diversity; InSAR; 2D deformation

1. Introduction

On 16 September 2015, an earthquake of Mw 8.3 [1] occurred 48 km from the west of Illapel, Chile, causing extensive damages to the area and leading to a large tsunami [2]. In this earthquake, at least 15 people died, 34 people injured, and tens of thousands of people became homeless. A total of 13 aftershocks $\geqslant$ Mw 6.0 soon afterwards occurred within 200 kilometers around the epicenter [1]. Chile is in one of the most earthquake-prone regions in the world and has been attacked by more than a dozen of quakes $\geqslant$ Mw 8.0 (see Figure 1) since 1900. The Nazca plate is moving at an average velocity of 74 mm/year east-northeast, and subducting beneath the continent at the Peru-Chile Trench of the South American plate [3]. The 2015 Mw 8.3 Illapel earthquake occurred as the result of thrust faulting on the interface between the Nazca and South American plates in central Chile [1,4]. The USGS report [4] shows that the epicenter of this event locates at 31.57°S, 71.654°W with a focal depth of about 25 km, and its dimension is typically about 230 × 100 km (length × width).

Currently, few geodesy measurements (such as GPS, spirit leveling) are available for scientists to further study this event. Thanks to a new generation of SAR systems, such as the Sentinel-1A (S1A) [5,6] equipped with the Terrain Observation by Progressive Scans (TOPS) [7], the continent-wide

Interferometric Synthetic Aperture Radar (InSAR) measurements becomes possible. Previous studies [8–12] have demonstrated the capability and great potential of S1A TOPS in ground surface deformation monitoring. In contrast to the conventional strip-map or ScanSAR data, multiple consecutive S1A TOPS frames (defined by ESA, where one zip file of S1A TOPS data distributed in Scihub [13] denotes a single frame) can generate long seamless strip data which is 250 km wide. Thus, it extremely benefits the study of large continent-wide event, such as the 2015 Mw 8.3 Illapel earthquake.

The S1A TOPS, as a new mode of InSAR, also has some key technical issues that need further study, especially for co-registering multiple S1A TOPS data frames. The Doppler Centroid of TOPS SAR data varies strongly along tracks, so a co-registration accuracy of at least 0.001 SLC pixel is required to eliminate the effect [7]. Currently, the Spectral Diversity (SD) [14] method and the enhanced SD (ESD) [15], implemented in GAMMA [16] or GMTSAR [17], could perform well for a single frame data case, but cannot resolve the problem of co-registering multiple S1A TOPS data frames well.

In this study, we firstly proposed an improved Spectral Diversity (ISD) method to co-register the multiple S1A TOPS Interferometric Wide Swath (IWS) data frames. We then map the co-seismic LOS deformation associated with the 2015 Mw 8.3 Illapel earthquake using both ascending and descending S1A data and the methods of D-InSAR and Offset Tracking, respectively. Subsequently, the D-InSAR and Offset Tracking measurements are assessed in terms of precision and used together to retrieve the east-west and vertical co-seismic deformations of this earthquake. Finally, we discussed implications of ISD co-registration method and co-seismic measurements, as well as the potential contribution of the S1A data to further continent-wide event studies.

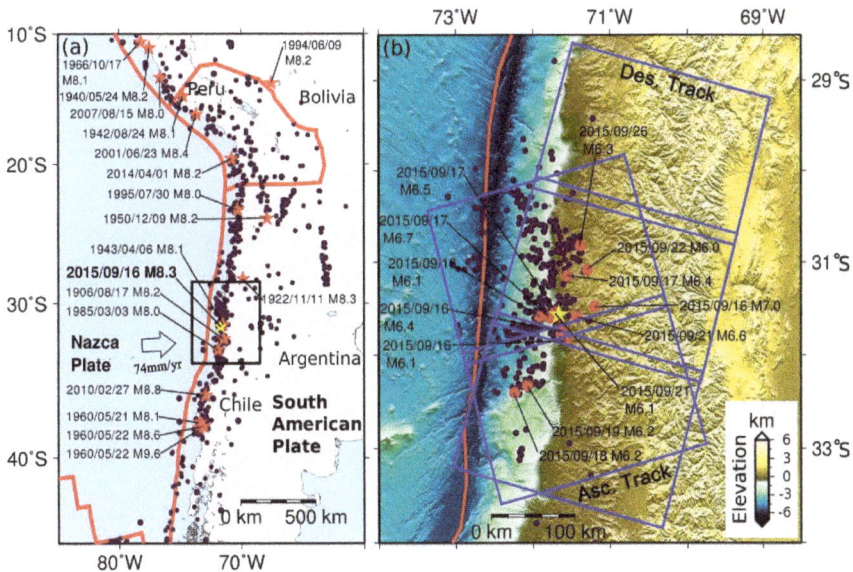

Figure 1. Tectonic settings of the 2015 Mw 8.3 Illapel earthquake and SAR data coverage. The main shock is presented by the yellow star. (**a**) shows earthquakes occurred along the interface between the Nazca and the South American plates (USGS, 2015a) since 1900. Red stars and dark purple dots represent earthquakes ≥Mw 8.0 and <Mw 8.0, respectively; (**b**) shows the topography of the area marked by black rectangle in (**a**) and the aftershocks occurred between 17 September 2015 and 1 October 2015. Red dots and dark purple dots represent earthquakes ≥Mw 6.0 and <Mw 6.0, respectively. The coverage of the S1A images is shown by blue rectangles. One rectangle denotes one S1A frame. The red line denotes the plate boundaries.

2. ISD Method for Co-Registering Multiple S1A TOPS Frames

2.1. ISD Method

Compared with the conventional strip-map InSAR, the TOPS mode InSAR has difficulty in azimuth co-registration due to a strong azimuth-variant Doppler centroid [7]. The interferometric phase bias caused by the Doppler centroid induced azimuth mis-registration (or azimuth offset) can be expressed as [7,18]:

$$\phi = 2\pi \cdot f_{DC} \cdot d_{azi}/PRF, \tag{1}$$

where f_{DC} is the Doppler centroid of the target point, d_{azi} is the azimuth co-registration error, and *PRF* is the Pulse Repetition Frequency of focused SLC data. We should note that f_{DC} varies linearly within a TOPS burst in the along-track dimension. Thus, a constant d_{azi} will produce a series of azimuth phase ramps imposed onto an interferogram, resulting in phase discontinuity (or phase jump) on the burst interface [19].

To suppress the phase jump at burst overlap to $\pm 1.5°$, a co-registration accuracy of at least 0.001 SLC pixel is required [7,14]. Currently, an overall mis-registration error is usually corrected with constant azimuth offset d_{azi} (see Equation (1)) based on the method of SD [18] or ESD [14]. For a single S1A IWS frame [13], which contains three sub-swaths and in each sub-swath there are typically nine burst, the number of burst is small (about $3 \times (9 - 1) = 24$ of burst overlaps), and these conventional methods can work well. However, conventional methods with a constant correction would be ineffective in the case of multiple consecutive S1A TOPS frames [20], because large variations of the Doppler centroid will result in considerable variations of azimuth mis-registration (or azimuth offset) for long imaging duration. More importantly, obvious phase jump would also occur on overlap area between burst, especially at the ends of the long data taken. In order to address this problem, an improved Spectral Diversity method is proposed in this study. Rather than treating d_{azi} as a constant, the new method compensates the variation of the azimuth mis-registration by a linear model:

$$d_{azi} = d_{azi}^0 + k \cdot t, \tag{2}$$

where d_{azi}^0 is the intercept, k is the change rate of azimuth mis-registration error, and t is the acquisition time difference between the current and the first SLC line. By substituting Equation (2) into Equation (1), one gets:

$$\phi \cdot s = d_{azi}^0 + k \cdot t, \tag{3}$$

where $s = \frac{PRF}{2\pi \cdot f_{DC}}$ is a scale factor converting phase to SLC azimuth offset. Given N observations (N is the quantity of burst overlaps) we get a linear system in matrix form:

$$L = B \cdot x, \tag{4}$$

where $x = [d_{azi}^0, k]^T$ is the unknown vector, and B is the design matrix:

$$B = \begin{bmatrix} 1 & t_1 \\ \vdots & \vdots \\ 1 & t_N \end{bmatrix}.$$

The observation vector $L = [L_1, L_2, \cdots, L_N]^T$, where $L_i = \phi_i \cdot s_i$ ($i = 1, \cdots, N$) is the azimuth offset of the i-th burst overlap, with ϕ_i being the average of the phase differences of the current and the next burst interferogram of the i-th burst overlap and $s_i = \frac{PRF}{2\pi \cdot f_{DC}^i}$ being the scale factor of i-th burst overlap.

When $N > 2$, Equation (4) becomes over-determined problem. To resolve the parameters x, we use an iterative weighted least square estimator [21–23], to account for possible gross error in observation vector L. When the estimation of x was obtained, the azimuth co-registration error for each azimuth line could be calculated by Equation (2) and used to refine the SLC resample look-up table in S1A TOPS SLC co-registration, which will be discussed in next section.

2.2. Procedure of S1A TOPS SLC Co-Registration with ISD

To meet the requirement and to counteract the topographic effect, the co-registration of wide-swath SLC images of S1A TOPS SLC [16] generally involves a sequentially DEM-assisted method (Step 1), conventional Cross-Correlation (CC) (Step 2) method, and a SD method (Step 3). Our ISD method can be easily embedded into the S1A SLC co-registration procedure for multiple S1A TOPS frames. Figure 2 shows the flow chart of the SLC co-registration, and the three key steps are summarized as follows.

- Step 1: initial co-registration. Estimate initial SLC images offset and set up SLC resampling look-up table [24] with orbital state vectors and an external DEM, e.g., SRTM DEM, and resample the slave SLC into master imaging geometry.
- Step 2: co-registration by the CC method. Determine the image offsets between master and the resampled slave SLC from Step 1 by the CC method, and use the image offsets to refine the look-up table. The original slave SLC is, again, resampled into master imaging geometry.
- Step 3: co-registration by ISD method. Calculate the mean phase differences of all bursts overlaps between the master and the resampled slave SLC from Step 2. Then the constant azimuth offset d_{azi}^0 and the change rate k are resolved by ISD method. The look-up table is refined with an azimuth offset calculated from Equation (2) and the estimated parameters and, finally, output the resampled slave SLC using the ultimately-refined look-up table.

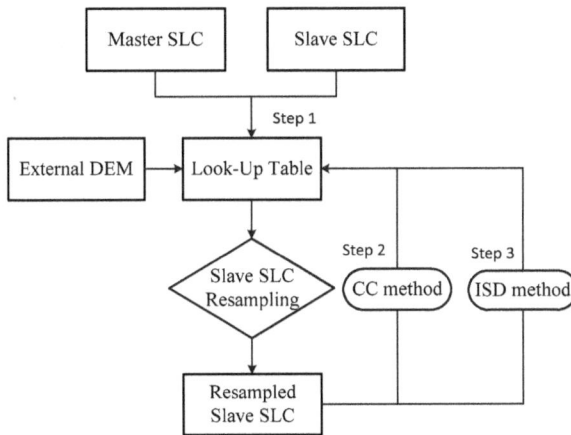

Figure 2. Flowchart of S1A TOPS SLC co-registration. The procedure of S1A co-registration is somewhat like a 3× iteration, which gradually improves the accuracy of SLC co-registration.

3. Data Processing

Ascending and descending tracks S1A data are used in this study (see Figure 1b). The details of the SAR data are shown in Table 1. The methods of Differential InSAR (D-InSAR) and Offset tracking will be exploited to estimate the Light of Sight (LOS) deformations. Finally, 2-D co-seismic displacement fields of this event will be derived.

Table 1. Information of SAR data.

Acquisition Date	Orbit Direction	Track	No. of Frame	No. of Total Burst	$B_\perp$ (m)
24 August 2015	Ascending	18	2	$36\,^1 = 2 \times 2 \times 9$	
17 September 2015	Ascending	18	2	$36 = 2 \times 2 \times 9$	70
26 August 2015	Descending	156	3	$84 = 3 \times 2 \times 9$	
19 September 2015	Descending	156	3	$84 = 3 \times 2 \times 9$	-116

[1] No. of Total Burst = No. of Frame × No. of Sub-Swath per Frame × No. of Burst per Sub-Swath.

3.1. Co-Registering Multiple Frames of S1A TOPS Data

The SAR co-registration follows the steps in Section 2.2. The 90 m SRTM DEM (v4.1) is used to estimate the initial look-up table [24] and remove the topographic phase. Thanks to the precise orbital state vector [6], an accuracy of a few pixels could be achieved [16] in the preliminary co-registration using orbital information. Then, the CC method is used to estimate the image offsets from evenly-distributed 64 × 192 (range × azimuth) tie-points with matching windows size of 128 × 256 (range × azimuth) pixels. All offsets are used to fit an offset polynomial that will subsequently be used to refine the look-up table. Due to limited accuracy of the CC method (~1/32 pixel) [25,26], the required co-registration accuracy of 0.001 SLC pixel could still not be achieved (see Table 2). Therefore, azimuth offset estimation with the proposed ISD is required for multiple S1A TOPS frames co-registration. We firstly calculated the observation vector L, the azimuth co-registration error, from each burst overlap. For descending data, three sub-swathes contain a number of $(3 \times (27 - 1) = 84)$ burst overlaps. Since some burst overlaps situate in low coherent area (average coherence < 0.75), such as sea or vegetation areas, 69 valid burst overlaps are used to fit the azimuth co-registration error. For ascending data, sub-swath 1 is located in the sea area, which is discarded directly. Sub-swathes 2 and 3 provide a total of 34 burst overlaps for fitting the azimuth co-registration error. The final fitted linear models are, respectively, $d_{azi}^{des} = 0.01320 - 2.1698 \times 10^{-4} \cdot t$ and $d_{azi}^{asc} = 0.01477 + 1.772 \times 10^{-4} \cdot t$ for descending and ascending tracks. The fitting results are also shown in Figure 3e. The azimuth co-registration accuracy with ISD is listed in Table 2, indicating that an accuracy of 0.001 SLC pixel has been achieved. Then the SLC resampling look-up table is refined by the fitted linear model. Finally, the slave SLC is precisely co-registered to the master SLC.

Table 2. SLC co-registration accuracies.

Interferometric Pair	CC Method		ISD Method	
	Azimuth (Pixel)	Range (Pixel)	Azimuth (Pixel)	Range (Pixel)
24 August 2015–17 September 2015	0.1170	0.0516	0.00104	—
26 August 2015–19 September 2015	0.0827	0.0425	0.00095	—

To demonstrate the effectiveness of our ISD method, the datasets are also co-registered with SD method implemented in GAMMA [16]. From comparisons in Figure 3a–d, we can see that obvious phase jumps still exist in the interferogram by the conventional SD method.

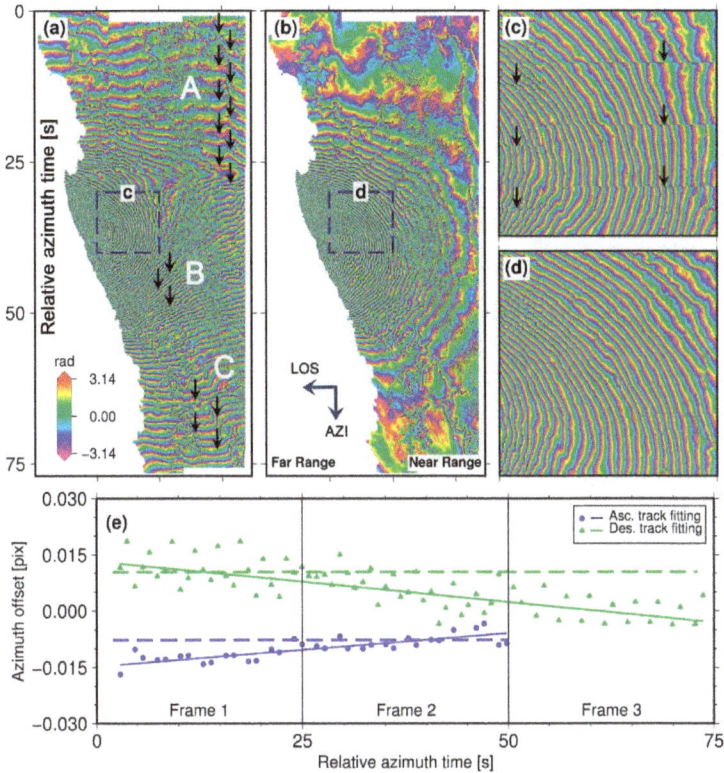

Figure 3. Comparison of the co-registration results by SD and ISD method. (a) Differential interferogram in the descending track co-registered by the SD method; (b) differential interferogram in the descending track co-registered by the ISD method; (c) enlarged view of the rectangle area marked by c in (a); (d) enlarged view of the rectangle area marked by d in (b); (e) variation of the azimuth co-registration error and its linear fitting. Black arrows in (a) and (c) mark the locations of phase jump; the solid green triangle, solid and dashed green line in (e) denote the samples of burst overlap azimuth offset, the fitted azimuth offset adopted by, the constant offset adopted by SD for the descending track, containing three S1A TOPS IWS frames with a number of 84 burst overlaps; the solid blue circles, solid and dashed blue lines in (e) represent the counterparts for the ascending track, containing two S1A TOPS IWS frames with 51 burst overlaps.

3.2. D-InSAR and Offset Tracking Processing

The TOPS D-InSAR processing, somewhat like traditional D-InSAR procedures [27], can be conducted after the SLC co-registration. As the Doppler centroid variations are different within bursts, we do not apply the azimuth common band filter in the interferogram [16]. The topographic phase is removed with 90 m SRTM, thus generating the differential interferogram, and multi-looking factors of 20 × 4 (range × azimuth) is applied to set the output result with a resolution of about 50 × 50 m. The improved adaptive phase noise filter [28] is utilized to smooth the interferogram. Subsequently, the smoothed interferogram is unwrapped by network programming-based phase unwrapping method [29]. Low coherent area (coherence ≤ 0.75) is masked out in the phase unwrapping process. Eventually, the LOS deformations from both the ascending and descending tracks are obtained (see Figure 4a,b).

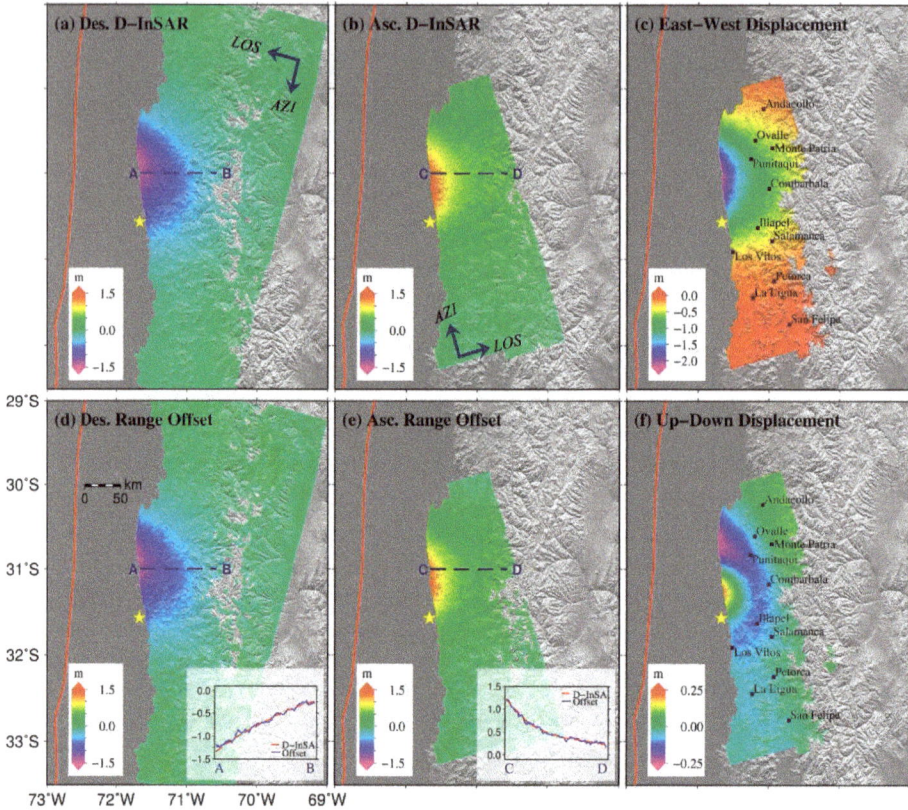

Figure 4. Co-seismic deformation of the earthquake. (**a**) and (**b**) are the D-InSAR LOS deformations from ascending and descending geometries, respectively. A positive sign means the ground moves towards the sensor; (**c**) east-west deformation. (+: towards east); (**d**) and (**e**) are the LOS deformations determined by Offset Tracking; (**f**) vertical deformation (+: Up). The yellow star indicates the location of the main shock. The red line denotes the boundary of the Nazca and South American plates; the inner figure in (**d**) shows the deformation profiles by D-InSAR and Offset Tracking along section A–B for comparisons, and the inner figure in (**e**) shows that of the profiles along section C-D.

Apart from the D-InSAR measurements, the Offset Tracking [19] measurements in the slant range and azimuth direction are also achieved by conventional CC method [25] using the co-registered SLC images. The sliding step size of offset estimate window is set to be 20 × 4 (range × azimuth) pixels to generate the same resolution result with D-InSAR. To reduce the noise, center weighted median filter [30] with a window size of 11 × 11 (range × azimuth) is employed to smooth the offset measurement, in which the correlation coefficient of the matching windows is used as the weight factor. From our offset tracking measurement implementation, we find that the azimuth deformation is difficult to be detected. This is also confirmed by Multiple-Aperture Interferometry (MAI) technology, whose expected accuracy for S1A TOPS IWS data is about 27 cm [25]. Thus, we do not use the azimuth offset measurements in our further study. The range offset measurements for ascending and descending orbits are shown in Figure 4d,e.

3.3. Precision Assessment and 2-D Displacement Retrieval

The precisions of both D-InSAR and Offset Tracking measurements are assessed by evaluating the local standard deviations (STDs) on sliding window with a size of about 1×1 km pixel-by-pixel [31]. The linear trend is fitted and subtracted from the sliding window to get stationary signals for calculating the STDs. The local STDs maps of the measurements from D-InSAR and Offset Tracking in the slant range are shown in Figure S1, which will be used as the weight factor to derive the 2D displacement of this earthquake.

Theoretically, three dimensional (3D) deformations [32] could be resolved with LOS and along-track measurements from two or more viewing geometries [33,34]. However, the north-south direction displacement is insensitive in D-InSAR measurements in general, and beyond the detectability of S1A TOPS IWS data in this earthquake in particular [25]. Thus, the north-south deformation is compromisingly neglected in this study. The unit projection vector of LOS deformation for ascending and descending orbits are $S_{asc} = [s_e, s_n, s_u]^T = [-0.607, -0.170, 0.755]^T$ and $S_{des} = [s_e, s_n, s_u]^T = [0.608, -0.168, 0.776]^T$, respectively. By assuming the north-south deformation is null and exploiting the projection vectors S_{asc} and S_{des} to establish observation equation, two dimensional (2D) deformations, i.e., east-west and up=down (or vertical) are derived by combing the D-InSAR and Offset Tracking measurements from ascending and descending orbits [35]. We acknowledge that neglecting the north-south deformation will induce errors in the east-west and vertical deformation estimation, but for the Illapel earthquake it might be very slim, because the azimuth deformation itself is very small. The final 2D deformations are shown in Figure 4c,f.

4. Results

We got the co-seismic LOS deformations and the 2-D, i.e., east-west and vertical displacements with the elaborated data processing strategy presented in previous section. Figure 4a,d show the co-seismic LOS deformations extracted by D-InSAR and Range Offset Tracking with descending track InSAR data, and Figure 4b,e are the co-seismic LOS deformations with ascending track InSAR data. Due to the effect of decorrelation, there are still some areas along the Andes Mountains that has been masked out.

Comparing the D-InSAR and the Offset Tracking measurements, we can find that they have the same LOS deformation pattern and the same deformation magnitude, showing remarkably good agreement between each other. To make more detailed comparisons, the profiles along A-B and C-D are displayed as insets in Figure 4d,e. These results illustrate quite good consistency in the LOS D-InSAR and Offset Tracking measurements for S1A TOPS mode data. Comparing the D-InSAR measurements from both descending and ascending track, we can see that they have quite a similar deformation pattern, which implies that the horizontal displacement dominates the ground deformation; we can also see that most of the deformation with large magnitude locates on the upper part along profile A-B near the coast in descending track, while it is opposite in ascending track, implying that obvious vertical displacement has occurred and the upper part sank and the lower part uplifted.

Figure 4c shows the east-west displacement, where up to 2 m and towards the west occurred in the coastal area. Vertical displacement (Figure 4f) associated with the ground uplift and subsidence ranging between $[-0.25, 0.25]$ m is also detected. It should be noted that east-west and vertical displacements are only calculated over the common area of all measurements (Figure 4a,b,d,e), which ensures that there are redundant observations for alleviating the effect of atmospheric [36] in the 2-D deformation inversion.

5. Discussion

To improve the accuracy of co-registering multiple S1A TOPS data frames, an ISD method is proposed and embedded into the S1A SLC co-registration procedure. Comparing with the conventional SD or ESD methods, the ISD method is more suitable for multiple S1A TOPS data frames co-registration by compensating the variation of azimuth co-registration error using a linear model. However, some issues shall be carefully settled to ensure the accuracy of ISD. Firstly, similar to the SD or ESD methods, the ISD method is subject to small azimuth offset measuring. The prerequisite for using ISD method is that the azimuth offset should be within ± 0.75 SLC pixels [19] Therefore, the azimuth offset shall firstly be truncated to this limit by other methods before applying the ISD method, e.g., the conventional CC method. Secondly, as shown in Equations (1) and (2), the observation, *i.e.*, azimuth co-registration error d_{azi} on each burst overlap, relates linearly to the phase offset ϕ. To obtain a high coherent phase offset, the burst overlap's phase difference of subsequent bursts shall be multi-looked with a large factor, e.g., 100×4 (range $\times$ azimuth) pixels and, subsequently, be filtered to reduce phase noise [28,31]. The multi-looking factor depends on the phase quality, and we found that a factor of 100×4 is empirically enough. Additionally, it should be noted that the north-south deformation would cause an azimuth offset [15] too, which will superimpose on the normal azimuth co-registration error. If only a part of burst overlaps have north-south deformation, our ISD method would automatically exclude those observations by an iterative weighted least square estimator [21], as those observations possibly manifest as gross errors with respect to the normal azimuth offset observations. Furthermore, given the prior information we could also manually remove the observations contaminated by north-south deformation, and the final ISD accuracy would be improved in such a situation.

The 2015 Mw 8.3 Illapel earthquake occurred on the interface between the Nazca and South American plates (see Figure 1). The co-seismic deformation of this earthquake is mapped with S1A TOPS data in this study. Comparing our results with those obtained by the European Space Agency's Project on Sentinel-1 INSAR Performance Study with TOPS Data [37], we found that ours are continuous and clear, while their D-InSAR interferograms still have obvious phase jumps, similar to the results in Figure 3c, for both ascending and descending tracks. Affecting by the discontinuities of the D-InSAR measurements, their decomposed vertical displacement has considerable leap. All of these errors would provide false information for the interpretation of earthquake mechanism. On the contrary, there are no phase jumps in our results (see Figure 4a,b). In addition to the D-InSAR measurements, we also extract the Offset Tracking measurements in the LOS direction. Both D-InSAR and Offset Tracking LOS measurements show good consistency (see Figure 4a,b,d,e). This confirms that the SLC images have been precisely co-registered by our ISD method. In addition, we have derived the 2-D deformations from the LOS D-InSAR and Offset Tracking measurements. The results show that the coastal area moved westwards by up to 2 m, and the vertical displacement (see Figure 4f) associated with ground uplift and subsidence ranging between $[-0.25, 0.25]$ m has also been detected. Additionally, by measuring the deformation area, we also found the dimension of the event is consistent with the result from USGS [4]. Additionally, the 2-D displacements, a recent study [38] has tried to recover the 3-D displacements of the event, in which they use the along-track interferometry [38] to measure the azimuth deformation. By comparing them, we found that our results are in line with theirs. However, the along-track measuring [38] can only be applied to the burst overlap region, which occupies only about 15% of the whole frame. While, for the non-burst overlap region, occupying about 85% of the frame, interpolation processes must be implemented, which will, of course, induce errors.

Chile is prone to earthquakes that have caused severe destruction in the past century. The Nazca plate moves to the east-northeast at a varying rate (approximately 80–65 mm/year from north to south) relative to a fixed South American plate, resulting in complex geologic processes along the Nazca subduction zone [1]. At the latitude of the 2015 Mw 8.3 Illapel earthquake, many massive earthquakes have occurred, including the 2010 Mw 8.8 Maule earthquake in central Chile [3,39]. The 2D results show that the coastal area has moved westwards by up to 2 m accompanying with a little vertical deformation. This implies the collision of the Nazca plate squeezed the coast and the east-west displacement dominating the co-seismic deformation. The result is in good accordance with the geological background of the region. All those measurements would benefit for the geophysical parameters inversion of the 2015 Mw 8.3 Illapel earthquake.

S1A satellite is now in normal on-orbit operation. As the first SAR satellite of Sentinel-1 constellation, S1A systematically acquires SAR image over land area with a revisit of 12 days in Europe and 24 days for other regions [40]. S1A is controlled to follow an orbital tube of 100 m radius [41], which guarantees short spatial perpendicular baseline. These two merits reduce the decorrelation, which has been demonstrated in previous studies [27]. As a C-band SAR system, S1A also has a limited detectable deformation gradient capability with D-InSAR technology [42]. However, the increasing slant range resolution (up to 2 m) enhances the detectability along the slant range with the SAR Offset Tracking method, and makes it a good supplement for D-InSAR measurement. Additionally, the S1A acquires data with a wide swath (250 kilometer wide for IWS mode), which would be more favorable to study continent-wide event than the conventional SAR satellite (<100 km), like the ENVISAT and the ALOS. The advantages of S1A for large earthquake study have been presented in this work. The advent of S1A will boost geo-hazard monitoring and crisis management [43,44].

6. Conclusions

In this study, we mapped the co-seismic deformation of the 2015 Mw 8.3 Illapel earthquake using multiple S1A TOPS frames data. In order to achieve a co-registration accuracy of 0.001 pixel, we proposed an ISD method by compensating the variation of azimuth co-registration error in the azimuth direction with a linear model, and successfully embedded it into the S1A TOPS SLC co-registration procedure. The co-registration results of the Illapel earthquake demonstrate that our ISD method can be used for co-registering multiple frames of S1A TOPS data, which will be beneficial for continent-wide event study with S1A TOPS data.

The co-seismic LOS deformation of the Illapel earthquake is measured by both the D-InSAR and the Offset Tracking method. Comparing the D-InSAR and the Range Offset measurements, we found that they are consistent with each other, which reveals the enhanced range deformation detectability of S1A TOPS data. Thus, the range offset tracking can act as a good supplement for D-InSAR in the situation where interferometric phases are turned to be decorrelated because of large displacement. The 2D deformations (east-west and vertical directions) are also resolved based on the D-InSAR and Offset Tracking LOS measurements. Results show that the coastal area has moved westward by up to 2 m, and the vertical deformation varies between $[-0.25, 0.25]$ m, implying the collision of the Nazca plate squeezed the coast.

Supplementary Materials: The following are available online at www.mdpi.com/2072-4292/8/5/376/s1, Figure S1: Local standard deviations of LOS measurements.

Acknowledgments: This work is supported by the National Natural Science Foundation of China (41574005, 41474007 and 41404013) and the Fundamental Research Funds of Central South University (2014zzts049). The S1A data is from the ESA project and is downloaded from the Sentinel-1 Scientific Data Hub. Bathymetry data used in Figure 1 is provided by TOPEX, UCSD. Figures are produced by General Mapping Tools (v5.1.2). The GAMMA software (v20150702) is partially used in InSAR processing of S1A data. The author would like to thank the four anonymous reviewers for their very constructive remarks and suggestions.

Author Contributions: Bing Xu and Zhiwei Li conceive the research work and Bing Xu wrote the first draft of the paper. Guangcai Feng, Zeyu Zhang, Qijie Wang, Jun Hu, Xingguo Chen contributed to experiment implementation and result interpretation. All authors contributed to paper writing and revision.

Conflicts of Interest: The authors declare no conflict of interest.

References

1. Usgs Earthquake Hazards Program, m8.3-48km w of Illapel, Chile, 2015. Available online: http://earthquake. Usgs.Gov/earthquakes/eventpage/us20003k7a#general_summary (accessed on 30 October 2015).
2. Pacific tsunami Warning Center (PTWC). Widespread Hazardous Tsunami Waves Are Possible along the Coast of Chile and Peru, 2015. Available online: http://ptwc.Weather.Gov/ptwc/text.Php?Id=pacific. Tsupac.2015.09.16.2301 (accessed on 30 October 2015).
3. Moreno, M.; Rosenau, M.; Oncken, O. 2010 maule earthquake slip correlates with pre-seismic locking of andean subduction zone. *Nature* **2010**, *467*, 198–202. [CrossRef] [PubMed]

4. Usgs Earthquake Hazards Program, m8.3 Coastal Chile Earthquake of 16 September 2015. Available online: http://earthquake.Usgs.Gov/earthquakes/eqarchives/poster/2015/20150916.Php (accessed on 30 October 2015).

5. Torres, R.; Snoeij, P.; Geudtner, D.; Bibby, D.; Davidson, M.; Attema, E.; Potin, P.; Rommen, B.; Floury, N.; Brown, M. Gmes sentinel-1 mission. *Remote Sens. Environ.* **2012**, *120*, 9–24. [CrossRef]

6. Schubert, A.; Small, D.; Miranda, N.; Geudtner, D.; Meier, E. Sentinel-1a product geolocation accuracy: Commissioning phase results. *Remote Sens.* **2015**, *7*, 9431–9449. [CrossRef]

7. De Zan, F.; Guarnieri, A.M. Topsar: Terrain observation by progressive scans. *IEEE Trans. Geosci. Remote Sens.* **2006**, *44*, 2352–2360. [CrossRef]

8. Grandin, R.; Vallée, M.; Satriano, C.; Lacassin, R.; Klinger, Y.; Simoes, M.; Bollinger, L. Rupture process of the Mw = 7.9 2015 gorkha earthquake (Nepal): Insights into himalayan megathrust segmentation. *Geophys. Res. Lett.* **2015**, *42*, 8373–8382. [CrossRef]

9. González, P.J.; Bagnardi, M.; Hooper, A.J.; Larsen, Y.; Marinkovic, P.; Samsonov, S.V.; Wright, T.J. The 2014–2015 eruption of fogo volcano: Geodetic modeling of Sentinel-1 TOPS interferometry. *Geophys. Res. Lett.* **2015**, *42*, 9239–9246. [CrossRef]

10. Salvi, S.; Stramondo, S.; Funning, G.; Ferretti, A.; Sarti, F.; Mouratidis, A. The Sentinel-1 mission for the improvement of the scientific understanding and the operational monitoring of the seismic cycle. *Remote Sens. Environ.* **2012**, *120*, 164–174. [CrossRef]

11. Wen, Y.; Xu, C.; Liu, Y.; Jiang, G. Deformation and source parameters of the 2015 Mw 6.5 earthquake in pishan, western China, from Sentinel-1a and ALOS-2 data. *Remote Sens.* **2016**, *8*, 134. [CrossRef]

12. Kim, J.-W.; Lu, Z.; Degrandpre, K. Ongoing deformation of sinkholes in wink, texas, observed by time-series Sentinel-1a SAR interferometry (preliminary results). *Remote Sens.* **2016**, *8*, 313. [CrossRef]

13. Esa, Scientific Data Hub. Avaliable online: https://scihub.Copernicus.Eu/ (accessed on 30 December 2015).

14. Prats-Iraola, P.; Scheiber, R.; Marotti, L.; Wollstadt, S.; Reigber, A. TOPS interferometry with TerraSAR-x. *IEEE Trans. Geosci. Remote Sens.* **2012**, *50*, 3179–3188. [CrossRef]

15. De Zan, F.; Prats-Iraola, P.; Scheiber, R.; Rucci, A. Interferometry with TOPS: Coregistration and azimuth shifts. In Proceedings of 10th European Conference on Synthetic Aperture Radar, Berlin, Germany, 3–5 June 2014; pp. 1–4.

16. Wegmüller, U.; Werner, C.; Strozzi, T.; Wiesmann, A.; Frey, O.; Santoro, M. Sentinel-1 support in the gamma software. In Proceedings of the Fringe 2015 Conference, ESA ESRIN, Frascati, Italy, 23–27 March 2015; pp. 23–27.

17. Sandwell, D.; Mellors, R.; Tong, X.; Wei, M.; Wessel, P. Open radar interferometry software for mapping surface deformation. *EOS Trans. Am. Geophys. Union* **2011**, *92*, 234. [CrossRef]

18. Scheiber, R.; Moreira, A. Coregistration of interferometric SAR images using spectral diversity. *IEEE Trans. Geosci. Remote Sens.* **2000**, *38*, 2179–2191. [CrossRef]

19. Scheiber, R.; Jager, M.; Prats-Iraola, P.; De Zan, F.; Geudtner, D. Speckle tracking and interferometric processing of TerraSAR-x TOPS data for mapping nonstationary scenarios. *IEEE J. Sel. Top. Appl. Earth Obs. Remote Sens.* **2015**, *8*, 1709–1720. [CrossRef]

20. Grandin, R. Interferometric processing of SLC Sentinel-1 TOPS data. In Proceedings of the 2015 ESA Fringe Workshop, Frascati, Italy, 23–27 March 2015.

21. Hooper, P.M. Iterative weighted least squares estimation in heteroscedastic linear models. *J. Am. Stat. Assoc.* **1993**, *88*, 179–184.

22. Hu, J.; Wang, Q.; Li, Z.; Zhao, R.; Sun, Q. Investigating the ground deformation and source model of the yangbajing geothermal field in Tibet, China with the WLS InSAR technique. *Remote Sens.* **2016**, *8*, 191. [CrossRef]

23. Li, Z.; Zhao, R.; Hu, J.; Wen, L.; Feng, G.; Zhang, Z.; Wang, Q. InSAR analysis of surface deformation over permafrost to estimate active layer thickness based on one-dimensional heat transfer model of soils. *Sci. Rep.* **2015**, *5*. [CrossRef] [PubMed]

24. Sansosti, E.; Berardino, P.; Manunta, M.; Serafino, F.; Fornaro, G. Geometrical SAR image registration. *IEEE Trans. Geosci. Remote Sens.* **2006**, *44*, 2861. [CrossRef]

25. Jung, H.-S.; Lee, W.-J.; Zhang, L. Theoretical accuracy of along-track displacement measurements from multiple-aperture interferometry (MAI). *Sensors* **2014**, *14*, 17703–17724. [CrossRef] [PubMed]

26. Feng, G.; Xu, B.; Shan, X.; Li, Z.; Zhang, G. Coseismic deformation and source parameters of the 24 September 2013 awaran, pakistan m (w) 7. 7 earthquake derived from optical Landsat 8 satellite images. *Chin. J. Geophys.* **2015**, *58*, 1634–1644.

27. Feng, G.; Li, Z.; Shan, X.; Xu, B.; Du, Y. Source parameters of the 2014 Mw 6.1 south Napa earthquake estimated from the Sentinel 1a, cosmo-skymed and GPS data. *Tectonophysics* **2015**, *655*, 139–146.

28. Li, Z.; Ding, X.; Huang, C.; Zhu, J.; Chen, Y. Improved filtering parameter determination for the goldstein radar interferogram filter. *ISPRS J. Photogramm. Remote Sens.* **2008**, *63*, 621–634. [CrossRef]

29. Costantini, M. A novel phase unwrapping method based on network programming. *IEEE Trans. Geosci. Remote Sens.* **1998**, *36*, 813–821. [CrossRef]

30. Ko, S.J.; Lee, Y.H. Center weighted median filters and their applications to image enhancement. *IEEE Trans. Circuits Syst.* **1991**, *38*, 984–993. [CrossRef]

31. Goldstein, R.M.; Werner, C.L. Radar interferogram filtering for geophysical applications. *Geophys. Res. Lett.* **1998**, *25*, 4035–4038. [CrossRef]

32. Jo, M.-J.; Jung, H.-S.; Won, J.-S. Detecting the source location of recent summit inflation via three-dimensional InSAR observation of Kīlauea volcano. *Remote Sens.* **2015**, *7*, 14386–14402. [CrossRef]

33. Li, Z.; Yang, Z.; Zhu, J.; Hu, J.; Wang, Y.; Li, P.; Chen, G. Retrieving three-dimensional displacement fields of mining areas from a single InSAR pair. *J. Geod.* **2015**, *89*, 17–32. [CrossRef]

34. Samsonov, S.; Tiampo, K.F.; Rundle, J.B. Application of DinSAR-GPS optimization for derivation of three-dimensional surface motion of the southern California region along the san andreas fault. *Comput. Geosci.* **2008**, *34*, 503–514. [CrossRef]

35. Hu, J.; Li, Z.; Ding, X.; Zhu, J.; Zhang, L.; Sun, Q. Resolving three-dimensional surface displacements from InSAR measurements: A review. *Earth Sci. Rev.* **2014**, *133*, 1–17. [CrossRef]

36. Zhan, W.J.; Li, Z.W.; Wei, J.C.; Zhu, J.J.; Wang, C.C. A strategy for modeling and estimating atmospheric phase of SAR. *Chin. J. Geophys.* **2015**, *58*, 2320–2329.

37. Insarapp Project, Chile Earthquake: Sentinel-1 InSAR Analysis. Available online: http://insarap.Org/ (accessed on 30 October 2015).

38. Grandin, R.; Klein, E.; Métois, M.; Vigny, C. Three-dimensional displacement field of the 2015 Mw 8.3 illapel earthquake (Chile) from across- and along-track Sentinel-1 TOPS interferometry. *Geophys. Res. Lett.* **2016**, *43*. [CrossRef]

39. Delouis, B.; Nocquet, J.M.; Vallée, M. Slip distribution of the 27 February 2010 Mw = 8.8 maule earthquake, central Chile, from static and high-rate GPS, InSAR, and broadband teleseismic data. *Geophys. Res. Lett.* **2010**, *37*, 1–7. [CrossRef]

40. Elliott, J.; Elliott, A.; Hooper, A.; Larsen, Y.; Marinkovic, P.; Wright, T. Earthquake monitoring gets boost from a new satellite. *Eos* **2015**, *96*, 14–18. [CrossRef]

41. Geudtner, D.; Torres, R.; Snoeij, P.; Ostergaard, A.; Navas-Traver, I. Sentinel-1 mission capabilities and SAR system calibration. In Proceedings of the 2013 IEEE Radar Conference (RadarCon), Ottawa, ON, Canada, 29 April–3 May 2013; pp. 1–4.

42. Jiang, M.; Li, Z.; Ding, X.; Zhu, J.-J.; Feng, G. Modeling minimum and maximum detectable deformation gradients of interferometric SAR measurements. *Int. J. Appl. Earth Obs. Geoinf.* **2011**, *13*, 766–777. [CrossRef]

43. Xu, W.; Burgmann, R.; Li, Z. An improved geodetic source model for the 1999 Mw 6.3 Chamoli earthquake, India. *Geophys. J. Int.* **2016**, *205*, 236–242. [CrossRef]

44. Sun, Q.; Zhang, L.; Hu, J.; Ding, X.; Li, Z.; Zhu, J. Characterizing sudden geo-hazards in mountainous areas by DinSAR with an enhancement of topographic error correction. *Nat. Hazard.* **2015**, *75*, 2343–2356. [CrossRef]

remote sensing

MDPI

Article

Coseismic Fault Model of Mw 8.3 2015 Illapel Earthquake (Chile) Retrieved from Multi-Orbit Sentinel1-A DInSAR Measurements

Giuseppe Solaro [1,*], Vincenzo De Novellis [1], Raffaele Castaldo [1], Claudio De Luca [1,2], Riccardo Lanari [1], Michele Manunta [1] and Francesco Casu [1]

[1]   Istituto per il Rilevamento Elettromagnetico dell'Ambiente, IREA-CNR, Via Diocleziano 328, 80124 Napoli, Italy; denovellis.v@irea.cnr.it (V.D.N.); castaldo.r@irea.cnr.it (R.C.); deluca.c@irea.cnr.it (C.D.L.); lanari.r@irea.cnr.it (R.L.); manunta.m@irea.cnr.it (M.M.); casu.f@irea.cnr.it (F.C.)

[2]   Department of Electrical and Information Technology Engeneering (DIETI), University of Naples, Federico II, via Claudio 21, 80124 Napoli, Italy

*   Correspondence: solaro.g@irea.cnr.it; Tel.: +39-0817620631

Academic Editors: Zhenhong Li, Zhong Lu and Prasad S. Thenkabail

Received: 28 January 2016; Accepted: 7 April 2016; Published: 12 April 2016

Abstract: On 16 September 2015, a Mw 8.3 interplate thrust earthquake ruptured offshore the Illapel region (Chile). Here, we perform coseismic slip fault modeling based on multi-orbit Sentinel 1-A (S1A) data. To do this, we generate ascending and descending S1A interferograms, whose combination allows us to retrieve the EW and vertical components of deformation. In particular, the EW displacement map highlights a westward displacement of about 210 cm, while the vertical map shows an uplift of about 25 cm along the coast, surrounded by a subsidence of about 20 cm. Following this analysis, we jointly invert the multi-orbit S1A interferograms by using an analytical approach to search for the coseismic fault parameters and related slip values. Most of the slip occurs northwest of the epicenter, with a maximum located in the shallowest 20 km. Finally, we refine our modeling approach by exploiting the Finite Element method, which allows us to take geological and structural complexities into account to simulate the slip along the slab curvature, the von Mises stress distribution, and the principal stress axes orientation. The von Mises stress distribution shows a close similarity to the depth distribution of the aftershock hypocenters. Likewise, the maximum principal stress orientation highlights a compressive regime in correspondence of the deeper portion of the slab and an extensional regime at its shallower segment; these findings are supported by seismological data.

Keywords: Illapel (Chile) earthquake; Sentinel 1-A; DInSAR; fault slip analytical model; 2D Finite Element model

1. Introduction

On 16 September 2015, at 22:54 UTC, an earthquake of Mw 8.3, at a depth of 25 km, occurred off the coast of Central Chile in Coquimbo area. The epicenter, located 46 km west of Illapel city (Figure 1), shook buildings in the capital city of Santiago and generated a tsunami that caused flooding in some coastal areas, such as the coastal town of Coquimbo, NW of Illapel, which were hit by waves up to 11 m high after the earthquake [1]. It has been estimated that more than 27,000 people were exposed to severe shaking from this earthquake [USGS PAGER].

The focal mechanism relevant to the main shock, calculated by USGS [2], indicates a N 4° striking sub-horizontal thrust fault. This rupture mechanism is in accordance with the convergence of the Nazca plate toward South American plate at an overall rate of about 6.6 cm/yr [1] (Figure 1). In the first hours, after the main shock event, there were 17 aftershocks in the same area with the strongest one having Mw 7.2 (16 September, 23:18 UTC); most of these aftershocks reveal the same rupture mechanism.

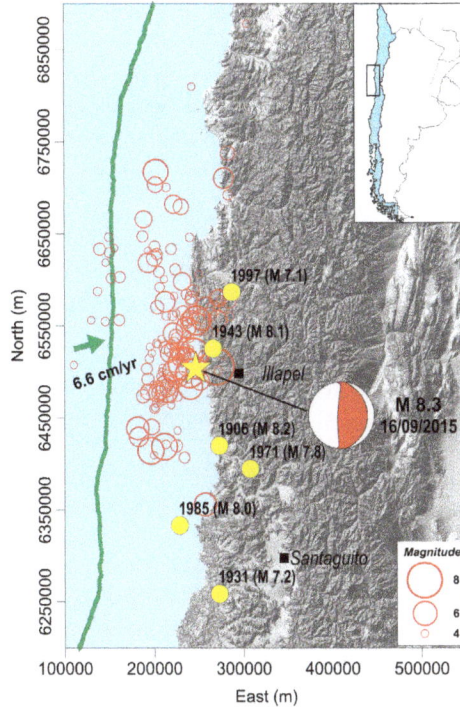

Figure 1. Map of seismicity in the Illapel region reporting the earthquakes from 30 August to 19 September 2015, with M > 4 (red circles): the epicenter of the main shock is outlined with the yellow star, and its focal mechanism is also represented. Yellow dots show the past most energetic earthquakes of the area. The solid green line indicates the trench axis position, while the green arrow shows the direction of motion of the Nazca Plate towards the South American Plate [3]. The upper right inset shows the location of the epicentral area.

The main event is categorized as an interplate subduction earthquake, localized at the interface between the subducting Nazca and overriding South American plates, supported also by a shallow depth (<50 km) and a nodal plane dipping with a low angle to the East. The South American subduction zone hosted a significant number of large earthquakes that provide details on strain accumulation and release during the earthquake cycle; another key feature of this tectonic setting is the trench-parallel variation in overriding plate shortening, which is maximum in the center and progressively decreases to the North and South [3]. Since 1900, numerous M 7 or larger earthquakes have occurred in this subduction zone (see Figure 1).

We remark that the 2015 Illapel earthquake occurs within the rupture zone of the 1943 M 8.1 seismic event. Since then, after many years of quiescence, the seismic activity of this plate interface suddenly increased in 1997; indeed, 7 events with M > 6 occurred between July 1997 and January 1998 along this shallow dipping subduction zone [4]. During this period, the rupture zone followed a cascade pattern propagating toward the location of the 15 October 1997 M 7.1 earthquake, which occurred at a depth of 68 km [5].

Recently, based on teleseismic data, Ye *et al.* [6] found that the Illapel earthquake occurred on a bilateral along-strike rupture zone with a larger slip north-northwest of the epicenter, with a peak slip of 7–10 m [6]; Tilmann *et al.* [7] determined a comprehensive rupture model through the inversion of seismic, GPS, and Sentinel 1-A (S1A) descending data; Melgar *et al.* [1] resolved a kinematic slip model by using GPS, strong motion and S1A descending and ascending data and jointly analyzed them with teleseismic backprojection. Grandin *et al.* [8] retrieved the full 3D displacement field by using S1A ascending and descending data, by jointly exploiting the surface deformation measurements components in the across image and the along-track satellite direction, retrieved through the DInSAR and Multiple Aperture Insar (MAI) techniques, respectively.

In this work, a detailed coseismic slip fault model is presented, obtained by taking advantage of the wide spatial coverage and reduced revisit time offered by multi-orbit S1A data, as well as of the high-accuracy measurement capability of the DInSAR technique. In particular, we proceed taking the following three steps: (1) we generate two S1A interferograms for both ascending and descending orbits, respectively, encompassing the main shock, in order to combine the displacements along the satellite Line Of Sight (LOS) for retrieving the EW and vertical components of deformation; (2) we jointly invert multi-orbit S1A deformation LOS measurements by using an analytical approach to search for the coseismic fault parameters; (3) we refine our modeling approach by exploiting the Finite Element (FE) method, which allows us to take into account geological and structural complexities of the region, and to simulate the slip along the subducted slab, the stress distribution, and the principal stress axes orientation.

2. Satellite Data

The used dataset consists of four SAR acquisitions that were taken on 26 August 2015 and 19 September 2015 along ascending orbits (Track 18), and 31 July 2015 and 17 September 2015 over descending ones (Track 156), by the C-Band S1A sensor acquiring data with the Terrain Observation with Progressive Scans (TOPS) mode, which is specifically designed for interferometric application and guarantees a very large spatial coverage [9]. Note that, similar to the ScanSAR, TOPS mode is burst-based; indeed, during the acquisition time, the antenna beam is switched cyclically among different sub-swaths, allowing a significant improvement of the range coverage, but with the detriment of the azimuth resolution. In particular, our SAR data are acquired through the Interferometric Wide Swath (IWS) TOPS mode configuration which is characterized by a swath extension of about 250 km [10]. Such an acquisition mode happens to be particularly effective in the framework of a very large area deformation analysis, as in the case of the Illapel earthquake, at the expense of a coarse spatial resolution (about 15 m and 4 m along azimuth and range, respectively). In our analysis, we jointly process 4 and 3 S1A slices for ascending and descending tracks, respectively, which correspond to an area of ~130,000 km^2.

By exploiting the available S1A Single Look Complex (SLC) images, we generate two DInSAR interferograms (characterized by a spatial baseline of 75 m and 7 m for ascending and descending orbits, respectively), following a burst-by-burst processing approach and working in radar coordinates (see Figure A1). In particular, we first co-register the SLC bursts through the use of precise orbits information and a 3-arcsec Shuttle Radar Topography Mission (SRTM) DEM of the investigated area [11]. Then, we calculate the DInSAR interferograms for each burst, by using the same DEM for the topographic residue component removal, with a multilook factor of 2 and 10 pixels along the azimuth direction and range, respectively. Subsequently, we compensate for the residual azimuth phase ramp due to possible azimuth mis-registration, through the Enhanced Spectral Diversity method [12] applied to the overlapping areas across consecutive bursts, both on azimuth and range directions. Finally, the compensated wrapped phases of each burst are combined into a single DInSAR interferogram to which the phase unwrapping [13], and the geocoding steps are then applied. Note that, for what concerns the phase unwrapping operation, which permits the retrieval of the interferometric phase (unwrapped interferogram) from its restriction to the $[-\pi, \pi]$ interval (wrapped interferogram), it is generally carried out by properly integrating the interferometric phase gradient directly estimated from the wrapped interferogram [14]. Because of the above mentioned integration step, the unwrapped interferogram is available only for the phase of one pixel. This pixel, usually referred to as a reference point, is typically located in a stable area, and its interferometric phase is set to zero (corresponding to the absence of deformation).

The so-generated coseismic DInSAR deformation maps are presented in Figure 2a,b and exhibit in both cases one main deformation lobe. Moreover, as clearly shown in Figure 2a,b, no phase ramps are present in the resulting interferograms. Such a result is also evident from the analysis of wrapped interferograms, provided in the Appendix (see Figure A1).

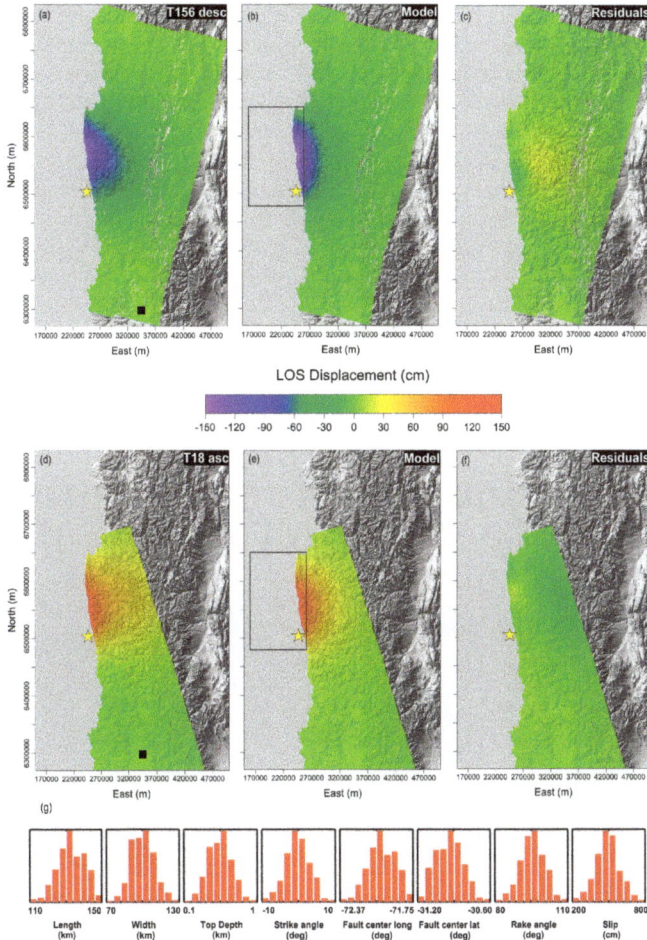

Figure 2. DInSAR data and analytical model. (**a**) Line-of-sight (LOS) coseismic displacement map computed by using Sentinel 1-A (S1A) data acquired from descending orbit (Track 156) on 31 July 2015 and 17 September 2015 (**b**) S1A descending LOS projected analytical model; (**c**) Descending model residual (**d**) Line-of-sight (LOS) coseismic displacement map computed by using Sentinel 1-A (S1A) data acquired from ascending orbit (Track 18) on 26 August 2015 and 19 September 2015 (**e**) S1A ascending LOS projected analytical model (**f**) Ascending model residual (**g**) Uncertainty analysis for the nonlinear inversion: standard deviation (red histograms) of the computed model fault parameters. The yellow star is the epicenter of the main shock. The black square in panels (**a**) and (**c**) represents the DInSAR reference point. The black rectangle in panels (**b**) and (**e**) represents the Okada fault boundary retrieved by the non-linear inversion.

In particular, the descending track shows negative LOS displacement values down to -146 cm, corresponding to a sensor-target range increase, while the ascending is characterized by positive values up to ~150 cm, indicating a decrease of the sensor-target distance; the ascending and descending LOS deformation maps reveal that the displacement is almost equal in magnitude and opposite in sign. These displacement values are in agreement with other works carried out with S1A data to study the Illapel earthquake [8], so phase unwrapping errors can be neglected. Accordingly, the main source of displacement uncertainty is mostly related to atmospheric artefacts, which can be considered in the order of a few phase fringes with a local extension, thus not significantly impacting the retrieved deformation signal induced by the main shock. The availability of both ascending and descending SAR data set allows us not only to detect the ground deformation in the corresponding LOS, but also to discriminate the vertical and east–west components of the displacement [15]. To achieve this task, we properly combine the geocoded displacement maps computed from the ascending and descending orbits on pixels common to both maps, taking into account the different acquisition geometries at each pixel [15,16]. Following the previous discussion, we present in Figure 3a,b the achieved east–west and vertical displacement maps, respectively, evaluated with respect to a pixel identified with a black square in Figure 3a. The east–west displacement map, shown in Figure 3a, clearly highlights a huge displacement to the west of more than 210 cm, while the vertical displacement map (Figure 3b) shows an uplift of about 25 cm along the coast, surrounded by an annular shaped subsidence of about 20 cm; these findings are consistent with GPS measurements reported in Tilmann *et al.* [7], which indicate an uplift of the coastline and a westward motion approximately radially towards a point offshore near 31.2° S, and with Grandin *et al.* [8], which highlight a shift from a coastal subsidence to a coastal uplift at 31.1° S.

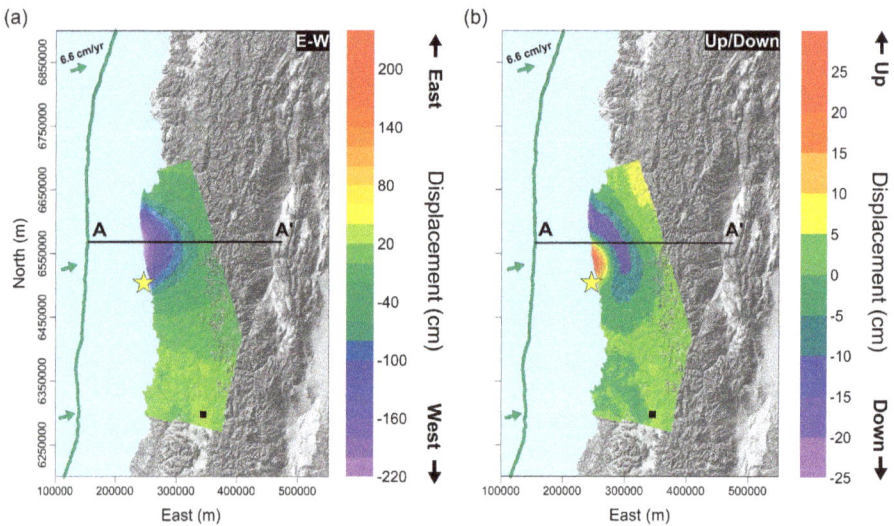

Figure 3. Displacement components computed from the descending and ascending displacement maps shown in Figure 2a,b, respectively, for the pixels that are common to both maps; (**a**) east–west deformation component; (**b**) vertical deformation component. The pixel identified by the black square represents the reference point of the common mask. AA' line is the trace of the SAR data profile used for the 2D Finite Element (FE) modeling.

3. Analytical Modeling

In order to retrieve the seismogenic fault parameters, we now jointly invert S1A DInSAR ascending and descending data following two main steps: a nonlinear inversion to constrain the fault geometries with uniform slip, followed by a linear inversion to retrieve the slip distribution on the fault plane. The observed data is modeled with a finite dislocation fault in an elastic and homogeneous half-space [17]. We search for 8 fault parameters by using a nonlinear inversion algorithm which is based on the Levenberg-Marquardt (LM) least-squares approach, a combination of a gradient descent and Gauss-Newton iteration [18]. Our implementation of the LM is modified with multiple random restarts to guarantee the catching of the global minimum in the optimization process. The dip angle is fixed at 22°, constrained by the aftershock spatial distribution and CGMT fault solution. The cost function is a weighted mean of the residuals, expressed as:

$$CF = \sqrt{\frac{1}{N} \sum_i^N \frac{(d_{i,obs} - d_{i,mod})^2}{\sigma_i}}$$

where $d_{i,obs}$ and $d_{i,mod}$ are the observed and modeled displacement of the i-th point, and σ_i is the standard deviation for the N points. The DInSAR data is sub-sampled through a QuadTree algorithm [19] over a mesh of about 17,200 points for the descending pass and 25,000 points for the ascending one.

The best fit solution consists of a reverse fault, whose parameters are summarized in Table 1.

Table 1. Analytical fault model parameters retrieved through non-linear inversion, with the relative standard deviation in parentheses.

Length (km)	Width (km)	Top Depth (km)	Strike (deg)	Dip (deg)	Rake (deg)	Lat (°)	Lon (°)	Average Slip (cm)
170 (0.4)	100 (0.7)	0.1 (0.1)	1 (2)	22 (1)	92 (2)	31.015S (0.01)	72.033W (0.01)	400 (15)

The modeled displacement maps (Figure 2c,d) show a good fit with the measured DInSAR data (Figure 2a,b), as highlighted by the residual maps in Figure 2e,f, where the residuals are generally below 25 cm. Figure 2g reports the uncertainty analysis for this non-linear inversion. Some residual deformation is still observed near the main rupture area, which perhaps mainly relates to some local deformation or a more complex geometry of the rupture plane. In order to have a more accurate estimate of the slip along the fault plane, a distributed slip on 20 × 20 patches (with dimensions of 15 × 7.5 km^2) is calculated. To this aim, a linear inversion is performed by fixing the parameters of the non-linear inversion (Table 1) and inverting the following system:

$$\begin{bmatrix} d_{InSAR} \\ 0 \end{bmatrix} = \begin{bmatrix} G \\ k^2 \cdot \nabla^2 \end{bmatrix} \cdot \begin{bmatrix} m \\ 0 \end{bmatrix}$$

where d_{InSAR} is the InSAR data vector, m is the vector of slip values, G is the Green's matrix with the point-source functions, and ∇^2 is a smoothing Laplacian operator weighted by an empirical coefficient k. The system solution is obtained by means of the Singular Value Decomposition technique of the kernel matrix. In this case, we find a maximum slip of about 7 m in the shallowest 20 km (Figure 4a,b); the RMS for this solution is about 4.6 cm. The standard deviation slip for each patch is reported in Figure A2. The fault length and width are extended to consider the border effects as negligible. Note that most of the aftershocks do not take place close to the areas with a larger slip, but they occur in their surroundings; this finding is very similar to what was found by USGS [20].

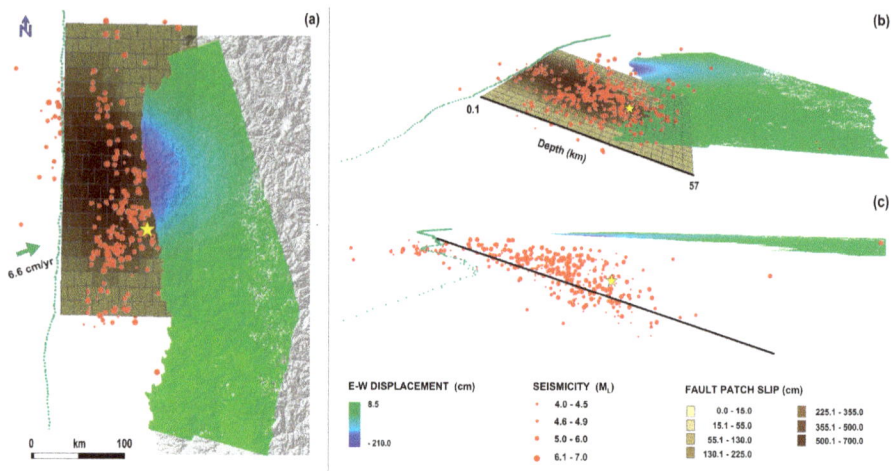

Figure 4. Coseismic slip distribution retrieved through the performed linear inversion. (**a**) Map view of the coseismic slip over 20 × 20 patches (with dimensions of 15 × 7.5 km^2); (**b**) and (**c**) report a 3D view of the fault patch slip. The aftershock distribution, with M > 4, spanning from 16 September (main shock) to 19 September (corresponding to the last day of temporal coverage of satellite data) are depicted with red circles. The yellow star is the main shock epicenter.

4. Finite Element Modeling

We extend our analysis by performing a 2D numerical modeling of the ground deformation pattern retrieved by the DInSAR measurements; our solution is based on the FE approach and allows us to account for the geological and structural information available over the considered area, as well as the seismicity distribution. More specifically, taking into account the geometric features of the active seismogenic slab and the mechanical heterogeneities, we apply a loading along the shallow segment of the slab (see black line in Figure 5a) in order to simulate the ground deformation pattern. We reproduce the retrieved displacements within a 2D structural mechanical context, under the linear elastic material approximation and by constraining the sub-domain setting with the geological and structural information reported in Tassara and Echaurren [21].

We define a simplified geometry of the study region, which extends 390 km and 100 km deep; such a very large area allows us to neglect the possible edge effects. Concerning the geological setting, we develop a heterogeneous model by considering five geological units having isotropic mechanical properties: (i) upper continental crust; (ii) lower continental crust; (iii) continental mantle; (iv) oceanic lithosphere; and (v) oceanic asthenosphere (Figure 5a). The elastic parameters values are reported in Table 2.

As boundary conditions, we apply a free constraint at the upper boundary domain, which corresponds to the topography of the considered area. The bottom boundary is fixed, while a roller condition at the two sides of the numerical domain is applied. The entire numerical domain is discretized in 17,712 tetrahedral elements with a higher refinement along the slab (Figure 5b).

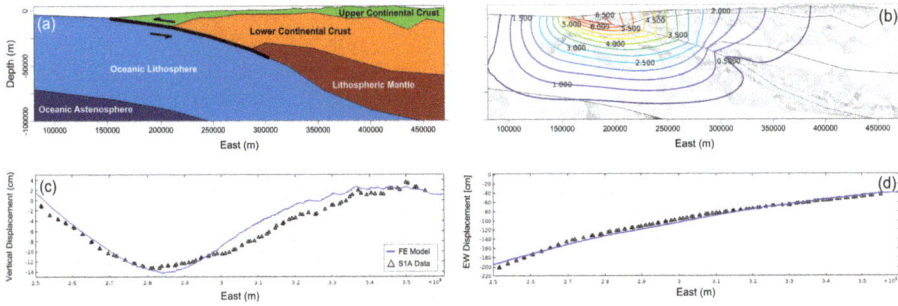

Figure 5. FE results. (**a**) Geological cross-section, modified from [18]. Black segment is the segment of the slab along which the loading is applied to simulate the earthquake thrusting. (**b**) Contour line of the retrieved slip model, superimposed on the discretized domains. (**c**) Comparison between the DInSAR vertical component and model. (**d**) Comparison between the DInSAR EW component and model.

Table 2. Parameters for each layer defined in the model: Density (kg/m^3), Young's Modulus (GPa) and Poisson Ratio. Modified from [7,21].

Parameters	Upper Crust	Lower Crust	Lithospheric Mantle	Ocean Lithospheric	Ocean Asthenosphere
Density	2700	3100	3300	3300	3400
Young's Modulus	100	110	150	150	160
Poisson's Ratio	0.25	0.26	0.26	0.25	0.26

In order to reproduce the coseismic displacement along the fault, we assume the stationarity and linear elasticity of the involved materials by considering the solution of the equilibrium mechanical equations [22]. The earthquake simulation is developed through two stages: During the first one (pre-seismic), the initial stress field was reproduced by applying gravity acceleration under elastic condition; note that this condition permits the model to compact under the weight of the rock successions (gravity loading) until it reached a stable equilibrium [23]. At the second stage, (coseismic), where the stresses are released through a non-uniform slip along the fault, we use an iterative analysis based on a trial-and-error approach [24,25]; it is performed by fixing the location and geometry of the dipping plane constrained by the available structural information and the hypocentral distribution and searching for the loading along the slab. In particular, to evaluate this parameter, we generate several forward structural mechanical models (up to 200) and compare the achieved model results with the EW and vertical component selected along the AA' profile shown in Figure 3a,b. This profile is selected as it passes through the area of maximum EW displacement.

Our best fit model, evaluated by the minimum RMS solution, is reported in Figure 5c,d for the vertical and EW component, respectively. We also report the retrieved 2D displacement along the modeled section, which provides a maximum slip of ~7 m at a depth of 10 km, in the shallower sector of the slab (Figure 5b). A comparison between the slip values retrieved from the FE model and Okada solution along the slab is shown in Figure A3. The achieved major discrepancy, located between depths of 20 and 30 km (hypocentral depth), is probably related to the geometric complexity of subducted slab, considered in the 2D FE model only.

5. Discussion

Our Okada model inferred from the multi-orbit S1A DInSAR measurements shows that almost all of the slip during the 2015 Illapel earthquake occurred northwest of the epicenter at a distance of about 60 km (Figure 4a). A large slip area of about 70 km along strike and 50 km along dip is found with a maximum slip located in the shallowest 20 km. The along-strike extent of the main shock rupture roughly coincides with the trench axis (see green line in Figure 4a). Moreover, the comparison between the slip model map and the aftershock distribution, spanning from 16 September (main shock) to 19 September (corresponding to the last day of temporal coverage of satellite data), reveals that a very small number of aftershocks occurred in the large-slip area (Figure 4a,b); this finding is consistent with what was found by other studies for the Illapel earthquake [6,7,26] and for similar large earthquakes [27]; indeed, the authors of the previous studies, by using the slip distributions for several moderate to large earthquakes, together with their aftershock distributions, concluded that the aftershocks occurred mostly outside or near the edges of the areas of large slip [28]. Our coseismic slip model suggests the slip direction is dominantly downdip and assuming the shear modulus to be 40 GPa, with an average slip of ~4.5 m; the total moment of the preferred model is 2.08×10^{21} Nm, corresponding to a moment magnitude 8.1, comparable to the seismic moment magnitude 8.3, calculated by the USGS-National Earthquake Information Center—NEIC [29].

We further improve our analysis by including in the modeling approach the structural and the geological information available for the study area; this allows us to simulate the slip along the subducted slab, stress distribution, and the principal stress axes orientation. To achieve this task, we perform the analysis via a FE approach. Therefore, we make several test in order to search for the loading along the subducted slab; the best model provides a maximum slip of ~7 m at a depth of 10 km, corresponding to the shallower sector of the slab (Figure 5b).

In addition, in order to confirm the validity of our model, we analyze the stress distribution along the subducted slab, in terms of von Mises scalar quantity and orientation of maximum principal stress, and compare them with seismological data. The von Mises stress expresses the difference between the principal components of stress and gives an indication of the amount of shear stress. At the same time, this parameter is proportional to the octahedral shear stress [30] by a constant factor $\sqrt{2}/3$ and thus can be directly compared with the yield strength of the materials to give an estimate of the possibility of failure. Indeed, the obtained von Mises stress distribution, reported in Figure 6a, shows a close similarity with the depth distribution of the aftershock hypocenters. Note that we report one month of aftershocks (with M > 4.3) located within 10 km north and south from the AA' profile and recorded by GEOFON network. Moreover, the maximum principal stress orientation highlight a compressive regime (horizontally-oriented σ_1) in correspondence of the deeper portion of the slab and an extensional regime (vertically-oriented σ_1) at its shallower segment (Figure 6b). This finding is supported by seismological data that show that at least 3 aftershocks (3 weeks after the main shock) exhibit a high-angle normal faulting close to the trench, and most of the aftershocks share low-angle thrust faulting mechanisms consistent with the megathrust geometry [1] (Figure 6b). Finally, we propose a conceptual model (not to scale) with the aim of synthesizing the kinematics of the megathrust faulting inferred from our modeling results and supported by observed data (Figure 6c). Such a model shows how the megathrust subduction induces a horizontal displacement toward the West, an uplift along the coast and a subsidence behind it of the overriding plate; in the same way, the motion along the subducted slab, considering the distribution of von Mises stress, could explain the occurrence of normal faulting earthquakes (extensional regime) across the trench axis and thrust faulting (compressive regime) along a deeper segment of the slab.

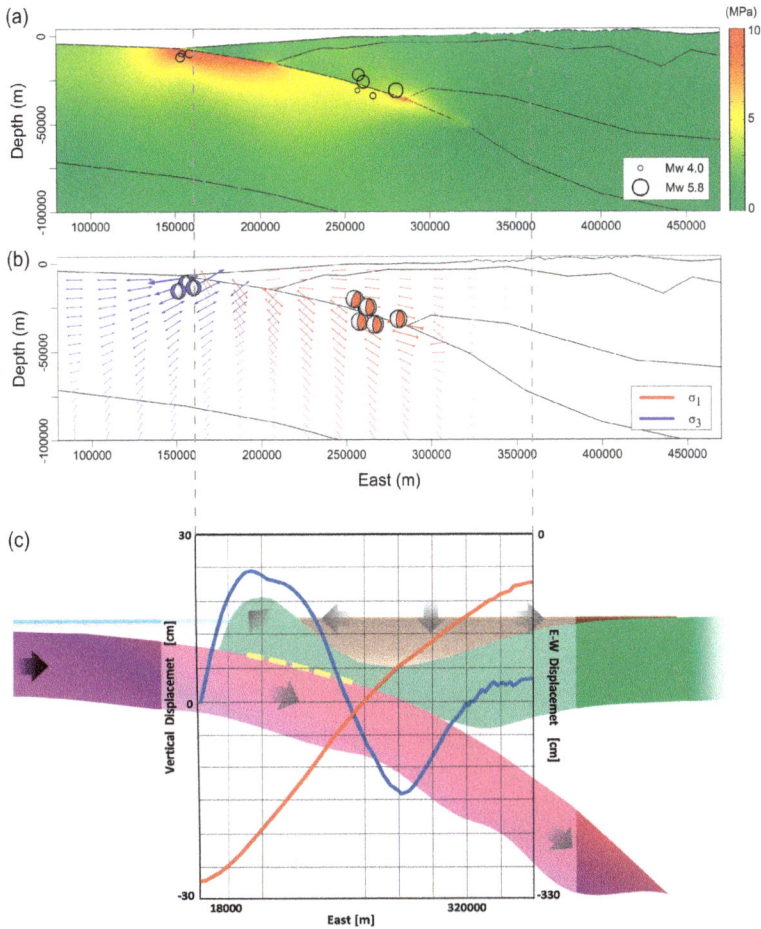

Figure 6. Stress distribution analysis inferred from the FE modeling. (**a**) Von Mises distribution. One month of aftershocks (with M > 4.3), located within 10 km north and south from the AA′ profile (see Figure 3 for AA′ trace profile) and recorded by the GEOFON network, is reported. (**b**) Principal stress orientation. (**c**) Conceptual model, modified after [21], on the kinematics of the megathrust faulting inferred from our modeling results, with the indication of the vertical (blue curve) and EW (red curve) DInSAR displacement along a portion of the subduction zone highlighted by the grey dashed lines.

6. Conclusions

Our main findings can be summarized as follows:

- S1A satellite measurements are a powerful tool to analyze the deformation induced by large mega-thrusting earthquakes, as in the case of the Illapel earthquake. More specifically, S1A peculiarities include wide ground coverage (250 km of swath), C-band operational frequency, and short revisit time (that will reduce from 12 to 6 days when the twin system Sentinel-1B will be placed in orbit during 2016). Such characteristics, together with the global coverage acquisition policy, make the Sentinel-1 constellation an extremely suitable region for studying high seismic hazard and monitoring worldwide, thus allowing the generation of both ground displacement information with increasing rapidity and new geological understanding.

- The east–west displacement map highlights a huge westward displacement of about 210 cm, while the vertical displacement map shows an uplifting area of about 25 cm along the coast, surrounded by an annular shaped subsidence of about 20 cm.

- The Okada modeling consists of a reverse fault, accounting for the main seismic event and corresponding to the shallow portion of the subducted slab. Most of the slip occurred northwest of the epicenter at a distance of about 60 km. A large slip area of about 70 km along strike and 50 km along dip is found with a maximum slip located at a depth ranging from 10 to 30 km.

- The FE modeling, obtained by also including our analysis geological and structural information, allows us to estimate values of maximum slip comparable with the analytical solution and to evaluate the von Mises distribution and axis stress orientation, which are in agreement both with the location and the type of faulting of the aftershocks.

Acknowledgments: This work has been partially supported by MIUR ("Progetto Bandiera RITMARE"), and the Italian Department of Civil Protection (DPC); manuscript contents reflect authors' positions that could be different from the DPC official statements. Part of the presented research has been carried out through the I-AMICA (Infrastructure of High Technology for Environmental and Climate Monitoring-PONa3_00363) project of Structural improvement financed under the National Operational Programme (NOP) for "Research and Competitiveness 2007–2013", co-funded with European Regional Development Fund (ERDF) and National resources. Sentinel-1A data are copyright of Copernicus (2015). The DEMs of the investigated zone were acquired through the SRTM archive.

Author Contributions: G.S. and F.C. conceived and organized the research activity. C.D.L., V.D.N., M.M., and F.C. processed the S1A SAR data. R.C., V.D.N., and G.S. performed the SAR inversion and fault slip modeling. All authors co-wrote the paper and G.S., R.L. and M.M. extensively reviewed the final version of the manuscript.

Conflicts of Interest: The authors declare no conflict of interest.

Appendix A

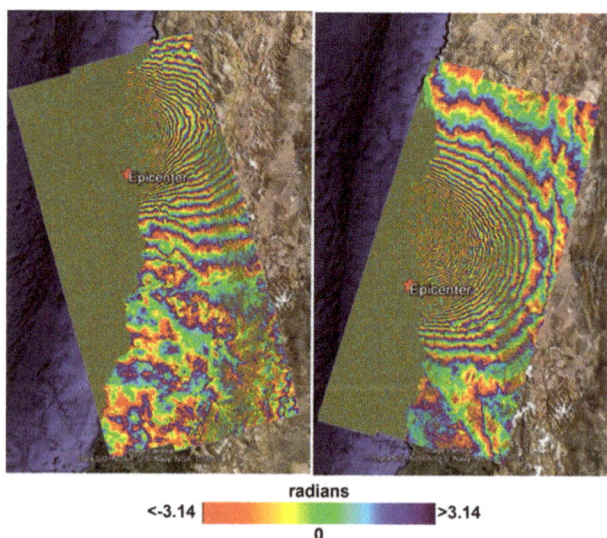

Figure A1. Wrapped interferograms superimposed on a satellite Google Earth image. (**left**) 26 August 2015–19 September 2015 inteferogram along the ascending orbit (Track 18) (**right**) 31 July 2015–17 September 2015 inteferogram along the descending orbit (Track 156). The red star is the main shock epicenter.

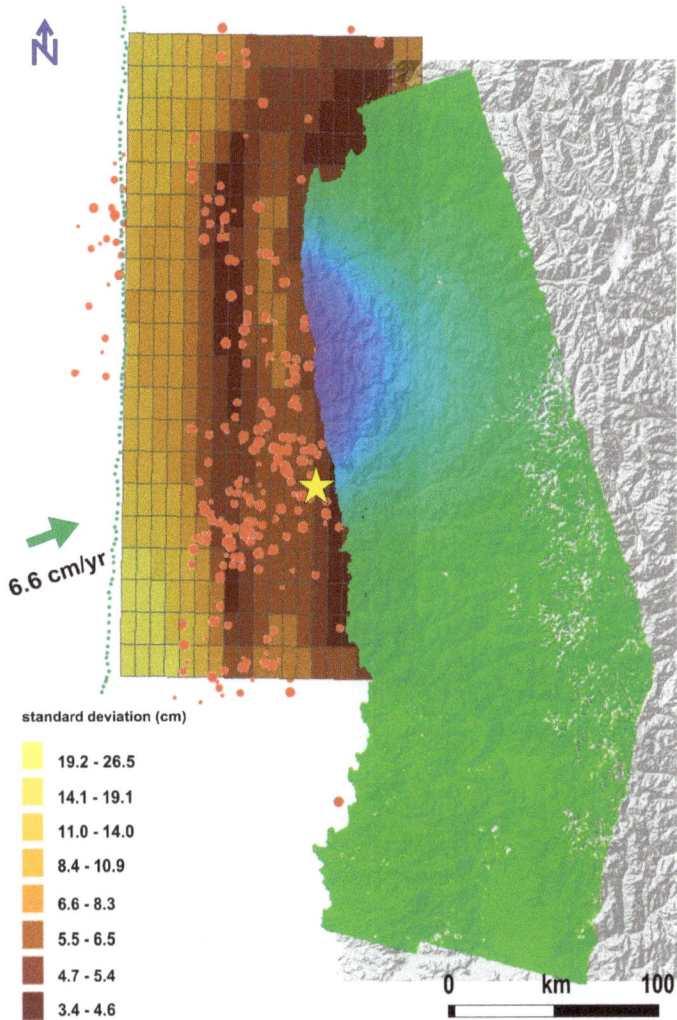

Figure A2. Standard deviation associated to the slip values of Figure 5a. The aftershock distribution, with M > 4, spanning from 16 September (main shock) to 19 September (corresponding to the last day of temporal coverage of satellite data) are depicted with red circles. The yellow star is the main shock epicenter.

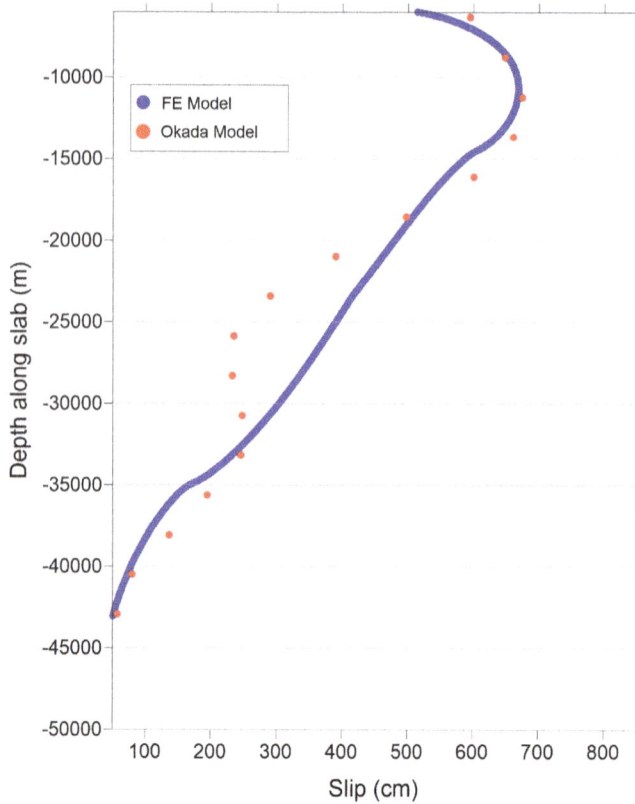

Figure A3. Comparison between the slip values retrieved from FE model and Okada solution. Note how the variation of the slip values with depth shows similar trend up to ~20 km depth, while they deviate from 22 to 33 km depth. This discrepancy may be ascribed to the curvature of the slab, which has been taken into account in the FE modeling only.

References

1. Melgar, D.; Fan, W.; Riquelme, S.; Geng, J.; Liang, C.; Fuentes, M.; Vargas, G.; Allen, R.M.; Shearer, P.M.; Fielding, E.J. Slip segmentation and slow rupture to the trench during the 2015, Mw 8.3 Illapel, Chile earthquake. *Geophys. Res. Lett.* **2016**. [CrossRef]
2. USGS. Available online: http://earthquake.usgs.gov/earthquakes/eventpage/us20003k7a#finite-fault (accessed on 8 April 2016).
3. Kendrick, E.; Bevisa, M.; Smalley, R.; Brooksa, B.; Barriga Vargasc, R.; Lauria, E.; Souto Fortese, L. The Nazca–South America Euler vector and its rate of change. *J. South Am. Earth Sci.* **2003**, *16*, 125–131. [CrossRef]
4. Lemoine, A.; Madariaga, R.; Campos, J. Evidence for earthquake interaction in Central Chile: The July 1997–September 1998 sequence. *Geophys. Res. Lett.* **2001**, *28*, 2743–2746. [CrossRef]
5. Gardi, A.L.; Lemoine, A.; Madariaga, R.; Campos, J. Modeling of stress transfer in the Coquimbo region of central Chile. *J. Geophys. Res.* **2006**, *111*, B04307. [CrossRef]
6. Ye, L.; Lay, T.; Kanamori, H.; Koper, K.D. Rapidly Estimated Seismic Source Parameters for the 16 September 2015 Illapel, Chile Mw 8.3 Earthquake. *Pure Appl. Geoph.* **2015**. [CrossRef]
7. Tilmann, F.; Zhang, Y.; Moreno, M.; Saul, J.; Eckelmann, F.; Palo, M.; Deng, Z.; Babeyko, A.; Chen, K.; Baez, J.C.; *et al.* The 2015 Illapel earthquake, central Chile: A type case for a characteristic earthquake? *Geophys. Res. Lett.* **2015**, *43*. [CrossRef]

8. Grandin, R.; Klein, E.; Metois, M.; Vigny, C. 3D displacement field of the 2015 Mw8.3 Illapel earthquake (Chile) from across- and along-track Sentinel-1 TOPS interferometry. *Geophys. Res. Lett.* **2016**, *43*. [CrossRef]

9. De Zan, F.; Monti Guarnieri, A. TOPSAR: Terrain Observation by Progressive Scans. *IEEE Trans. Geosci. Remote Sens.* **2006**, *44*, 2352–2360. [CrossRef]

10. ESA. Available online: https://earth.esa.int/web/sentinel/technical-guides/sentinel-1-sar/sar-instrument/acquisition-modes (accessed on 8 April 2016).

11. Sansosti, E.; Berardino, P.; Manunta, M.; Serafino, F.; Fornaro, G. Geometrical SAR Image Registration. *IEEE Trans. Geosci. Remote Sens.* **2006**, *44*, 2861–2870. [CrossRef]

12. Prats-Iraola, P.; Scheiber, R.; Marotti, L.; Wollstadt, S.; Reigber, A. TOPS Interferometry With TerraSAR-X. *IEEE Trans. Geosci. Remote Sens.* **2012**, *50*, 3179–3188. [CrossRef]

13. Costantini, M. A novel phase unwrapping method based on network programming. *IEEE Trans. Geosci. Remote Sens.* **1998**, *36*, 813–821. [CrossRef]

14. Fornaro, G.; Franceschetti, G.; Lanari, R.; Rossi, D.; Tesauro, M. Interferometric SAR phase unwrapping using the finite element method Radar, Sonar and Navigation. *IEE Proc.* **1997**, *144*, 266–274.

15. Manzo, M.; Ricciardi, G.P.; Casu, F.; Ventura, G.; Zeni, G.; Borgstrom, S.; Berardino, P.; Del Gaudio, C.; Lanari, R. Surface deformation analysis in the Ischia island (Italy) based on spaceborne radar interferometry. *J. Volc. Geoth. Res.* **2006**, *151*, 399–416. [CrossRef]

16. Wright, T.J.; Parsons, B.E.; Lu, Z. Toward mapping surface deformation in three dimensions using InSAR. *Geoph. Res. Lett.* **2004**, *31*. [CrossRef]

17. Okada, Y. Internal deformation due to shear and tensile faults in a half-space. *Bull. Seismol. Soc. Am.* **1985**, *75*, 1135–1154.

18. Marquardt, D. An algorithm for least-squares estimation of nonlinear parameters. *SIAM J. Appl. Math.* **1963**, *11*, 431–441. [CrossRef]

19. Jónsson, S.; Zebker, H.; Segall, P.; Amelung, F. Fault slip distribution of the 1999 Mw 7.2 Hector Mine earthquake, California, estimated from satellite radar and GPS measurements. *Bull. Seismol. Soc. Am.* **2002**, *92*, 1377–1389. [CrossRef]

20. USGS. Available online: http://earthquake.usgs.gov/earthquakes/eqarchives/poster/2015/20150916.pdf (accessed on 8 April 2016).

21. Tassara, A.; Echaurren, A. Anatomy of the Andean subduction zone: Three-dimensional density model upgraded and compared against global-scale models. *Geophys. J. Int.* **2012**, *189*, 161–168. [CrossRef]

22. Fagan, M.J. *Finite Elements Analysis: Theory and Practice*; Prentice Hall: Upper Saddle River, NJ, USA, 1992; pp. 1–311.

23. Apuani, T.; Corazzato, C.; Merri, A.; Tibaldi, A. Understanding Etna flank instability through numerical models. *J. Volc. Geoth. Res.* **2013**, *251*, 112–126. [CrossRef]

24. Tizzani, P.; Castaldo, R.; Solaro, G.; Pepe, S.; Bonano, M.; Casu, F.; Manunta, M.; Manzo, M.; Pepe, A.; Samsonov, S.; *et al.* New insights into the 2012 Emilia (Italy) seismic sequence through advanced numerical modeling of ground deformation InSAR measurements. *Geophys. Res. Lett.* **2013**, *40*, 1971–1977. [CrossRef]

25. Tarantola, A. Inverse Problem Theory and Methods for Model Parameter Estimation. In *SIAM Society for Industrial and Applied Mathematics*; SIAM: Philadelphia, PA, USA, 2005; pp. 1–343.

26. Heidarzadeh, M.; Murotani, S.; Satake, K.; Ishibe, T.; Riadi Gusman, A. Source model of the 16 September 2015 Illapel, Chile Mw 8.4 earthquake based on teleseismic and tsunami data. *Geoph. Res. Lett.* **2015**, *43*. [CrossRef]

27. Mendoza, C.; Hartzell, S.H. Aftershock patterns and main shock faulting. *Bull. Seismol. Soc. Am.* **1988**, *78*, 1438–1449.

28. Das, S.; Henry, C. Spatial relation between main earthquake slip and its aftershock distribution. *Rev. Geoph.* **2003**, *41*. [CrossRef]

29. USGS. Available online: http://earthquake.usgs.gov (accessed on 8 April 2016).

30. Jaeger, J.C.; Cook, N.G.W. *Fundamentals of Rocks Mechanics. Science Paperbacks*; Chapman-Hall: New York NY, USA, 1971; pp. 1–515.

remote sensing

MDPI

Article

Source Parameters of the 2003–2004 Bange Earthquake Sequence, Central Tibet, China, Estimated from InSAR Data

Lingyun Ji [1,*], Jing Xu [1], Qiang Zhao [1] and Chengsheng Yang [2]

[1]   Second Monitoring and Application Center, China Earthquake Administration, 316 Xiying Rd., Xi'an 710054, China; xjinggis@163.com (J.X.); zq_gke1990@163.com (Q.Z.)
[2]   College of Geology Engineering and Geomatics, Chang'an University, 126 Yanta Rd., Xi'an 710054, China; yangchengsheng@chd.edu.cn
*   Correspondence: dinsar010@hotmail.com; Tel.: +86-29-8550-6645

Academic Editors: Zhenhong Li, Roberto Tomas, Zhong Lu and Prasad S. Thenkabail
Received: 23 March 2016; Accepted: 14 June 2016; Published: 18 June 2016

Abstract: A sequence of Ms $\geq$ 5.0 earthquakes occurred in 2003 and 2004 in Bange County, Tibet, China, all with similar depths and focal mechanisms. However, the source parameters, kinematics and relationships between these earthquakes are poorly known because of their moderately-sized magnitude and the sparse distribution of seismic stations in the region. We utilize interferometric synthetic aperture radar (InSAR) data from the European Space Agency's Envisat satellite to determine the location, fault geometry and slip distribution of three large events of the sequence that occurred on 7 July 2003 (Ms 6.0), 27 March 2004 (Ms 6.2), and 3 July 2004 (Ms 5.1). The modeling results indicate that the 7 July 2003 event was a normal-faulting event with a right-lateral slip component, the 27 March 2004 earthquake was associated with a normal fault striking northeast–southwest and dipping northwest with a moderately oblique right-lateral slip, and the 3 July 2004 event was caused by a normal fault. A calculation of the static stress changes on the fault planes demonstrates that the third earthquake may have been triggered by the previous ones.

Keywords: radar interferometry; satellite geodesy; earthquake source observations; deformation; earthquake sequence; Bange earthquakes

1. Introduction

From July 2003 through July 2004, a complex earthquake sequence occurred in Bange County on the border between Qinghai province and Tibet, China (Figure 1). According to the China Earthquake Networks Center's (CENC) catalogue [1], the sequence started with an Ms = 6.0 earthquake on 7 July 2003 (Table 1). The National Earthquake Information Center's (NEIC, United States Geological Survey) catalogue indicated a normal faulting mechanism, whereas the Global Centroid Moment Tensor's (GCMT) database showed a strike-slip mechanism (Figure 1). On 27 March 2004, approximately eight months later, three earthquakes occurred approximately 70 km south of the 2003 event: a large shock (Ms 6.2, 18:47 GMT) was preceded by two Ms $\geq$ 5.0 shocks (Figure 1; Table 1). Five events with Ms $\geq$ 5.0 were reported through July (Table 1). We list all of the earthquakes with Ms $\geq$ 5.0, and whether the focal mechanism solutions are subject to normal-faulting or strike-slip mechanisms, in Table 1.

On a regional scale, the epicentral area of the 2003–2004 Bange earthquake sequence is situated in the central Tibetan Plateau. Tectonically, the earthquake sequence occurred in the northern Qiangtang block, approximately 100 km south of the boundary between the Bayan Har block and Qiangtang block. Generally, active north–south shortening and an east–west extension in central Tibet are accommodated by north–south trending normal faults and conjugate strike-slip faults (e.g., [2]). Previous studies indicate that recently active faults within the Qiangtang block range from strike-slip to normal faulting kinematics [2].

Table 1. Catalogue of the 2003–2004 Bange earthquake sequence from CENC (shown as stars in Figure 1).

Date (yyyymmdd)	Time (hh:mm)	Latitude (°)	Longitude (°)	Magnitude (Ms)	Depth (km)	Focal Mechanism	
						GCMT	NEIC
20030707	06:55	34.51	89.37	6.0	13	◕	O
20040327	18:45	33.92	89.20	5.8	13	—	—
20040327	18:47	34.01	89.22	5.5	10	—	—
20040327	18:47	33.95	89.37	6.2	9	O	O
20040406	10:30	33.93	89.13	5.0	14	O	—
20040422	10:02	33.87	89.12	5.1	8	O	—
20040523	02:22	34.00	89.30	5.1	10	O	—
20040523	07:38	34.08	89.28	5.3	9	◕	—
20040703	14:10	34.00	89.20	5.1	6	◕	—

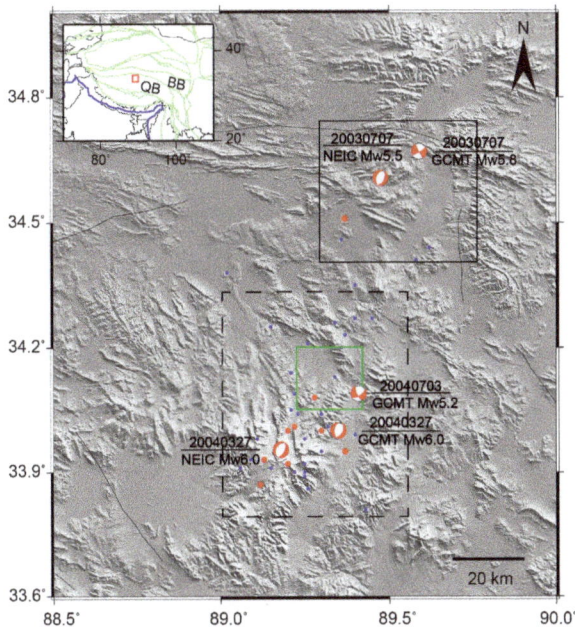

Figure 1. Topographic map of Bange County in central Tibet, China, with the location shown in the inset. Green lines in inset represent block boundaries [3]: BB, Bayan Har Block; QB, Qiangtang Block. Shaded relief topography is SRTM DEM at 90 m resolution. Black thin lines are fault traces [4]. Earthquakes listed in Table 1 are shown as red circles. Blue circles are aftershocks with Ms ⩾ 3.0 through 2015. Earthquake catalogue is from China Earthquake Networks Center (CENC) [1]. Black box with solid line marks areas covered by interferograms of the 7 July 2003 event. Dashed box marks areas covered by interferograms of the 27 March 2004 event. Green box marks areas covered by interferograms of the 3 July 2004 event. Focal mechanisms from NEIC and GCMT for 7 July 2003 Ms 6.0, 27 March 2004 Ms 6.2, and 3 July 2004 Ms 5.1 events are shown.

According to the China Earthquake Networks Center's (CENC) catalogue (2000 to 2015), earthquakes larger than magnitude 5 were common around the 2003–2004 Bange sequence area; and all of these events were smaller than magnitude 6, except for the two aforementioned events, that is, the 7 July 2003 Ms 6.0 and 27 March 2004 Ms 6.2 events. Thus, identification and characterization of the seismic sources of the 2003–2004 sequence can provide an important contribution to the understanding of the deformation style of this seismically active area. Additionally, the modeling of the displacement fields provides new insights into the seismotectonic setting and the seismic hazard for the central Tibetan Plateau.

Coseismic deformation fields caused by the 2003–2004 Bange earthquake sequence are unknown because of difficult logistics and persistently inclement weather in this remote area. Because of the sparseness of the geodetic arrays, the earthquakes responsible for the sequence were not identified previously. Moreover, the absence of any seismic rupture at the surface after the earthquake does not allow for direct field identification of the seismogenic fault. This makes the application of satellite-based monitoring techniques, such as interferometric synthetic aperture radar (InSAR), highly desirable. InSAR combines two or more SAR images of the same area acquired at different times from nearly the same position in space to map any surface deformation that might occur during the time interval spanned by the images (e.g., [5,6]). InSAR has been well known for imaging coseismic displacements and estimating source parameters since the June 1992 Mw = 7.3 Landers earthquake (e.g., [7–12]). In this study, we measure the ground deformation due to the 2003–2004 Bange earthquake sequence using InSAR. We invert the source geometries based on the observed surface deformation patterns and subsequently perform linear inversions to retrieve the slip distributions. Then, we evaluate the static stress drop of the 2003–2004 Bange earthquake sequence. Finally, Coulomb failure function (CFF) analysis is used to study the interactions among the earthquakes.

2. InSAR Data and Analysis

We collect SAR images covering the 2003–2004 Bange earthquake sequence from the Envisat satellite, operating on the C band. The data are processed using the GAMMA InSAR processing software [13]. We use the two-pass InSAR approach (e.g., [5,6]) to form deformation interferograms. The effects of topography are removed from the interferograms using a filled 3 arc·s (~90 m) resolution Shuttle Radar Topography Mission (SRTM) digital elevation model (DEM) [14] obtained from the Consultative Group on International Agricultural Research's Consortium for Spatial Information (CGIAR-CSI) [15]. To improve the signal-to-noise ratio, interferograms are downsampled to 4 looks in range and 20 looks in azimuth (80 m × 80 m) and are filtered twice using an adaptive filter function based on the local fringe spectrum [16], with the dimensions of the windows being 128 × 128 and 32 × 32 pixels. This filtering strategy efficiently removes the high frequency noise [17] and makes the phase unwrapping much easier. To remove residual orbit errors, a fine estimation of the interferogram baseline is obtained by a nonlinear least-square adjustment of the observed phase over presumably stable areas [18,19]. The time chart in Figure 2 shows the temporal coverage of the coseismic differential interferograms, along with their perpendicular baselines. In addition, marked are the times of seismic events, for reference.

2.1. The 7 July 2003 Event

Figure 3 shows two interferograms that span the 7 July 2003 event, mapping the coseismic deformation pattern. The reasonably good coherence permits a clear view of the surface deformation field associated with the earthquake. As the bull's-eye pattern signal is consistent between the two interferograms, which are calculated using two independent pairs of images, we rule out the possibility that the observed signals are strongly affected by atmospheric artifacts. Additionally, the bull's-eye pattern signal cannot be attributed to DEM error because the baselines of these interferograms are short, making them insensitive to any plausible errors in the DEM (Figure 2).

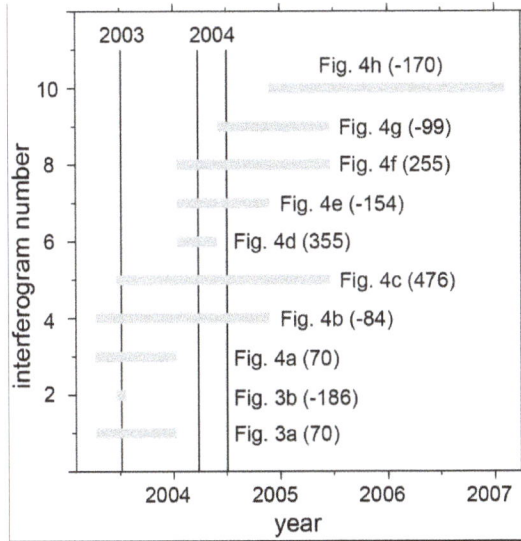

Figure 2. Time intervals covered by each of the interferograms shown in Figures 3 and 4. Values in parentheses are the perpendicular baseline of each interferogram. Grey bars show time intervals. Black solid vertical lines mark the times of the 7 July 2003 Ms 6.0, 27 March 2004 Ms 6.2, and 3 July 2004 earthquakes (see Table 1 for details).

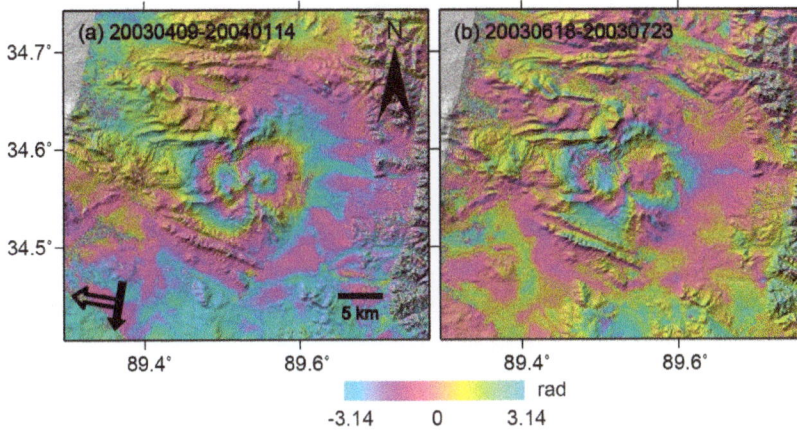

Figure 3. Coseismic interferograms of the 7 July 2003 earthquake. Location of the interferograms is shown in Figure 1 as a box with solid line. Start and end dates are provided above each image using the format yyyymmdd. (**a**) 20030409–20040114. Satellite flight direction and radar look direction are labeled as a solid arrow and open arrow, respectively; (**b**) 20030618–20030723. Each fringe, *i.e.*, full color cycle from red through yellow to blue, represents 28 mm of range increase between the ground and the satellite.

The deformation fields show two approximately symmetric lobes with two color cycles, indicating an approximately 5-cm range change in the radar's LOS (Line Of Sight) direction. The left lobe moved away from the satellite, whereas the right lobe moved towards the satellite.

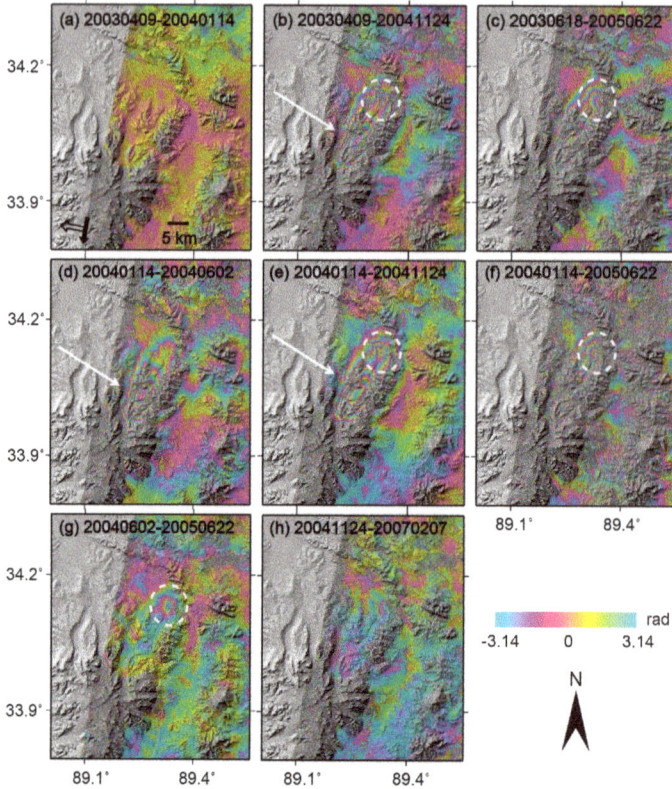

Figure 4. Interferograms related to the 2004 earthquakes. Location of the interferograms is shown in Figure 1 as a dashed box. Start and end dates are provided above each image using the format yyyymmdd. White solid arrows point to the oval pattern caused by the 27 March 2004 event, whereas the white dashed circles delineate the circular pattern caused by the 3 July 2004 event. Satellite flight direction and radar look direction are labeled as short solid arrow and open arrow, respectively. Each fringe, *i.e.*, full color cycle from red through yellow to blue, represents 28 mm of range increase between the ground and satellite. (**a**) 20030409–20040114; (**b**) 20030409–20041124; (**c**) 20030618–20050622; (**d**) 20040114–20040622; (**e**) 20040114–20041124; (**f**) 20040114–20050622; (**g**) 20040602–20050622; (**h**) 20041124–20070207.

2.2. The 2004 Earthquakes

Figure 4 shows interferograms related to the 2004 earthquakes. Before 14 January 2004 and after 24 November 2004, no obvious deformation is detected in the corresponding interferograms (Figure 4a,h). In addition, similar to the analysis of the 7 July 2003 event, we confirm that the large-scale oval pattern signal that elongated NE–SW, as shown in Figure 4b–e, was primarily caused by the 2004 earthquakes. The oval signal shows approximately five red-yellow-blue color cycles, indicating a negative displacement reaching up to 14 cm. However, we note another small-scale circular signal that persisted in several interferograms (Figure 4b,c,e–g), located northeast of the oval. The oval fringes and the circular fringes partially overlap. However, the signal is not detected before 2 June 2004 (Figure 4d). Referring to the catalogue (Table 1), we infer that the local small-scale subsidence signal may be caused by the earthquake that occurred on 3 July 2004.

3. Source Modeling and Analysis

Using the InSAR surface displacements, we can potentially place constraints on the fault orientation according to the depth and the spatial extent through modeling. To consider the high spatial correlation of pixels and expedite the modeling process, we first attempt to down-sample the InSAR data using the quadtree method (e.g., [20]). However, this fails to capture the main deformation pattern with a high enough resolution. Then, we sample the near-field area at a dense regular spacing grid and the far-field area at a sparse regular spacing grid. The modeling is executed in two steps. First, we perform an exhaustive search for the best-fit fault parameters assuming uniform slip. Then, we divide the faults into sub-faults and estimate the slip on each patch.

3.1. Uniform Slip Model

3.1.1. The 7 July 2003 Event

We choose Figure 3b for modeling because it has a shorter time interval than Figure 3a, containing less possible pre- and post-seismic deformation. Qualitatively, the coseismic deformation pattern shown in Figure 3 is consistent with a NW-SE fault plane for the 7 July 2003 earthquake. However, because all of the available interferograms are in the same viewing geometry, we cannot easily identify the hanging wall of the causative fault. Therefore, we perform two inversions on the 7 July 2003 earthquake. The first model constrains the strike to be within 270° and 360°, and the second model constrains the strike to be within 90° and 180°. The first model produces a solution with an approximately horizontal fault plane, which is inconsistent with the deformation pattern and is physically unrealistic for a causative fault. Hence, we consider the first model to be unlikely. In contrast, the second model achieves reasonable fault parameters. Referring to the pattern of the observed deformation, we assume that the causative fault could be interpreted as a single rectangular plane with a uniform slip embedded in a homogeneous, isotropic, elastic half-space [21]. Nine parameters define the rectangular dislocation: length, width, depth, strike, dip, slip magnitude (dip- and strike-slip along the fault), and location (two parameters). In the model, we introduce linear terms to account for any possible phase ramp due to uncertainties in satellite positions [5]. We use the downhill simplex method and Monte Carlo simulations [22] to estimate the optimal parameters and their uncertainties, and the root mean square errors (RMSE) between the observed and modeled interferograms as the prediction-fit criterion. We randomly choose the starting parameters within broad bounds to generate 1000 uniformly distributed samples. For each of the nine parameters in the inversion, the histogram of the set of best-fit solution parameters is a Gaussian from which we select the mean value as the optimal solution and estimate the standard deviation. Figure 5a shows the distributions of the solution parameters, indicating they are well retrieved.

Figure 6 shows the observed (a), modeled (b), and residual (c) interferograms of the model. The model fits the observed interferogram reasonably well. The displacement profile in Figure 6d–f does not show a discontinuity, implying that the fault does not break the surface. Table 2 shows the optimal parameters of the best-fit fault. All of the model parameters are well constrained according to their uncertainties (Table 2). We attribute the goodness of fit to the reasonably good coherence of the observed interferogram. The causative fault is located at a depth of 5 km, dipping 80° to the northeast. The slip is a dominantly dip slip, with a small amount of dextral slip. In other words, the 7 July 2003 earthquake is supposed to be a normal-faulting event with a right-lateral slip component. This focal mechanism is consistent with that in the NEIC's catalogue, but different with from that in the GCMT's catalogue (Figure 1).

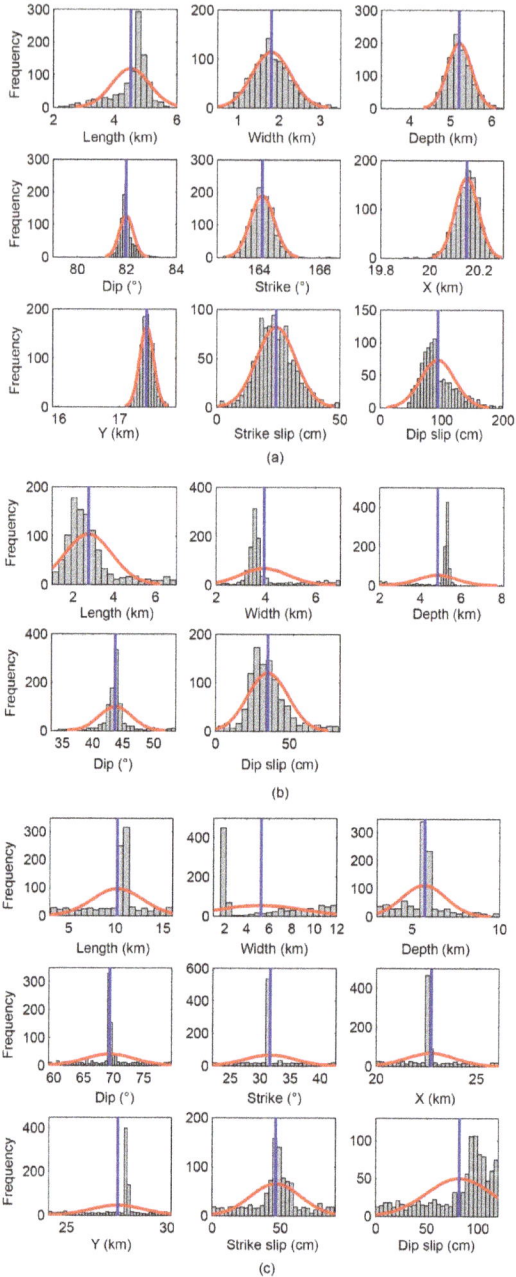

Figure 5. Frequency histograms of modelled parameters determined from 1000 independent runs of the inversion algorithm. Histograms represent the 1000 best-fit solution parameters (**black** bins) obtained from inversions of InSAR coseismic deformation maps. The optimal solution for the parameters is estimated from the mean value (**blue** vertical line) of the best-fit Gaussian (**red** curve). (**a**) 7 July 2003 earthquake; (**b**) 3 July 2004 earthquake; (**c**) 27 March 2004 earthquake.

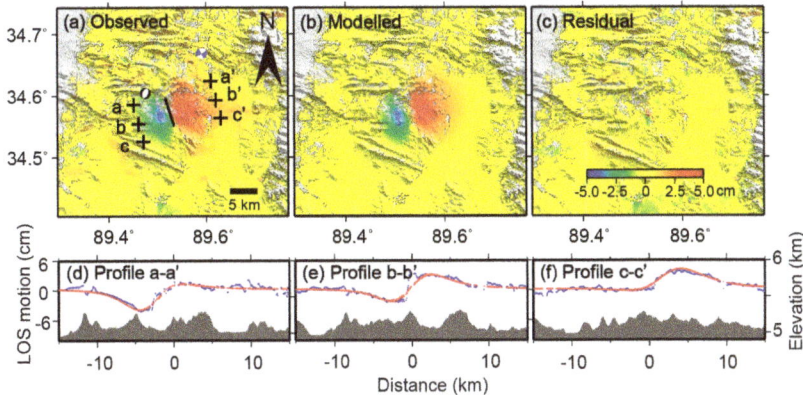

Figure 6. Coseismic deformation (range displacement–negative away from the satellite) and model for uniform slip inversion of the 7 July 2003 earthquake. (**a**) Observed interferogram spanning 20030618–20030723. Black and blue beach balls show focal mechanisms of NEIC and GCMT catalogues, respectively; (**b**) Synthetic interferogram for a uniform slip elastic dislocation model; (**c**) Residual interferogram, which is the difference between observed (**a**) and modeled (**b**) interferograms; (**d**,**e**,**f**) are profiles of line-of-sight (LOS) displacements (**blue** dots), model LOS displacements (**red** dots) and topography (**grey**), respectively. Crosses in (**a**) indicate profile locations. Black line in (**a**) represents the modeled fault trace.

Table 2. Source fault parameters and their 2σ uncertainties of the 7 July 2003 earthquake.

Parameter (Unit)	20030707 Ms 6.0
Length (km)	4.5 ± 0.5
Width (km)	1.8 ± 0.6
Depth (km)	5.2 ± 0.4
Strike (°)	164.0 ± 0.5
Dip (°)	81.9 ± 0.5
Strike slip (cm)	25.1 ± 11.0
Dip slip (cm)	88.0 ± 20.0
Longitude [1] (°)	89.5239 ± 0.001
Latitude [1] (°)	34.5901 ± 0.001

[1] The latitude/longitude location is the top left corner of the modeled fault projected to the surface.

3.1.2. The 2004 Earthquakes

Based on the analysis in Section 2.2, we infer that the observed large-scale oval signal and the small circular pattern were created by two separate earthquakes, probably occurring on separate faults. In other words, we assume that the 27 March 2004 Ms 6.2 earthquake produced the oval pattern displacement and the 3 July event with Ms 5.1 was related to the small circular displacement. To determine the two sources of observed displacements, we employ the following modeling strategy. We choose Figure 4e,g for modeling because they have the best coherence and least apparent atmospheric contaminations. First, we model Figure 4g to determine the fault parameters of the 3 July event because it only includes deformation caused by that event (Figures 4 and 6b; Table 3). Because only a single lobe is shown in the interferogram, we fix the location of the presumed fault during modeling (Figure 7a; Table 3). The deformation fringes shown in Figure 7a change gradually from maximum subsidence to zero toward the west, whereas the fringes change suddenly to zero toward the east. This deformation pattern supports a west dipping normal fault. Therefore, we fix the strike of the model fault to be approximately N–S and constrain the rake angle to a purely normal faulting mechanism. Then, we remove Figure 7b from Figure 4e, and model the residual interferogram to estimate the source of the 27 March event (Figures 6c and 8; Table 3). Figure 8 shows the modeling result. The model fault produces a first order fit to the observed deformation pattern. The northern remaining signal in the residual interferogram is most probably due to atmospheric artifacts. The near-fault fringes of the 3 July 2004 event are unmodeled due to a simple uniform slip elastic dislocation model (Figure 7d), which has also been evidenced by several previous InSAR studies (e.g., [23,24]). The best-fitting model for the 27 March 2004 Ms 6.2 earthquake indicates a normal fault striking southwest–northeast, dipping to the southeast with a moderately oblique right-lateral slip. This mechanism is consistent with the NEIC and GCMT catalogues (Figure 1). The 3 July Ms 5.1 earthquake was caused by a normal fault, which is consistent with the GCMT's catalogue (Figure 1).

Table 3. Source fault parameters and their 2σ uncertainties of the 27 March 2004 Ms 6.2 and 3 July Ms 5.1 earthquakes.

Parameter (Unit)	20040327 Ms 6.2	20040703 Ms 5.1
Length (km)	10.2 ± 2.1	2.8 ± 1.1
Width (km)	5.3 ± 2.9	3.9 ± 0.8
Depth (km)	5.7 ± 0.8	4.9 ± 0.9
Strike (°)	31.7 ± 2.1	182.3 *
Dip (°)	69.4 ± 3.9	43.8 ± 2.2
Strike slip (cm)	46.8 ± 14.0	0.0 *
Dip slip (cm)	82.1 ± 21.0	34.0 ± 10.0
Longitude (°)	89.2004 ± 0.01 [1]	89.3613 [1,*]
Latitude (°)	34.0157 ± 0.01 [1]	34.1443 [1,*]

[1] The latitude/longitude locations are the top left corner of the modeled faults projected to the surface; * The parameters denoted by an asterisk are fixed in modeling.

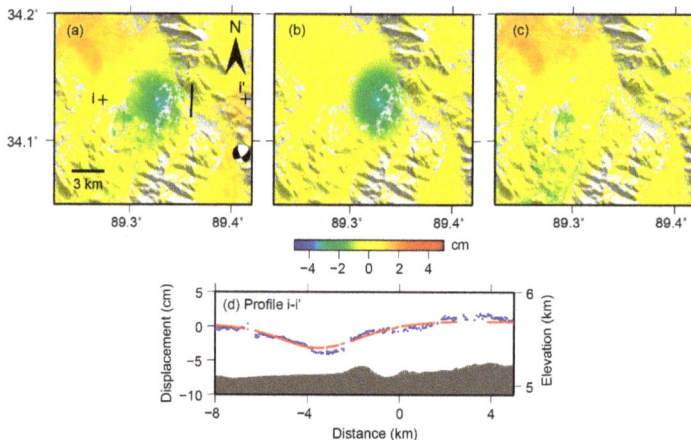

Figure 7. Coseismic deformation (range displacement—negative away from the satellite) and model for uniform slip inversion of the 3 July 2004 earthquake. Location of the interferograms is shown in Figure 1 as a green box. (**a**) Observed interferogram spanning 20040602–20050622. Focal mechanism from GCMT catalogue is shown; (**b**) Synthetic interferogram for a uniform slip elastic dislocation model; (**c**) Residual interferogram, which is the difference between observed (**a**) and modeled (**b**) interferograms; (**d**) Profile of line-of-sight (LOS) displacements (**blue** dots), model LOS displacements (**red** dots) and topography (**grey**). Crosses in (**a**) indicate profile locations. Black lines in (**a**) represent modeled fault trace.

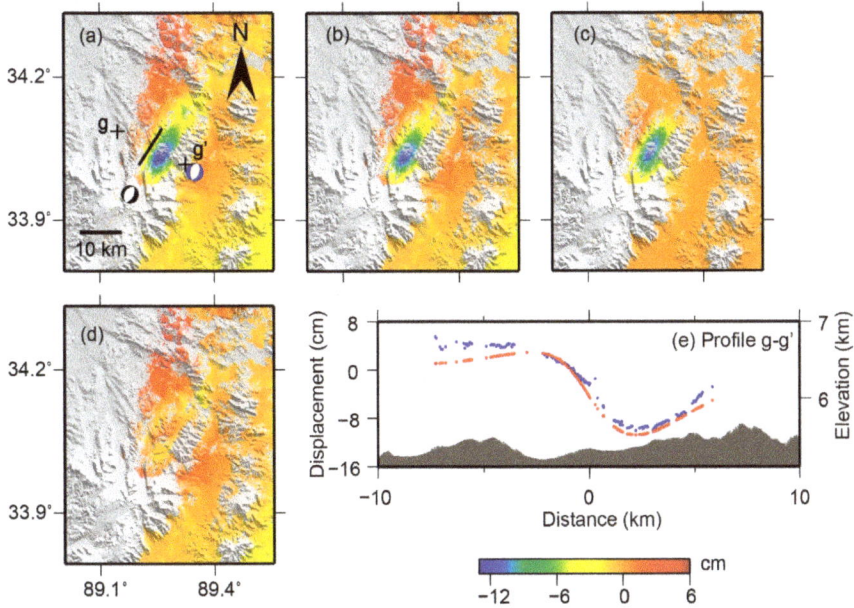

Figure 8. Coseismic deformation (range displacement—negative away from the satellite) and model for uniform slip inversion of the 27 March 2004 earthquake. (**a**) Observed interferogram spanning 20040114–20041124. Black and blue beach balls show focal mechanisms from NEIC and GCMT catalogues, respectively; (**b**) Observed interferogram spanning 20040114–20041124 obtained by subtracting Figure 7b; (**c**) Synthetic interferogram for uniform slip elastic dislocation model; (**d**) Residual interferogram, which is the difference between observed (**b**) and modeled (**c**) interferograms; (**e**) Profile of line-of-sight (LOS) displacements (**blue** dots), model LOS displacements (**red** dots), and topography (**grey**). Crosses in (**a**) indicate profile locations. Black line in (**a**) represents modeled fault trace.

3.2. Distributed Slip Model

Although uniform slip models can provide a reasonable fit to the data, we know that a homogeneous slip on a sharply bounded fault plane is not physically reasonable. Moreover, the simple uniform slip dislocation lacks the capability to model near-fault processes of the 3 July 2004 event (Figure 7d). Thus, we obtain more realistic models by discretizing the fault planes into sub-faults and solve for the slip on each patch, thereby allowing the slip to smoothly taper to zero towards the edges of the fault plane. We use the inversion code SDM [25] based on the constrained least-squares method, which has been used in a number of recent publications for analyzing GPS and InSAR coseismic deformation data (e.g., [26–29]). To overcome the problem of the non-uniqueness and instability of the inversion result, a smoothing constraint is applied to the slip distribution. An optimal smoothing factor is determined by analyzing the trade-off curve between the data misfit and slip roughness (Figure 9).

3.2.1. The 7 July 2003 Event

Using the fault geometry determined in the uniform slip modeling, we extend the fault plane along the strike and down-dip by increasing its total length to 17 km and its down-dip width to 10 km. The fault is discretized into patches that are 1 km in both the along-strike and down-dip directions. Then, the slip on all of the small patches is estimated using the SDM code [25]. Figure 10 shows the modeling result. Compared to the uniform slip dislocation modeling result (Figure 6), the modeling result shows a slight improvement in the near fault fit (Figure 10). The correlation coefficient between the observation and prediction is 97.2%. However, the results are not substantially improved from the uniform to the distributed slip model. The calculated slip distribution is shown in Figure 11. Most of the slip occurs at depths of 3 to 9 km, with a maximum slip of approximately 24 cm at a depth of 6 km. The geodetic moment based on the distributed slip is 2.19×10^{17} Nm, corresponding to Mw 5.56, which is comparable to the seismological estimates ranging from Mw 5.5 (from NEIC) to Mw 5.8 (from GCMT) (Figure 1). Error analysis shows that the maximum standard deviation is approximately 1.5 cm, indicating the slip is well retrieved (Figure 10c).

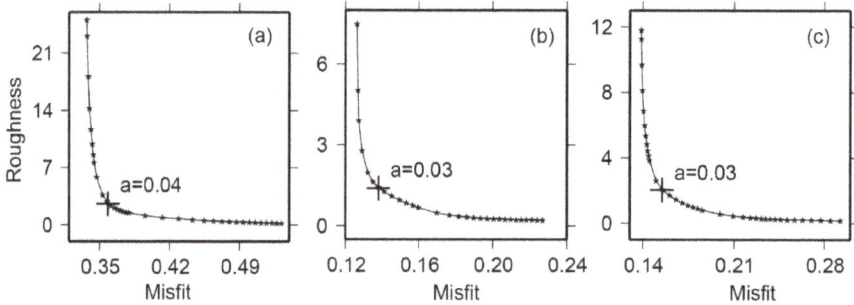

Figure 9. Trade-off curves between misfit and model roughness. The roughness is the normalized value. Pluses indicate locations of optimal smoothing parameters where balances between model misfit and smoothness is achieved. (**a**) 7 July 2003 earthquake; (**b**) 3 July 2004 earthquake; (**c**) 27 March 2004 earthquake.

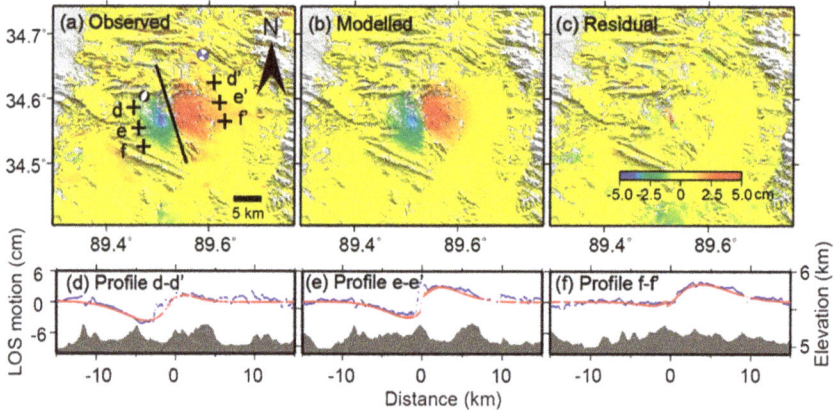

Figure 10. Coseismic deformation (range displacement—negative away from the satellite) and model for distributed slip inversion for the 7 July 2003 earthquake. (**a**) Observed interferogram spanning 20030618–20030723. Black and blue beach balls show focal mechanisms from NEIC and GCMT catalogues, respectively; (**b**) Synthetic interferogram and (**c**) residual interferogram based upon the fault plane (**black** line in (**a**)) slip distribution shown in Figure 11; (**d,e,f**) are profiles of line-of-sight (LOS) displacements (**blue** dots), model LOS displacements (**red** dots), and topography (**grey**), respectively. Crosses in (**a**) indicate profile locations.

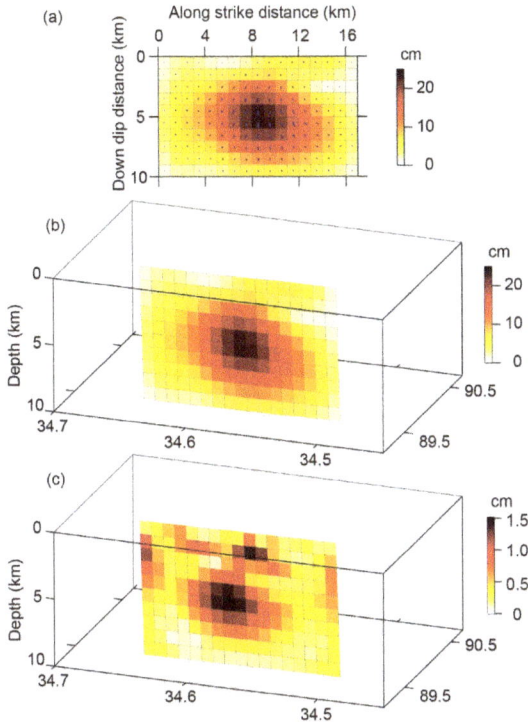

Figure 11. Slip distributions for modeled seismic source of the 7 July 2003 event. (**a**) Perpendicular view of the fault, with slip vectors plotted in addition to the slip magnitudes shown in color; (**b**) 3-D view from WSW; (**c**) 1σ uncertainty for slip distribution as shown in (**a**) and (**b**), estimated from performing 100 inversions.

3.2.2. The 2004 Earthquakes

Similar to the procedure described in Section 3.2.1, we use the fault geometry of the parameters from the best-fitting uniform-slip model (Table 3) and estimate the slip distribution for the 2004 earthquakes. Figure 12 shows the modeling result of the 7 July event. Compared to the uniform slip dislocation modeling result (Figure 7d), the modeling result based on distributed-slip model shows a significant improvement in the near fault fit (Figure 12d). The correlation coefficient between the observation and prediction is 98.9%. The calculated slip distributions are shown in Figure 13. Figure 13 shows that most of the slip occurs at depths of 4 to 9 km, with a maximum slip of approximately 14 cm at a depth of 7 km. The geodetic moment is 1.11×10^{17} Nm, resulting in a Mw of 5.33. Figure 13c shows small slip uncertainties on the fault, demonstrating that the slip distributions shown in Figure 13a,b are reliable. Figure 14 shows the modeling result of the 27 March 2004 Ms 6.2 event, showing improvement compared to the uniform slip model. Figure 15 shows the slip distribution. The coseismic slip is concentrated at depths of 1 to 6 km, with a maximum slip of approximately 55 cm at a depth of 4 km. The geodetic moment is 6.92×10^{17} Nm, resulting in a Mw of 5.86, which is slightly lower than the seismological estimates, that is, Mw 6.0 (from NEIC and GCMT) (Figure 1). Error analysis shows that the maximum standard deviation is approximately 7 cm, on the southern part (Figure 15c). Slip on the central part (where major slip occurs) is well retrieved, with the 1σ uncertainty generally ≤4 cm.

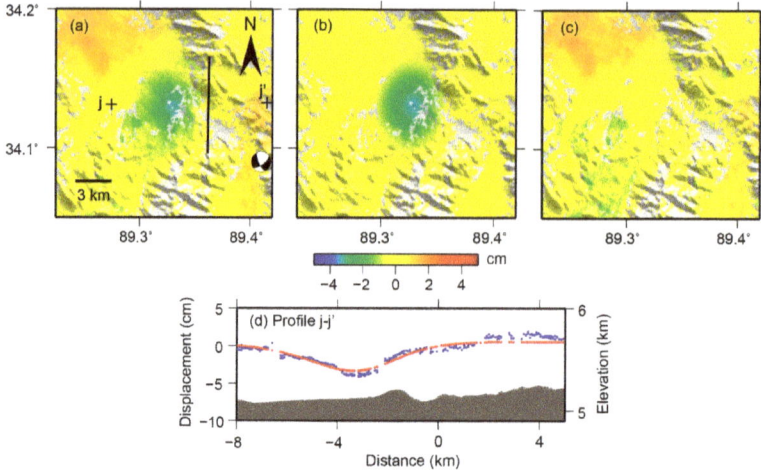

Figure 12. Coseismic deformation (range displacement—negative away from the satellite) and model for distributed slip inversion of the 3 July 2004 earthquake. (**a**) Observed interferogram spanning 20040602–20050622. Focal mechanism from GCMT catalogue is shown; (**b**) Synthetic interferogram and (**c**) residual interferogram based upon the fault plane' (black line in (**a**)) slip distribution shown in Figure 13; (**d**) Profile of line-of-sight (LOS) displacements (**blue** dots), model LOS displacements (**red** dots), and topography (**grey**). Crosses in (**a**) indicate profile locations.

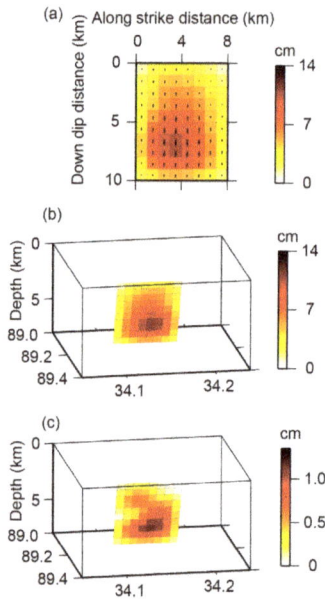

Figure 13. Slip distributions for the modeled seismic sources of the 3 July 2004 event. (**a**) Perpendicular view of the fault, with slip vectors plotted in addition to the slip magnitudes shown in color; (**b**) 3-D view from ENE; (**c**) 1σ uncertainty for the slip distribution as shown in (**a**,**b**), estimated from performing 100 inversions.

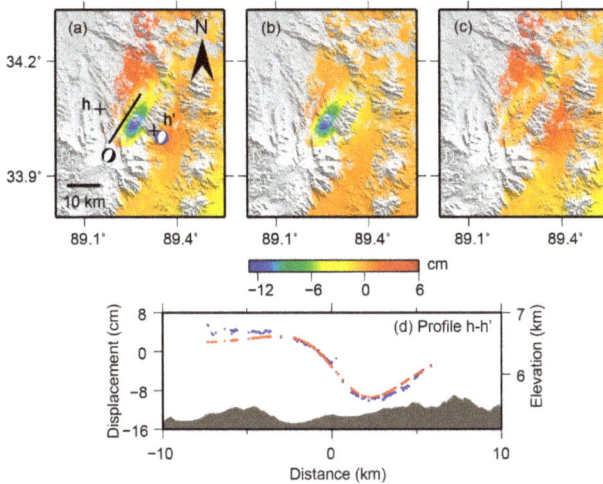

Figure 14. Coseismic deformation (range displacement—negative away from the satellite) and model for distributed slip inversion of the 27 March 2004 earthquake. (**a**) Observed interferogram spanning 20040114–20041124 obtained by subtracting Figure 4g. Black and blue beach balls show focal mechanisms from NEIC and GCMT catalogues, respectively; (**b**) Synthetic interferogram and (**c**) residual interferogram based upon the fault plane's (**black** line in (**a**)) slip distribution shown in Figure 15; (**d**) Profile of the line-of-sight (LOS) displacements (**blue** dots), model LOS displacements (**red** dots), and topography (**grey**). Crosses in (**a**) indicate profile locations.

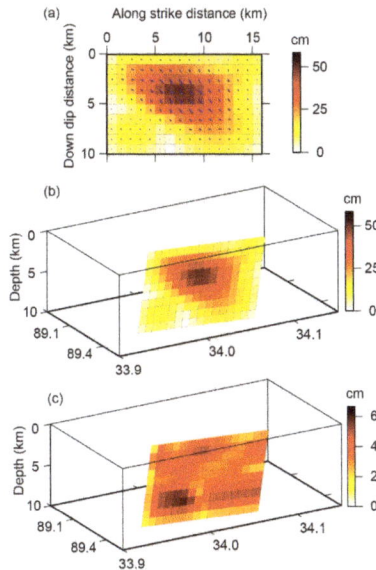

Figure 15. Slip distribution for modeled seismic source of the 27 March 2004 event. (**a**) Perpendicular view of the fault, with slip vectors plotted in addition to the slip magnitudes shown in color; (**b**) 3-D view from WSE; (**c**) 1σ uncertainty for the slip distribution as shown in (**a,b**), estimated from performing 100 inversions.

4. Discussion

4.1. Static Stress Drop

The static stress drop provides hints on the scaling of the static parameters (such as the source size or final displacement) characterizing an earthquake [30]. In this study, we evaluate the static stress drop of the 2003–2004 Bange earthquake sequence using the relationship [31–33]:

$$\Delta\sigma = 7 \times M_0/16/r^3 \qquad (1)$$

where M_0 is the seismic moment and r is the source radius. This relationship assumes a circular rupture, and the parameter r was modeled as $r = \text{sqrt}(LW/\pi)$, where L and W are the rupture length and width, respectively.

Table 4 provides the calculated stress drop for the best-fitting uniform slip model. The average stress drops of the 2003 and 2004 earthquakes are approximately 19, 13, and 6 Mpa, which are consistent with the stress drops of moderate earthquakes that occur in the Tibetan Plateau (e.g., [34]). Though the stress drops are larger than the median of an intraplate earthquake's stress drop of 5.95 MPa [33], they fall within the typical range of 0.3 to 50 MPa [33].

Table 4. Average static stress drop for the 2003–2004 earthquake sequence inferred from InSAR observations.

Earthquake	Seismic Moment (Nm × 10^17) [1]	Inferred Source Radius (m)	Stress Drop (MPa)
20030707 Ms 6.0	2.35	1748	19.2
20040327 Ms 6.2	7.11	2891	12.9
20040703 Ms 5.1	4.19	3116	6.1

[1] The adopted shear modulus μ was 30 GPa.

4.2. Coulomb Stress Change Analysis

To test whether more recent earthquakes in the Bange earthquake sequence may have been triggered by previous ones, we calculate the progression of the Coulomb stress changes by using the PSGRN/PSCMP code based on the distributed slip source [35] (e.g., [36]). First, we calculate the static Coulomb stress change induced by the 7 July 2003 event on the fault plane of the 27 March 2004 event (Figure 16a). Then, the joint effect of the first two large events (*i.e.*, the 7 July 2003 event and 27 March 2004 event) on the July 3 2004 event is calculated (Figure 16b). The Coulomb failure criterion used is [37]:

$$\Delta\text{CFF} = \Delta\tau + \mu' \times \Delta\sigma_n$$

where ΔCFF is the Coulomb stress change, $\Delta\tau$ is the change in shear stress on the receiver fault (positive in the slip direction), μ' is the apparent coefficient of friction and $\Delta\sigma_n$ is the change in normal stress acting on the receiver fault (with extension positive). The value μ' is treated as a constant with a value of 0.4 [38]. A positive ΔCFF implies that the effect of previous events advanced subsequent shocks toward failure, whereas a negative ΔCFF represents stress release and a delayed failure time.

Figure 16a shows that the stress change induced by the 7 July 2003 event on the fault plane of the 27 March 2004 event is around 0 MPa. The low stress change indicates the 27 March 2004 event may not be enhanced by the 7 July 2003 event, possibly because of the long distance (approximately 70 km) between these two earthquakes. The effect of the 7 July 2003 and 27 March 2004 events on the July 3 2004 event is positive on most part of the fault (Figure 16b), indicating that the receiver fault are brought closer to failure (e.g., [39,40]). Therefore, the 3 July 2004 event is positively triggered by the 7 July 2003 and 27 March 2004 events.

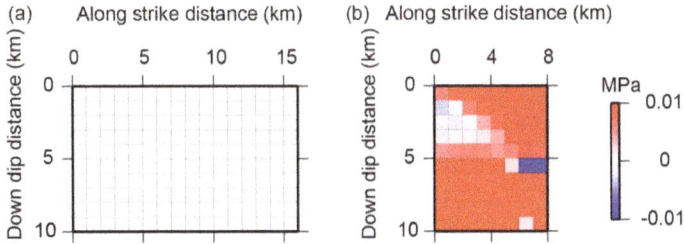

Figure 16. (**a**) Coseismic Coulomb stress change on the fault plane of the 27 March 2004 event triggered by the 7 July 2003 event; (**b**) Stress change induced on the 3 July 2004 earthquake triggered by 7 July 2003 and 27 March 2004 events.

4.3. Normal Faulting Earthquakes in Tibetan Plateau

Figure 17 indicates that moderate-sized (Mw 5.5–7.0) normal faulting earthquakes are common in the Tibetan Plateau. For instance, the 2008 Yutian earthquake in the northwest part of the plateau [41,42] was the largest normal faulting earthquake ever recorded instrumentally in northern Tibet. Elliott *et al.* [41] studied a series of eight Mw 5.9–7.1 normal faulting earthquakes and inferred that the extension of the Tibetan Plateau is driven primarily by gravitational forces. Ryder *et al.* [34] took the 2004–2008 Zhongba earthquake sequence as an example, and studied an extensional earthquake sequence that occurred on the Tibetan Plateau.

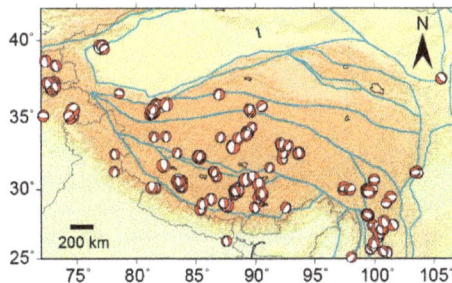

Figure 17. Normal faulting earthquakes with Mw ⩾ 5.5 in Tibetan Plateau (1976–2015). Focal mechanisms are based on GCMT catalogue.

Tectonically, east–west extension in the central Tibetan Plateau is accommodated by normal faults and rift systems that trend north–south (e.g., [2,43]), which are active structures that rupture in some normal faulting earthquakes. Therefore, moderate-sized normal faulting earthquakes will probably occur on the Tibetan Plateau in the future to account for the continuing east–west extension.

5. Conclusions

The 2003–2004 Bange earthquake sequence involves a series of normal faulting events with magnitudes larger than 5.0, which occurred on previously unknown faults. InSAR observations provide important constraints on the source parameters and slip distributions. The results indicate that the 2003 earthquake was a normal-faulting event with a right-lateral slip component. The 27 March 2004 Ms 6.2 earthquake was associated with a normal fault striking southwest–northeast and dipping southeast with a moderately oblique right-lateral slip. The 3 July Ms 5.1 event was caused by a normal fault with a slight normal slip component. A calculation of the static stress changes on the fault planes demonstrates that the third earthquake may have been triggered by the previous ones.

This study indicates that InSAR can provide reliable source parameters of shallow, moderate-sized earthquakes in areas that lack dense seismic networks. Earthquakes with M ⩾ 5.0 are common on the Tibetan Plateau. However, because of the sparse seismic station distribution, most of the focal mechanisms that are important for understanding local tectonic activity are unknown. Nevertheless, it is possible to learn more about moderate–sized earthquakes using InSAR, as more SAR data are available from new satellite missions (e.g., Sentinel-1A and Advanced Land Observing Satellite-2).

Acknowledgments: This study was supported by the Spark Programs of Earthquake Sciences granted by the China Earthquake Administration (XH14069Y) and the Special Earthquake Research Project granted by the China Earthquake Administration (201508009). Envisat SAR data are copyrighted by ESA and were provided by ESA under the 'Group on Earth Observations (GEO) Geohazard Supersite initiative' (http://www.earthobservations.org/gsnl.php) with the Category 1 project (C1F.28413). The author would like to thank the editors and three anonymous reviewers for their very constructive comments and suggestions that greatly improved this paper.

Author Contributions: All of the authors participated in editing and reviewing the manuscript. Lingyun Ji led the research and processed the InSAR data. Jing Xu calculated the Coulomb stress change. Lingyun Ji and Chengsheng Yang inverted the coseismic slip distribution. Lingyun Ji and Qiang Zhao analyzed and interpreted the results. All contributed to the writing of this manuscript.

Conflicts of Interest: The authors declare no conflict of interest.

References

1. The China Earthquake Networks Center's (CENC) Catalogue. Available online: http://data.earthquake.cn/datashare/csn_catalog_p001_new.jsp (accessed on 1 March 2016).
2. Taylor, M.; Yin, A.; Ryerson, F.J.; Kapp, P.; Ding, L. Conjugate strike-slip faulting along the Bangong-Nujiang suture zone accommodates coeval east-west extension and north-south shortening in the interior of the Tibetan Plateau. *Tectonics* **2003**, *22*, 1–18. [CrossRef]
3. Zhang, P.Z.; Deng, Q.D.; Zhang, G.M.; Ma, J.; Gan, W.; Min, W.; Mao, F.; Wang, Q. Active tectonic blocks and strong earthquakes in the continent of China. *Sci. China Ser. D Earth Sci.* **2003**, *46* (Suppl. S2), 13–24. (In Chinese)
4. Tapponnier, P.; Zhiqin, X.; Roger, F.; Meyer, B.; Arnaud, N.; Wittlinger, G.; Jingsui, Y. Oblique stepwise rise and growth of the Tibet Plateau. *Science* **2001**, *294*, 1671–1677. [CrossRef] [PubMed]
5. Massonnet, D.; Feigl, K. Radar interferometry and its application to changes in the Earth's surface. *Rev. Geophys.* **1998**, *36*, 441–500. [CrossRef]
6. Rosen, P.A.; Hensley, S.; Joughin, I.R.; Li, F.K.; Madsen, S.N.; Rodriguez, E.; Goldstein, R.M. Synthetic aperture radar interferometry. *Proc. IEEE* **2000**, *88*, 333–380. [CrossRef]
7. Massonnet, D.; Rossi, M.; Carmona, C.; Adragna, F.; Peltzer, G.; Feigl, K.; Rabaute, T. The displacement field of the Landers earthquake mapped by radar interferometry. *Nature* **1993**, *364*, 138–142. [CrossRef]
8. Wright, T.J.; Lu, Z.; Wicks, C. Constraining the slip distribution and fault geometry of the Mw 7.9, 3 November 2002, Denali fault earthquake with interferometric synthetic aperture radar and global positioning system data. *Bull. Seismol. Soc. Am.* **2004**, *94*, S175–S189. [CrossRef]
9. Biggs, J.; Nissen, E.; Craig, T.; Jackson, J.; Robinson, D.P. Breaking up the hanging wall of a rift-border fault: The 2009 Karonga earthquakes, Malawi. *Geophys. Res. Lett.* **2010**, *37*. [CrossRef]
10. Li, Z.; Elliott, J.; Feng, W.; Jackson, J.; Parsons, B.; Walters, R. The 2010 MW 6.8 Yushu (Qinghai, China) earthquake: Constraints provided by InSAR and body wave seismology. *J. Geophys. Res.* **2011**, *116*. [CrossRef]
11. Elliott, J.; Nissen, E.; England, P.; Jackson, J.; Lamb, S.; Li, Z.; Oehlers, M.; Parsons, B. Slip in the 2010–2011 Canterbury earthquakes, New Zealand. *J. Geophys. Res.* **2012**, *117*. [CrossRef]
12. Wen, Y.; Xu, C.; Liu, Y.; Jiang, G. Deformation and source parameters of the 2015 Mw 6.5 earthquake in Pishan, western China, from Sentinel-1A and ALOS-2 Data. *Remote Sens.* **2016**, *8*, 1–14. [CrossRef]
13. Werner, C.; Wegmüller, U.; Strozzi, T.; Wiesmann, A. GAMMA SAR and interferometric processing software. In Proceedings of the ERS-Envisat Symposium, Gothenburg, Sweden, 16–20 October 2000.
14. Farr, T.G.; Rosen, P.A.; Caro, E.; Crippen, R.; Duren, R.; Hensley, S.; Kobrick, M.; Paller, M.; Rodriguez, E.; Roth, L.; *et al.* Shuttle radar topography mission. *Rev. Geophys.* **2007**, *45*. [CrossRef]
15. The Consultative Group on International Agricultural Research's Consortium for Spatial Information. Available online: http://srtm.csi.cgiar.org (accessed on 1 March 2016).

16. Goldstein, R.M.; Werner, C.L. Radar interferogram filtering for geophysical applications. *Geophys. Res. Lett.* **1998**, *25*, 4035–4038. [CrossRef]

17. Nof, R.N.; Ziv, A.; Doin, M.P.; Baer, G.; Fialko, Y.; Wdowinski, S.; Eyal, Y.; Bock, Y. Rising of the lowest place on Earth due to Dead Sea water-level drop: Evidence from SAR interferometry and GPS. *J. Geophys. Res.* **2012**, *117*. [CrossRef]

18. Rosen, P.A.; Hensley, S.; Zebker, H.; Webb, F.H.; Fielding, E.J. Surface deformation and coherence measurements of Kilauea Volcano, Hawaii, from SIR-C radar interferometry. *J. Geophys. Res.* **1996**, *101*, 23109–23125. [CrossRef]

19. Lu, Z.; Dzurisin, D. *InSAR Imaging of Aleutian Volcanoes: Monitoring a Volcanic Arc from Space*; Springer: Chichester, UK, 2014; p. 390.

20. Jonsson, S.; Zebker, H.; Segall, P.; Amelung, F. Fault slip distribution of the Mw 7.2 Hector Mine earthquake estimated from satellite radar and GPS measurements. *Bull. Seismol. Soc. Am.* **2002**, *92*, 1377–1389. [CrossRef]

21. Okada, Y. Surface deformation due to shear and tensile faults in a half-space. *Bull. Seismol. Soc. Am.* **1985**, *75*, 1135–1154.

22. Press, W.; Teukolsky, S.; Vetterling, W.; Flannery, B. *Numerical Recipes in C, the Art of Scientific Computing*; Cambridge University Press: New York, NY, USA, 1992; p. 994.

23. Funning, G.; Parsons, B.; Wright, T.J. The 1997 Manyi (Tibet) earthquake: Linear elastic modelling of coseismic displacements. *Geophys. J. Int.* **2007**, *169*, 988–1008. [CrossRef]

24. Lohman, R.B.; Simons, M. Some thoughts on the use of InSAR data to constrain models of surface deformation: Noise structure and data downsampling. *Geochem. Geophys. Geosyst.* **2005**, *6*. [CrossRef]

25. Wang, R.; Diao, F.; Hoechner, A. SDM—A geodetic inversion code incorporating with layered crust structure and curved fault geometry. In Proceedings of the EGU General Assembly 2013, Vienna, Austria, 7–12 April 2013.

26. Wang, L.; Wang, R.; Roth, F.; Enescu, B.; Hainzl, S.; Ergintav, S. Afterslip and viscoelastic relaxation following the 1999 M7.4 Izmit earthquake from GPS measurement. *Geophys. J. Int.* **2009**, *178*, 1220–1237. [CrossRef]

27. Xu, C.; Liu, Y.; Wen, Y.; Wang, R. Coseismic slip distribution of the 2008 Mw 7.9 Wenchuan earthquake from joint inversion of GPS and InSAR data. *Bull. Seismol. Soc. Am.* **2010**, *100*, 2736–2749. [CrossRef]

28. Wen, Y.; Xu, C.; Liu, Y.; Jiang, G.; He, P. Coseismic slip in the 2010 Yushu earthquake (China), constrained by wide-swath and strip-map InSAR. *Nat. Hazards Earth Syst. Sci.* **2013**, *13*, 35–44. [CrossRef]

29. Motagh, M.; Bahroudi, A.; Haghighi, M.H.; Samsonov, S.; Fielding, E.; Wetzel, H.U. The 18 August 2014 Mw 6.2 Mormori, Iran, Earthquake: A thin-skinned faulting in the Zagros Mountain inferred from InSAR measurements. *Seismol. Res. Lett.* **2015**, *86*, 775–782. [CrossRef]

30. Cotton, F.; Archuleta, R.; Causse, M. What is sigma of the stress drop? *Seismol. Res. Lett.* **2013**, *84*, 42–48. [CrossRef]

31. Brune, J.N. Tectonic stress and the spectra of seismic shear waves from earthquakes. *J. Geophys. Res.* **1970**, *75*, 4997–5009. [CrossRef]

32. Scholz, C.H. *The mechanics of Earthquakes and Faulting*; Cambridge University Press: New York, NY, USA, 2002.

33. Allmann, B.P.; Shearer, P.M. Global variations of stress drop for moderate to large earthquakes. *J. Geophys. Res.* **2009**, *114*. [CrossRef]

34. Ryder, I.; Burgmann, R.; Fielding, E. Static stress interactions in extensional earthquake sequences: An example from the South Lunggar Rift, Tibet. *J. Geophys. Res.* **2012**, *117*. [CrossRef]

35. Wang, R.; Lorenzo-Martín, F.; Roth, F. PSGRN/PSCMP—A new code for calculating co-and post-seismic deformation, geoid and gravity changes based on the viscoelastic-gravitational dislocation theory. *Comput. Geosci.* **2006**, *32*, 527–541. [CrossRef]

36. Nissen, E.; Elliott, J.R.; Sloan, R.A.; Craig, T.J.; Funning, G.J.; Hutko, A.; Parsons, B.E.; Wright, T.J. Limitations of rupture forecasting exposed by instantaneously triggered earthquake doublet. *Nat. Geosci.* **2016**, *9*, 330–336. [CrossRef]

37. Harris, R. Introduction to special section: Stress triggers, stress shadows, and implications for seismic hazard. *J. Geophys. Res.* **1998**, *103*, 24347–24358. [CrossRef]

38. Freed, A.M. Earthquake triggering by static, dynamic, and postseismic stress transfer. *Annu. Rev. Earth Planet. Sci.* **2005**, *33*, 335–367. [CrossRef]

39. King, G.C.P.; Stein, R.S.; Lin, J. Static stress changes and the triggering of earthquakes. *Bull. Seismol. Soc. Am.* **1994**, *84*, 935–953.

40. Lin, J.; Stein, R.S. Stress triggering in thrust and subduction earthquakes and stress interaction between the southern San Andreas and nearby thrust and strike-slip faults. *J. Geophys. Res.* **2004**, *109*. [CrossRef]

41. Elliott, J.R.; Walters, R.J.; England, P.C.; Jackson, J.A.; Li, Z.; Parsons, B. Extension on the Tibetan plateau: Recent normal faulting measured by InSAR and body wave seismology. *Geophys. J. Int.* **2010**, *183*, 503–535. [CrossRef]

42. Furuya, M.; Yasuda, T. The 2008 Yutian normal faulting earthquake (Mw 7.1), NW Tibet: Non-planar fault modeling and implications for the Karakax Fault. *Tectonophysics* **2011**, *511*, 125–133. [CrossRef]

43. Taylor, M.; Yin, A. Active structures of the Himalayan-Tibetan orogen and their relationships to earthquake distribution, contemporary strain field, and Cenozoic volcanism. *Geosphere* **2009**, *5*, 199–214. [CrossRef]

remote sensing

MDPI

Article

Space Geodetic Observations and Modeling of 2016 Mw 5.9 Menyuan Earthquake: Implications on Seismogenic Tectonic Motion

Yongsheng Li [1,2], Wenliang Jiang [1,2,*], Jingfa Zhang [1,2] and Yi Luo [1,2]

[1]   Institute of Crustal Dynamics, China Earthquake Administration, Beijing 100085, China;
      whlys@163.com (Y.L.); Zhangjingfa@hotmail.com (J.Z.); luoyi1983@126.com (Y.L.)
[2]   Key Laboratory of Crustal Dynamics, China Earthquake Administration, Beijing 100085, China
*   Correspondence: jiang_wenliang@163.com; Tel.: +86-10-8284-6721

Academic Editors: Zhenhong Li, Roberto Tomas, Zhong Lu and Prasad S. Thenkabail
Received: 3 May 2016; Accepted: 14 June 2016; Published: 22 June 2016

Abstract: Determining the relationship between crustal movement and faulting in thrust belts is essential for understanding the growth of geological structures and addressing the proposed models of a potential earthquake hazard: A Mw 5.9 earthquake occurred on 21 January 2016 in Menyuan, NE Qinghai Tibetan plateau. We combined satellite interferometry from Sentinel-1A Terrain Observation with Progressive Scans (TOPS) images, historical earthquake records, aftershock relocations and geological data to determine fault seismogenic structural geometry and its relationship with the Lenglongling faults. The results indicate that the reverse slip of the 2016 earthquake is distributed on a southwest dipping shovel-shaped fault segment. The main shock rupture was initiated at the deeper part of the fault plane. The focal mechanism of the 2016 earthquake is quite different from that of a previous Ms 6.5 earthquake which occurred in 1986. Both earthquakes occurred at the two ends of a secondary fault. Joint analysis of the 1986 and 2016 earthquakes and aftershocks distribution of the 2016 event reveals an intense connection with the tectonic deformation of the Lenglongling faults. Both earthquakes resulted from the left-lateral strike-slip of the Lenglongling fault zone and showed distinct focal mechanism characteristics. Under the shearing influence, the normal component is formed at the releasing bend of the western end of the secondary fault for the left-order alignment of the fault zone, while the thrust component is formed at the restraining bend of the east end for the right-order alignment of the fault zone. Seismic activity of this region suggests that the left-lateral strike-slip of the Lenglongling fault zone plays a significant role in adjustment of the tectonic deformation in the NE Tibetan plateau.

Keywords: Menyuan earthquake; interferometry; Sentinel-1A TOPS; Lenglongling fault; characteristics of the tectonic environment

1. Introduction

A Mw 5.9 earthquake struck the Menyuan county, Qinghai (101.641°E, 37.67°N) on 21 January 2016. Moment tensor solution from teleseismic data suggests that the Menyuan earthquake occurred on a 43° southern dipping thrust fault at about 10 km depth with a strike of 134° [1,2]. The hypocenter was located at the intersection of Lenglongling fault and Tuolaishan fault. Since 1927, more than five earthquakes with Ms > 6 have occurred within a 100 km range from the epicenter of the event according to the U.S. Geological Survey (USGS). The largest one, with a magnitude Ms 8.0, occurred in May 1927 at Gulang, the closest earthquake with a magnitude Ms 6.5 occurred on 26 August 1986. Both the 1986 and 2016 events occurred near the secondary fault of the Lenglongling fault; the distance between the epicenters of the earthquakes is about 15 km (Figure 1a). The focal mechanism solution of 2016 event indicated that seismogenic fault is a thrust fault, with a strike slip component according to

USGS [1]. Since the late Quaternary, the activities of the Lenglongling fault have been characterized by the left-lateral slip and a minor component of the dip slip at some segment. Thus, there is a series of large-scale sinistral slip fault geomorphology along the Lenglongling fault [3]. The activity behavior of the Lenglongling main fault and the focal mechanisms of two earthquakes (1986 and 2016) show great differences in mechanical properties [4], which indicates the complexity of the tectonic stress field and structural styles in this area (Figure 1b). The characteristics of the rupture process of the 2016 Menyuan earthquake offer an outstanding occasion to better constrain and resolve the fault geometry of the northeast margin of NE Tibetan plateau. In the paper, we will report the deformation patterns of the 2016 Menyuan Earthquake from the ascending and descending track Sentinel-1A data. Joint analysis with the 1986 Menyuan earthquake, and the background and mechanical properties of the seismogenic fault will be analyzed to reveal the tectonic relationship between the seismogenic fault and the main fault of Lenglongling zone. It will enhance our understanding of the implications on seismogenic tectonic motion of the NE Tibetan plateau.

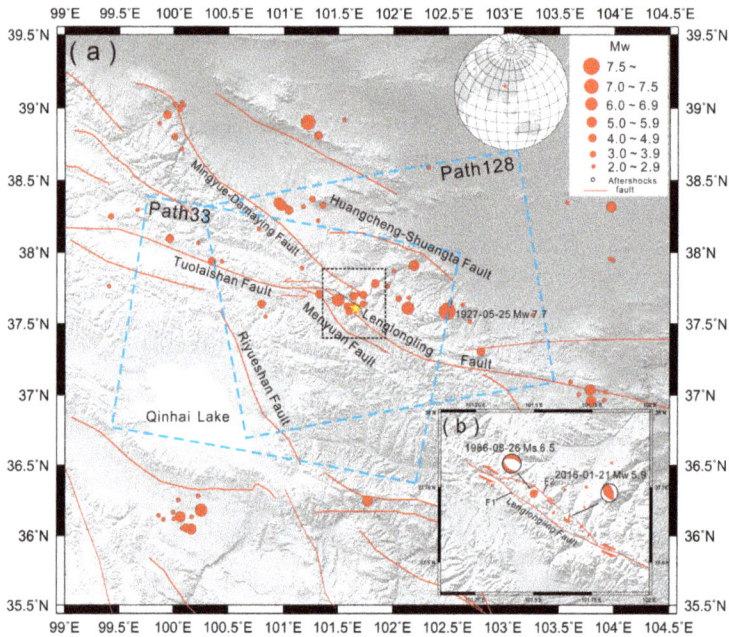

Figure 1. (**a**) Tectonic background of the 21 January 2016 Menyuan Earthquake superimposed on topographic relief. The star is location of the 2016 Menyuan event. The red lines denote the active faults. The blue frames are the coverage of the Sentinel-1A data. The red dots show the historic events since 1927; (**b**) The partially enlarged view of the black dotted frames in (**a**). F1: The main fault of Lenglongling; F2: The secondary fault of Lenglongling; the circles express the aftershocks location. Both ends of the secondary fault of Lenglongling are bent to converge to the main fault.

2. Tectonic Settings

The stress environment of seismic activity and tectonic deformation in the northeastern margin of the Tibetan Plateau is mainly derived from the northward push from the Indian plate. The continental collision of the India and Eurasia plates causes the plate convergence at a relative rate of 40–50 mm/year [5]. The northward thrusting of India beneath Eurasia led to the development of the Altyn-Tagh and Qilianshan orogens in the northern margin of the Tibetan Plateau and generated numerous earthquakes which consequently make this area one of the most seismically hazardous regions [6–10].

A large number of active faults are widely distributed, and control the activities of the strong earthquakes in this region. The earthquakes in the region have the characteristics of high frequency, high intensity, shallow hypocenter and wide distribution. It is one of the most active regions in China [11–13]. Geological research and GPS observation results show that the basic activities of the main boundary zone in the northeastern margin of the Qinghai Tibetan Plateau are left-lateral torsion and reverse thrust. Recently, the left-lateral activity has also been quite significant [11,14]. The GPS convergence rate shows that the northeastern margin of the Tibetan plateau is accumulating strain. From the compilation of historical records, it can certainly be shown that this region is capable of large magnitude events. The most recent of the 1986 Mw 6.0 Menyuan earthquake was in the western region [15].

The 2016 Menyuan earthquake occurred in the northeastern margin of the Tibetan plateau. This region is one of the most tectonically active areas. The Lenglongling left-lateral strike-slip fault is located in the front margin of the NE Tibetan plateau (Figure 1a), which plays an important role in adjustment and conversion of the tectonic deformation of this region [3]. This fault is also considered to be an important segment of the Qilian-Haiyuan fault [3,13,16]. Under regional structural stress, the crustal block is undergoing NE-oriented compression and shortening, clockwise rotation and extrusion along the SSE direction [12,17,18]. The overall strike of the Lenglongling fault is NE 110°~115°, and the length is approximately 120 km. The strike slip rate of the Lenglongling fault zone is within the range of 4–19 mm/a [3,13,16–19]. A seismic gap (Tianzhu seismic gap) has been observed on the Qilian-Haiyuan fault which is mainly composed of the Lenglongling fault, Jingqianghe fault, Maomaoshan fault and Laohushan fault [3]. Therefore, it is better to pay more attention to the tectonic deformation and seismic activity in this region.

3. InSAR Coseismic Measurements and Geodetic modeling

3.1. InSAR Coseismic Deformation

The coseismic deformation due to the 2016 Menyuan earthquake was mapped using both ascending and descending tracks of the Sentinel-1A TOPS (Terrain Observation with Progressive Scans) mode (paths 33 and 128). The ascending coseismic interferogram was generated from 13 January 2016 to 6 February 2016 (13 January 2016–6 February 2016), and the descending one was generated from 18 January 2016 to 11 February 2016 (18 January 2016–11 February 2016). The parameters of the interferometric pairs are shown in Table 1.

Table 1. Sentinel-1A pairs used in this study.

No.	Acquisition Time	Path	Mode	Orbit	$B_\perp$	Incidence
Ifg1	20160113–20160206	128	ScanSAR	Asc	15	30°–46°
Ifg2	20160118–20160211	33	ScanSAR	Des	6.5	30°–46°

Due to a steep azimuth spectra ramp in each burst and a small overlap between consecutive bursts, conventional interferometry with TOPS SAR data is challenging [20]. GAMMA software supplies a new coregistration strategy to process TOPS SAR pairs, which uses a method that considers the effects of the scene topography and then uses a spectral diversity method that considers the interferometric phase of the burst overlap regions between any two adjacent bursts. The topographic phase is removed using a simulated phase from the 1-arc (~30 m) DEM from SRTM (Shuttle Radar Topography Mission). The phase filtering [21], phase unwrapping (e.g., using a minimum cost-flow approach [22]), phase to displacement conversion and coherence estimation are the same as those of the conventional stripmap interferometry. Finally, we obtain surface deformation maps (Figure 2a,b). The maps completely recorded the ground deformation field caused by the earthquake.

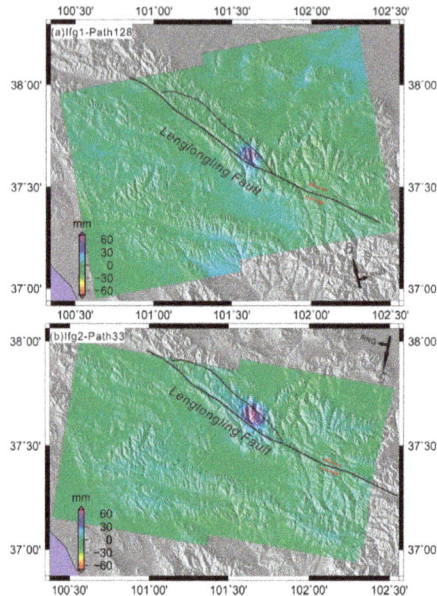

Figure 2. The light of sight (LOS) deformation maps of the 2016 Menyuan earthquake. Each map is labeled and has a background of shaded topography. The black lines indicates the main and the secondary fault of Lenglongling. (**a**) Ascending LOS deformation map with pairs 13 January 2016–6 February 2016; (**b**) Descending LOS deformation map with pairs 18 January2016–11 February 2016. The main and the secondary faults of Lenglongling are labeled consistently with Figure 1b.

The deformation maps (Figure 2a,b) suggest an uplift of about 7 cm along the light of sight (LOS) direction of the satellite. The temporal and spatial baselines are relatively small (Table 1), and limited vegetation coverage exists in the epicenter region, thus the coherence is high. The patterns of the earthquake epicenter are smooth and distinct. Identifying the location of coseismic displacement from the interferograms is crucial to understanding the relationship between the ground motions detected by InSAR and the fault planes that caused the earthquakes [23,24]. Additionally, we compared the descending and ascending interferograms; the difference of deformation patterns can be observed in Figure 2. The deformation map of the descending track shows that the location is shifted relatively eastward and a little larger in magnitude. This is because the different SAR viewing geometric parameters could lead to different LOS deformation patterns.

3.2. Geodetic Modeling for Earthquake Rupture

After detailed analysis of coherence, orbital ramp effects and atmospheric artifacts of the coseismc interfergrams, we use the coseismic deformation map of the Menyuan earthquake from Sentinel-1A satellite as the geodetic inversion constraints to determine the detailed slip distribution pattern on the causative fault. The main and secondary faults of Lenglongling are well constrained by fault distribution data, and the approximate location is guided by new and existing geologic mapping (Figure 2). From the fault geometric parameters, we can infer that the secondary fault of the Lenglongling was responsible for this earthquake. The secondary fault of the Lenglongling was derived from the geological mapping. From the whole geological structure background, a series of shovel like faults have been developed. Although the interferometric patterns in Figure 2 are very simple and can be inverted easily using the one-segment model, this event was triggered by a mid-dip angle fault (40°–45°) and the inferred junction between fault and surface is further north than the secondary fault. Thus, a two-segment fault model can be simpler and more effective at fitting the faults.

In order to constrain the possible geometric configurations, we model the surface geodetic displacements due to slip on the secondary fault of Lenglongling for varying fault dips and ramp location. We produced a two-segment shovel-shaped reverse fault model that can agree with InSAR observations of ground deformation. Note that the fault surface trace is fixed during the inversion based on the geological features. The two segments ruptured with different dip angles. The first fault segment extends toward the southwest with a dip of 85° down to 6.5 km. The second segment is relatively flat and deep with a dip of ~40°; it extends farther southwest, and it is responsible for this earthquake. The earthquake was mainly triggered by the deep parts, so we primarily inverted the deep segment; the shallow part is inferred from the supplementary information, such as the location of the fault and the aftershock distribution.

The source parameters and variable slip distribution were determined by using the geodetic inversion package PSOKINV [25], we conduct a global nonlinear inversion to determine the fault geometry of the 2016 Menyuan earthquake, which uses a random search method with Particle Swarm Optimization (PSO) based on an improved group cooperation algorithm [25,26]. First, we determined the fault location and principal focal mechanisms based on a uniform fault. From the structural features of the seismogenic faults, we approximate the geometry of the ruptured fault with two connected fault segments. The Okada elastic dislocation model [27] and PSO nonlinear optimization algorithm are employed to automatically compare and determine the optimum parameters of the simulation results, *i.e.*, finding the minimum solution of the adaptation function in the whole parameter space. The best-fit uniform slip model suggests that the earthquake occurred at (101.65°, 37.64°) at a depth of 10.5 km. The fault had a strike of 134°, a dip of 40° and a slip angle of 65°. The magnitude could be up to Mw 5.9, which is consistent with the Global Centroid Moment Tensor Catalogue (GCMT) and USGS solutions, as shown in Table 2.

Table 2. Optimal fault geometric parameters determined with Sentinel-1A coseismic deformation.

Source	Location		Epicenter Depth (km)	Focal Mechanisms	Fault Dimensions (km)				Magnitude
	Lon (°)	Lat (°)		Strike (°)	Dip (°)	Rake (°)	Length	Width	
GCMT	101.76	37.65	13.9	134	43	68	-	-	5.9
USGS	101.641	37.67	9.0 ± 1.6	134	43	68	-	-	5.9
This study	101.65	37.64	10.5	134	40	65	20	10	5.9 [a]
	101.65	37.64	10.5	134	43	68	24	20	5.9 [b]

[a] Fault parameters derived from uniform slip model; [b] Fault parameters derived from distributed slip model.

To determine the distribution of the coseismic slip, a linear inversion was used to estimate the slip distribution along a fixed fault plane which is determined from the uniform solutions. To prevent a physically impossible oscillatory slip, a Laplacian smoothing was employed to constrain the slip roughness [28]. The optimal dip angle and smoothing coefficient can be determined simultaneously by the log function [25,29]. We fixed a given dip angle and applied different smoothing coefficients, and then analyzed the variation trends. Figure 3 shows that a dip angle ranging from 40° to 45° and smoothing factor as in reference [16] allowed us to obtain the global minimum point. By applying the above methods, the dip angle and smoothing coefficients were determined as 43° and 2.5°, respectively. Figure 4 shows the simulated results derived from the optimal slip model. Figure 4a,d represents the InSAR observations. Figure 4b,e shows the simulation interferogram based on InSAR inversion. Figure 4c,f show the residuals relative to Figure 4b,e respectively. It is clear that the general patterns of the both Sentinel-1A observations can be sufficiently explained by the distributed slip model and there are no notable residual fringes left in the residuals (Figure 4c,f). The correlation coefficient between the observations and simulation is 95.4%.

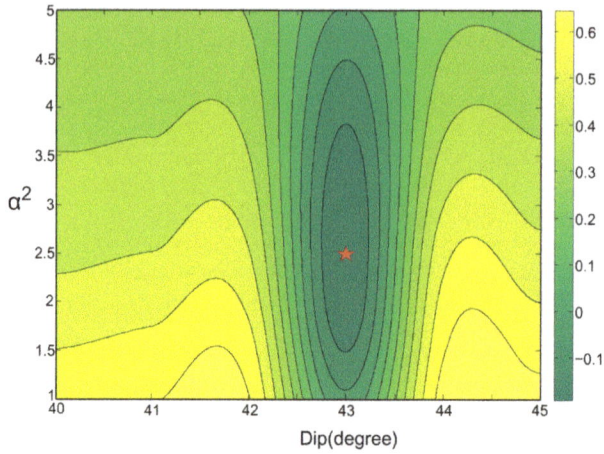

Figure 3. Contour map of log function with variations in dips and smoothing coefficients (α^2). The red star indicates the global minimum point.

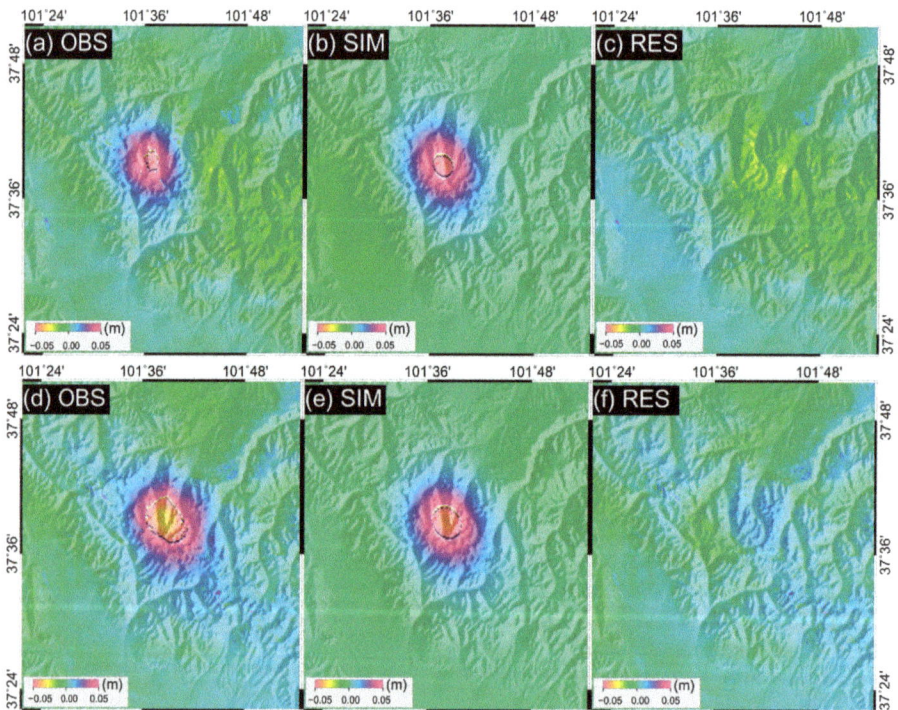

Figure 4. Original, modeled and residual interferograms for InSAR-derived slip models. (**a–c**) are from SENTINEL-1A Path 128 while (**d–f**) are from SENTINEL-1A Path 33.

Our optimal slip model suggests that the earthquake nucleation was initiated at the deeper ramp portion of the rupture plane (Figures 5 and 6). The thrust propagation along the updip rift would have caused breaking of the shallow part of the fault, resulting in coseismic surface uplift [30]. Using the double-difference relocation algorithm, 647 aftershocks were relocated within 60 h of the main shock [31]. The aftershocks occurred along the maximum regional slip, and the aftershocks take on a shovel-shaped structure at a depth of 7–15 km (Figure 6). The relocation results of this event show that the mainshock and the aftershock sequence are mainly distributed in the southwest plane of the secondary fault.

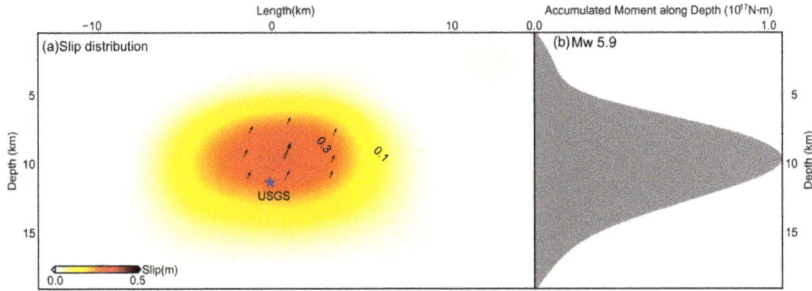

Figure 5. Slip distribution of the Menyuan earthquake. (**a**) Coseismic slip model of the 2016 Menyuan earthquake derived using two paths of Sentinel-1A IW deformation maps; (**b**) The seismic moment distribution along the depth in the slip model.

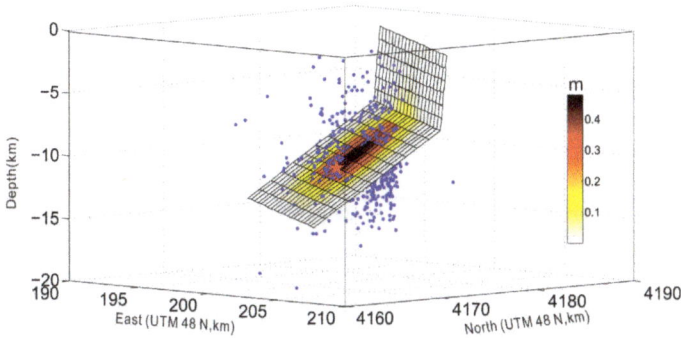

Figure 6. Three-dimensional illustration of Slip distribution of the Menyuan earthquake. The blue points represent the relocation of aftershocks, the red star denotes the epicenter.

Finally, the best-fit slip inversion model shown in Figures 5 and 6 suggests that the major seismogenic fault is a thrust fault with a strike of ~134°, a dip of ~43° and an average rake angle of ~68°. The inferred optimal slip model suggests that the coseismic slip is concentrated at depths of 8–11 km. A maximum slip of ~0.45 m appears at a depth of 9.5 km. The cumulative seismic moment is up to 9.9×10^{17} N·m, equivalent to a magnitude of Mw 5.9. This reveals that the seismic distribution was under the control of the secondary fault of Lenglongling and of the extrusion force in the NE direction of the region.

4. Discussion

4.1. Lateral Variation of the Motion along the Lenglongling Fault

The Lenglongling fault zone is located at the frontal margin of the NE Tibetan Plateau. The overall stress field in this area is under the SW-NE extrusion. A great compressive nappe structure zone is formatted between the two predominant left-lateral strike-slip faults (Altyn Tagh fault and Haiyuan-Qilian fault) in the NE Tibetan Plateau (Figure 7). A series of active structural zones are thrust-nappe movements from southwest to northeast, which are mainly characterized by compressional thrust faults according to the results of GPS velocity field [15]. With the transition in space from the west to the Lenglongling fault zone, the movement of the blocks causes the clockwise rotation of the stress direction [12].

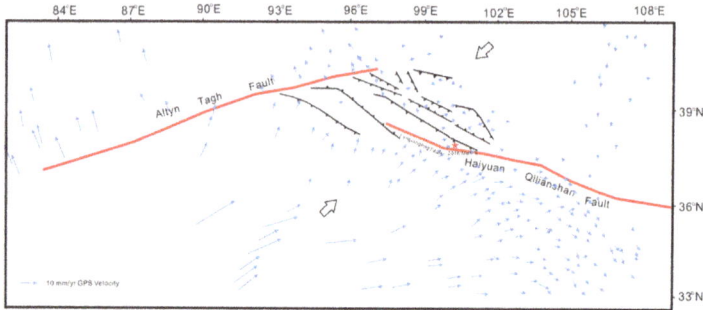

Figure 7. Tectonic transformation mode in the Lenglongling fault.

The movement mode of the Qilian-Haiyuan fault zone gradually changes to left-lateral strike slip from the west to the east. Many obduction faults in the north margin of the Lenglongling fault zone are in the southwest dipping direction and overthrusted successively from the south to the north [32]. These faults were turned into the shovel-shaped by the strong nappe structure under the continental driving [12]. Therefore, the 2016 earthquake was dip-thrusting with a limited strike slip component, which is consistent with the inversion results determined in Section 3.

4.2. Regional Active Tectonic Features Inferred from Two Earthquakes

Although the overall motion features of the Lenglongling fault is left-lateral strike-slip, due to the complexity of fault structure styles, different segments of the fault show different movement characteristics. The most direct representation is the obvious difference between the focal mechanism solution of the two earthquakes in 1986 and 2016 (Figure 1b). According to the previous research, focal mechanism solutions imply that the 1986 earthquake originated on a normal fault which was dominated by the dip slip component [4]. Its fault tensile activity is in a SEE direction. Regional crust extension induced this seismic activity and the tensile rupture zone of surface in the meizoseismal area also confirmed this conclusion [4]. The two earthquake epicenters are located in the secondary faults of western part of the Lenglongling fault zone, 5 km north of the main fault. The two ends of the secondary fault converge to the main faults which compose the left-order alignment in the west and right-order alignment in the east of the left-lateral strike-slip fault, respectively [33–35]. Therefore, the structural style and stress environment are complicated at the curved part of the two ends of the secondary fault, which may mean that the local faulting activity has a diversified performance.

The fault activity characteristics at the curved parts of the strike-slip fault can be changed due to the shear extension or compression (Figure 8). The curved part of the left-order alignment along the left-lateral strike-slip fault is shown as a shear extension and could produce a normal fault and pull-apart basin. In contrast, the right-order alignment along the left-lateral strike-slip fault is shown as a shear compression and could produce folding structure and reverse fault. The epicenter of the 1986 Menyuan earthquake is located in the NW section of the secondary faults, where the segment is in a shear tensile environment due to the left-lateral strike slip. Therefore, the focal mechanism of the earthquake shows a normal fault with a tiny shear effect. In contrast, the epicenter of the 2016 Menyuan earthquake is located in the SE section of the secondary fault and this segment is in a shear compression environment. In consequence, this event shows a very strong compression effect and caused significant uplift signals near the earthquake zone in the InSAR coseismic deformation map.

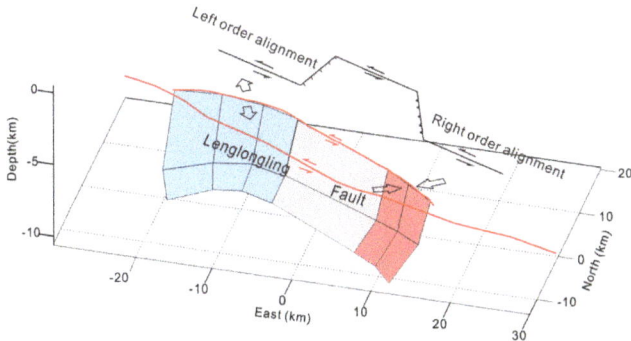

Figure 8. Three-dimensional block diagram of the geometry proposed for the Lenglongling fault. Red rectangles indicate the compression regions of the secondary fault of Lenglongling, while the blue ones indicate the tensile regions.

5. Conclusions

The surface displacements caused by the 2016 Mw 5.9 Menyuan earthquake have been derived from both the ascending and descending Sentinel-1A TOPS data. Two Sentinel-1A IW interferometric pairs show that the significant ground displacements have a maximum uplift of 7 cm in the satellite light-of-sight (LOS). A two-step inversion strategy was used to determine the fault geometry and slip distribution. The results show a two-segment shovel-shaped reverse fault model that can explain InSAR observations very well. The secondary fault of the Lenglongling fault zone should be responsible for this event. The best-fit slip model suggests that the coseismic slip is concentrated on the deeper segment of thrust fault with a strike of 134°, a dip of 43° and an average rake angle of 68°. The maximum slip of ~0.45 m occurred at a depth of ~9.5 km. The cumulative seismic moment is approximately up to 9.9×10^{17} N· m, which is equivalent to a magnitude of M_w 5.9. A joint analysis combined with the1986 Menyuan Earthquake near Lenglongling faults suggests that the western section of the secondary fault is in a shear tensile environment due to the left-lateral strike-slip of the Lenglongling fault zone, while the eastern section of the secondary fault is in a shear compression environment. The two earthquake sequences exactly reflected the left-lateral strike-slip characteristics of the Lenglongling faults zone. This showed us how to accommodate the regional tectonic deformation of the Qilian-Haiyuan tectonic zone and the stress variation characteristics of the NE Tibetan plateau.

Acknowledgments: This work was supported by Research grant from Institute of Crustal Dynamics, China Earthquake Administration under Grant (Grant numbers ZDJ2015-16 and ZDJ2015-15), National Natural Science Foundation of China under Grant (Grant number 41374050) and Major Projects of High Resolution Earth Observation of China (31-Y30B09-9001-13/15-02). We thank Wanpeng Feng for his constructive comments. The Sentinel-1A data we used were provided by the ESA. The relocation results of the aftershocks are supplied by Fan Lihua from Institute of Geophysics, China Earthquake Administration. We thank five anonymous reviewers for their detailed and thoughtful comments.

Author Contributions: Yongsheng Li Led the research work, designing the experiments and writing the first draft; Wenliang Jiang contributed to experiment implementation and geosciences result interpretation. Jingfa Zhang and Yi Luo contributed to paper writing and revision.

Conflicts of Interest: The authors declare no conflict of interest.

References

1. U.S. Geological Survey (USGS). Available online: http://earthquake.usgs.gov/earthquakes/search/ (accessed on 16 June 2016).
2. Global Centroid Moment Tensor Catalogue (GCMT). Available online: http://www.globalcmt.org/ CMTsearch.html (accessed on 16 June 2016).

3. Gaudemer, Y.; Tapponnier, P.; Meyer, B.; Peltzer, G.; Shunmin, G.; Zhitai, C.; Huagung, D.; Cifuentes, I. Partitioning of crustal slip between linked, active faults in the eastern Qilian Shan, and evidence for a major seismic gap, the 'Tianzhu gap', on the western Haiyuan Fault, Gansu (China). *Geophys. J. Int.* **1995**, *120*, 599–645. [CrossRef]

4. Xu, J.; Yao, L.; Wang, J. Earthquake source mechanisms of Menyuan Earthquake (Ms = 6.4, on 26 August 1986) and its strong aftershocks. *Northwest. Seismol. J.* **1986**, *8*, 84–86.

5. Liu, J.; Ji, C.; Zhang, J.Y.; Zhang, P.Z.; Zeng, L.S.; Li, Z.F.; Wang, W. Tectonic setting and general features of coseismic rupture of the 25 April 2015 Mw 7.8 Gorkha, Nepal earthquake. *Chin. Sci. Bull.* **2015**, *60*, 2640–2655. (In Chinese) [CrossRef]

6. England, P.; Houseman, G. Finite strain calculations of continental deformation: 2. Comparison with the India-Asia collision zone. *J. Geophys. Res.* **1986**, *91*, 3664–3676. [CrossRef]

7. England, P.; McKenzie, D. A thin viscous sheet model for continental deformation. *Geophys. J. Int.* **1982**, *70*, 295–321. [CrossRef]

8. Dewey, J.F.; Shackleton, R.M.; Chang, C.; Yiyin, S. The tectonic evolution of the Tibetan Plateau. *Philos. Trans. R. Soc. Lond. Ser. A Math. Phys. Sci.* **1988**, *327*, 379–413. [CrossRef]

9. Tapponnier, P.; Shu, Z.; Roger, F.; Meyer, B.; Arnaud, N.; Wittlinger, G.; Jingsui, Y. Oblique stepwise rise and growth of the Tibet Plateau. *Science* **2011**, *294*, 1671–1677. [CrossRef] [PubMed]

10. Staisch, L.M.; Niemi, N.A.; Clark, M.K.; Chang, H. Eocene to late Oligocene history of crustal shortening within the Hoh Xil Basin and implications for the uplift history of the northern Tibetan Plateau. *Tectonics* **2016**, *35*. [CrossRef]

11. Wang, S.; Zhang, X.; Zhang, S.; Zhang, X.; Xue, F. Characteristics of recent tectonic deformation and seismic activity in the northeastern margin of Tibetan Plateau. *Acta Geosci. Sin.* **2005**, *26*, 209–216.

12. Yuan, D.Y.; Zhang, P.Z.; Liu, B. Geometrical imagery and tectonic transformation of the late Quaternay active tectonics in northeast margin of Qinghai-Xizang plateau. *Acta Geol. Sin.* **2004**, *78*, 270–278. (In Chinese)

13. Zheng, W.; Zhang, P.; He, W. Transformation of displacement between strike-slip and crustal shortening in the northern margin of the Tibetan plateau: Evidence from decadal GPS measurements and late Quaternary slip rates on faults. *Tectonophysics* **2013**, *584*, 267–280. [CrossRef]

14. Liang, S.; Gan, W.; Shen, C.; Xiao, G.; Liu, J.; Chen, W.; Ding, X.; Zhou, D. Three-dimensional velocity field of present-day crustal motion of the Tibetan Plateau derived from GPS measurements. *J. Geophys. Res. Solid Earth* **2013**, *118*, 5722–5732. [CrossRef]

15. Gan, W.; Zhang, P.; Shen, Z.-K.; Niu, Z.; Wang, M.; Wan, Y.; Zhou, D.; Cheng, J. Present-day crustal motion within the Tibetan Plateau inferred from GPS measurements. *J. Geophys. Res.* **2007**, *112*, B08416. [CrossRef]

16. Lasserre, C.; Gaudemer, Y.; Tapponnier, P.; Mériaux, A.-S.; van der Woerd, J.; Daoyang, Y.; Ryerson, F.J.; Finkel, R.C.; Caffee, M.W. Fast late Pleistocene slip rate on the Leng Long Ling segment of the Haiyuan fault, Qinghai, China. *J. Geophys. Res.* **2002**, *107*, 2276. [CrossRef]

17. He, W.; Liu, B.; Yuan, D.; Yang, M. Research on the slip rate of the Lenglongling fault zone. *Northwest. Seismol. J.* **2000**, *22*, 90–97. (In Chinese)

18. He, W.; Yuan, D.; Ge, W.; Luo, H. Determination of the slip rate of the Lenglongling fault in the middle and eastern segments of the Qilian mountain active fault zone. *Earthquake* **2010**, *30*, 131–137. (In Chinese)

19. Li, Q.; Jiang, Z.; Wu, Y. Present-day tectonic deformation characteristics of Haiyuan-Liupanshan Fault Zone. *J. Geodesy Geodyn.* **2013**, *33*, 18–22.

20. Wegmüller, U.; Werner, C.; Strozzi, T.; Wiesmann, A.; Frey, O.; Santoro, M. Sentinel- 1 a support in the GAMMA software. In Proceedings of the Fringe 2015 Conference, ESA ESRIN, Frascati, Italy, 23–27 March 2015.

21. Goldstein, R.M.; Werner, C.L. Radar interferogram filtering for geophysical applications. *Geophys. Res. Lett.* **1995**, *25*, 4035–4038. [CrossRef]

22. Chen, C.W. Statistical-Cost Network-Flow Approaches to Two-Dimensional Phase Unwrapping for Radar Interferometry. Ph.D. Thesis, Stanford University, Stanford, CA, USA, 2001.

23. Copley, A.; Karasozen, E.; Oveisi, B.; Elliott, J.R.; Samsonov, S.; Nissen, E. Seismogenic faulting of the sedimentary sequence and laterally variable material properties in the Zagros Mountains (Iran) revealed by the August 2014 Murmuri (*E. Dehloran*) earthquake sequence. *Geophys. J. Int.* **2015**, *203*, 1436–1459. [CrossRef]

24. Solaro, G.; De Novellis, V.; Castaldo, R.; De Luca, C.; Lanari, R.; Manunta, M.; Casu, F. Coseismic fault model of Mw 8.3 2015 Illapel Earthquake (Chile) retrieved from multi-orbit Sentinel1-A DInSAR measurements. *Remote Sens.* **2016**, *8*, 323. [CrossRef]

25. Feng, W.; Li, Z.; Hoey, T.; Zhang, Y.; Wang, R.; Samsonov, S.; Li, Y.; Xu, Z. Patterns and mechanisms of coseismic and postseismic slips of the 2011 M W 7.1 Van (Turkey) earthquake revealed by multi-platform synthetic aperture radar interferometry. *Tectonophysics* **2014**, *632*, 188–198. [CrossRef]

26. Feng, W.; Lindsey, E.; Barbot, S.; Samsonov, S.; Dai, K.; Li, P.; Li, Z.; Almeida, R.; Chen, J.; Xu, X. Source characteristics of the 2015 M W 7.8 Gorkha (Nepal) earthquake and its M W 7.2 aftershock from space geodesy. *Tectonophysics* **2016**. [CrossRef]

27. Okada, Y. Surface deformation due to shear and tensile faults in a half-space. *Bull. Seismol. Soc. Amer.* **1985**, *75*, 1135–1154.

28. Harris, R.A.; Segall, P. Detection of a locked zone at depth on the Parkfield, California, segment of the San Andreas fault. *J. Geophys. Res.* **1987**, *92*, 7945–7962. [CrossRef]

29. Bürgmann, R.; Ayhan, M.E.; Fielding, E.J.; Wright, T.J.; McClusky, S.; Aktug, B.; Demir, C.; Lenk, O.; Türkezer, A. Deformation during the 12 November 1999 Düzce, Turkey, earthquake, from GPS and InSAR data. *Bull. Seismol. Soc. Am.* **2002**, *92*, 161–171. [CrossRef]

30. Sreejith, K.M.; Sunil, P.S.; Agrawal, R.; Saji, A.P.; Ramesh, D.S.; Rajawat, A.S. Coseismic and early postseismic deformation due to the 25 April 2015, Mw 7.8 Gorkha, Nepal, earthquake from InSAR and GPS measurements. *Geophys. Res. Lett.* **2016**, *43*. [CrossRef]

31. Fang, L.; Wu, J.; Wang, W. Relocation of mainshock and aftershock sequences of Ms7.0 Sichuan Lushan earthquake. *Chin. Sci. Bull.* **2013**, *58*, 1901–1909. [CrossRef]

32. Zhao, W.; Mechie, J.; Feng, M.; Si, D.; Xue, G.; Su, H.; Song, Y.; Yang, H.; Liu, Z. Cenozoic orogenesis of the Qilian Mountain and the lithosphere mantle tectonic framework beneath it. *Geol. China* **2014**, *41*, 1411–1423. (In Chinese)

33. Taylor, M.; Yin, A. Active structures of the Himalayan-Tibetan orogen and their relationships to earthquake distribution, contemporary strain field, and Cenozoic volcanism. *Geosphere* **2009**, *5*, 199–214. [CrossRef]

34. Wang, J.; Qin, B.; Dang, Q. A study on the fracture process of the Menyuan M6.4 earthquake occurred on 26 August 1986. *North China Earthq. Sci.* **1992**, *10*, 25–33.

35. Li, Y.; Liu, X. Activity analysis of faults around Qilianshan before the 2016 Menyuan Ms6.4 earthquake. *J. Geodesy Geodyn.* **2016**, *36*, 288–293.

remote sensing

MDPI

Article

Deformation and Related Slip Due to the 2011 Van Earthquake (Turkey) Sequence Imaged by SAR Data and Numerical Modeling

Elisa Trasatti *, Cristiano Tolomei, Giuseppe Pezzo, Simone Atzori and Stefano Salvi

Istituto Nazionale di Geofisica e Vulcanologia, Rome 00143, Italy; cristiano.tolomei@ingv.it (C.T.);
giuseppe.pezzo@ingv.it (G.P.); simone.atzori@ingv.it (S.A.); stefano.salvi@ingv.it (S.S.)
* Correspondence: elisa.trasatti@ingv.it; Tel.: +39-06-5186-0349

Academic Editors: Zhenhong Li, Magaly Koch and Prasad S. Thenkabail
Received: 5 February 2016; Accepted: 14 June 2016; Published: 22 June 2016

Abstract: A Mw 7.1 earthquake struck the Eastern Anatolia, near the city of Van (Turkey), on 23 October 2011. We investigated the coseismic surface displacements using the InSAR technique, exploiting adjacent ENVISAT tracks and COSMO-SkyMed images. Multi aperture interferometry was also applied, measuring ground displacements in the azimuth direction. We solved for the fault geometry and mechanism, and we inverted the slip distribution employing a numerical forward model that includes the available regional structural data. Results show a horizontally elongated high slip area (7–9 m) at 12–17 km depth, while the upper part of the fault results unruptured, enhancing its seismogenic potential. We also investigated the post-seismic phase acquiring most of the available COSMO-SkyMed, ENVISAT and TERRASAR-X SAR images. The computed afterslip distributions show that the shallow section of the fault underwent considerable aseismic slip during the early days after the mainshock, of tens of centimeters. Our results support the hypothesis of a seismogenic potential reduction within the first 8–10 km of the fault through the energy release during the post-seismic phase. Despite non-optimal data coverage and coherence issues, we demonstrate that useful information about the Van earthquake could still be retrieved from SAR data through detailed analysis.

Keywords: SAR interferometry; multi aperture interferometry; Van earthquake; remote sensing; numerical modelling; inverse methods; coseismic deformation; post-seismic deformation

1. Introduction

A Mw 7.1 earthquake struck the Eastern Anatolia at 10:41 a.m. on 23 October 2011 (Figure 1). The epicenter was approximately located at 38.76°N, 43.36°E (Turkish Kandilli Observatory and Earthquake Research Institute, KOERI) at a depth of about 16 km, with considerable spatial variations of centroid and aftershocks solutions among different international seismological institutions, of the order of tens of kilometers. The earthquake, located close to the Tabanli village, about 20 km N-NE the city of Van (about 400,000 inhabitants), caused significant losses and casualties. The focal mechanism indicates an ENE-WSW thrust fault, consistent with the trend observed in eastern Turkey, SW of the Karliova junction along the Arabian plate boundary [1], with an additional minor left-lateral component. The epicenters of the ~1400 M > 3 (5300 events M > 1) aftershock registered until the end of January 2012 are ~NE-SW aligned. The focal mechanisms of the Mw > 5 aftershocks confirm the dominant thrust component (Figure 1) apart from the M 5.1 event of 29 October 2011 which shows a right-lateral strike-slip mechanism. Another earthquake of Mw 5.7 (labeled Edremit-Van earthquake according to KOERI) took place on 9 November 2011. It was located offshore, near the town of Edremit, South of Van, and increased the level of structural damage.

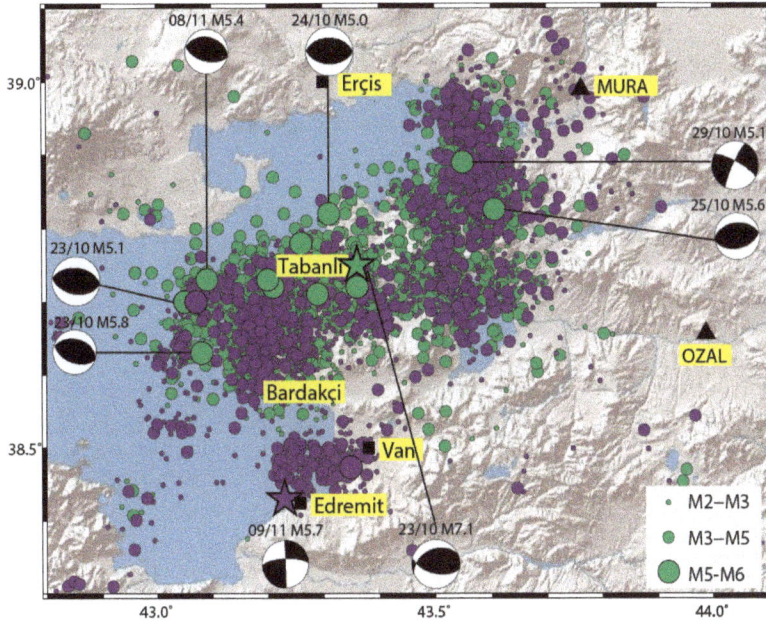

Figure 1. Map of the Mw 7.1 Van earthquake (green star) and related aftershocks until the end of 2011. The Edremit-Van earthquake epicenter is indicated with the purple star. The aftershocks (green before the Edremit-Van earthquake, purple after) are from [2] and KOERI. The available focal mechanisms of M > 5.0 events from Global-CMT earthquake catalog and [3] are shown. Near field GPS stations (MURA and OZAL) are indicated by black triangles.

The Van earthquake occurred along a fault that was not previously mapped among the active faults of the region [4], but has been included afterwards [5]. The rupture induced aftershocks on secondary structures, several surface tensional cracks of tens of centimeters, landslides and liquefactions (e.g., [6,7]). Since the length of the surface ruptures does not justify the mainshock magnitude, part of the main fault of the Van earthquake should be considered as blind. Furthermore, along with the main thrust plane, secondary back thrust and left/right lateral fault ruptures occurred [7]. It is not clear whether these fault ruptures are directly connected to the main rupture at depth or not.

Initially, the seismic fault was imaged by [8] in a preliminary seismic data inversion. Results show a maximum slip patch of 4 m at the hypocentral depth (16–20 km), mainly elongated up-dip toward SW. Irmak *et al.* [9] used 35 teleseismic stations to obtain a rupture with bilateral propagation nucleating at greater depths, ~20 km. Other teleseismic inversions described a rupture dominated by failure of a major asperity located up-dip and SW of the hypocenter [10]. The rupture history results indicated above are not in accordance with each other, due to the few seismic recordings in the region and the limited spatial coverage of the seismic stations. The main consequence is a poor estimate of the seismic hazard of the earthquake area. Gallovič *et al.* [11] presented different models to illustrate the broad variability of possible rupture propagation, depending on the unfavorable seismic data constraints. Akinci & Antonioli [12] used a stochastic approach to overcome the described limitations in order to study the characteristics of ground motion. Fielding *et al.* and Elliott *et al.* [3,13] presented two comprehensive studies on the Van earthquake using geodetic, seismological and field observations. The rupture propagated along the dip direction of the fault from the waveform modeling, with two maximum slip zones close and above the hypocenter [3]. The seismological and geodetic inversions both constrain the slip in an area extending about 30 km along strike. Elliott *et al.* [13] used a pair of en echelon NW, 40°–54° dipping faults, finding the slip distributed within two separated large concentration zones (one for each fault). Results show the lack of significant slip above 8 km depth, implying a potential future rupture in the shallower part of the faults [3,13–15]. However, GPS data covering 1.5 years after the mainshock to model afterslip, suggested a lower likelihood for a large earthquake to occur in the SW shallow sector of the fault [16].

In this paper, we have analyzed Synthetic Aperture Radar (SAR) data in order to image the coseismic and post-seismic deformation of the Van earthquake, mapping also coseismic fractures and landslides. We have constrained the slip distribution on the main fault using Line Of Sight (LOS) and azimuth (~NS) components of ground displacement obtained from SAR data. We used a numerical fault modeling approach (by means of Finite Elements (FE)) in order to consider the elastic heterogeneities of the Van province. The 3D model was derived from tomography and receiver functions studies, and allowed us to investigate the possible structural control on the retrieved fault-slip source. We focused on the contribution of the ground deformation observations to the understanding of the main features of the Van earthquake sequence. Some of the SAR data employed in our work were previously published (e.g., [3,13–15]), but in this study we performed a detailed analysis of most of the available SAR data in order to give insights to the seismic hazard of the region despite non-optimal data coverage and coherence issues.

2. Tectonic Settings and Structural Data

The Anatolian plateau is geologically complex, and is dissected into numerous seismogenic faults. The compound lithospheric structure is accompanied by large seismic wave velocities variations, and the seismic activity is intense along highly heterogeneous zones (e.g., [17]). Unfortunately, little information is available at smaller scale, e.g., for the epicentral area of the Van earthquake. The Lake Van basin is located near the Karliova triple junction between the Anatolian microplate and the Eurasian and the Arabian plates, and this allowed materials upwelling from the Earth's mantle to accumulate in Lake Van and in the nearby volcanic area of Nemrut volcano [18]. Lake Van was the drill site of an International Continental scientific Drilling Program (ICDP) called PALEOVAN [19,20]. The seismic survey related to ICDP PALEOVAN revealed tephra deposits due to the historical activity of Nemrut volcano, and very localized features, such as clinoforms extending few hundreds of meters. However, the physical properties reported are related to very shallow depths, and therefore unable to characterize the whole seismogenic zone. Salah *et al.* [20] computed a 3D tomography of the crust beneath the eastern Anatolia (between latitudes 37.0°N–41.0°N and longitudes 38.0°E–44.5°E) from *P*-waves and *S*-waves (*Vp* and *Vs*, respectively). The Lake Van area is associated with a heterogeneous velocity structure. High *Vp* and *Vs* anomalies are constrained near the surface, while low velocity zones are widely present at 20–30 km depth. The Moho depth in the Van Province is estimated to be at 42–44 km depth [17]. The low velocity anomalies are interpreted as being caused by hot lithosphere resulting from the Arabian-European plates collision, while high *Vp* and low *Vs* (corresponding to a high Poisson ratio anomaly) imply the presence of fluids ascending upward from the hot lithosphere.

We took advantage from the numerical modeling approach to include in our models the elastic structure as imaged by the tomographic study [20]. In addition to these data, we considered *Vs* values in the epicentral area resulting from joint inversion of teleseismic receiver functions and surface waves by [17] at the stations VANB and CLDR.

3. Coseismic Ground Deformation and Modeling

3.1. Coseismic Geodetic Data

We processed a total of 32 SAR images (Table S1 in the Supplementary Data) from both X-Band (COSMO-SkyMed and CSK) and C-Band (ASAR ENVISAT, and ENV) satellites along the descending orbit using the software packages Sarscape [21] and Gamma [22]. Unfortunately, the ascending orbit data could not be considered since the first pre-seismic image was too far in time, *ca.* 3 years before the mainshock, and was not suitable to produce coherent interferograms. It is also worth to mention that an ascending coseismic TERRASAR-X (TSX) pair exists [23], but it shows serious coherence issues due to the very large temporal baseline (pre-seismic image acquired on the 2009).

We adopted two different interferometric techniques to retrieve the coseismic displacement field, the classical Differential SAR Interferometry (InSAR) and Multi Aperture Interferometry (MAI) [24]. These techniques allow for the cross-validation of independent results, although with different spatial resolution and accuracy. The InSAR images were generated with 10 m pixel ground resolution for CSK (multilooking factor equal to 5 in azimuth and range) and 90 m for ENV (multilooking factor equal to 20 for the azimuth and 4 for the range), using a 30 m DEM from the ASTER mission to remove the topographic phase contribution and for geocoding. A set of ground control points was chosen from highly coherent regions located outside the high displacement area, in order to estimate and remove the contribution of orbital uncertainties. A filtering step was performed, based on the Goldstein algorithm [25]. Finally, the phase unwrapping was performed applying the Delauney minimum cost flow algorithm [26] to minimize possible phase jumps. The MAI analysis was carried out using a cross-correlation window size of 400 m × 400 m. The characteristics of the coherent SAR interferograms are sketched in Figure 2 and listed in Table 1, while the measured coseismic displacement field is shown in Figure 3. The range measurements (CSK1, Figure 3a) cover a large part of the coseismic displacement field and show maximum LOS values of less than 1 m towards the satellite in the North, decreasing southwards. The azimuth measurements (Figure 3b) confirm the compressional kinematics (northern sector negative and southern sector positive) with extreme values located in the same area of the LOS highest values. The displacement maps obtained from the two ENV interferograms from adjacent tracks show very different coherence. The first one (ENV1, Figure 3c) captures the eastern part of the ground displacement with a good spatial coverage in spite of the large baselines (Figure S1), showing 44 cm maximum LOS displacement. The second interferogram is much less coherent (Figure S2). The phase unwrapping was very problematic and we eventually resolved to unwrap the clearest fringes using a manual procedure involving visual interpretation of the fringe continuity in three different areas of the image. The analysis was carried out given the important constraint provided by the coverage of the very near field coseismic area. This procedure generated ground displacement values (ENV2, Figure 3d) affected by a somewhat higher uncertainty, estimated as 1–2 fringe uncertainty. The superposition with CSK1 in the western part of the image was used to choose an offset for the unwrapped phase, and it was useful for a comparison.

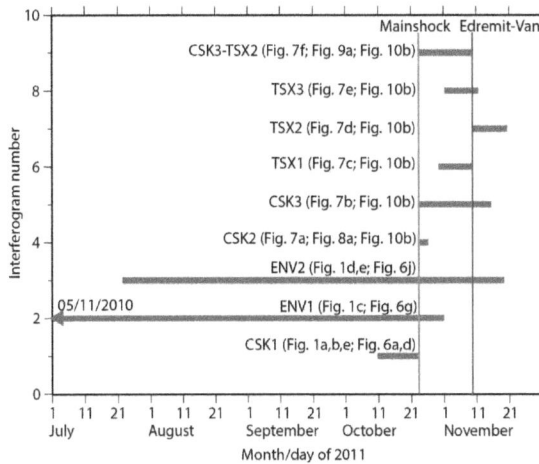

Figure 2. Graphical representation of the interferograms' time span during the coseismic and post-seismic phases.

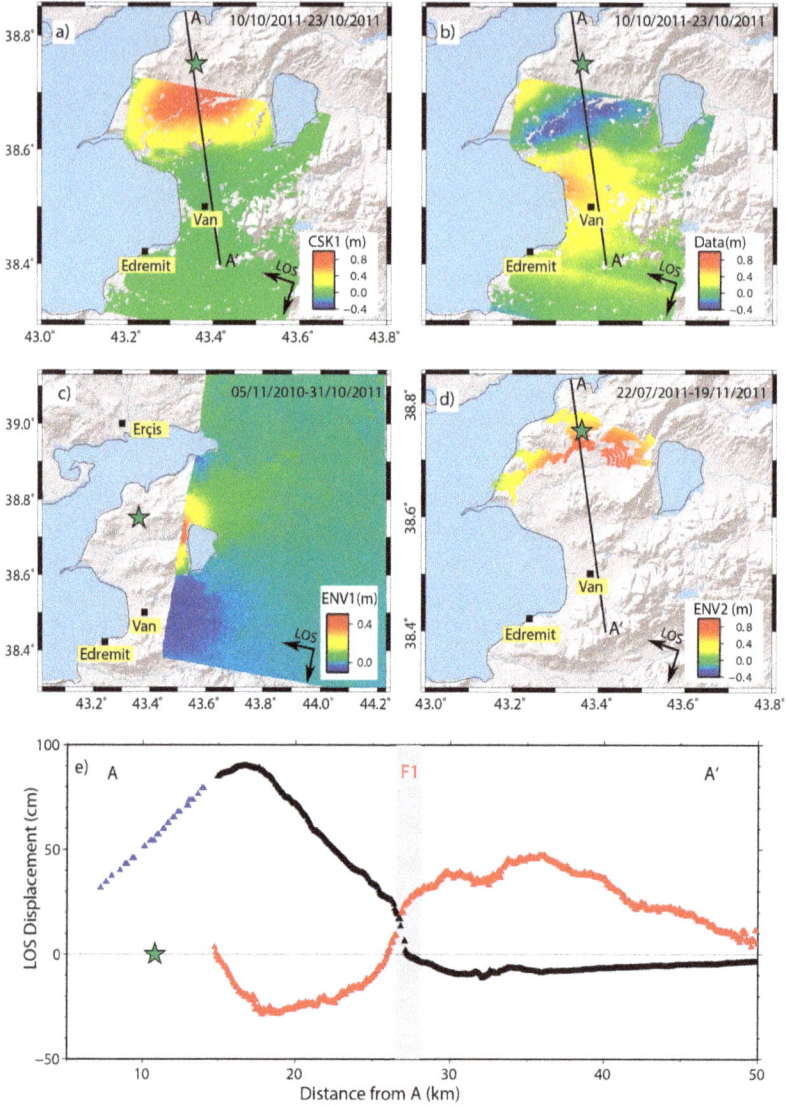

Figure 3. InSAR results: (**a**) CSK range results (*i.e.*, LOS direction); CSK1; (**b**) CSK azimuth results (positive Northwards); (**c**) ENV1, LOS displacements; and (**d**) ENV2, LOS displacements. SAR data details can be found in Table 1; (**e**) Data along the profile AA': CSK1 (range measurements, black), CSK (azimuth measurements, red), and ENV2 (LOS measurements, blue). The grey vertical band indicates the fault trace (F1) and its uncertainty (~1 km). The green star is the hypocenter.

Table 1. Characteristics of the interferograms used to map the coseismic and post-seismic phases.

Mission	Orbit	Acquisition Dates	Phase [a]	Perpendicular Baseline (m)	Temporal Baseline (days)	Incidence Angle (°)	Fringe Rate (cm)
COSMO-SkyMed (CSK1)	Descending	10 October 2011 23 October 2011	Co	192	13	29	1.6
ENVISAT (ENV1)	Descending	5 November 2010 31 November 2011	Co	633	360	41	2.8
ENVISAT (ENV2)	Descending	22 July 2011 19 November 2011	Co	270	120	41	2.8
COSMO-SkyMed (CSK2)	Descending	23 October 2011 26 October 2011	Post	307	3	29	1.6
COSMO-SkyMed (CSK3)	Descending	23 October 2011 15 November 2011	Post	79	23	29	1.6
Terrasar-X (TSX1)	Descending	29 October 2011 9 November 2011	Post	292.2	11	26.3	1.6
Terrasar-X (TSX2)	Descending	9 November 2011 20 November 2011	Post	11.5	11	26.3	1.6
Terrasar-X (TSX3)	Ascending	31 October 2011 11 November 2011	Post	190	11	33.2	1.6

[a] Co is referred to the coseismic phase, Post is the post-seismic phase.

The profile AA' (Figure 3e), across the extreme LOS values and close to the epicenter, shows a continuous pattern of ENV2 and CSK1 datasets. The grey vertical band (F1) reports the fault trace position (within errors) resulting from our following geophysical inversion and corresponds to the steepest ground displacement gradient. Both the range and azimuth results from CSK show a compressive regime and a halfway discontinuity. The spatial continuity between the CSK1 and ENV1 unwrapped interferograms is shown in Figure S3.

3.2. Fractures and Landslides from InSAR Data

The Van earthquake caused some surface ruptures, mapped during field work and summarized by [7] and references therein. We analyzed the high resolution CSK wrapped InSAR image in order to identify phase jumps and fringe discontinuities corresponding to surface fractures. We distinguished two families of subparallel discontinuities ~ENE oriented and located at about 38.57°N (FA) and 38.64°N (FB), as shown in Figure 4. The group FA is approximately located along the surface projection of the mainshock fault plane, and it corresponds to the main surface rupture, approximately 8 km long, as described in [7]. Another fringe discontinuity group, FB, occurred close to the lobe of largest coseismic LOS deformation. These discontinuities are compatible with a set of probable ruptures related to secondary left-lateral strike-slip structures activated during the Van seismic sequence, as suggested by [7] and shown in their Figure 4.

Additional anomalous fringe patterns and/or decorrelation areas can be found in Figure 4A. Some of these patterns could be due to surface phenomena related to the earthquake occurrence, like landslides. Indeed, the coseismic shaking could activate landslides characterized by a deep or shallow rupture surface (e.g., [27] and references therein). By analyzing the CSK wrapped interferogram, we identified a landslide type called *Deep-seated Gravitational Slope Deformation* (DGSD), located about 43.5123°E, 38.6244°N (landslide 1 in Figure 4A–C). This type of phenomena is normally characterized by variable deformation rates from less than 1 mm to few centimeters per year, and they could be triggered by seismic events or heavy rainfall. Moreover, DGSDs are often revealed with geomorphological evidences, such as double crest ridges, counterscarps or gravitational half-grabens. Some of these features, such as crest ridges and counterscarps, were also recognized for the landslide

1 of this work. In the CSK interferogram, the area involved by the DGSD shows a different fringe pattern. In particular, the spacing between the fringes into the landslide body appears to be larger (and chaotic) with respect to the area nearby the landslide, where the fringe spacing is regular and follows the pattern due to the coseismic displacement (Figure 4B). This is also observable in the along-slope displacement profile (green stripe in Figure 4C) from the CSK unwrapped phase, where we observe the regular coseismic displacement distorted by a well-localized displacement pattern related to the coseismic activation of the DGSD. Indeed, the signal corresponds to the landslide body that moves downstream as an independent mass, and it is characterized by a different dislocation of the upper part of the landslide body with respect to the accumulation zone. In addition, North of landslide 1 there are three small areas where a loss of coherence occurs (Figure 4A,B). The regular fringe pattern is sharply interrupted in correspondence of these areas: this phenomenon could be associated to sudden motion temporally occurred between two SAR acquisitions. By comparing the InSAR fringes with optical imagery, we identify three landslides (labeled 2, 3 and 4 in Figure 4A,B,E) located near the epicenter area, suggesting the occurrence of additional coseismic displacement due to the seismic shaking.

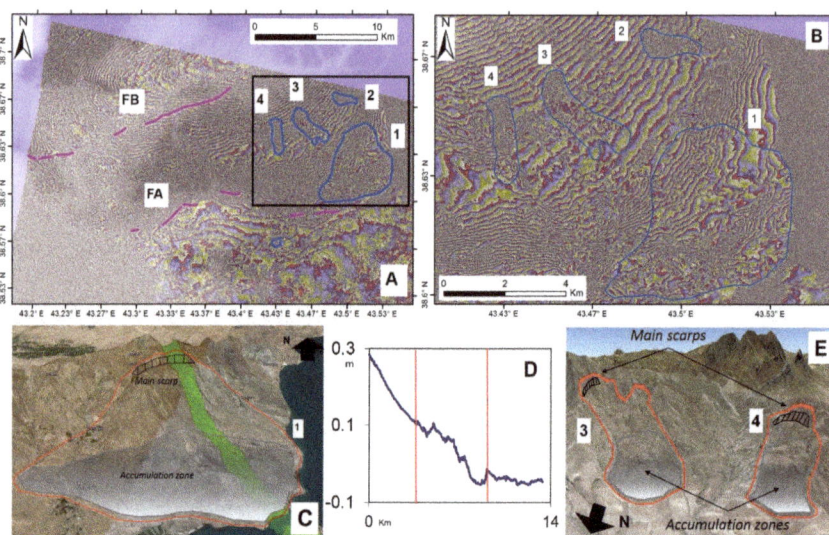

Figure 4. Fractures and landslides from InSAR: (**A**) wrapped phase of the CSK interferogram, where some crack lineaments (FA and FB, purple lines) and landslides (blue lines) are recognized; (**B**) zoomed view of the landslides; (**C**) Google Earth view from South of landslide 1 (red lines); and in green the displacement profile shown in (**D**), where the two red bars indicate the DGSD body; (**E**) Google Earth view from NNW of landslides 3 and 4, indicated in red.

3.3. Model of the Mainshock Source

The geodetic data collected were used to retrieve the coseismic slip distribution of the mainshock fault through an inversion procedure. We adopted a two-step inversion framework, initially optimizing the source parameters [28] using a non-linear inversion algorithm, followed by a Bayesian study on the inverted parameters (Neighborhood Algorithm (NA) [29,30]). This procedure settles the most probable ensemble of solutions, instead of a single best-fit model. In the second step, the defined fault plane is accepted for FE modeling with the commercial code [31]. The slip distribution was constrained through a linear inversion in the heterogeneous FE forward model, adopting a procedure already tested in seismic source inversions [32].

The SAR datasets were subsampled in a regular grid with a step of 1 km (except for ENV1, subsampled with a step increasing according with the epicentral distance), obtaining a total of 5314 data points. In addition to the SAR data, we considered the coseismic GPS data belonging to the Turkish CGPS network. The stations are spread in a region of 350 km × 450 km, and only two stations (MURA and OZAL) out of 16 in total are included in the largest ENV1 frame, *i.e.*, within 50 km from the epicenter (Figure 1).

We carried out several tests during the non-linear inversions in order to balance the weight of the SAR and GPS data (e.g., [32]), and to limit the trade-offs among fault parameters (e.g., top left corner coordinates and strike, fault width and depth). In the preferred non-linear inversion configuration, the GPS misfit weighs 5% respect to the SAR misfit, due to the very high signal-to-error ratio of the GPS dataset and the presence of only two near-field GPS benchmarks. The fault width and dip were fixed at 18 km and 50°, respectively (e.g., [8,9], and within the ranges proposed by [3,13]). Both the fault width and dip could be estimated from accurate aftershock locations in 3D, but even in the relocated catalogue from [2], the seismicity is too scattered for this purpose (Figure 5b). A down-dip rupture width of 20 km for a M 7.1 reverse slip earthquake was estimated [33].

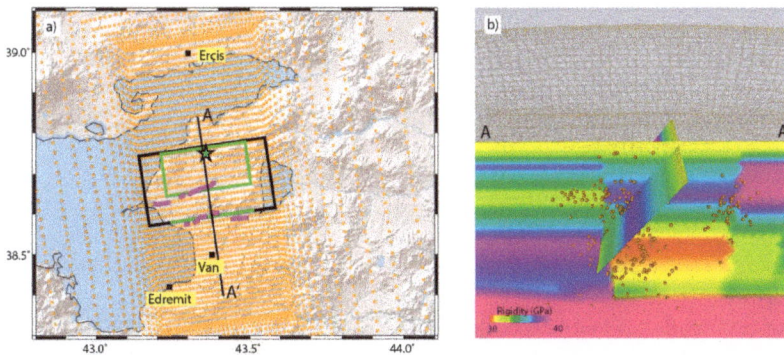

Figure 5. FE model of the Van Earthquake: (**a**) Top view of the model. The uniform slip fault is reported by green line while the fault plane used to retrieve the slip distribution is indicated by the black line. Surface ruptures constrained by InSAR are indicated by purple lines. The green star is the hypocenter and the orange dots are the surface nodes of the FE model; (**b**) Rigidity distribution on the section AA′ (see Figure 3) and on the fault plane within the heterogeneous FE model, view from West. The external edges of the FE model are shown with ochre lines. The seismicity within few kilometers from the section is reported by ochre spheres.

The resulting values of the free parameters are reported in Table 2, while the Posterior Probability Density (PPD) functions are shown in Figures S4 and S5. The source parameters are well constrained since the PPDs have narrow bell-shaped distributions. The uncertainties of the obtained parameters are taken from the half-widths of the distributions themselves. The length (L = 24 km) is the worst constrained parameter and shows trade-offs with the Easting, Northing and slip of the fault. This is not surprising since minor variations of the fault position may be accommodated by small adjustments in the fault length and slip amount, in order to reproduce the surface displacement pattern. A minor trade-off between the strike angle and Northing is also to be mentioned. Data inversion shows that the thrust fault mechanism has a minor left-lateral component, being the rake equal to 72°, as expected from the focal mechanism (Figure 1). The fault trace, obtained extending the fault plane to the free surface, follows the surface fractures (Figure 5a) and corresponds to the displacement discontinuity shown in the profile in Figure 3e (fault F1). The computed scalar seismic moment is 5.5×10^{19} Nm, using a rigidity value of 35 GPa (average of fault rigidities depicted in Figure 5b), corresponding to Mw 7.1. The error associated to the scalar moment is estimated from its PPD distribution as 10%–15%

its mean value (Figure S5). All the source parameters constrained are in accordance, within their uncertainties, with previous findings, e.g., the InSAR inversions by [13]. Residuals are reported in Figures S6 and S7.

Table 2. Fault parameters retrieved by non-linear inversion.

Latitude [a,b]	Longitude [a,b]	Depth [a,b] (km)	Length [b] (km)	Width [c] (km)	Strike [b] (°)	Dip [c] (°)	Slip [b] (m)	Rake [b] (°)
38.676	43.506	8.7 ± 0.5	24.0 ± 1.1	18	263 ± 4	50	3.6 ± 0.5	72 ± 5

[a] The fault position (latitude, longitude and depth) is referred to the top left corner; [b] The standard deviation of every parameter is estimated from the half-width of the PPD distributions (Figure S4). The standard deviations retrieved are 800 m for the latitude and 600 m for the longitude; [c] The parameter is fixed from literature (see text for details).

We then proceeded to constrain the slip distribution in a 3D numerical model, employing the fault geometry determined above. The fault dimensions were enlarged up to L = 36 km and width W = 30 km, maintaining the same dip, strike and trace center, and extending the plane till the surface. The 3D FE model is made of ~200,000 8-node brick elements (partial layout depicted in Figure 5). The whole FE cylindrical domain has a diameter of 440 km and height 170 km, to avoid undesirable boundary effects. The grid resolution is 1 km in the fault near field, and increases up to 10-20 km in the far field (bottom and edges). The fault plane is subdivided into patches of 2 km side. The elastic structure of the FE model was computed from the Vp and Vs data resulting from tomography [20] and receiver functions studies [17]. Although the tomography by [20] is one of the few tomographic studies of the region, it is rather coarse for our purpose since it is referred to a regional scale, with horizontal resolution of 20–30 km and layers at 4/12/25/40/55 km. The velocities were converted into elastic parameters using a density profile linearly increasing with depth, as the Vp increases, $\rho = 541 + 360Vp$ [34]. Each element of the grid was characterized by independent constants without layering approximation [32]. The FE domain assumed the following values: rigidity 20 GPa $< \mu <$ 63 GPa, Poisson coefficient 0.17 $< \nu <$ 0.33 and density 2400 kg/m^3 $< \rho <$ 3300 kg/m^3. The fault plane was characterized by a smaller variability of elastic constants: 30 GPa $< \mu <$ 40 GPa, 0.25 $< \nu <$ 0.31 and 2700 kg/m^3 $< \rho <$ 2900 kg/m^3. Similar results were also obtained along the AA′ profile perpendicular to the fault strike (Figure 5b). Once the FE model was set up, elementary Green's Functions were computed by applying unitary slips on each patch separately, and the obtained surface displacements were recorded in a matrix. In this way, the slip distribution was optimized in the heterogeneous medium. The full procedure was described in [32], while the slip distribution was obtained through a linear inversion procedure based on the singular value decomposition (e.g., [35]). The slip uncertainty was calculated according to the standard rules for uncertainty propagation $cov_m = G^{-g}cov_d G^{-gT}$, where cov_m and cov_d are the variance/covariance matrices of the observed data and model parameters, respectively, and G^{-g} is the generalized inverse of the linear system. The standard deviation of the slip is the root of the cov_m diagonal values.

The slip distribution is characterized by a main zone of slip concentration expanding along strike between 12 km and 17 km depth (Figure 6a). The high slip zone is horizontally drop-shaped and extends up-dip from the hypocenter. It becomes thinner and shallower in the SW part of the fault. The maximum slip amounts to 8.4 m, and no significant slip is retrieved at shallow depths, above 7–9 km depth, in accordance with previous findings (e.g., [13]). The fractures of the group FB detected in Section 3.2 correspond to the vertical projection of the upper limit of the slip area in the SW part of the fault. The high slip gradients at the edges of this area may have concurred to the generation of these fractures. The errors associated to the slip distribution amount to few centimeters at shallow depths and increase up to more than 1 m at the fault bottom (Figure 6b). The area of high slip is characterized by uncertainties of 20–30 cm. The scalar seismic moment, determined by taking into account the actual rigidity associated to each patch within the FE model, is 5.0×10^{19} Nm, equivalent to Mw 7.1.

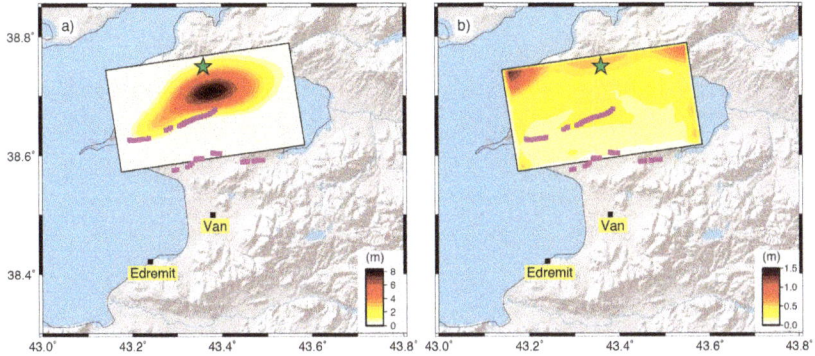

Figure 6. Fault slip distribution retrieved from geodetic data: (**a**) fault slip distribution; and (**b**) error associated. The surface fractures are reported by purple lines, and the Van earthquake epicenter by the green star.

Figure 7 shows comparisons between data and model. The datasets are rather heterogeneous, being composed by four coseismic displacement maps having different characteristics in terms of spatial coverage and associated errors, plus the GPS displacements. The whole pattern of the LOS displacements is generally reproduced and the agreement between data and model is within the related errors with percentages of 83.4% (CSK1), 99.8% (ENV1), and 95.6% (ENV2). Larger differences are observed between the azimuth direction data and model (71.9% of residuals are within errors) and GPS (only 52% within errors, full coverage residual map in Figure S8). The bad fit with GPS data is due to the regional scale of the GPS network, extending behind the coseismic far field, and their reduced weight in the misfit function. Even if positive residuals are found near Van, the NS data help to reduce the single orbit constraint of CSK1, ENV1 and ENV2. This is particularly compelling due to the specific fault orientation (~EW) of the Van earthquake fault.

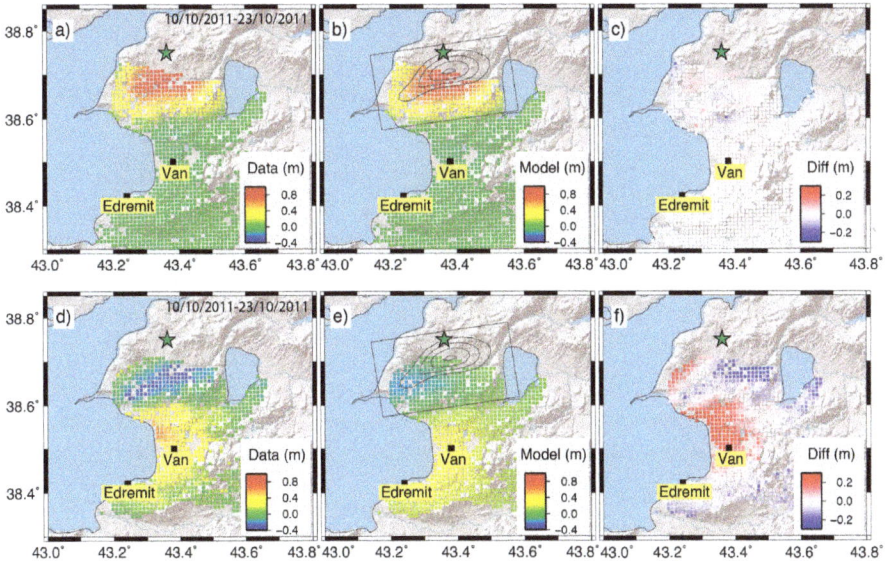

Figure 7. *Cont.*

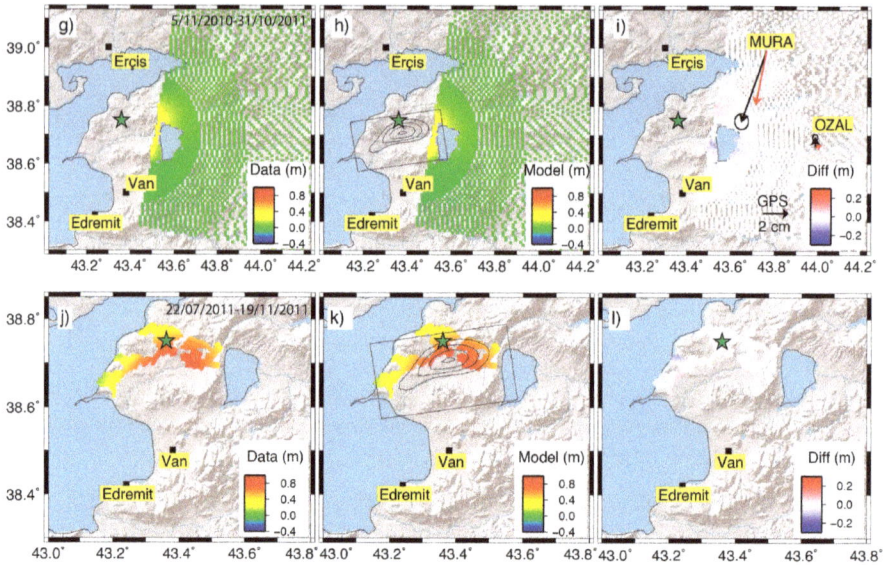

Figure 7. Comparisons between observed data (first column) and predictions (second column) related to the coseismic displacements of the Van earthquake. The residuals are observed minus modeled data (third column): (**a–c**) CSK1 (1210 data points); (**d–f**) CSK azimuth data (891 points); (**g–i**) ENV1 (2317 data points), near field GPS displacement vectors are also reported (black, observed; red, computed); and (**j–l**) ENV2 (896 data points). The green star is the hypocenter while the fault embedded in the FE model is indicated in black, with slip contour each 2 m.

4. Post-Seismic Ground Deformation and Modeling

4.1. Post-Seismic Geodetic Data

Table 1 and Figure 8 report the CSK and TSX image pairs used to measure the post-seismic displacements, selected from a larger number of images, including ENV (Table S1). We exploited the SAR dataset (Table S1) by computing several post-seismic pairs. Complete multi-temporal processing was not possible because of the low number of available acquisitions. The main problem with the post-seismic InSAR results was the interferograms' decorrelation due to the snow coverage and atmospheric effects. To reduce the influence of these limitations, we selected a CSK pair with a temporal baseline of only three days (and about 300 m for the normal baseline), whose master is 23 October 2011. In this way, we minimized the decorrelation contribution due to the temporal baseline. Moreover, we reduced the decorrelation noise by multilooking with a factor 11 in azimuth and in range, in order to get a pixel ground dimension of 25 m. Once the differential interferogram was computed, filtering and phase unwrapping were performed similarly to the coseismic data analysis. The obtained descending interferogram (CSK2, Figure 8a) shows a belt of positive values (up to 11 cm LOS) in the hanging wall, close to the fault trace. Some residuals due to atmospheric artifacts are present in the southern part of the interferogram, quite far from the high displacement area. Few centimeters of negative deformation are found in the northern part of the image. Another CSK interferogram is between the mainshock and 15 November 2011 (CSK3, Figure 8b), with longer temporal baseline and showing values up to 16 cm LOS. However, the Edremit-Van earthquake (9 November 2011) is included in this temporal span, and the displacement pattern located South of Van may be attributed to this Mw 5.7 event.

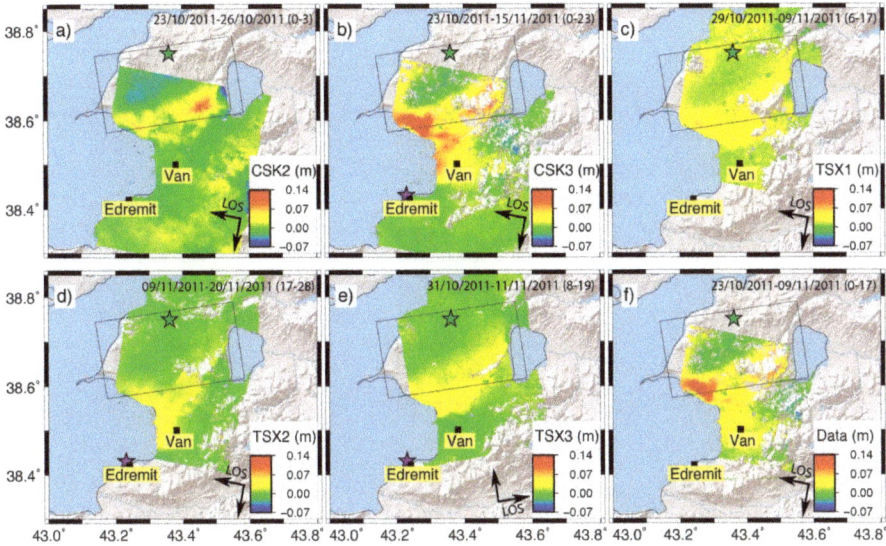

Figure 8. Post-seismic InSAR results from CSK and TSX satellites: (**a**) CSK2; (**b**) CSK3; (**c**) TSX1; (**d**) TSX2; (**e**) TSX3; and (**f**) difference between CSK3 and TSX2 (see text and Table 1 for details). Master/slave dates are indicated, along with the days after the mainshock in brackets. The green star is the mainshock epicenter and the purple star is the Edremit-Van earthquake epicenter (shown only if included in the temporal baseline).

From the exploitation of the TSX dataset, we obtained two descending orbit interferograms spanning from few days after the Van earthquake to few hours before the Edremit-Van earthquake (TSX1, Figure 8c) and from before the Edremit-Van earthquake to 20 November 2011 (TSX2, Figure 8d). We use the same multilook factor as CSK, 11, in range and azimuth. TSX1 shows a pattern similar to CSK2 and CSK3, with highest LOS displacements close to the shore of Lake Van, North of Van. The signal in TSX2 between Van and the shore may be associated to the displacement due to the Edremit-Van earthquake. Only one pair of ascending TSX images provided good results in terms of coherence (TSX3, Figure 8e). The map shows displacements strongly affected by both the post-seismic deformation of the Van earthquake and the deformations due to the Edremit-Van earthquake. As a final remark, we exclude that the observed signals are due to atmospheric artifacts since the post-seismic deformation is found in all the interferograms in Figure 8.

In order to obtain a better representation of the temporal evolution of the Van earthquake post-seismic deformation, we attempted to remove the geodetic signal of the Edremit-Van earthquake by subtracting the TSX2 displacements from the CSK3 displacements. The CSK3 image spans from the mainshock to 15 November 2011, while the TSX2 master image was acquired few hours before the Edremit-Van earthquake and the slave image is dated 20 November 2011. In this process we assumed that the CSK and TSX satellites had the same LOS (there are few degrees of difference in the looking angles) and that the five-day difference between the two slave images carried negligible post-seismic displacement. In this way, we obtained a map of 17-day displacements from the mainshock to the Edremit-Van earthquake, excluding the latter. The result, shown in Figure 8f, has an oblique belt of positive displacements near the fault trace, similarly to the three-day post-seismic interferogram CSK2, to TSX1 and to the original CSK3 (Figure 8a–c). High positive displacements are located close to the Lake Van shoreline, near the town of Bardakçi, for a maximum amount of 14 cm. The area corresponds to the place of surface ruptures as documented by [7], and confirmed by this work.

4.2. Afterslip Modeling

To study the post-seismic phase of the Van earthquake we assumed that the ground displacements were generated by the afterslip on the coseismic fault. The afterslip distributions are computed by means of the FE model adopted in the coseismic phase. The three-day afterslip constrained by CSK2 (Figure 9) shows a shallow distribution that extends from Lake Erçek, where it reaches ~30 cm, toward the WSW close to Lake Van. The high slip patch is shallow, between 1 km and 7 km, in opposition to the coseismic results in Figure 6. The slip distribution uncertainty is shown in Figure S9. The geodetic moment, computed using the actual rigidity values on the fault, amounts to 9.8×10^{17} Nm, corresponding to an equivalent Mw 5.9 (similarly to [13]). A band of positive residuals is found near and across the fault trace (Figure 9c), while in the northern area (hanging wall) the residuals are lower and negative. The residuals are at 78.9% within the data error (1.5 cm). These residuals may imply different fault geometry at shallow depths.

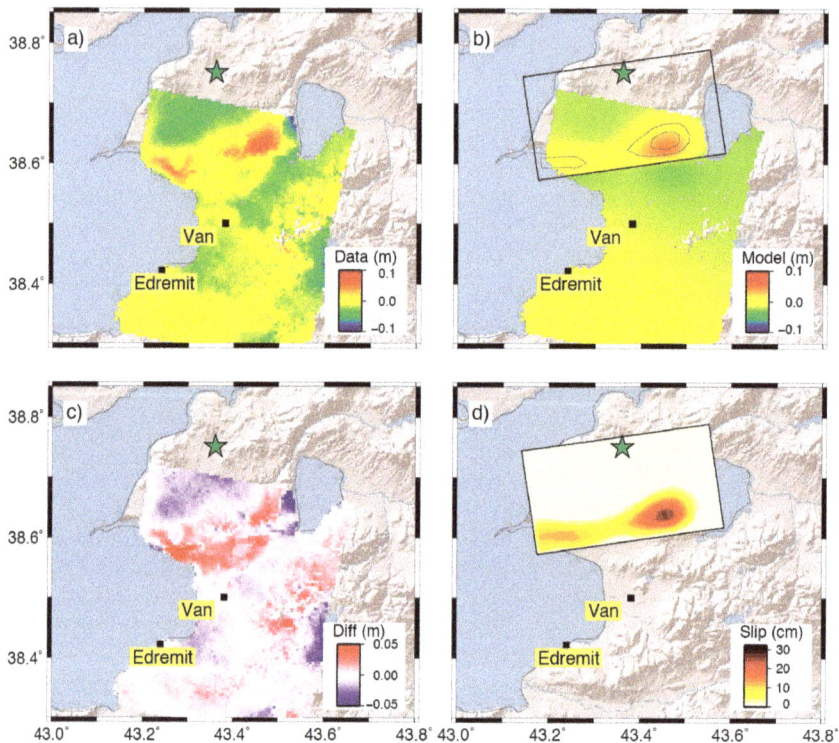

Figure 9. Post-seismic InSAR data and modeling (three-day temporal baseline): (**a**) subsampled CSK2 data (step of 500 m, 5660 data points); (**b**) modeled LOS displacements and the fault afterslip contour each 10 cm; (**c**) residuals (observed minus modeled data); and (**d**) related afterslip distribution. The green star is the mainshock epicenter.

Results from the afterslip computations related to the 17-day displacements are shown in Figure 10. The afterslip distribution shows two separate areas of slip concentration, both in the upper part of the fault, above 9 km depth. The larger one is located in the eastern part of the fault, and broadens the slip pattern already found in the first three days, reaching about 40 cm of slip. The second is located in the very shallow fault, at 3–4 km depth in the western sector of the fault, close to Lake Van. Data and model show quite similar patterns, but an area of positive residuals is found between Bardakçi and

Van. Unfortunately, the inverted data have low coherence in the high slip area (due to loss of coherence of both satellites from which it is computed), where higher values were expected, according to the inferred slip. The residuals are at 88.6% within data errors, considering a larger error (3 cm) due to the procedure used to isolate the displacement for the 17-day temporal baseline. The computed geodetic moment in the FE heterogeneous medium is 1.6×10^{18} Nm, corresponding to Mw 6.1. The three- and 17-day post-seismic energy releases amount to $1/50$ and $1/30$ of the coseismic energy release, respectively. From the total energy released in 17 days post mainshock, ~60% was released in the first three days. A further aseismic release of ~10^{19} Nm (equivalent to Mw 6.6) from the end of November 2011 for 1.5 years was computed [16].

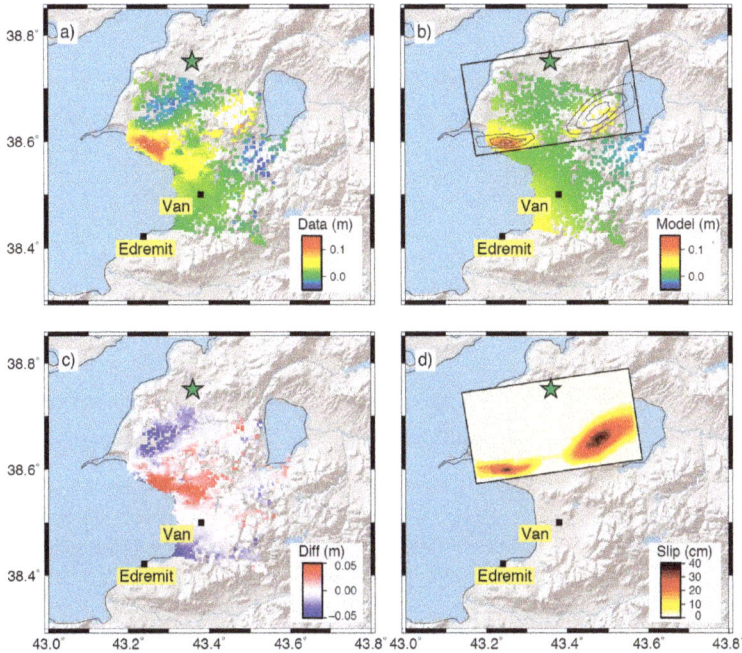

Figure 10. Post-seismic InSAR data and modeling (17-days temporal baseline): (**a**) subsampled data (step of 500 m, 1590 data points) computed from the difference between images CSK3 and TSX2 (see text for details); (**b**) modeled LOS displacements and the fault afterslip contour each 10 cm; (**c**) residuals (observed minus modeled data); and (**d**) related afterslip distribution. The green star is the mainshock epicenter.

5. Discussion

5.1. The 9 November 2011 Edremit-Van Earthquake

The Mw 5.7 Edremit-Van earthquake occurred offshore the town of Edremit, 15 km SW of Van. This event caused 40 fatalities and further collapse of tens of already damaged buildings. The earthquake, similarly to the event of 23 October 2011, took place on a fault that was not previously mapped (e.g., [4]). This second mainshock originated a sequence with a M 5.0 aftershock on 30 November 2011. The Edremit-Van earthquake had a dominantly dextral strike-slip focal mechanism (Figure 1). The conjugate planes' ambiguity (~EW and ~NS oriented) was discussed by various authors [2,10,12,15], generally endorsing the North dipping EW orientation, similarly to the mainshock fault [36]. Furthermore, the aftershocks of the Edremit-Van earthquake were fairly distributed with EW trend (Figure 1). The observation of the combined patterns due to different SAR orbits (TSX2,

Figure 8d and TSX3, Figure 8e) contributes to resolve the ambiguity, supporting the EW orientation hypothesis. Indeed, despite the ascending displacements (TSX3) includes part of the post-seismic data of the mainshock, they show opposite patterns South of Van, confirming the ~EW dextral mechanism (Figure 8 and profiles in Figure S10). The hypothesis of a NS strike slip fault at the epicenter longitude would provide negligible displacement in this area.

The role of the 23 October earthquake in promoting the Edremit-Van earthquake of 9 November was debated, discussing whether the latter is an aftershock of the mainshock or not. Computations of the changes in Coulomb stress (Coulomb Failure Function, CFF, e.g., [37]) endorsed the active role of the Van earthquake in promoting the Edremit-Van earthquake [12,15]. We simulated the variation of CFF projected on the presumed Edremit-Van fault (oriented EW at Lat. 38.45°N and extending 8 km × 5.75 km, [33]), by taking into account the coseismic and post-seismic slip distributions (Figure S11a). The whole Edremit-Van fault plane undergoes an increment of the CFF, supporting the possibility that the Van mainshock may have promoted the earthquake of 9 November 2011.

5.2. Geophysical Insights and Hazard Implications from SAR Analysis

From the comparison between slip and rigidity distributions (Figure 11), we observe that the high coseismic slip area expands from the hypocenter towards depths with higher rigidity. This is in accordance with previous findings (e.g., [32,38]), since the slip concentrates in asperities zones. However, in this case the rigidity shows long wavelength variations of only 22% along the fault plane and the tomographic data is too coarse compared to the fault dimensions to quantitatively affect the retrieved slip distribution. On the other side, good resolution structural data often evidence strong heterogeneities at the fault scale, with short wavelength variations that impact on the obtained slip distributions, either in case of continental earthquakes of moderate magnitude [32] or megathrust events [38]. As a final remark, we may also notice that the aftershocks are not particularly dense in the area of high coseismic and post-seismic slip but concentrate on the edges and on the bottom of the coseismic slip.

The coseismic slip distribution shows a large concentration area at hypocentral depths (12–17 km) with highest peak at about 14 km depth (Figure 11), and the rupture reaches only the depth of 8–10 km. These results are common to other authors [11,13,15]. Based on similar coseismic results, it was argued by [13] that the change of the Coulomb stress brought the upper, un-ruptured part of the fault closer to failure. Our post-seismic data and modeling show that the shallow section of the fault underwent considerable aseismic slip during the early days after the mainshock. Yet during the first 3 days after the mainshock we retrieved few dozens of centimeters of afterslip in the upper part of the fault close to Lake Erçek. During the following weeks, the slip has continued to increase in the same area, and a slip concentration patch appeared close to the western corner of the fault, near the Lake Van shore. Therefore, we argue that all the upper part of the fault accommodated aseismic slip. The shallow aseismic slip has continued also for the following 1.5 years, based on GPS data [16]. The afterslip in the upper section of the fault released an equivalent Mw 6.1 in the first 17 days after the mainshock and a further Mw 6.6 in the following 1.5 years (computed from the end of November 2011). This partially compensate the shallow slip deficit observed right after the mainshock, and overall, our results show the lowering of the seismogenic potential within the first 8–10 km of the fault through the release of a significant amount of energy during the post-seismic phase. We have also shown that shallow afterslip occurred along the full length of the fault and not only in the western sector as suggested by [16], whose results were probably biased by the partial coverage of the GPS network in the area near Lake Erçek, and by [13,14] using the three-day post-seismic image CSK2.

We simulated the variation of CFF by taking into account the coseismic and post-seismic slip distributions. The Coulomb stress change within the first 10 km of the fault observed in section AA' (e.g., Figure 3) results to be positive (*i.e.*, earthquake occurrence enhanced), even including the post-seismic afterslip (Figure S11b). This is due to the difference of one order of magnitude between the coseismic and post-seismic scalar moments, which implies that the stress drop due to the aseismic slip cannot shadow the positive coseismic stress changes. However, we may account for several factors that may contribute to reduce the seismic hazard at shallow depths. As an example, the velocity strengthening characteristics of the shallow crust promotes faster falloff of slip velocity behind the rupture front and a decrease of slip towards the free surface (e.g., [39]). Therefore, limited slips are expected in the first kilometers below the free surface, further reducing the seismic potential. In our computations the shallow velocity strengthening characteristics, as feasible for the entire area east of Lake Van consisting of accretionary complex materials [40], are not taken into account. Furthermore, other post-seismic mechanisms such as viscoelastic and poroelastic behaviors may contribute to the post-seismic response, reducing the seismic hazard of that sector. Unfortunately, there is no detailed knowledge of the local structure and these possibilities cannot be fully investigated.

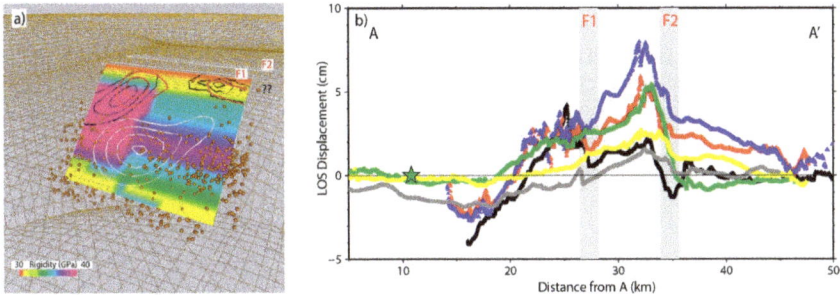

Figure 11. Coseismic and post-seismic results: (**a**) 3D view of the Van earthquake fault from the NW. Coseismic slip (white lines, contour every 2 m) and the post-seismic slip (red and black lines are three and 17 days after the mainshock, respectively, contour every 10 cm) superimposed on the rigidity heterogeneities of the fault plane within the FE model. Aftershocks hypocenters within 4 km from the fault are shown by ochre spheres; (**b**) AA' profile (see Figure 5a) across the traces of the main fault (F1) and the splay fault (F2). The colors are: CSK2, black; CSK3, blue; TSX1, grey; TSX2, yellow; TSX3, green; and the difference between CSK3 and TSX2, red. The green star is the Van earthquake epicenter.

We adopted only one fault in our inversions. Elliott *et al.* [13] resolved the slip on a pair of en echelon fault planes, based on the local morphology and the fault strike detectable in the CSK coseismic interferogram (Figure 4a). Their slip distribution shows two lobes, one on each fault, decreasing at the central border due to the sharp change of the fault dip. The existence of a reactivated aseismic fault, constrained by the GPS post-seismic data, was supposed [16]. Such splay fault extends from the western edge to the middle of the coseismic fault trace, shifted 7–8 km to the South. This secondary fault is supposed to join the main rupture at about 500 m below the surface, and using only GPS data the afterslip is found to be distributed both on the main and splay faults [16]. As mentioned above, even the relocated aftershocks [2] are too scattered and cannot be used to visualize more detailed features such as main fault segmentation and/or splay faults, especially those very shallow. Field observations and InSAR fringe patterns do identify minor surface fractures, as it occurs in many earthquakes but their direct relation to the mainshock rupture is far from clear. To better understand the InSAR contribution to the presence of the splay fault, Figure 11b reports all the post-seismic InSAR data along the AA' profile (as shown in Figure 3), where F2 is the trace of the presumed splay fault (as depicted by [16]) with ±1 km uncertainty. Most of the patterns show a high gradient close to F2. This secondary plane seems actually to cause a discontinuity on the surface displacements, but nor the seismicity and other geophysical data constrain its depth. Given the limited independent knowledge about the local fault system, we did not attempt to set up inversions of complex faults (*i.e.*, double and/or listric faults), and eventually we adopted a single fault plane.

6. Conclusions

We have analyzed a large dataset of SAR images in order to study the coseismic and post-seismic phases of the Van earthquake sequence. Only few good quality interferograms resulted to be fruitful to constrain the slip distributions. Following the approach outlined in [32], we built a FE model of the Van earthquake that includes the structural information of the Van region. We computed coseismic and post-seismic slip distributions from the numerical model-based inversions.

We have here attempted to improve previous findings (e.g., [3,13–15]) employing most of the available geodetic data (InSAR and GPS) to constrain the slip characteristics of the Van Earthquake. Our approach, along with the employment of new-generation FE models, was aimed to contribute to the knowledge of the seismic hazard of the region. Our method allowed us to disclose further characteristics of the Van sequence, and to provide new insights for the regional/local risk assessment. The main new outcomes of our study are that the shallow part of the fault (above 7–9 km depth), unruptured during the coseismic phase, underwent afterslip in the post-seismic phase that may have reduced the seismic potential in its whole length from NW to SE. Furthermore, from the analysis of the InSAR data we were able to discuss the existence of a reactivated aseismic fault, that actually caused a discontinuity in the SAR profiles, previously hypothesized by means of few GPS [16].

As a conclusive remark, despite non-optimal data coverage (only single-orbit good-quality data available in the coseismic phase and for which coherence issues in the X-band post-seismic phase prevented the use of time series processing), we demonstrated that useful information could still be retrieved from SAR data through a detailed analysis.

Supplementary Materials: The following are available online at www.mdpi.com/2072-4292/8/6/532/s1. Figures S1 and S2: Wrapped InSAR images from ENVISAT satellite, Figure S3: Unwrapped adjacent coseismic InSAR data, Figure S4: PPD functions obtained from the non-linear fault inversion, Figure S5: Two-dimensional PPD distributions, Figure S6: Comparisons between observed and computed data of the coseismic displacements in the non-linear inversion, Figure S7: Full-scale map with GPS as computed in the non-linear inversion, Figure S8: Full-scale map with GPS as computed in the linear inversion, Figure S9: slip uncertainty distributions for the post-seismic analysis, Figure S10: Profiles of the InSAR data across the faults, Figure S11: CFF computations for the Edremit-Van earthquake, Table S1: Characteristics of all the SAR images acquired.

Acknowledgments: TERRASAR-X images and GPS data are available through the Group on Earth Observations (GEO) Geohazard Supersite initiative (http://www.earthobservations.org/gsnl.php). R. Cakmak (TUBITAK MRC EMSI) is acknowledged for the GPS coseismic field. ENVISAT data were provided by the European Space Agency (ESA) in the framework of the Category-1 project n. 5605. COSMO-SkyMed data were provided by the Italian Space Agency (ASI) in the framework of the SiGRiS project (ASI n. I/024/07/0), that partially funded this work. This work was also partially funded by the EU's H2020 EVER-EST Project (grant agreement 674906). We thank J. P. Merryman Boncori for taking part to the COSMO-SkyMed data processing. P. De Gori provided the relocated seismicity and M.K. Salah the tomographic data. A. Akinci and A. Antonioli are acknowledged for useful discussions. We thank three anonymous referees that helped us to improve the first version of this manuscript. The figures were generated using the commercial softwares ArcGis 10.0, Amira 5.4 and the open source software GMT (http://gmt.soest.hawaii.edu/).

Author Contributions: All the coauthors conceived and designed the present study. C.T. processed the InSAR data; G.P. worked on the fractures and landslides; E.T. performed the modeling and inversions; S.A wrote the codes for linear inversions; and E.T., C.T. and G.P. wrote the paper.

Conflicts of Interest: The authors declare no conflict of interest.

References

1. Örgülü, G.; Aktar, M.; Türkelli, N.; Sandvol, E.; Barazangi, M. Contribution to the seismotectonics of eastern turkey from moderate and small size events. *Geophys. Res. Lett.* **2003**, *30*. [CrossRef]

2. De Gori, P.; Akinci, A.; Lucente, F.P.; Kılıç, T. Spatial and temporal variations of aftershock activity of the 23 October 2011 mw 7.1 Van, Turkey, earthquake. *B Seismol. Soc. Am.* **2014**, *104*, 913–930. [CrossRef]

3. Fielding, E.J.; Lundgren, P.R.; Taymaz, T.; Yolsal-Cevikbilen, S.; Owen, S.E. Fault-slip source models for the 2011 m 7.1 van earthquake in Turkey from sar interferometry, pixel offset tracking, gps, and seismic waveform analysis. *Seismol. Res. Lett.* **2013**, *84*, 579–593. [CrossRef]

4. Şaroğlu, F.; Emre, Ö.; Kuşçu, I. *Active Fault Map of Turkey, 1:2,000,000 Scale*; General Directorate of Mineral Research and Exploration: Ankara, Turkey, 1992.

5. Emre, Ö.; Duman, T.Y.; Özalp, S.; Elmaci, H.; Olgun, Ş.; Şaroğlu, F. *Active Fault Map of Turkey with an Explanatory Text, 1:250,000 Scale*; General Directorate of Mineral Research and Exploration: Ankara, Turkey, 2013.

6. Cetin, K.O.; Turkoglu, M.; Ünsal Oral, S.; Nacar, U. Van-Tabanli Earthquake (mw = 7.1) 23 October 2011 Preliminary Reconnaissance Report. Available online: supersites.earthobservations.org/Van_EQ_Preliminary_Report_KOC.pdf (accessed on 15 March 2012).

7. Doğan, B.; Karakaş, A. Geometry of co-seismic surface ruptures and tectonic meaning of the 23 October 2011 mw 7.1 Van earthquake (east anatolian region, Turkey). *J. Struct. Geol.* **2013**, *46*, 99–114. [CrossRef]

8. Hayes, G. USGS. Available online: earthquake.usgs.gov/earthquakes/eqinthenews/2011/usb0006bqc/finite_fault.php (accessed on 25 October 2011).

9. Irmak, T.S.; Doğan, B.; Karakaş, A. Source mechanism of the 23 October, 2011, Van (Turkey) earthquake (m-w = 7.1) and aftershocks with its tectonic implications. *Earth Planets Space* **2012**, *64*, 991–1003. [CrossRef]

10. Utkucu, M. 23 October 2011 Van, eastern anatolia, earthquake (mw 7.1) and seismotectonics of lake van area. *J. Seismol.* **2013**, *17*, 783–805. [CrossRef]

11. Gallovič, F.; Ameri, G.; Zahradník, J.; Janský, J.; Plicka, V.; Sokos, E.; Askan, A.; Pakzad, M. Fault process and broadband ground-motion simulations of the 23 October 2011 Van (eastern turkey) earthquake. *B Seismol. Soc. Am.* **2013**, *103*, 3164–3178. [CrossRef]

12. Akinci, A.; Antonioli, A. Observations and stochastic modelling of strong ground motions for the 2011 October 23 m-w 7.1 Van, Turkey, earthquake. *Geophys. J. Int.* **2013**, *192*, 1217–1239. [CrossRef]

13. Elliott, J.R.; Copley, A.C.; Holley, R.; Scharer, K.; Parsons, B. The 2011 mw 7.1 Van (eastern Turkey) earthquake. *J. Geophys. Res. Sol.* **2013**, *118*, 1619–1637. [CrossRef]

14. Feng, W.P.; Li, Z.H.; Hoey, T.; Zhang, Y.; Wang, R.J.; Samsonov, S.; Li, Y.S.; Xu, Z.H. Patterns and mechanisms of coseismic and postseismic slips of the 2011 m-w 7.1 Van (Turkey) earthquake revealed by multi-platform synthetic aperture radar interferometry. *Tectonophysics* **2014**, *632*, 188–198. [CrossRef]

15. Moro, M.; Cannelli, V.; Chini, M.; Bignami, C.; Melini, D.; Stramondo, S.; Saroli, M.; Picchiani, M.; Kyriakopoulos, C.; Brunori, C.A. The 23 October 2011, Van (Turkey) earthquake and its relationship with neighbouring structures. *Sci. Rep.* **2014**, *4*. [CrossRef] [PubMed]

16. Dogan, U.; Demir, D.Ö.; Çakir, Z.; Ergintav, S.; Ozener, H.; Akoğlu, A.M.; Nalbant, S.S.; Reilinger, R. Postseismic deformation following the mw 7.2, 23 October 2011 Van earthquake (Turkey): Evidence for aseismic fault reactivation. *Geophys. Res. Lett.* **2014**, *41*, 2334–2341. [CrossRef]

17. Gök, R.; Mellors, R.J.; Sandvol, E.; Pasyanos, M.; Hauk, T.; Takedatsu, R.; Yetirmishli, G.; Teoman, U.; Turkelli, N.; Godoladze, T.; *et al.* Lithospheric velocity structure of the anatolian plateau-caucasus-caspian region. *J. Geophys. Res. Sol.* **2011**, *116*. [CrossRef]

18. Kipfer, R.; Aeschbachhertig, W.; Baur, H.; Hofer, M.; Imboden, D.M.; Signer, P. Injection of mantle type helium into lake Van (Turkey)—The clue for quantifying deep-water renewal. *Earth Planet. Sci. Lett.* **1994**, *125*, 357–370. [CrossRef]

19. Litt, T.; Krastel, S.; Sturm, M.; Kipfer, R.; Örcen, S.; Heumann, G.; Franz, S.O.; Ülgen, U.B.; Niessen, F. 'Paleovan', international continental scientific drilling program (ICDP): Site survey results and perspectives. *Quaternary Sci. Rev.* **2009**, *28*, 1555–1567. [CrossRef]

20. Salah, M.K.; Sahin, Ş.; Aydin, U. Seismic velocity and poisson's ratio tomography of the crust beneath east anatolia. *J. Asian Earth Sci.* **2011**, *40*, 746–761. [CrossRef]

21. Sarscape Sarmap sa, 5.0. Available online: http://www.sarmap.ch (accessed on 25 March 2013).

22. Werner, C.; Wegmüller, U.; Strozzi, T.; Wiesmann, A. Interferometric point target analysis for deformation mapping. *Geosci. Remote Sens. Symp.* **2003**, *7*, 4362–4364.

23. Wang, T.; Jonsson, S. Improved sar amplitude image offset measurements for deriving three-dimensional coseismic displacements. *IEEE J. Stars* **2015**, *8*, 3271–3278. [CrossRef]

24. Bechor, N.B.D.; Zebker, H.A. Measuring two-dimensional movements using a single insar pair. *Geophys. Res. Lett.* **2006**, *33*. [CrossRef]

25. Goldstein, R.M.; Werner, C.L. Radar interferogram filtering for geophysical applications. *Geophys. Res. Lett.* **1998**, *25*, 4035–4038. [CrossRef]

26. Costantini, M. A novel phase unwrapping method based on a network programming. *Geosci. Remote Sens. IEEE Trans.* **1998**, *36*, 813–821. [CrossRef]

27. Jibson, R.W. Methods for assessing the stability of slopes during earthquakes-a retrospective. *Eng. Geol.* **2011**, *122*, 43–50. [CrossRef]

28. Okada, Y. Internal deformation due to shear and tensile faults in a half-space. *B Seismol. Soc. Am.* **1992**, *82*, 1018–1040.

29. Sambridge, M. Geophysical inversion with a neighbourhood algorithm—II. Appraising the ensemble. *Geophys. J. Int.* **1999**, *138*, 727–746. [CrossRef]

30. Sambridge, M. Geophysical inversion with a neighbourhood algorithm—I. Searching a parameter space. *Geophys. J. Int.* **1999**, *138*, 479–494. [CrossRef]

31. Abaqus *Dassault Systèmes Simulia Corp. Providence*, 6.11. Available online: www.simulia.com (accessed on 1 February 2012).

32. Trasatti, E.; Kyriakopoulos, C.; Chini, M. Finite element inversion of dinsar data from the mw 6.3 l'aquila earthquake, 2009 (Italy). *Geophys. Res. Lett.* **2011**, *38*. [CrossRef]

33. Wells, D.L.; Coppersmith, K.J. New empirical relationships among magnitude, rupture length, rupture width, rupture area, and surface displacement. *B Seismol. Soc. Am.* **1994**, *84*, 974–1002.

34. Christensen, N.I.; Mooney, W.D. Seismic velocity structure and composition of the continental-crust—A global view. *J. Geophys. Res. Sol.* **1995**, *100*, 9761–9788. [CrossRef]

35. Atzori, S.; Hunstad, I.; Chini, M.; Salvi, S.; Tolomei, C.; Bignami, C.; Stramondo, S.; Trasatti, E.; Antonioli, A.; Boschi, E. Finite fault inversion of dinsar coseismic displacement of the 2009 l'aquila earthquake (central Italy). *Geophys. Res. Lett.* **2009**, *36*. [CrossRef]

36. Görgün, E. The 23 October 2011 m-w 7.2 Van-ercis, Turkey, earthquake and its aftershocks. *Geophys. J. Int.* **2013**, *195*, 1052–1067. [CrossRef]

37. Harris, R.A. Introduction to special section: Stress triggers, stress shadows, and implications for seismic hazard. *J. Geophys. Res. Sol.* **1998**, *103*, 24347–24358. [CrossRef]

38. Romano, F.; Trasatti, E.; Lorito, S.; Piromallo, C.; Piatanesi, A.; Ito, Y.; Zhao, D.; Hirata, K.; Lanucara, P.; Cocco, M. Structural control on the tohoku earthquake rupture process investigated by 3d fem, tsunami and geodetic data. *Sci. Rep.* **2014**, *4*. [CrossRef] [PubMed]

39. Kaneko, Y.; Lapusta, N.; Ampuero, J.P. Spectral element modeling of spontaneous earthquake rupture on rate and state faults: Effect of velocity-strengthening friction at shallow depths. *J. Geophys. Res. Sol.* **2008**, *113*. [CrossRef]

40. Şengör, A.M.C.; Özeren, M.S.; Keskin, M.; Sakınç, M.; Özbakır, A.D.; Kayan, İ. Eastern turkish high plateau as a small turkic-type orogen: Implications for post-collisional crust-forming processes in Turkic-type orogens. *Earth Sci. Rev.* **2008**, *90*, 1–48. [CrossRef]

Chapter 2:
Landslide Hazards

remote sensing

Article

Remote Sensing for Characterisation and Kinematic Analysis of Large Slope Failures: Debre Sina Landslide, Main Ethiopian Rift Escarpment

Jan Kropáček [1,*], Zuzana Vařilová [2,3], Ivo Baroň [4], Atanu Bhattacharya [5], Joachim Eberle [1] and Volker Hochschild [1]

[1] Department of Geosciences, University of Tuebingen, Rümelinstr. 19–23, 72070 Tübingen, Germany; joachim.eberle@uni-tuebingen.de (J.E.); volker.hochschild@uni-tuebingen.de (V.H.)
[2] Municipal Museum of Ústí nad Labem, Masarykova 1000/3, 400 01 Ústí nad Labem, Czech Republic; varilova@muzeumusti.cz
[3] Geo-Tools, U Mlejnku 128, Přemyšlení, 250 66 Zdiby, Czech Republic
[4] Department of Geology and Paleontology, Natural History Museum Vienna, Burgring 7, 1010 Vienna, Austria; Ivo.Baron@nhm-wien.ac.at
[5] Institute for Cartography, TU Dresden, Helmholzstr. 10, 01062 Dresden, Germany; atanudeq@gmail.com
* Correspondence: jan.kropacek@uni-tuebingen.de; Tel.: +49-7071-297-8940; Fax: +491-7071-295-378

Academic Editors: Zhenhong Li, Roberto Tomas, Richard Gloaguen and Prasad S. Thenkabail
Received: 10 July 2015; Accepted: 19 November 2015; Published: 2 December 2015

Abstract: Frequently occurring landslides in Ethiopia endanger rapidly expanding settlements and infrastructure. We investigated a large landslide on the western escarpment of the Main Ethiopian Rift close to Debre Sina. To understand the extent and amplitude of the movements, we derived vectors of horizontal displacements by feature matching of very high resolution satellite images (VHR). The major movements occurred in two phases, after the rainy seasons in 2005 and 2006 reaching magnitudes of 48 ± 10.1 m and 114 ± 7.2 m, respectively. The results for the first phase were supported by amplitude tracking using two Envisat/ASAR scenes from the 31 July 2004 and the 29 October 2005. Surface changes in vertical direction were analyzed by subtraction of a pre-event digital elevation model (DEM) from aerial photographs and post-event DEM from ALOS/PRISM triplet data. Furthermore, we derived elevation changes using satellite laser altimetry measurement acquired by the ICESat satellite. These analyses allowed us to delineate the main landslide, which covers an area of 6.5 km^2, shallow landslides surrounding the main landslide body that increased the area to 8.5 km^2, and the stable area in the lower part of the slope. We assume that the main triggering factor for such a large landslide was precipitation cumulated over several months and we suspect that the slope failure will progress towards the foot of the slope.

Keywords: Ethiopian rift; Tarmaber area; Debre Sina; large landslide; feature tracking; amplitude tracking; DEM differencing; ICESat

1. Introduction

Deformation of the Earth's surface caused by various geodynamic processes represents a serious hazard for settlements and infrastructure. In the last few decades, remote-sensing techniques have proven to be an effective tool to identify and monitor surface deformations of glaciers, permafrost and landslides. Correlation techniques of consecutive optical space-borne images results in displacement vector fields, which provide valuable information on the character of the movement [1–3]. These techniques are based on image windows correlation in space domain [4] or frequency domain [5,6]. The accuracy of these approaches is by the rule of thumb on order of pixel size of the correlated image data [2]. However, recent studies have shown that an accuracy of one-fourth to one-fifth of a pixel size can be achieved provided VHR stereo pairs are available [7,8]. Furthermore, Differential SAR interferometry (DInSAR) techniques were successfully used for investigation of landslides, e.g., [9–11]. The use of this technique is, however, limited by temporal decorrelation, terrain setting and orientation and by relation of the movement velocity to the used radar frequency [12]. Amplitude tracking is another technique for movement detection based on SAR (Synthetic Aperture Radar) data. Unlike DInSAR it is based purely on the amplitude information. This technique is often used for glacier monitoring [13–15]. In this approach, the movement velocities can be measured independently from the movement direction with respect to the range direction and additionally, the coherence of the image pair is not required [13]. A time series of DEMs extracted form historical photographs were used to capture morphological change caused by a landslide by [16,17]. The difficulties caused by the limited ground control and missing camera calibration protocol can be solved by the application of self-calibrating bundle adjustment methods [17]. It was demonstrated by [18] that DEM differencing using DEMs derived from VHR satellite data can be effectively used to investigate mass displacement of large landslides. This technique is also effective for a delineation of large landslides [18].

Slopes of the Ethiopian Highlands are frequently affected by landslides of various types, which often lead to eviction of inhabitants, damage to housing, infrastructure and arable land and even loss of human lives [19,20]. Most of the landslides in this region, including the largest ones, are triggered by heavy precipitation occurring at the end of the rainy periods in July and August [21], whereas earthquakes mainly trigger fast moving slope failures such as rock slides, topples and falls [19]. An extraordinarily large slope failure occurred in the Yizaba locality of the Tarmaber area north of the town of Debre Sina (Figure 1). The landslide has been studied by [22] who mapped susceptibility zones, analyzed precipitation, described basic geological site settings and pointed out earthquakes in Afar Rift as the probably main triggering factor. Field hydro-geological investigations and geophysical sounding were reported by [23]. There are some substantial discrepancies in the reported extent and evolution of the Debre Sina landslide. Several opinions regarding the dating of the major sliding phases can be found in literature. The occurrence of the first cracks was dated as 21 August 2005 by [24], while [25] reports September 2005. Based on eyewitness accounts, [23] states that the formation of tension cracks had already occurred in August 2004. The major sliding event was dated as 13–14 September 2005 by the report of Action by Churches released in 2006 (in [22,26]). This is in accordance to [24] who reports 13 September 2005, while September 2006 is reported by [24,25]. We suppose that the major movements occurred in two phases as [23] reports two major movements in September 2005 and in September 2006 based on interviews with local inhabitants.

In this study, we aim at demonstration of the potential of optical and microwave remote sensing techniques in combination with DEMs for investigation of the extent, evolution and properties of a large landslide.

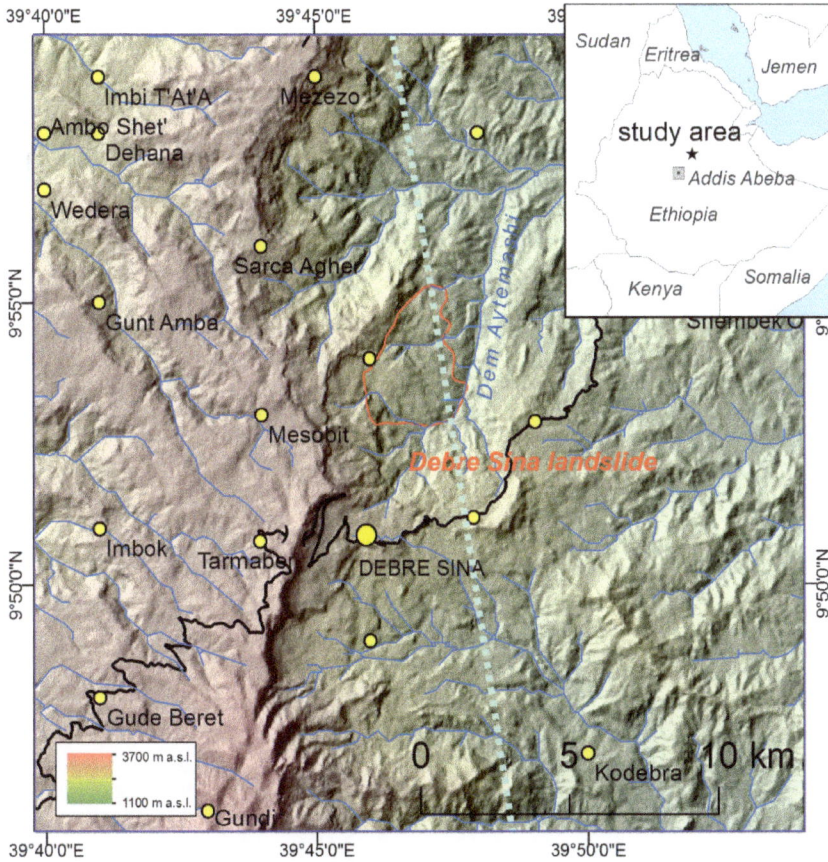

Figure 1. The study area is located on the western escarpment of the Main Ethiopian Rift close to the trunk road that connects Addis Ababa with the northern regions of the country. The nominal ground track of ICESat is shown as a light blue dotted line.

2. Study Area

The Debre Sina landslide is located on the western escarpment of the Main Ethiopian Rift (MER) (Figure 1). The rift escarpment is formed by sequences of tertiary volcanic rock formations. The study area is characterized by rugged relief, with rock outcrops, deeply dissected creeks and channels but also by less steep areas with terraced arable land (Figure 2).

Figure 2. Panoramic view of the Debre Sina landslide from the SE (above) and E (below) with examples of characteristic geodynamic features within the main landslide body and its close surroundings: rockfalls (**a**); gully erosion along the Dem Aytemashi River and its tributaries (**b**); rotational landslides (**c**); debris flows (**d**); shallow landslide and earthflow (**e**); and displaced rock blocks with well-marked scarp lines (**f**). The estimated position of the shear plane outcrop is shown as a blue line. All photos were taken in March 2015 by Zuzana Vařilová.

The bedrock in the lower part of the slope consists of alternating layers of basalt, rhyolitic or trachytic ignimbrites as well as tuffs and agglomerates of different volcanic material (Alaje formation), which is overlaid by basalts of the Tarmaber formation in the head scarp area of the Debre Sina landslide. The ignimbrites and tuffs of the Alaje formation, in particular, are highly altered and intensely weathered [22,23]. The slopes with lower inclination are covered by Quaternary sediments (alluvial, colluvial-eluvial deposits and residual soils). The landslide is located in a tectonically active area with an extension character [24]. The predominant direction of discontinuity (faults and lineaments) orientation is WSW-ENE (NW-SE) [23]. The major faults on the western boundary of the rift can have occasional earthquake tremors leading to activation of unstable ground [23]. The study area is located in a high seismic risk zone [27–29]. Fifteen earthquakes with magnitudes ranging from 4.1 to 5.9 were registered in the surrounding area within 200 km of distance since 1980, while in the relatively small area of the Affar depression, which is located 280 km North from the study area, 170 shakes with magnitude > 4.0 have occurred since 2005 [30]. Historical records from the past 150 years show that no large magnitude earthquake occurred in MER [31]. The precipitation in the area follows the general pattern of the Ethiopian Highlands. After a long dry period the precipitation increases in March, April and May followed by the main rainy season in July, August and September. The mean annual precipitation measured in Debre Sina station for the period 1990–2013 is 1750 mm, while the mean precipitation in the period from July to September is 1035 mm. Due to the high elevation gradient in the study area the climate in the upper part is considerably colder and wetter [23]. The landslide area is drained by the Dem Aytemashi River towards the north and later to the east towards the Awash River. The drainage pattern of the basin probably follows tectonic predisposition.

3. Methods

3.1. Estimation of Vertical Changes by Subtraction of Two DEMs

To understand changes in mass distribution due to the land sliding pre- and post-event DEMs were compared. A cell-by-cell subtraction of post and pre-event DEMs provides a straightforward way of obtaining positive and negative elevation differences corresponding to accumulation or depletion, respectively.

As we could not find VHR satellite images older than 2005, we used aerial photographs from 1986 to build a pre-event DEM. Black and white contact copies of three photographs covering the study area were scanned at a resolution of 1025 dpi, which resulted in a ground resolution of approximately 1 m, and were supplied on a DVD by the Ethiopian Mapping Agency. As no calibration protocol and accurate GCPs were available to allow us to calculate the internal and external orientation, we could not carry out the processing using traditional photogrammetry. Instead we used the Structure From Motion (SFM) approach [32,33] implemented in the PhotoScan v 1.0.4 software package. The acquisition geometry was reconstructed by an iterative bundle adjustment without reference to the cartographic coordinate system. The resulting point cloud was georeferenced *a posteriori* using 25 GCP identified in VHR imagery in Google Earth as no Differential Global Positioning System (DGPS) measurements were possible. The root mean square error (RMSE) calculated from residuals of 7 check points, also collected in the Google Earth, were 4.9, 5.5 and 10.6 m in x, y and z directions, respectively. The advantage of this approach is that the inaccuracy of GCPs does not affect the internal geometry of the model. In the next step, the georeferenced point cloud with the mean density 0.83 points/m^2 was converted to a raster DEM with a resolution of 2.5 m. Furthermore, an ortho-image with a resolution of 1 m was generated to complete the time series of VHR images with an image capturing the pre-event situation (Table 1). The high resolution pre-event DEM allowed us to analyze the morphology of the original terrain including theoretical surface runoff. We generated a Topography Wetness Index (TWI) (Figure 3) [34] representing potential infiltration, which is a relevant parameter for addressing the development of a slope failure [35,36].

To build a post-event DEM we used an image triplet acquired by the PRISM instrument carried by the Japanese satellite ALOS on 19 November 2008. This satellite was launched in 2006 and is dedicated to cartography and disaster monitoring. The triplet consists of three images taken by backward, nadir and forward pointing cameras which enable stereo-processing [37]. We processed the data using the Leica Photogrammetry Suite version 9.3. The resulting PRISM DEM has a resolution of 10 m. The resulting RMSE on nine check points collected from Google Earth 6.1, 6.5 and 7.4 m in the x, y and z directions, respectively. This value is, however, strongly influenced by the accuracy of the reference satellite imagery. Additionally, an ortho-image with grid spacing of 2.5 m was generated employing the SRTM DEM as the elevation reference. The use of the PRISM DEM led to an identical result even over the area of the landslide. We opted for the SRTM DEM as the PRISM DEM includes some artifacts in the areas of cloud cover.

Table 1. High-resolution satellite images covering the study area. RMSE after co-registration of the orthorectified images with respect to the Ikonos-2 image from 14 December 2005.

Satellite/Sensor	Acquisition Date	Corresponding Rainy Season	Resolution PAN/MS (m)	Off-Nadir Angle (Degrees)	Sun Elevation. (Degrees)	RMSE of Co-Registration (m); (No. Points)
Air photos	27 November 1986	1986	1.0	-	-	6.3; (16 points)
Ikonos-2	14 December 2005	2005	0.8/4.0	17.5	52.0	-
Ikonos-2	05 June 2007	2006	0.8/4.0	10.6	65.9	0.88; (12 points)
Kompsat-2	27 January 2008	2007	1.0/4.0	0.0		0.67; (88 points)
WorldView-1	22 December 2008	2008	0.5	19.5	51.7	1.60; (123 points)
QuickBird-2	18 May 2010	2009	0.6	10.2	66.7	1.90; (62 points)
GeoEye	25 November 2012	2012	0.46	20.4	54.9	1.94; (143 points)
WorldView-2	27 May 2014	2013	0.46/1.84	9.6	69.5	1.75; (31 points)

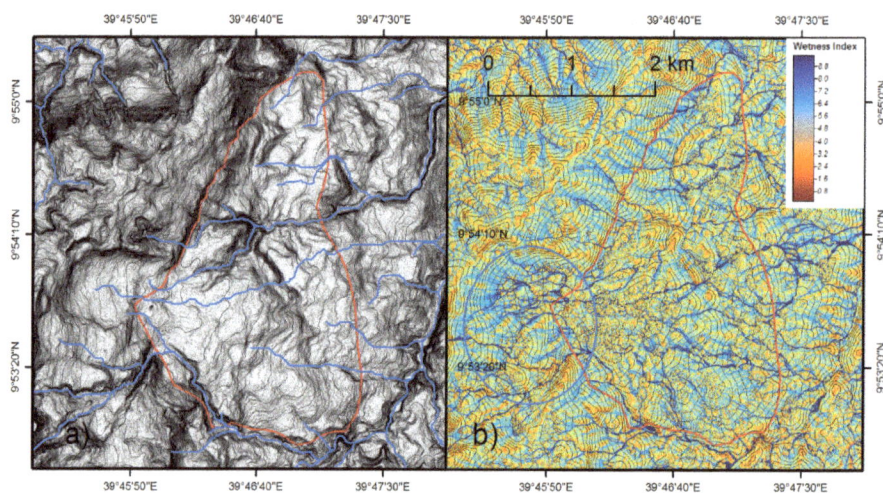

Figure 3. The pre-event DEM extracted from aerial photographs is shown as a slope inclination image overlaid with drainage (**a**); Topographic Wetness Index derived from the pre-event DEM shows large zones of potential infiltration above the southern part of the landslide, marked by the blue ellipse (**b**). The contour of the active main landslide body is in red.

The two models were accurately co-registered in a horizontal direction using piecewise linear transformation in AutoSync Workstation. In this approach, the triangular areas between the tie points are fitted using the first order polynomial. The RMSE of the co-registration in horizontal direction calculated from residuals on the used tie points was 4.6 m. This figure indicates the amount of the residual discrepancy in horizontal direction between the two DEMs. The uncertainty of the DEM differencing was assessed as the RMSE of vertical differences on 43 randomly spaced points in the off landslide area. This resulted in the value of 3.6 m. The tie points were automatically identified in the ortho-images produced during the derivation of the two DEMs, which provide abundant image structures for point identification. This way the problem of tie point identification between two DEMs, which provide no suitable patterns for feature matching, was circumnavigated. Both ortho-images and both the DEMs were resampled to 2.5 m beforehand which is the resolution of the DEM from the aerial photographs and also approximately of the PRISM ortho-image. Residual distortions of the pre-event DEM in a vertical direction were modeled by a second order polynomial fitted to a regular grid of 20 × 20 points representing the elevation differences between the two DEMs. Points falling onto the landslide as well as points with outlying values of elevation difference were excluded from the fitting. The higher polynomial order was needed to account for a non-linear distortion of the pre-event DEM probably due to an inaccuracy of the fitted camera model.

3.2. Vertical Elevation Differences Derived by ICESat

The study area is crossed by one nominal ground track of ICESat (Ice, Cloud, and land Elevation Satellite) (see Figure 1). The single repeat tracks do not exactly match the nominal nadir track because the ICESat's precision spacecraft pointing control was not used in the mid latitudes (NSIDC 2015). The maximum spread of ground tracks reaches 2.0 km over the study area. The GLAS instrument (Geoscience Laser Altimeter System) on-board ICESat measured the surface elevation of signal footprints with a diameter of 70 m along the nadir tracks each 172 m. The data were acquired every 3–6 months during 18 one-month campaigns between 2002 and 2009. The elevation is derived from the two way travel time of the emitted laser pulse and from the position of the satellite. Data records containing the elevation are provided by NSIDC (National Snow and Ice Data Center). We used the ICESat product L2 Global Land Surface Altimetry Data denoted as GLA14, release 34 [38]. This satellite mission was primarily dedicated to monitoring atmospheric aerosol and ice sheets with monotonous terrain and low inclination. We followed an approach that has been successfully used for mountain glaciers with a similar size and topography to large landslides [39–41].

To derive the elevation changes over the landslide body we calculated differences of the ICESat elevation measurements with respect to the pre-event SRTM DEM for each ICESat point. The elevation corresponding to the ICESat measurements was obtained using bi-linear interpolation of elevations of the four neighboring cells in the SRTM DEM. To account for the effect of clouds we discarded all points with the elevation difference > 200 m. Furthermore, we selected points acquired after the initial sliding event given by several authors as being on 13 September 2005 over the study area. This resulted in 40 point measurements distributed mainly over the accumulation area of the landslide. For a reliable estimation of the elevation differences the two datasets were accurately co-registered following [42]. The horizontal shift of the SRTM DEM with respect to the ICESat dataset was estimated as being 30 meters with an azimuth of 31 degrees. This shift was removed by an adjustment of the reference coordinated from the SRTM DEM dataset. As the ICESat elevation is referred to the TOPEX/Poseidon Ellipsoid we subtracted the geoid height provided in the ICESat data records and applied a conversion to the WGS84 ellipsoid following [43] to obtain an elevation coherent with the elevation contained in the SRTM DEM dataset. The error ranges were estimated following [44]. The resulting elevation differences on the ICESat points were compared to the results of the DEM differencing. To assess the uncertainty of the vertical offsets the differences between ICESat measurements and the SRTM were averaged for the same ICESat track over a distance of 70 km stretching over both plains and slopes including 4600 points. This resulted in a mean difference of −1.8 m and standard deviation of 11.6. A threshold of the elevation difference of 100 m was applied beforehand to sort out ICESat measurements affected by clouds and atmospheric noise [41].

3.3. Mapping of Surface Features from the Time Series of Optical Satellite Data

Surface features and patterns identified on remote sensing images of landslides can reveal information on their origin and controlling mechanisms [7,45]. A time series of VHR images was used to analyze the evolution of the slope failure (Table 1). These features are commonly mapped using aerial or drone photographs [45–47], however, the large scale of the Debre Sina landslide allowed us to use also satellite images (Figure 4). We assembled a time series of VHR scene subsets for the landslide and surroundings from the archive of DigitalGlobe™ and from the European Space Agency (the entire scene from Kompsat-2). The images document slope deformations in the period from

2005 to 2014. Only the WorldView-2 image from 2014 was partially cloud-covered (11%). Apart from the WorldView-1 image, which has only a panchromatic band, all of the images are composed of a panchromatic band, three visible bands, and one near infrared band.

The first image was acquired only three months after the presumed initial slope failure that occurred in September 2005 [22]. The images were delivered in a pre-processed form, which means that standard radiometric and geometric corrections have already been applied. The images were ortho-rectified in Leica Photogrammetry Suite (LPS) applying the Ratio Polynomial Coefficients approach using a set of 6 GCPs and 6 check points identified in GoogleEarth in the areas around the body of the landslide. We used a post-event DEM derived from ALOS/PRISM data from 19 November 2008 as the elevation reference. To check the influence of the DEM on the spatial accuracy of the ortho-images, we generated two ortho-images from QuickBird-2 data for 2010 using the post-event PRISM DEM and the global SRTM DEM with grid spacing of one arc second corresponding to 30 meters [48,49]. The ortho-images were compared by feature matching in AutoSync, which is included in the Erdas Imagine software package. The influence of the choice of the DEM appears to be low as the RMSE of the identified homologous points was 4.42 m for the area of the landslide and 2.28 m for an adjacent area.

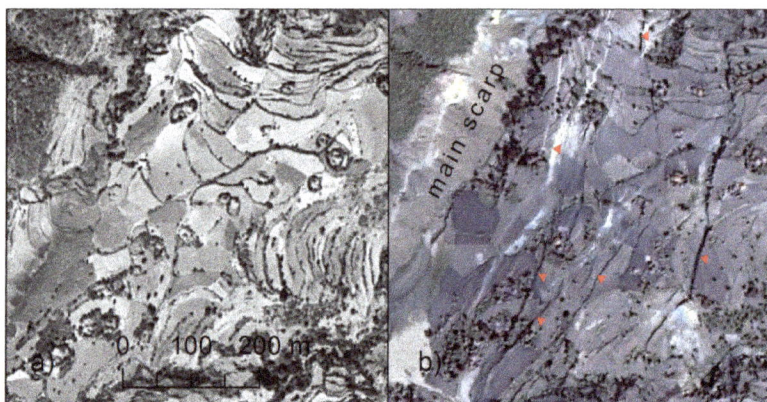

Figure 4. Comparison of ortho-images from aerial photographs taken on 27 November 1986 (**a**) and from Kompsat-2 taken on 27 January 2008 (**b**) for an area in the upper part of the landslide (For its position see the Section 4.2). The main scarp, secondary scarps and open cracks are clearly visible. The scarps facing the SE appear as bright lines, whereas the open cracks and scarps facing the E and NE (back-scarps) are dark. The position of the subset is indicated in the Figure 7.

The ortho-image of Ikonos-2 from 2005 is affected with a higher uncertainty in position as there is no DEM available capturing the landslide surface between the two major phases. Since we used the post-event DEM for the orthorectification, the error in position can in theory reach up to 3.1 m per 10.0 m meters of elevation difference as the off-nadir angle of the acquisition was 17.5°. Assuming the maximal horizontal difference of 25 m, which is the half of the maximal difference resulting from both the major movement phases, we can estimate the uncertainty of the Ikonos-2 image from 2005 due to the DEM as 6.0 m.

As we used ground control with unknown accuracy, we could not guarantee a high absolute positional accuracy of the ortho-images. However, our aim was to derive information on the relative movement of the landslide surface with respect to the stable surrounding terrain. This could be achieved by an accurate co-registration of the ortho-images before the application of feature tracking. We applied an automatic tie point identification and co-registration using polynomial adjustment in AutoSync. The Ikonos-2 image from 2005 was used as the master image (Table 1).

To assess the residual error a set of tie points between each image and the Ikonos-2 image from the year 2005 was identified in the off-landslide area. These tie points were selected independently from the previously used ground control points (GCPs). The RMSE of the residuals on the tie points was <2 m for all of the ortho-images (Table 1). The ortho-image from the aerial photographs that captured the situation before the first movements (described below), and is thus the first image of the time series, was co-registered in the same way. In this case, the RMSE was higher (6.3 m).

The accurately co-registered ortho-images allowed us to map in detail a number of surface features indicating the mass movement. The main scarp, minor scarps, small water bodies and a large number of cracks were identified and mapped. Contrast manipulation and RGB combinations of the visible and near infrared bands were used to fully exploit the potential of the images. The cracks already identified in a previous image and shifted to a new position were not mapped twice but only in the image of the first occurrence. The mapping was validated during a field trip in March 2015 (Figure 5).

Figure 5. The main scarp of the Debre Sina landslide, which is bordered from above by an old scarp now covered by shrubs (**a**); A large block of weathered volcanic rock, which was displaced by approximately 90 meters (**b**); Back-scarps in the main landslide body can reach up to 15 m in height (**c**). All photographs were taken by Zuana Vařilová in March 2015.

3.4. Estimation of Horizontal Displacement by Feature Tracking in Optical Satellite Data

Following [7,33,50] we applied an automatic approach for the extraction of vectors of horizontal surface movement of a landslide from a time series of remote sensing images. We used feature tracking in a frequency domain implemented in the COSI-Corr tool [5] to automatically extract the magnitude and azimuth of the surface displacements. The images of the time series (Table 1) were down-sampled to 4 m to minimize the effects of small noisy features and each two consecutive images were matched. Feature matching in frequency domain with initial and final window sizes of 32 and 128 pixels, respectively, was applied in a regular grid with spacing of 128 m. The resulting movement vectors have to be checked for outliers [7]. The vectors of excessive length and direction not following the slope inclination were checked against the satellite images and in the event of a discrepancy were manually discarded. The error of displacement magnitudes due to the image co-registration was estimated as the RMS of amplitudes in the off-landslide area, which yields 4.0 m for 2004 (Kompsat-2 and Ikonos-2) and 8.9 m for 2005 (aerial photographs and Ikonos-2). The total error of the displacement magnitudes was calculated taking into the account also the error due to the used DEM (6.0 m), which yielded 10.1 m and 7.2 m, respectively.

The matching of Ikonos-2 images from 2005 and 2007 provided poor results. This was due to the difference in phenology and illumination conditions as the 2007 image was acquired in June while most of the other images were taken in winter. The next image of the time series was used instead (Kompsat-2 image from 2008). As a visual check of the shifts between the Ikonos-2 from 2007 and Kompasat-2 images revealed no changes, the resulting vectors represent the movements after the rainy season in 2006. We calculated the gradient of the displacement fields to better understand the kinematics of the landslide. We applied a simple approach based on plane fitting to the values of displacement in a floating window 3 by 3 pixels using the Least Square technique. To account for the noise, the displacement images were median filtered beforehand. The gradient was estimated as the range of the fitted values and converted to m/m units.

3.5. Estimation of Horizontal Displacement by Amplitude Tracking in Microwave Domain

Two ascending scenes, acquired by ASAR (Advances Synthetic Aperture Radar) instrument on Envisat on 31 July 2004 and 29 October 2005, were obtained from the archive of the ESA. The first represents the situation before the slope failure while the second corresponds to the time period between the initial and the second movement. To account for the inaccuracies in the orbital parameters a co-registration with sub-pixel accuracy was applied to the image pair using cross-correlation function in 32×64 window. Surface displacements in slant and azimuth directions between the scenes were detected using the amplitude tracking module of the GAMMA software package. SAR amplitude tracking is a suitable method for measuring surface displacement over landslides [51]; however, its results are sensitive to the search window size, which in theory should not be less than the estimated ground displacement [15]. Different window sizes for different SAR sensors were tested and discussed for glacier studies by [15]. In this study, a window size of 32×64 single-look pixels was used, which corresponds to approximately 640 m in the ground-range direction and 256 m in the azimuth direction. Following [51] a high signal to noise ratio (SNR) value of 11 was applied to ensure high coherence. The resulting displacements in range and azimuth directions were projected to the horizontal plane using the local incidence angle and then converted to magnitude using Euclidean geometry. Further, the magnitudes were checked for outliers and manually corrected. The gaps were closed using ordinary kriging [52] interpolation which resulted in a regular grid of displacement values with grid spacing of 120 m. The arrows in the off-landslide area is probably due to the difference of velocities as the size of the processing window is optimized to certain displacement magnitude and can therefore produce noise in the area of no movements [10].

4. Results

4.1. Structural Predispositions for the Mass Movements in the Area

An expert-based morphostructural analysis of the MER escarpment around Debre Sina using SRTM DEM revealed major morpholineaments striking SSW-NNE (Figure 6). These topographic features most probably represent traces and scarps of tectonic (normal) faults related to the African rift, which also provide evidence of an extensional regime in the area of the landslide. Several minor morpholineaments striking NE-SW, NW-SE, E-W, and N-S are dissected by the major ones and are probably of older age.

Areas affected by slope failures were indicated by large concave scarps, irregular instead of strata-controlled topography, and a convex slope foot, *etc.*, and they were also mapped by the morphostructural analysis of the SRTM DEM. The results revealed an extremely large extent of gravitational slope failures in the broader area of the MER escarpment (Figure 6); slope failures covered approximately 126 km^2 of study area, which is approximately 12 % of the area shown in the Figure 6. These morphologically significant slope failures are mostly represented by rotational deep-seated rockslides and large earthflows.

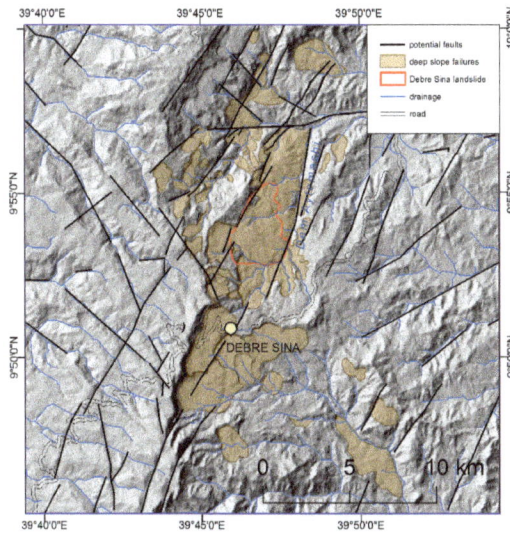

Figure 6. Morphostructural analysis of the SRTM DEM of the surroundings of the Debre Sina landslide revealed a number of areas affected by old deep seated slope failures of the MER escarpment.

4.2. Surface Morphology and Extent of the Landslide

A detailed look at the Debre Sina landslide using satellite images (Table 1) revealed a pattern of topographic features on the main body of the slide, as well as the relatively shallower subsequent slope failures (Figure 7). The field survey revealed that the majority of the mapped cracks featured a vertical displacement, *i.e.*, they represent scarps or back-scarps (Figure 5). Most of the scarps and open cracks were found on the first and second images of the time series (Table 1), corresponding to movements after the rainy seasons of 2005 and 2006. Several cracks appeared after the rainy season of 2007. No cracks were found on the later images indicating that no major re-activation has occurred since 2007. The shape of the cracks indicates the rotational character of the movement in the upper part of the affected slope.

According to our observations, especially on zones of high displacement gradients, the extent of the main landslide is approximately 6.5 km^2. The lower margin of this zone marks the presumed outcrops of the shear plane of the main landslide. The total area of the Debre Sina landslide, including the relatively shallow subsequent landslides, is 4.7 km in an N-S direction and 3.5 km in an E-W direction, which is approximately 8.5 km^2. The elevation difference between the crown and the toe is approximately 500 m, taking into the account only the main landslide. The subsequent landslides reached the bottom of the valley, which is 150 m lower. The maximum estimated thickness of the active main landslide body is approximately 150–200 m according to the interpretation of the topographic profiles, whereas the maximum thickness of the old slope failure is up to 300–400 m (Figure 8). To estimate the landslide volume we used an empirical formula for rotational landslides by [53]: V_{Ls}

$= 1/6 \, \pi \, D_d \cdot W_d \cdot L_d$, where V_{Ls} is the landslide volume after the movement including loosening of the mass, D_d is the landslide depth, W_d is the width of the landslide, and L_d is the length of the landslide. Assuming a thickness of 200 m, the estimated volume of the active main landslide body is approximately 1.7 km^3.

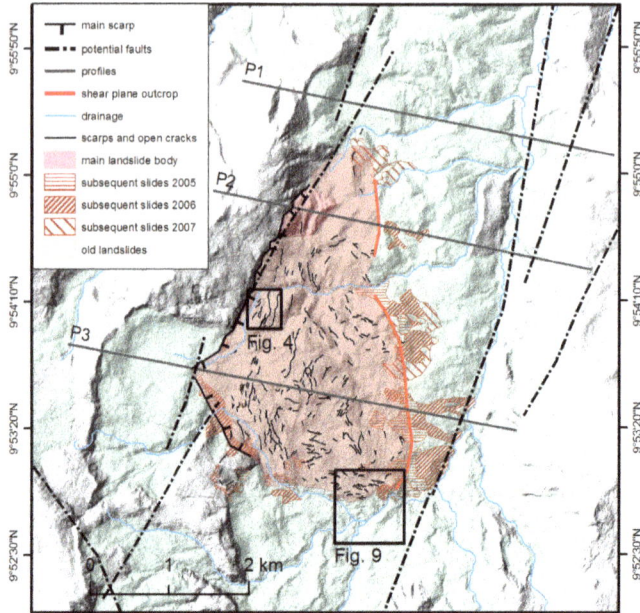

Figure 7. The main scarp coincides with the tectonic lineament. The main landslide body is followed by subsequent relatively shallow landslides below the presumed outcrop of the shear plane.

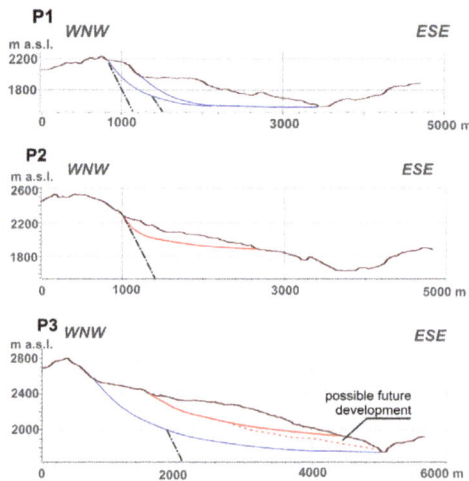

Figure 8. Topographic profiles and interpreted cross-sections of the slope failure under study: the estimated basal shear zones of old slope failures are presented in blue, the active slope failures are in red, and tectonic faults are marked as purple dash-and-dot lines.

The time series of the satellite images revealed that the southern part of the landslide was more active and the deep-seated mass movements reached the bottom of the valley in the SE. The river was blocked temporarily in 2005 leading to the emergence of a small lake (1.1 ha) and to a diversion of the river course approximately 150 m to the south (Figure 9).

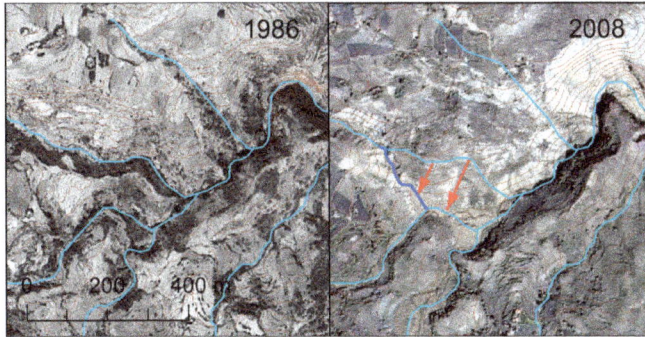

Figure 9. Change of the drainage pattern on the southern margin of the Debre Sina landslide. The position of the subset is indicated in the Figure 7.

4.3. Horizontal Displacements Revealed by Feature Tracking

The displacement fields derived from the satellite images clearly document that the main movements occurred in two phases after the rainy seasons of 2005 and 2006 (Figure 10). The amplitude of horizontal movements reached up to 48 ± 10.1 m and 114 ± 7.2 m in the first and second phase, respectively. It can be seen that the movement vectors are divergent (Figure 10), which means that lateral spreading took place in both phases. Both the displacement fields show that the lower part of the slope between the landslide body and the river remained stable during both of the major movement phases. Furthermore, it can be observed that the displacement magnitude increases towards the lower part of the slope within the main landslide. This indicates surface extension in both the lateral and transversal extension. The displacements also increase in an N-S direction, which could be due to higher infiltration above the southern part of the main landslide indicated by the high values of TWI (Figure 3).

The displacement fields of all of the other subsequent image pairs covering the period 2007–2014 do not show any coherent pattern over the landslide area. Furthermore, the magnitudes of these displacement fields do not exceed the RMS of the displacements in the off-landslide area, which means that they represent noise. This implies that there were no displacements or their magnitude was lower than the level of noise. As no cracks were detected in the VHR images after 2007 and no reactivation was reported by the local inhabitants interviewed by [23], we conclude that the slope stayed stable after 2007.

The displacement field from the amplitude tracking is in good agreement with the feature tracking in the optical domain (Figure 10). Its theoretical accuracy was calculated following [13] as being 0.3 m; however, the real error is much higher as the vectors identified in the off-landslide area result in an RMSE of 6.6 m.

The gradients derived from the displacement field for movements in 2005 and 2006 (Figure 11) provide detailed information on the kinematics of the slope failure. The gradient images feature spatial patterns of belts of high gradient that are bordered by zones of low gradient which correspond to areas of similar movement magnitude. The belts of high gradient partially correspond to the zones of open cracks and secondary scarps.

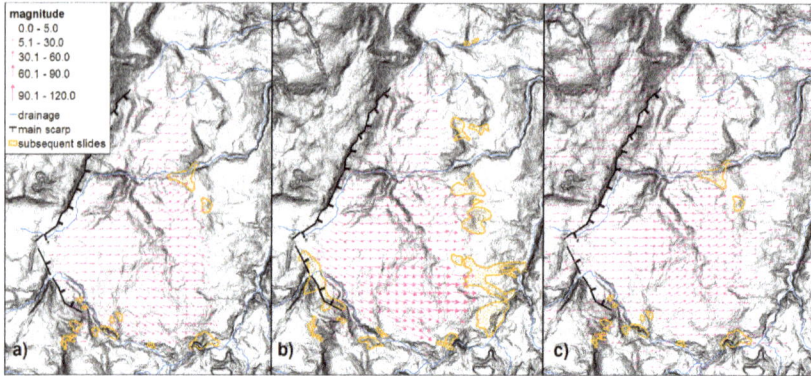

Figure 10. Vectors of horizontal displacements of the two major phases of activity from the feature tracking in the optical domain in 2005 (**a**) and 2006 (**b**). Aerial photographs from 1986, Ikonos-2 image from 14 December 2005 and Kompsat-2 image from 27 January 2008 were used for the feature tracking. No vectors could be derived for the relatively shallow subsequent landslides as their surface was disturbed during the sliding. The displacement field from Envisat/ASAR images from 31 July 2004 and 29 October 2005 indicates that the major movements in 2005 took place before the end of October (**c**). The vectors in the off-landslide area (also above the main scarp) are considered to be noise. The slope inclination map based on the pre-event DEM is shown in the background.

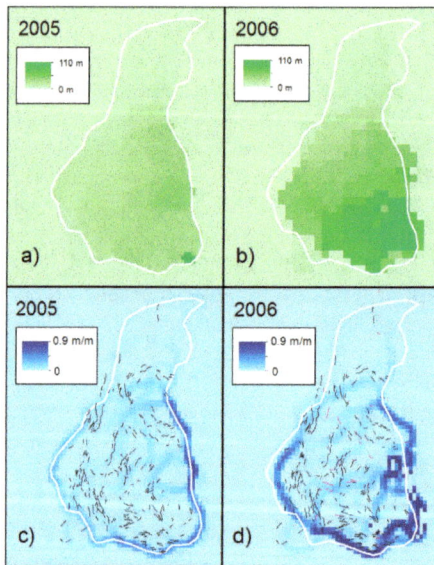

Figure 11. Magnitude of horizontal movements calculated by feature tracking from VHR images for the movement phases in 2005 (**a**) and 2006 (**b**). Gradients of displacement for 2005 (**c**) and for 2006 (**d**) represent surface deformation patterns. The extent of the main landslide body activated in 2005 is shown as a white outline in all sub-images showing the change in the extent of the landslide area in 2006. Scarps and open cracks longer than 50 meters are shown as black (2005) and magenta (2006) lines.

4.4. Vertical Displacements Identified by Differential DEMs

The map of vertical displacements (Figure 12) shows the total change in vertical direction in the period from 1986 to 2008 and thus comprises both the 2005 and 2006 events. The map shows zones of elevation decrease in the upper part of the main body reaching −42 ± 3.6 m and zones of elevation increase in its lower part showing the displacement of the mass with vertical differences reaching 49 ± 3.6 m (Figure 12). This pattern corresponding to depletion and accumulation distinguishes the deep seated movement of the main body from the subsequent shallower landslides. Furthermore, detailed information such as the thickness of sediment filling of the valleys by mobilized unconsolidated material can be retrieved from the difference image. The depth of the filling along the southern margin of the main landslide reaches around 47 ± 3.6 m. Another pattern clearly visible in Figure 12 is an interleaving of zones of positive and negative elevation differences in the middle part of the main landslide. These zones correspond to a replacement of surface undulations already present in the pre-event DEM, leaving areas of negative elevation differences in their original position and creating areas of positive differences in the new position. One of these ridges oriented in a NW-SE direction is formed by deeply weathered volcanic rock surrounded by unconsolidated colluvial material (Figure 5b). The DEM difference also captured a large rock fall that transformed the main scarp in the SW part of the landslide (Figure 2a).

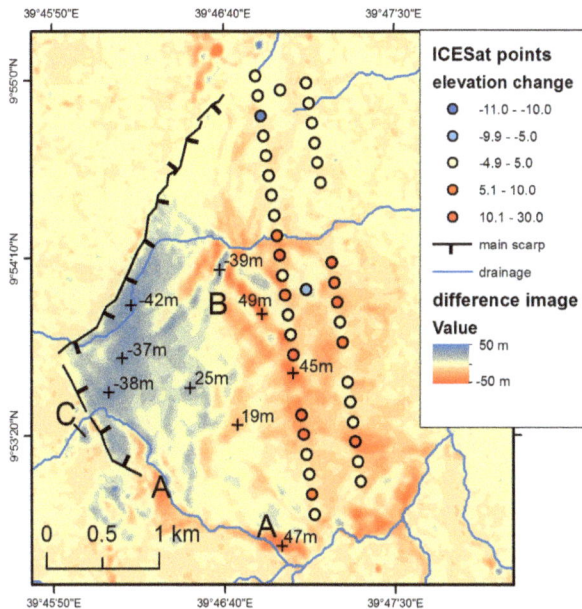

Figure 12. Elevation changes calculated from the pre-and post-event DEMs. The filling of valleys by mobilized sediment can be clearly seen (**A**). The stripes of negative values followed by stripes of positive values in the central part of the landslide are due to a shift of terrain ridges. Negative and positive areas mark their original and new positions, respectively (**B**). A large rock fall beyond the main scarp is marked by (**C**). The point rows across the accumulation area represent elevation changes derived as a difference between the ICESat measurements and the pre-event DEM. The points follow three satellite ground tracks from 17 June 2006, 2 April 2009 and 8 December 2008 (from W to E).

Three usable ICESat ground tracks that cross the accumulation part of the main landslide body (Figure 12) contain 39 point measurements yielding elevation differences ranging from −11.0 m to 24.6 m. One track was acquired between the two major phases of the movement (17 June 2006) while the two remaining tracks were acquired after the second phase. Over land, single elevation differences can be affected by inaccuracies due to various influences including local surface slope, roughness and errors in fitting the return waveform [54]. However, the elevation changes extracted by the two different methods feature a high correlation (r = 0.92). Only the points of the two tracks acquired after the second movement phase (19 points) were used for the correlation. To our knowledge this is the first time that a combination of ICESat data and DEM was used to measure vertical elevation changes of a landslide surface.

5. Discussion

5.1. Character of the Movement

The displacement fields revealed an extension in a longitudinal direction and lateral spread. The lateral spread is further confirmed by the existence of scarps and open cracks in a longitudinal direction (Figure 7). In addition, a higher movement magnitude in the southern part of the main landslide can be seen in both the displacement fields (Figure 10). This irregularity may be due to higher infiltration above this part, as indicated by the TWI (Figure 3).

The 3D reconstruction based on the aerial imagery resulted in detailed DEM but the lower geometrical quality affected the accuracy of feature tracking. Nevertheless, the accuracy was sufficient to map the movements as its amplitude reached values one order of magnitude higher than the error. The huge elevation differences detected by DEM differentiating were a good indication of the depth of the slope failure. The differential DEM allowed us to delineate the main deep-seated landslide body from the subsequent shallower landslides. We could also identify the largely unaffected lower part of the slope. It is remarkable that large areas (Figure 7) of the main landslide body feature little change to the original surface despite significant horizontal movements.

5.2. Evolution of the Slope Failure

The analysis of the VHR images of the Debre Sina slope failure showed that there were large movements before 14 December 2005 and between 14 December 2005 and 5 June 2007. The displacement field from Envisat/ASAR shows that the first phase took place after 31 July 2004 and before 29 October 2005. Based purely on the VHR image analysis, we could not decide whether a continuous or abrupt movement took place. However, taking into account that the main triggering factor of landslides on the rift margin is rainfall [19,21,55] we can suppose that the main movement of the Debre Sina landslide occurred in two phases after the rainy periods of 2005 and 2006. This agrees with the findings of [24] who carried out interviews with local inhabitants. We could also see that the extent of the main landslide was almost identical in both phases. In addition, we could measure the amplitude of the horizontal movements, which revealed that the second phase had higher amplitude than the first one. Furthermore, we could assess the position of the outcrop of the shear plane. The shear plane outcrop was presumed to be much deeper by [23,26] but the feature tracking and the DEM differencing clearly showed that the lower part of the slope was only partially affected by relatively shallow movements. These subsequent landslides were caused by a collapse of pushed out material along the shear plane of the main body. It can be expected that the landslide will prograde in this area and it will eventually reach the river [55].

Large shallow landslides were still occurring on the slope below the shear plane outcrop in 2007. The present processes mainly affect the disturbed surfaces of the shallow landslides and they have a prevalently water erosion character. Some of these disturbed areas are also the subject of small scale remediation measures by local farmers such as terracing on the denudated surface of shallow landslides.

5.3. Predisposition and Triggering

Regarding the predisposition of the Debre Sina landslide, the active tectonics, seismic activity and thick sediment mantle on the long slopes of the MER clearly provide favorable conditions for the occurrence of large mass movements. The morphostructural analysis of the wider surroundings identified a number of old landslides in similar settings to the Debre Sina landslide (Figure 6) and clearly showed the tectonic predisposition of the landslides which agrees with the findings of [22]. The occurrence of the Debre Sina landslide has a very close resemblance to the rejuvenation of part of a boundary fault system in the Tarmaber area (Ankober border fault) [23,56]. Although it may be difficult to distinguish the origin of many of the terrain forms in a tectonically highly active area [57] the fault that is followed by the main scarp could be clearly identified in both pre- and post-event DEMs. The convex form of the landslide body, which is evident in both the pre- and post-event DEMs (Figure 3), indicates that the studied slope failure is probably a re-activation of an old large landslide. This finding is further supported by the existence of a steep slope covered with vegetation above the main scarp that has almost the same inclination and forms a continuous stripe above the present scarp (Figure 5a).

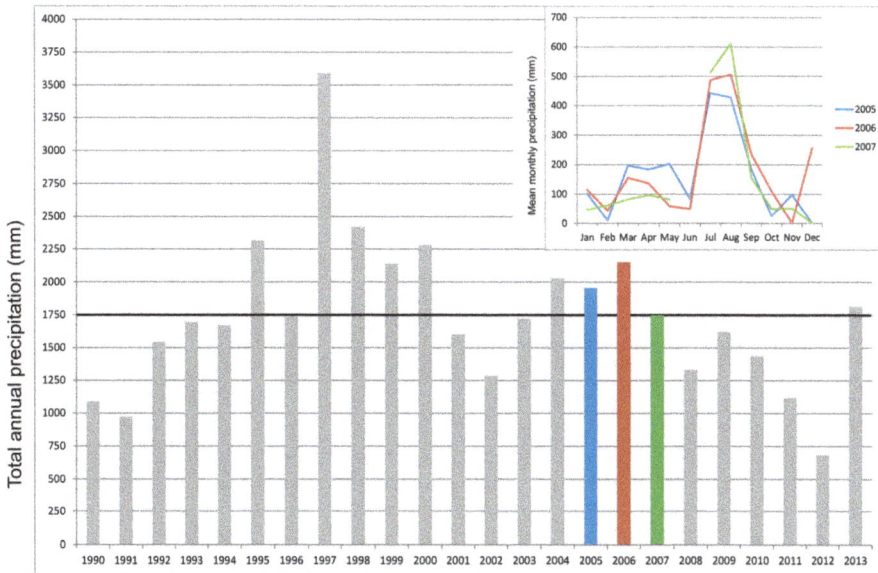

Figure 13. More than 20 years (1990 to 2013) of rainfall records from the Debre Sina station operated by the National Meteorology Agency of Ethiopia. The mean annual precipitation for the whole period marked by a horizontal black line is 1750 mm. The inlet graph of mean monthly precipitation in 2005, 2006 and 2007 shows periods of intensive rainfall in July and August.

A triggering of the Debre Sina landslide by an earthquake was suggested by several authors. A series of 162 earthquakes with magnitude > 4 connected to volcanic exposures in the Affar region in the period between 20 September and 4 October 2005 [58,59] and was identified as a trigger of the 2005 movement by [22]. However, the main movement in 2005 occurred sooner on 13 and 14 September [22,24,26] Furthermore, we were unable to find any seismic record corresponding to the largest 2006 event that also occurred in September [20]. Taking into account the huge volume, we presume that the main triggering factor is precipitation cumulated over a period of several months. However, the initiation of the movement can be due to a ground tremor. A single precipitation event listed in [22] as the daily maximum rainfall is unlikely to trigger a slope failure of such a large extent.

The precipitation records taken directly in Debre Sina show that the years 2005 and 2006 were above average but not extreme (Figure 13). We suggest that the sliding events were driven by a combination of geologic and tectonic predispositions together with external factors such as long-term water saturation and/or seismic events.

6. Conclusions

This study demonstrated the high potential of remote sensing techniques for the investigation of a large landslide with difficult accessibility. Using a number of approaches, some of them rarely used for landslides, we were able to carry out a new and detailed interpretation of the Debre Sina landslide. A combination of these techniques with the support of a limited field survey provided us with information about the landslide concerning its extent, kinematics, zonation and evolution over time.

The displacement fields derived by feature tracking in an optical domain provided us with quantitative information related to the particular phases of the slope failure development. Furthermore, we could distinguish the main landslide body from the subsequent shallower slides and identify the stable lower part of the slope affected only by the shallower movements. The amplitude tracking using archive Envisat data provided a lower resolution than the optical VHR images and appeared to be sensitive to processing parameters; nevertheless, its results allowed us to narrow the time window of the first landslide phase. In addition, it provided an independent validation of the results from the optical domain.

The DEMs derived by different techniques from historical aerial photographs and from modern stereo acquisitions appeared to show high potential for the general analysis of the extent and type of movements as well as for detailed feature identification. The derivatives of the detailed pre-event DEM such as TWI or slope inclination appeared to be useful tools for understanding the fine morphology and processes leading to the development of the slope failure. Our only high resolution pre-event dataset relies on an old aerial image from the EMA. This highlights the importance of national archives of aerial photographs reaching far before the era of VHR satellites. The good management of such archives ensuring easy accessibility and digitalization of analogous media is inevitable for the evaluation of this precious data source.

The analysis of ICESat measurements over the accumulation part of the main landslide body matches the results of the DEM differencing. Furthermore, it provides an independent validation. To our knowledge, this was the first time that a landslide accumulation was measured by a satellite laser altimeter. This indicates the potential of future satellite altimetry missions for the measurement of slope deformations.

Detailed terrain information representing the juxtaposition of the DEMs with the VHR images allowed us to identify a number of detailed features such as the appearance of temporary water bodies or changes of the river course and the type of the subsequent landslides. The occurrence and in particular the dimensions of the Debre Sina landslide confirm the high susceptibility of volcanic terrains to large gravitational mass movements in accordance with other published studies worldwide.

Acknowledgments: This study was carried out in the framework of the German Science Foundation (DFG) project "Integrated assessment of geomorphological process dynamics on different spatio-temporal scales in the Ethiopian Highlands using remote sensing and advanced modelling approaches" (HO1840/11-1) and the Czech Science Foundation (GAČR) project "Mass wasting and erosion as an indicator of morphotectonic activity in the Ethiopian Highlands based on remote sensing approaches" (P209/12/J068). The ALOS/PRISM data were provided by the ESA in the framework of the project ID13160. The aerial photographs were provided by the Ethiopian Mapping Agency. We would like to thank the Gesellschaft für Erd- und Völkerkunde, Stuttgart, Germany for supporting the field campaign in 2015. Last but not least, we would like to thank Jiří Šíma from AQUATEST a.s., Prague, Czech Republic and Leta Alemayehu from the Geological Survey of Ethiopia for providing valuable information and useful advice.

Author Contributions: Jan Kropáček carried out the remote sensing data analyses and wrote a substantial part of the manuscript. Zuzana Vařilová took care of the literature, evaluated the field investigations and took part in the writing of the text. Ivo Baroň performed the morphostructural analysis and contributed to the text.

Remote Sens. **2015**, 7, 16183–16203

Atanu Battacharya carried out the amplitude tracking. Joachim Eberle contributed by performing the field data interpretation. Volker Hochschild managed the project activities and reviewed the text.

Conflicts of Interest: The authors declare no conflict of interest.

References

1. Kääb, A. Photogrammetry for early recognition of high mountain hazards: New techniques and applications. *Phys. Chem. Earth Part B: Hydrol. Ocean. Atmos.* **2000**, *25*, 765–770. [CrossRef]
2. Kääb, A. Monitoring high-mountain terrain deformation from repeated air- and spaceborne optical data: Examples using digital aerial imagery and ASTER data. *ISPRS J. Photogramm. Remote Sens.* **2002**, *57*, 39–52. [CrossRef]
3. Delacourt, C.; Raucoules, D.; Le Mouélic, S.; Carnec, C.; Feurer, D.; Allemand, P.; Cruchet, M. Observation of a large landslide on La Reunion Island using differential SAR interferometry (JERS and Radarsat) and correlation of optical (Spot5 and Aerial) images. *Sensors* **2009**, *9*, 616–630. [CrossRef] [PubMed]
4. Kääb, A.; Vollmer, M. Surface Geometry, Thickness changes and flow fields on creeping mountain permafrost: Automatic extraction by digital image analysis. *Permafr. Periglac. Process.* **2000**, *11*, 315–326. [CrossRef]
5. Leprince, S.; Barbot, S.; Ayoub, F.; Avouac, J.-P. Automatic and precise orthorectification, coregistration, and subpixel correlation of satellite images, application to ground deformation measurements. *IEEE Trans. Geosci. Remote Sens.* **2007**, *45*, 1529–1558. [CrossRef]
6. Scherler, D.; Leprince, S.; Strecker, M.R. Glacier-surface velocities in alpine terrain from optical satellite imagery—Accuracy improvement and quality assessment. *Remote Sens. Environ.* **2008**, *112*, 3806–3819. [CrossRef]
7. Stumpf, A.; Malet, J.P.; Allemand, P.; Ulrich, P. Surface reconstruction and landslide displacement measurements with Pléiades satellite images. *ISPRS J. Photogramm. Remote Sens.* **2014**, *95*, 1–12. [CrossRef]
8. Lacroix, P.; Berthier, E.; Maquerhua, E.T. Earthquake-driven acceleration of slow-moving landslides in the Colca valley, Peru, detected from Pléiades images. *Remote Sens. Environ.* **2015**, *165*, 148–158. [CrossRef]
9. Fruneau, B.; Achache, J.; Delacourt, C. Observation and modelling of the Saint-Etienne-de-Tinee landslide using SAR interferometry. *Tectonophysics* **1996**, *265*, 181–190. [CrossRef]
10. Rott, H.; Scheuchl, B.; Siegel, A.; Grasemann, B. Monitoring very slow slope motion by means of SAR interferometry: A case study from a mass waste above a reservoir in the Ötztal Alps, Austria. *Geophys. Res. Lett.* **1999**, *26*, 1629–1632. [CrossRef]
11. Strozzi, T.; Farina, P.; Corsini, A.; Ambrosi, C.; Thüring, M.; Zilger, J.; Wiesmann, A.; Wegmüller, U.; Werner, C. Survey and monitoring of landslide displacements by means of L-band satellite SAR interferometry. *Landslides* **2005**, *2*, 193–201. [CrossRef]
12. Delacourt, C.; Allemand, P.; Berthier, E.; Raucoules, D.; Casson, B.; Grandjean, P.; Pambrun, C.; Varel, E. Remote-sensing techniques for analysing landslide kinematics: A review. *Bull. Soc. Geol. France* **2007**, *178*, 89–100. [CrossRef]
13. Strozzi, T.; Luckman, A.; Murray, T.; Wegmüller, U.; Werner, C.L. Glacier motion estimation using SAR offset-tracking procedures. *IEEE Trans. Geosci. Remote Sens.* **2002**, *40*, 2384–2391. [CrossRef]
14. Pritchard, H.; Murray, T.; Luckman, A.; Strozzi, T.; Barr, S. Glacier surge dynamics of Sortebræ, east Greenland, from synthetic aperture radar feature tracking. *J. Geophys. Res.: Earth Surf.* **2005**, *110*, 1–13. [CrossRef]
15. Jiang, Z.-L.; Liu, S.-Y.; Peters, J.; Lin, J.; Long, S.-C.; Han, Y.-S.; Wang, X. Analyzing Yengisogat Glacier surface velocities with ALOS PALSAR data feature tracking, Karakoram, China. *Environ. Earth Sci.* **2012**, *67*, 1033–1043. [CrossRef]
16. Weber, D.; Herrmann, A. Contribution of digital photogrammetry in spatio-temporal knowledge of unstable slopes: the example of the Super-Saute landslide (Alpes-de-Haute-Provence, France). *Bull. Soc. Geol. France* **2000**, *171*, 637–648. [CrossRef]
17. Walstra, J.; Chandler, J.H.; Dixon, N.; Dijkstra, T.A. In time for change-quantifying landslide evolution using historical aerial photographs and modern photogrammetric methods. In Proceedings of the 20th ISPRS, Istanbul, Turkey, 12–23 July 2004; pp. 475–480.

18. Tsutsui, K.; Rokugawa, S.; Nakagawa, H.; Miyazaki, S.; Chin-Tung, C.; Shiraishi, T.; Shiun-Der, Y. Detection and volume estimation of large-scale landslides based on elevation-change analysis using DEMs extracted from high-resolution satellite stereo imagery. *IEEE Trans. Geosci. Remote Sens.* **2007**, *45*, 1681–1696. [CrossRef]

19. Abebe, B.; Dramis, F.; Fubelli, G.; Umer, M.; Asrat, A. Landslides in the Ethiopian highlands and the Rift margins. *J. Afr. Earth Sci.* **2010**, *56*, 131–138. [CrossRef]

20. Woldearegay, K. Review of the occurrences and influencing factors of landslides in the highlands of Ethiopia: With implications for infrastructural development. *Momona Ethiop. J. Sci.* **2013**, *5*, 3–31.

21. Ayalew, L. The effect of seasonal rainfall on landslides in the highlands of Ethiopia. *Bull. Eng. Geol. Environ.* **1999**, *58*, 9–19. [CrossRef]

22. Abay, A.; Barbieri, G. Landslide Susceptibility and Causative Factors Evaluation of the Landslide Area of Debresina, in the Southwestern Afar Escarpment, Ethiopia. *J. Earth Sci. Eng.* **2012**, *2*, 133–144.

23. Alemayehu, L.; Gerra, S.; Zvelebil, J.; Šíma, J. *Landslide Investigations in Tarmaber, Debre Sina, North Shewa Zone*; Amhara Regional State; AQUATEST a.s.: Prague, Czech Republic, 2012.

24. Woldearegay, K. Characteristics of a large-scale landslide triggered by heavy rainfall in Tarmaber area, central highlands of Ethiopia. *Geophys. Res. Abstr.* **2008**, *10*. EGU2008-A-04506-EGU02008.

25. Schneider, J.; Woldearegay, K.; Atsbah, G. Reactivated large-scale landslides in Tarmaber district, central Ethiopian Highlands at the western rim of afar triangle. In Proceedings of the International Geological Congress, Oslo, Sweeden, 6–14 August 2008.

26. Hagos, A.A. Remote sensing and GIS-based mapping on landslide phenomena and landslide susceptibility evaluation of Debresina Area (Ethiopia) and Rio San Girolamo basin (Sardinia). Ph.D. Thesis, Universita degli Studi di Cagliari, Cagliari, Italy, 2012.

27. Kebede, F.; Kulhánek, O. Recent seismicity of the East African Rift system and its implications. *Phys. Earth Planet. Inter.* **1991**, *68*, 259–273. [CrossRef]

28. Keir, D.; Ebinger, C.J.; Stuart, G.W.; Daly, E.; Ayele, A. Strain accommodation by magmatism and faulting as rifting proceeds to breakup: Seismicity of the northern Ethiopian rift. *J. Geophys. Res.: Solid Earth* **2006**, *111*. [CrossRef]

29. Midzi, V.; Hlatywayo, D.J.; Chapola, L.S.; Kebede, F.; Atakan, K.; Lombe, D.K.; Turyomurugyendo, G.; Tugume, F.A. Seismic hazard assessment in Eastern and Southern Africa. *Ann. Geophys.* **1999**, *42*.

30. USGS. Earthquake Hazards Program. Available online: http://earthquake.usgs.gov/earthquakes/search/ (accessed on 2 September 2015).

31. Gouin, P. *Earthquake History of Ethiopia and the Horn of Africa*; International Development Research Centre: Ottawa, ON, Canada, 1979.

32. Westoby, M.J.; Brasington, J.; Glasser, N.F.; Hambrey, M.J.; Reynolds, J.M. Structure-from-Motion photogrammetry: A low-cost, effective tool for geoscience applications. *Geomorphology* **2012**, *179*, 300–314. [CrossRef]

33. Lucieer, A.; de Jong, S.M.; Turner, D. Mapping landslide displacements using Structure from Motion (SfM) and image correlation of multi-temporal UAV photography. *Prog. Phys. Geogr.* **2013**, *38*, 97–116. [CrossRef]

34. Beven, K.J.; Kirkby, M.J.; Seibert, J. A physically based, variable contributing area model of basin hydrology. *Hydrol. Sci. Bull.* **1979**, *24*, 43–69. [CrossRef]

35. Conoscenti, C.; Maggio, C.D.; Rotigliano, E. GIS Analysis to assess landslide susceptibility in a fluvial basin of NW Sicily (Italy). *Geomorphology* **2008**, *94*, 325–339. [CrossRef]

36. Yilmaz, I. Landslide susceptibility mapping using frequency ratio, logistic regression, artificial neural networks and their comparison: A case study from Kat landslides (Tokat-Turkey). *Comput. Geosci.* **2009**, *35*, 1125–1138. [CrossRef]

37. Takaku, J.; Futamura, N.; Iijima, T.; Tadono, T.; Shimada, M. High resolution DSM generation from ALOS PRISM–Perform-ance Analysis. In Proceedings of the IEEE IGARSS, Barcelona, Spain, 23–27 July 2007.

38. Zwally, H.J.; Schutz, R.; Bentley, C.; Bufton, J.; Herring, T.; Minster, J.; Spinhirne, J.; Thomas, R. GLAS/ICESat L2 Global Land Surface Altimetry Data, Version 33, National Snow and Ice Data Center, Boulder, CO, USA, 2003.

39. Kääb, A.; Berthier, E.; Nuth, C.; Gardelle, J.; Arnaud, Y. Contrasting patterns of early twenty-first-century glacier mass change in the Himalayas. *Nature* **2012**, *488*, 495–498. [CrossRef] [PubMed]

40. Neckel, N.; Kropáček, J.; Bolch, T.; Hochschild, V. Glacier mass changes on the Tibetan Plateau 2003–2009 derived from ICESat laser altimetry measurements. *Environ. Res. Lett.* **2014**, *9*, 014009. [CrossRef]

41. Kropáček, J.; Neckel, N.; Bauder, A. Estimation of mass balance of the Aletsch Glacier, Swiss Alps, from ICESat laser altimetry data. *Remote Sens.* **2014**, *6*, 5614–5632. [CrossRef]

42. Nuth, C.; Kääb, A. Co-registration and bias corrections of satellite elevation data sets for quantifying glacier thickness change. *Cryosphere* **2011**, *5*, 271–290. [CrossRef]

43. Bhang, K.J.; Schwartz, F.W.; Braun, A. Verification of the vertical error in C-band SRTM DEM using ICESat and Landsat-7, Otter Tail County, MN. *IEEE Trans. Geosci. Remote Sens.* **2007**, *45*, 36–44. [CrossRef]

44. Koblet, T.; Gärtner-Roer, I.; Zemp, M.; Jansson, P.; Thee, P.; Haeberli, W.; Holmlund, P. Reanalysis of multi-temporal aerial images of Storglaciären, Sweden (1959–99)—Part 1: Determination of length, area, and volume changes. *The Cryosphere* **2010**, *4*, 333–343. [CrossRef]

45. Parise, M. Observation of surface features on an active landslide, and implications for understanding its history of movement. *Nat. Hazards Earth Syst. Sci.* **2003**, *3*, 569–580. [CrossRef]

46. Niethammer, U.; James, M.R.; Rothmund, S.; Travelletti, J.; Joswig, M. UAV-based remote sensing of the Super-Sauze landslide: Evaluation and results. *Eng. Geol.* **2012**, *128*, 2–11. [CrossRef]

47. Stumpf, A.; Malet, J.-P.; Kerle, N.; Niethammer, U.; Rothmund, S. Image-based mapping of surface fissures for the investigation of landslide dynamics. *Geomorphology* **2013**, *186*, 12–27. [CrossRef]

48. Farr, T.G.; Kobrick, M. Shuttle radar topography mission produces a wealth of data. *Eos Trans. AGU* **2000**, *81*, 583–583. [CrossRef]

49. Rabus, B.; Eineder, M.; Roth, A.; Bamler, R. The shuttle radar topography mission–A new class of digital elevation models acquired by spaceborne radar. *ISPRS J. Photogramm. Remote Sens.* **2003**, *57*, 241–262. [CrossRef]

50. Booth, A.M.; Lamb, M.P.; Avouac, J.-P.; Delacourt, C. Landslide velocity, thickness, and rheology from remote sensing: La Clapière landslide, France. *Geophys. Res. Lett.* **2013**, *40*, 4299–4304. [CrossRef]

51. Raucoules, D.; de Michele, M.; Malet, J.P.; Ulrich, P. Time-variable 3D ground displacements from high-resolution synthetic aperture radar (SAR). Application to La Valette landslide (South French Alps). *Remote Sens. Environ.* **2013**, *139*, 198–204. [CrossRef]

52. Krige, D.G. A statistical approach to some basic mine valuation problems on the Witwatersrand. *J. Chem., Metall. Min. Soc. South Afr.* **1951**, *52*, 119–139.

53. Turner, A.K.; Schuster, L.R. *Landslides: Investigation and Mitigation*; Transportation Research Board: Washington, WA, USA, 1996.

54. Hilbert, C.; Schmullius, C. Influence of surface topography on ICESat/GLAS forest height estimation and waveform shape. *Remote Sens.* **2012**, *4*, 2210–2235. [CrossRef]

55. Záruba, Q.; Mencl, V. *Landslides and Their Control*; Elsevier: New York, NY, USA, 1972.

56. Wolfenden, E.; Ebinger, C.; Yirgu, G.; Deino, A.; Ayalew, D. Evolution of the northern Main Ethiopian rift: Birth of a triple junction. *Earth Planet. Sci. Lett.* **2004**, *224*, 213–228. [CrossRef]

57. Baroň, I.; Kernstocková, M.; Faridi, M.; Bubík, M.; Milovský, R.; Melichar, R.; Sabouri, J.; Babůrek, J. Paleostress analysis of a gigantic gravitational mass movement in active tectonic setting: The Qoshadagh slope failure, Ahar, NW Iran. *Tectonophysics* **2013**, *605*, 70–87. [CrossRef]

58. Wright, T.; Ebinger, C.; Biggs, T.; Ayele, A.; Yirgu, G.; Keir, D.; Stork, A. Magma maintained rift segmentation at continental rupture in the 2005 Afar diking episode. *Nature* **2006**, *442*, 291–294. [CrossRef] [PubMed]

59. Yirgu, G.; Ababa, A.; Ayele, A. Recent seismovolcanic crisis in northern Afar, Ethiopia. *Eos, Trans. Am. Geophys. Union* **2006**, *87*, 325–329. [CrossRef]

remote sensing

MDPI

Article

Hybrid-SAR Technique: Joint Analysis Using Phase-Based and Amplitude-Based Methods for the Xishancun Giant Landslide Monitoring

Tengteng Qu, Ping Lu, Chun Liu *, Hangbin Wu, Xiaohang Shao, Hong Wan, Nan Li and Rongxing Li

Center for Spatial Information Science and Sustainable Development Applications, College of Surveying and Geo-Informatics, Tongji University, Shanghai 200092, China; 1989tengteng@tongji.edu.cn (T.Q.); luping@tongji.edu.cn (P.L.); hb@tongji.edu.cn (H.W.); 1533336@tongji.edu.cn (X.S.); 8lovehappy@tongji.edu.cn (H.W.); 123linan@tongji.edu.cn (N.L.); rli@tongji.edu.cn (R.L.)
* Correspondence: liuchun@tongji.edu.cn; Tel.: +86-21-6598-4460

Academic Editors: Zhenhong Li, Roberto Tomas, Randolph H. Wynne and Prasad S. Thenkabail
Received: 29 June 2016; Accepted: 17 October 2016; Published: 23 October 2016

Abstract: Early detection and early warning are of great importance in giant landslide monitoring because of the unexpectedness and concealed nature of large-scale landslides. In China, the western mountainous areas are prone to landslides and feature many giant complex landslides, especially following the Wenchuan Earthquake in 2008. This work concentrates on a new technique, known as the "hybrid-SAR technique", that combines both phase-based and amplitude-based methods to detect and monitor large-scale landslides in Li County, Sichuan Province, southwestern China. This work aims to develop a robust methodological approach to promptly identify diverse landslides with different deformation magnitudes, sliding modes and slope geometries, even when the available satellite data are limited. The phase-based and amplitude-based techniques are used to obtain the landslide displacements from six TerraSAR-X Stripmap descending scenes acquired from November 2014 to March 2015. Furthermore, the application circumstances and influence factors of hybrid-SAR are evaluated according to four aspects: (1) quality of terrain visibility to the radar sensor; (2) landslide deformation magnitude and different sliding mode; (3) impact of dense vegetation cover; and (4) sliding direction sensitivity. The results achieved from hybrid-SAR are consistent with in situ measurements. This new hybrid-SAR technique for complex giant landslide research successfully identified representative movement areas, e.g., an extremely slow earthflow and a creeping region with a displacement rate of 1 cm per month and a typical rotational slide with a displacement rate of 2–3 cm per month downwards and towards the riverbank. Hybrid-SAR allows for a comprehensive and preliminary identification of areas with significant movement and provides reliable data support for the forecasting and monitoring of landslides.

Keywords: hybrid-SAR technique; joint analysis; phase-based SAR; amplitude-based SAR; giant complex landslide monitoring

---

1. Introduction

Landslides are one of the major geo-hazards that pose great threats to many areas around the world. Landslides are widely distributed in the mountainous areas of western China [1,2]. Especially after the Wenchuan Earthquake in 2008 in China (Mw 7.9 or Ms 8.0), a large number of landslides were triggered and received considerable attention [3]. Numerous villages are scattered throughout this large-scale landslide-prone area, which raises great importance to identify potential active landslides. It is quite common for a large storm to produce new landslides in this area since the earthquake occurred. A positive and effective monitoring tool that can help find the hidden nature of large-scale landslides and minimize the unexpectedness is of great importance for landslide early warning and early recognition.

Measurements of the ground surface deformation over large regions can be carried out by using the Spaceborne Synthetic Aperture Radar (SAR) techniques. In particular, Differential Interferometric Synthetic Aperture Radar (DInSAR) is an effective method to measure deformation in landslides. The successful application of this technique in landslide monitoring has been widely documented [4,5]. However, several problems hinder the exploitation of the DInSAR technique in landslide monitoring. These limitations include spatial decorrelation due to long perpendicular baselines between SAR acquisitions, decorrelation caused by vegetation coverage changes, large deformation gradients, errors resulting from atmospheric phase screen (APS) and phase unwrapping errors. Advanced DInSAR methods have also been developed to address some of the aforementioned issues, including Permanent Scatterer-InSAR [6,7], SqueeSAR [8], Small Baseline Subset (SBAS) [9,10], the Stanford method for Persistent Scatterers (StaMPS) [11,12] and interferometric point target analysis (IPTA) [13,14]. These techniques use phase shift analysis of long time-series SAR images to investigate landslides with low displacement velocities (mm/year to a few decimeters/year) [15].

However, there is difficulty for the DInSAR techniques to retrieve deformation information in relatively fast-moving areas. For example, when rapid landslides occur in densely-vegetated areas, the low spatial density of persistent scatterers (PS) makes phase unwrapping extremely difficult, resulting in the unsuccessful detection of fast movement. To identify rapid movement with velocities exceeding the limits of DInSAR and the associated techniques, the exploitation of amplitude information from the SAR data using the pixel offset tracking technique demonstrated its advantage in landslide monitoring. For example, XiaoFan et al. [16] applied a sub-pixel offset technique to TerraSAR Spotlight data to monitor the Shuping landslide in the Three Gorges of China. Singleton et al. [17] used sub-pixel offset techniques to monitor episodic landslide movements in vegetated terrain. Shi et al. [18] used multi-mode high-resolution TerraSAR-X data to monitor landslide deformation with point-like target offset tracking. Raspini et al. [19] exploited the amplitude information in SAR images to map the Montescaglioso landslide. Bhattacharya et al. [20] evaluated the potential of SAR intensity tracking to estimate the displacement rate in a landslide-prone area in India.

In general, a giant complex landslide consists not only of slow-moving areas, such as creeping and deep-seated gravitational slope deformation, but also of fast-moving areas exhibiting non-linear slope movement, such as toppling and rotational landslides. Thus, the movement of the entire landslide may vary considerably and exhibit non-uniform behavior, indicating that no single method would be sufficient for such a complex task. Moreover, a quick monitoring response may be necessary before a long series of a SAR dataset would be accumulated. Additionally, the monitoring method should benefit from an integrated analysis of phase-based and amplitude-based methods.

In this article, a hybrid-SAR technique is proposed and applied to a representative giant complex landslide in southwestern China using a high-resolution TerraSAR-X Stripmap dataset. Both phase-based and amplitude-based techniques are applied to obtain the landslide displacements. Then, the application circumstances and influential factors of phase-based and amplitude-based methods are evaluated according to four aspects: (1) quality of terrain visibility to the radar sensor; (2) landslide deformation magnitude and different sliding mode; (3) impact of dense vegetation cover; and (4) sliding direction sensitivity. Specifically, the surface displacements measured by the in situ sensors of four boreholes were used in the evaluation of the hybrid-SAR technique. The deformation tendencies from both the SAR data and the in situ data showed consistency. The applicability of this new hybrid-SAR technique in complex giant landslide research is demonstrated and evaluated.

2. Methodology Comparison of Phase-Based and Amplitude-Based Techniques

Phase-based InSAR techniques and amplitude-based offset tracking methods have their advantages and limitations. Table 1 provides a brief comparison of these two methods according to their methodological differences. The following section provides a more comprehensive methodological analysis of the two techniques.

Table 1. Comparison of the phase-based InSAR technique with the amplitude-based pixel offset technique.

Methodological Comparisons	Phase-Based InSAR Technique	Amplitude-Based Pixel Offset Technique
Accuracy	Higher accuracy, proportional to wavelength	Lower accuracy, proportional to pixel size
Phase unwrapping errors	Phase unwrapping errors	No need for phase unwrapping
Detectable deformation rate	Suitable for slow rate of deformation	Suitable for high rate of deformation
Sensitivity to temporal decorrelation	More sensitive to temporal decorrelation	Less sensitive to temporal decorrelation
Sensitivity to atmospheric phase screen	Significant signal delays caused by atmospheric phase screen	Not affected by atmospheric phase screen
Measurement direction	One-dimensional line-of-sight direction	Two-dimensional measurements involving range and azimuth directions

2.1. Phase Based: DInSAR and Advanced DInSAR Techniques

The phase-based InSAR techniques have high accuracy in monitoring displacements, and the accuracy is proportional to the wavelength. Thus, these techniques are suitable for monitoring slow rates of deformation. The traditional DInSAR technique has been applied to monitor slow-moving landslides on the order of cm/year, while the advanced time-series DInSAR technique can detect extremely slow-moving landslides on the order of mm/year.

Because of its methodological limitations, the DInSAR technique is unable to derive fast-moving displacements with high spatial gradients. In a wrapped interferogram, the maximum displacement between neighboring pixels cannot exceed $\lambda/2$, where λ is the wavelength [21]. Furthermore, when considering phase unwrapping, the maximum displacement between neighboring pixels cannot exceed $\lambda/4$, that is the highest deformation gradient should be less than 0.5 interferometric fringes per pixel [22]. DInSAR also has a serious limitation related to dense vegetation, which results in rapid decorrelation between SAR acquisitions. For high-resolution TerraSAR-X satellites, the wavelength is smaller and more sensitive to vegetation cover. Moreover, its sensitivity to atmospheric variability also hinders its exploitation in landside monitoring.

To mitigate the limitations of DInSAR, advanced DInSAR techniques have been developed and include permanent scatterer interferometry (PSInSAR) and small baseline subset (SBAS) [23]. PSInSAR makes use of stable permanent scatterer pixels that show high coherence during long time intervals in a stacking of multi-temporal co-registered images. SBAS chooses image combinations with short temporal and spatial baselines to reduce decorrelation effects. However, in thickly-vegetated areas, a low density of PS and the loss of coherence can produce unreliable phase unwrapping errors, which may result in the failure of these techniques.

2.2. Amplitude Based: Pixel Offset Technique

Because of its methodological limitations, the amplitude-based pixel offset technique has lower accuracy compared with phase-based InSAR methods. The key processing step of the pixel offset tracking method is to obtain the peak locations for a two-dimensional cross-correlation function of two SAR image patches. Signal-to-noise ratio (SNR) estimates are then calculated by comparing the height of the correlation peak relative to the average level of the correlation function. Serving as indicators of the confidence level of each offset, the SNR values should be set with a threshold, and only the SNR values above the threshold should be used to calculate the offsets. Furthermore, the orbital ramp errors, the topographical errors and the ionospheric effects should be eliminated. At the end, the final range and azimuth offsets can be estimated by measuring the row and column offsets between two acquisitions. During the processing, various parameters, such as the cross-correlation window size and oversampling factor, should be carefully evaluated to adjust to the size of deformation features and SAR image pixel size [24].

The achievable accuracy could be theoretically expressed as the following expression [25]:

$$\sigma = \sqrt{\frac{10}{3N}} \frac{\sqrt{2 + 5\gamma^2 - 7\gamma^4}}{\pi\gamma^2} \tag{1}$$

where γ is the coherence and N is the number of pixels within the estimation window.

The offset tracking technique is suitable for monitoring high rates of deformation, e.g., rapid landslides on the order of m/year, and the accuracy is proportional to the pixel size of the SAR image. This method does not require phase unwrapping, is less sensitive to temporal decorrelation and is not affected by atmospheric artifacts. Furthermore, pixel offset measurements can provide both range and azimuth vectors as the InSAR technique is only sensitive to the line-of-sight (LOS) direction.

3. Study Area and SAR Dataset

3.1. Geological Setting of the Xishancun Landslide

The Xishancun Landslide (Figure 1) is a giant landslide located on the northern bank of the Zagunao River in Li County (Sichuan Province), which lies to the east of the Tibetan Plateau. The region is featured by active tectonics similar to the other areas surrounded by a river network on the Tibetan Plateau [26]. The landslide is approximately 22 km from Wenchuan City, and several studies have pointed out that Li County is among the most severe geohazard regions triggered by the Wenchuan earthquake in 2008 [27,28]. The Xishancun Landslide is considered to be influenced and accelerated by the Wenchuan Earthquake. Dai et al. [29] found that several main types of landslides, including shallow disrupted landslides, rock falls, deep-seated landslides and rock avalanches, were triggered by the Wenchuan Earthquake. The Xishancun Landslide can be considered a complex mixture of the above landslide types, involving both slow-moving and fast-moving movements. This landslide poses severe threats to the 317 National Road and those villages both on the slope and at the foot of the mountain.

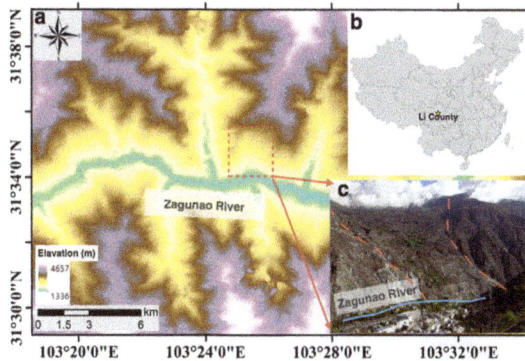

Figure 1. (a) Location of the Xishancun Landslide outlined in a red rectangular overlaid by an ASTER DEM; (b) location of Li County in China; (c) m of the Xishancun Landslide as seen from a terrestrial photo taken from the opposite bank of Zagunao River.

The Xishancun Landslide is a south-facing slope with erosional textures developed. The geomorphology of this landslide is relatively complex and forms a "V-shaped" valley. The giant thick accumulation body is bounded along the trailing edge by nearly vertical cliffs and along the leading edge by the Zagunao River. Both the eastern and western sides feature gullies that bound these two parts of the landslide. The elevation of the leading edge is approximately 1510 m, and the elevation of the trailing edge is approximately 3300 m. The elevation difference is 1790 m. The landslide length is approximately 3800 m; the minimum width is 680 m; and the maximum width is 980 m. The average thickness of the sliding body is 55 m. Thus, the volume of the landslide is approximately 1.7×108 m^3. Hence, it is considered as an oversized landslide.

There are many natural terraces in the landslide, and the terraces on the trailing edge and middle parts are much larger. The slope of the landslide body ranges from 25° to 55°, and steep slopes are mainly developed along the front part. Vegetation on the majority of the landslide body is scarce. However, due to the catchment function of the "V-shaped" geomorphology in the lower part and abundant water resources, groundwater is relatively accessible, and vegetation is present, with fruit trees and crops flourishing here, as well. The dense vegetation cover in the western part significantly limits the application of DInSAR in this area. Because of the construction of village roads on the mountain, the slopes cut into the toe along the leading edge are significant and form severe scarps that lead to collapses in some areas. The complexity of the topographic conditions and landslide slopes are considered to be the major factors resulting in the occurrence of geological hazards and were aggravated by the Wenchuan Earthquake.

This landslide is also less sensitive to displacement measurements along LOS due to the south-facing orientation. The complex landslide behaviors provide a good opportunity for the joint application of both the phase-based InSAR method and the amplitude-based pixel offset method.

3.2. Geotechnical Monitoring

To monitor the landslide with geotechnical equipment, a Spatial Sensor Network similar to the MUNOLD (Multi-Sensor Network for Observing Landslide Disaster) raised by Lu et al. [30] is currently being deployed in Xishancun Landslide. There are four boreholes (Bh1, Bh2, Bh3 and Bh4) located on the lower and middle parts of the landslide and distributed evenly along the slope (Figure 2). Four artificial corner reflectors are installed just beside the corresponding boreholes, which enable further comparisons (Section 5.5) between hybrid-SAR-derived borehole surface displacements and in situ sensor measurements. Four inclinometers with tilt sensors monitoring displacement along the main sliding direction have been installed inside the boreholes. The field data technically showed good performance, with a high rate of data return and reliable data access. The noise in the data collected was small and stable, thereby allowing reliable data support for further analysis.

Figure 2. Landslide map with locations of the geotechnical monitoring instrumentation and landslide boundary outlined in red. Red Dots 1–4 indicate the existing boreholes. Areas 1–6 will be introduced in the following sections.

3.3. TerraSAR-X Test Dataset

The acquisition of Stripmap TerraSAR-X data started in 2014. In total, six images of the study area were collected. The parameters of five image pairs used in later hybrid-SAR processing are listed in Table 2. Basic information on this TerraSAR-X dataset is provided in Table 3.

Table 2. The parameters of the five image pairs used in hybrid-SAR processing.

Master Image	Slave Image	Perpendicular Baseline (m)	Temporal Baseline (Day)
21 November 2014	13 December 2014	54	22
13 December 2014	15 January 2015	−190	33
15 January 2015	17 February 2015	−17	33
17 February 2015	11 March 2015	64	22
21 November 2014	11 March 2015	51	110

Table 3. Basic information on the acquired TerraSAR-X Stripmap (SM) datasets.

SM Data	
Orbit direction	Descending
Look angle (degree)	33.0
Heading (degree)	−169.7
Polarization	HH
Azimuth Spacing (m)	1.83
Range Spacing (m)	1.36

4. Analyses and Experimental Results of Hybrid-SAR

4.1. DInSAR Results from Representative Interferometric Pairs

Although limited datasets are not enough to carry out robust time series analyses, rapid identification of actively-deforming areas and early warnings of active regions are urgently needed. In this study, four interferograms with baselines shorter than 190 m and acquisition time intervals shorter than 33 days have been computed. To carry out two-pass differential interferometry, an external digital elevation model (DEM) covering the whole body of this landslide with a posting of 0.5 m produced by terrestrial laser scanning (TLS) was used (Figure 3). A multi-look factor of two was applied both in the range and in the azimuth directions.

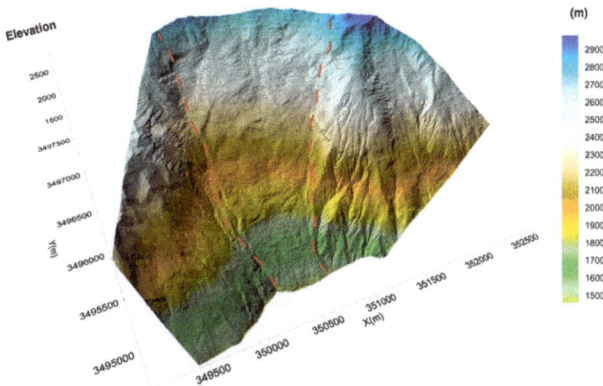

Figure 3. TLS-derived DEM with a spatial resolution of 0.5 m. The middle area surrounded by the red dashed line is the corresponding landslide location.

During the winter periods, the interferograms are less noisy, and the fringes caused by displacement can be determined in the interferograms. For example, in Figure 4a–d, the interferometric signals in the middle part of the landslide (Area 3) are evident. In addition, the eastern border of landslide (Area 2) and a region close to the toe of landslide (Area 1) also preserve good quality fringes in the interferograms. However, in the western part of the lower landslide body (Area 6), the coherence is not as well preserved due to the coverage of high-density vegetation. Additionally, the mountain to the west of the landslide is seriously affected by the layover effect of SAR images. These four interferograms with neighboring time spans show uniform fringe locations and indicate that long-lasting and slow deformation exists in these active regions.

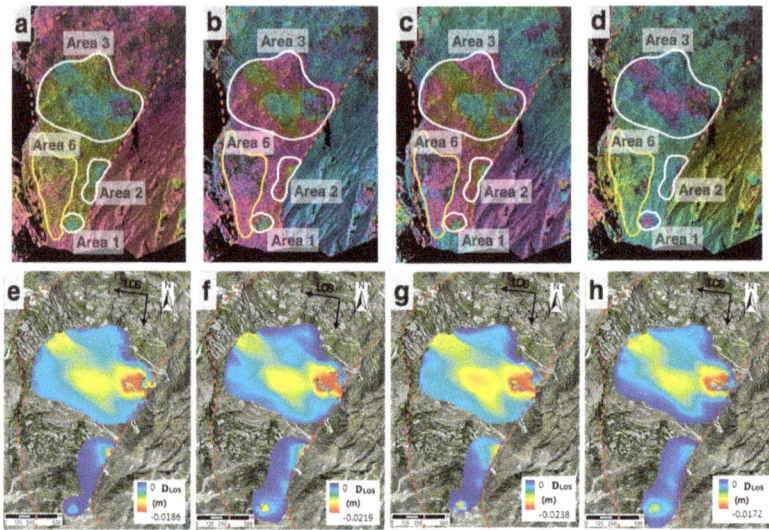

Figure 4. Geocoded differential interferograms (**a–d**) and geocoded displacement maps (**e–h**) in the LOS direction with temporal baselines and perpendicular baselines noted below. Significant landslide displacement signals are highlighted with white ellipses in the interferograms, which also correspond to Areas 1–3 in Figure 2; (**a,e**) 21 November 2014–13 December 2014 (22 days, 54 m); (**b,f**) 13 December 2014–15 January 2015 (33 days, −190 m); (**c,g**) 15 January 2015–17 February 2015 (33 days, −17 m); (**d,h**) 17 February 2015–11 March 2015 (22 days, 64 m).

The associated displacement maps are then derived along the LOS direction after performing the phase unwrapping of all interferograms. Because of the dense vegetation cover in the western lower part of this landslide, the coherence became quite low, and this area was masked during the unwrapping procedure in order to obtain a robust result for other parts of the landslide. Figure 4e–h shows that the maximum displacement in a 33-day time interval can reach up to 2.2 cm in the middle part of the landslide (Area 3) where three neighboring deformation areas can be identified. Moreover, the eastern borderline of the landslide body (Area 2) is quite evident in the displacement map, which provides good support for the interpretation of the landslide division. In the small deformation area close to the landslide toe (Area 1), the displacements showed an extremely slowly increasing tendency. A further detailed interpretation and analysis of deformation modes in Areas 1–3 will be carried out in Section 5.2.

Although unable to carry out an advanced time series DInSAR application at this stage, the quick response of conventional DInSAR technique over a short time span and its ability to outline the boundary of landslide-prone areas demonstrate the successful application of the phased-based InSAR technique in our landslide research.

4.2. Offset Tracking Results between SAR Acquisitions

In the offset tracking procedure, if the selected perpendicular baselines are quite low and the displacement rate is rather high, the range offsets due to local topography can be neglected with respect to ground displacement. The conventional co-registration procedure is sufficient for most cases where offsets due to topography are less than 0.2 pixels. However, for higher-resolution systems combined with large baselines, these topographical errors can exceed one pixel. In particular, in this study, the displacement gradients should not be this high, and the TerraSAR-X dataset exhibited high resolution and relatively large baselines. During the processing, offsets due to topography are incorporated into a co-registration look-up table that links the geometries of two images based on a DEM of the area. Thus, topography-related offsets were considered and removed.

Using the above approach, we derived two obvious deformation areas on the landslide body from two images in the TerraSAR-X dataset (spanning from 21 November 2014 to 11 March 2015). The results of azimuth and slant range deformation measurements are rendered in Figure 5. In the slant range displacement map, the geocoded DInSAR displacement result derived from 21 November 2014 to 13 December 2014 is also overlaid in order to make a clear comparison with the pixel offset result. It should be noted that only points with SNR values greater than 10.0 in pixel offset processing are highlighted. These values represent relatively high coherence and high reliability of the measurements.

From Figure 5, it can be seen that the whole landslide body demonstrates a significant movement magnitude from north to south downwards on the slope. In the vegetated area of the landslide (Area 5), phased-based InSAR failed to derive good quality fringe due to the decorrelation; however, some deformation signals are extracted successfully because of the existence of contrasting features (for example, buildings and corner reflectors) during the pixel offset processing. In the azimuth displacement map, this zone is active and moved a distance of 8–16 cm in a time span of four months. In the range displacement map, the displacements, i.e., approximately 2–4 cm per month, cannot be ignored either. The significant movement in Area 5 was also reflected from in situ measurements in this region, which will be analyzed in Section 5.5. Another active area (Area 4) that should receive more attention is the middle part of the Xishancun Landslide, which features obvious azimuth and slant range displacement. For this area, both phase and amplitude information are extracted successfully from SAR datasets and show good consistency. The slant range displacement is consistent with the results achieved from the DInSAR technique, being 2 cm per month. The azimuth displacement of this area amounts to 2–3 cm per month. A rotational failure mechanism is deduced in a later analysis in Section 5.2.

For our TerraSAR-X Stripmap dataset, the azimuth pixel size is 1.83 m (Table 2), which is also the minimum pixel dimension. The maximum detectable displacement between neighboring pixels in SAR images (a quarter of the TerraSAR wavelength) divided by the minimum pixel dimension is the maximum displacement gradient derived from the phase-based InSAR technique, being approximately 0.004 m/m. Therefore, even if we derive the original resolution interferogram, the maximum detectable difference between two points over a distance of 10 m will be only 0.04 m. In this case, a multi-look factor of two is applied for the range and azimuth direction; thus, the maximum InSAR detectable difference is only 0.02 m. However, in the Xishancun Landslide, regions with fast-moving phenomena exist, and their displacement gradients exceed the measurable limit (for example, Area 5). The results of the pixel offset technique demonstrate a good performance of this technique in detecting larger deformation magnitudes.

Theoretically, the range displacement map from the pixel offset technique is expected to contain the same information as the differential interferogram [24]. In order to quantitatively evaluate the combination of the two techniques, the differences between LOS displacements of ten points in Area 4 derived from DInSAR processing in a time span of 22 days and offset tracking processing in a time span of 110 days are calculated in Table 4. To make a uniform comparison of the consistency between two techniques, offset tracking displacements in 110 days are linearly converted to 22 days' displacements. The mean difference and RMS error of the LOS displacement differences of ten points in Table 4 derived from two techniques are 0.003 m and 0.0031 m, respectively, indicating an order of magnitude lower than the accumulated landslide displacements. Although there are regions that have quite low SNR values and were excluded from robust displacement calculation, the successful extraction of both phase and amplitude information in Area 4 confirms the pixel offset processing to be a good comparison with DInSAR results given enough time span and deformation magnitudes.

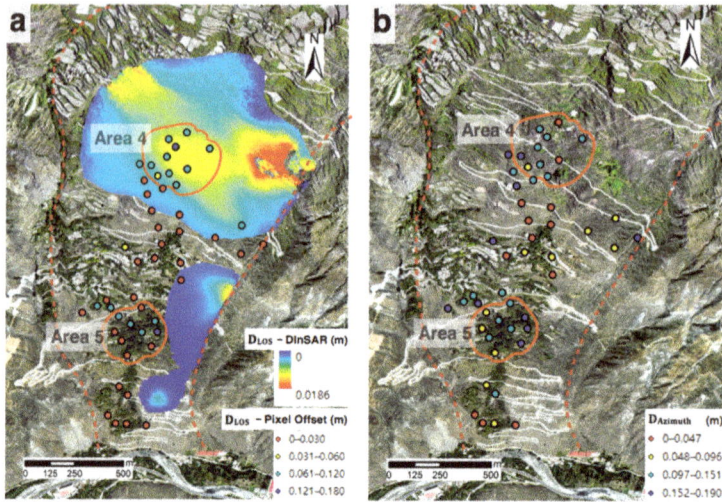

Figure 5. Slant range displacements (**a**) and azimuth displacements (**b**) measured from 21 November 2014 to 11 March 2015.

Table 4. LOS displacements and corresponding differences between ten points in Area 4 derived from DInSAR processing and offset tracking processing.

Point ID	Offset Tracking LOS Displacements (110 Days)/m	Offset Tracking LOS Displacements (22 Days)/m	DInSAR Displacements (22 Days)/m	Displacement Differences (22 Days)/m
1	0.09	0.016	0.013	0.003
2	0.10	0.018	0.014	0.004
3	0.06	0.011	0.009	0.002
4	0.10	0.018	0.014	0.004
5	0.06	0.011	0.010	0.001
6	0.07	0.013	0.010	0.003
7	0.09	0.016	0.012	0.004
8	0.06	0.011	0.008	0.003
9	0.08	0.014	0.012	0.002
10	0.07	0.013	0.009	0.004

5. Discussion

The methodological differences between phase-based and amplitude-based techniques presented in Section 2 led to their different application circumstances and influential factors in giant landslide research. As giant complex landslides always have diverse topographic features, complicated deformation patterns, as well as different orientations with respect to radar satellites, the applications of phase-based and amplitude-based methods in practical research may show unique advantages and disadvantages. In this section, the application circumstances and influential factors of phase-based and amplitude-based methods are evaluated according to four aspects: (1) quality of terrain visibility to the radar sensor; (2) landslide deformation magnitude and different sliding mode; (3) impact of dense vegetation cover; and (4) sliding direction sensitivity. The consistency and difference between phase-based and amplitude-based techniques demonstrated during their applications in the Xishancun Landslide research are fully exploited to evaluate the applicability of the hybrid-SAR technique in our case study.

5.1. Quality of Terrain Visibility to the Radar Sensor

The topography strongly influences the performance of both interferometric and non-interferometric techniques. The terrain visibility to the radar sensor depends on the satellite acquisition geometry and landslide terrain slope geometry.

The R-index (RI) represents a ratio between the pixel size in the slant range (radar geometry distance) and the ground range (Earth surface distance). To calculate the RI, the following parameters are needed: a DEM with slope (β) and aspect angles (α) and the LOS parameters, including the incidence angle (θ) and satellite ground track angle (γ). Notti et al. [31] proposed a simplified version of the formula to calculate the R-Index:

$$R = \sin\left[\theta - \beta * \sin\left(A\right)\right] \tag{2}$$

Here, A is the aspect correction factor. For descending data, A is computed as $A = \alpha - \gamma$ for descending and as $A = \alpha + \gamma + 180°$ for ascending data. The R-index ranges from -1 to 1. The meanings of the R-index values are listed below:

1 $R \leq 0$: The areas are affected by layover, foreshortening and shadow effects.
2 $0 < R < 0.4$: The pixel in this area exhibits strong compression.
3 $0.4 < R < 1$: The slope has good orientation, and the main factor that influences the following processing will be the land use.
4 $R = 1$: The slope is parallel to the LOS.

Figure 6 shows the R-index spatial distribution for descending geometry for the Xishancun Landslide. The calculated DEM was derived from TLS with a posting of 0.5 m, the same as that used in the previous processing. The figure shows that the Xishancun Landslide is mostly south oriented and has relatively high R-index values, indicating a good orientation. However, on the western slope of the landslide (Area 6), the R-index values are below 0.4, indicating the presence of compressed pixels. The overall R-index values of the whole landslide are higher than zero, indicating that the Xishancun Landslide has relatively good terrain visibility to the radar sensor.

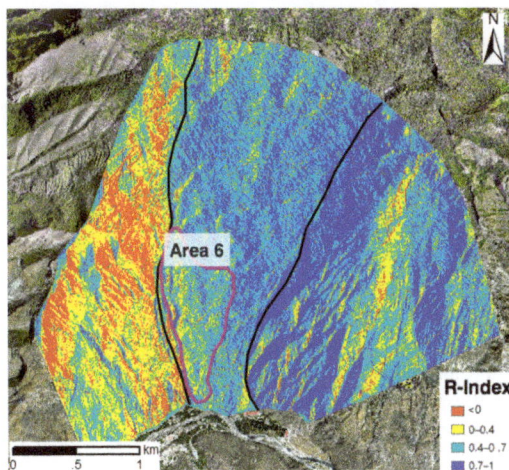

Figure 6. The R-index map of the Xishancun Landslide for the descending geometry.

The R-index can be used to identify areas of good terrain visibility and geometrical distortions, as well. This presence of image distortions may seriously hinder the exploitation of InSAR processing. From the interferograms in Figure 4, low R-index values in Area 6 could be one of the reasons that lead to its unclear interferometric fringes. Moreover, the presence of layover and shadowing not only prevents the application of the interferometric technique, but also limits the non-interferometric technique.

The good orientation of the landslide body and good terrain visibility to the radar sensor should be a prerequisite for the application of the hybrid-SAR technique. In practical cases, the calculation of the R-index spatial distribution of the landslide body with respect to the specific orbital geometry should be performed in advance to evaluate the application possibility of the hybrid-SAR technique.

5.2. Landslide Deformation Magnitude and Different Sliding Modes

During the in situ investigation, the Xishancun Landslide is shown to be a quite complex landslide exhibiting different types of movement. The application of the hybrid-SAR technique to the Xishancun Landslide successfully helps infer the existence of a rotational slide, an extremely slow earthflow and a creeping area.

Area 4 in Figure 7 is considered to be a rotational slide corresponding to the region in the middle part of the landslide. From the DInSAR results, obvious fringe also appeared in this area, with significant vertical gravitational movements at the head of the slide. From the offset tracking results, the deformation magnitudes in the azimuth and slant range direction are both relatively large, indicating significant downwards and northwards movements. From the high-resolution ortho-images, large main scarps have formed an obvious trailing edge of this rotational slide, and the slide boundaries show distinct terrain discontinuities with the surrounding areas. A close examination of the slope map of this area reveals that very high slope angles exist here. The sharp mountain trend and large height difference may be drivers of the rotation at the head of this slide. Both the texture of this area and hybrid-SAR results indicate that a rotational slide is a reasonable first-order interpretation.

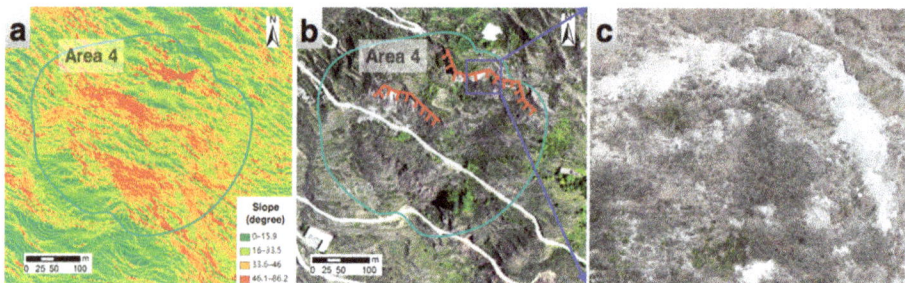

Figure 7. (**a**) The slope information of a rotational slide developed in the middle part of the Xishancun Landslide. Its location could be referred to Area 4 in Figure 2. (**b**) The rotational slide with a main scarp and a minor scarp both clearly visible in the DEM. (**c**) The enlarged view of the trailing edge earmarked by the rectangle in (**b**) from high-resolution ortho-images.

To be detected successfully by DInSAR, an earthflow motion has to be fast enough to be monitored over a short time span and slow enough to avoid radar decorrelation [32]. The earthflow corresponding to Area 2 is an ideal example and has a displacement rate of approximately 1 cm per month (Figure 8). This extremely slow earthflow corresponds to the fringe in DInSAR interferograms (Figure 4) along the eastern boundary of the landslide. The main earthflow body acts as a conveyor for material from the head of the slide and moves debris downslope through the transport zone to the depositional lobe and toe zone. During the in-situ investigation, many fractures have developed along the border scarp, and shallow surface flows of soil blocks were observed. As the slide material is not covered by foliage or tree canopy in this region, the identification from UAV-based low-altitude aerial photography could be easily accomplished. Especially from Figure 8b, the detailed enlarged view clearly verified the special textures and slide morphology of this earthflow.

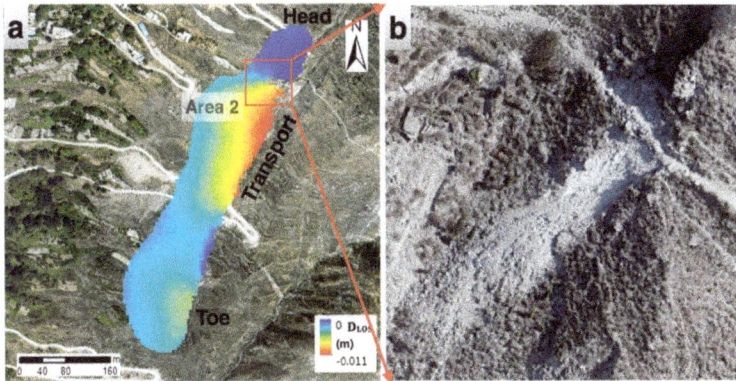

Figure 8. (**a**) An earthflow (Area 2) close to the eastern boundary of the landslide in the lower part. The DInSAR displacements derived from 13 December 2014–15 January 2015 are overlaid. (**b**) The enlarged view of the red square region obtained from the UAV-based high-resolution aerial photo.

For Area 1, the DInSAR results in Figure 4 (a uniform displacement velocity of no more than 1 cm/month) could help infer a very slow-moving landslide mode in this region. This may be creeping behavior related to the shallow sliding surface underground, which means the ground movements are mainly translational with the same slope angle as the topographic surface. The terrain slopes gently in this region; however, human activities are quite active, such as mountain road construction and farmland reclaiming. From Figure 9, the field survey found evidence of creeping mechanisms, such as ruptures in mountain roads, downward sliding of the turf on the rock beside the road and curved tree trunks on both sides of the road. The creeping in this area resulted in the fringe close to the landslide toe in DInSAR interferograms (Figure 4). The phase-based DInSAR could successfully detect the slow creeping behavior of the landslide, providing that the movement does not exceed the detectable gradient.

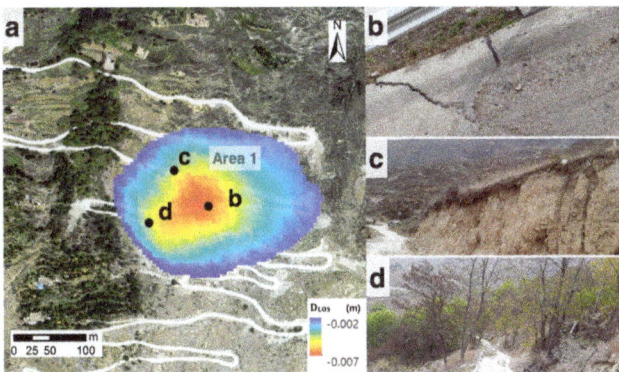

Figure 9. (**a**) A creeping region (Area 1) close to the toe of the landslide. The DInSAR displacements derived from 17 February 2015–11 March 2015 are overlaid. (**b**) Example of a rupture in the road. (**c**) Example of the turf sliding downwards. (**d**) Example of curved tree trunks in this area.

With the use of the high-resolution phased-based DInSAR technique, slow-moving landslides can be reliably detected and monitored. Slowly creeping sections and extremely slow earthflows show obvious fringes in interferograms, and rotational slides are evidenced by mainly vertical gravitational movement at the head of the slide. The offset tracking technique provides a robust method for measuring high gradient displacements, such as typical rotational slides and episodic movements, and a more reliable interpretation of landslide types providing both range and azimuth offsets. Pooling the strengths of the two methods, hybrid-SAR can provide more comprehensive information on the movement of giant complex landslides to help understand different landslide mechanisms.

5.3. Impact of Dense Vegetation Cover

The InSAR technique is tightly associated with the land cover at the regional scale [33]. In this study, the western part of the Xishancun Landslide body (Area 6) is covered by dense vegetation, which can be easily seen in Figure 2. This densely-vegetated region has low coherence, which seriously hinders the use of the InSAR technique. Decorrelation between the TerraSAR acquisitions caused by vegetation could be recognized on the interferometric coherence map in Figure 10. Although this interferometric pair has just 22 days of temporal baseline and 64 m of perpendicular baseline, the decorrelation due to dense vegetation cover still exerted quite an obvious effect. As the wavelength of TerraSAR-X is relatively small, its sensitivity to vegetated surfaces is more significant.

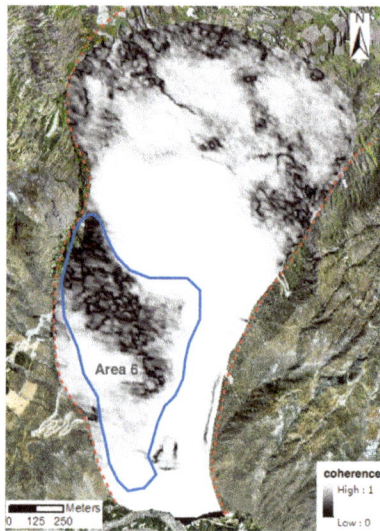

Figure 10. The interferometric coherence map for the pair of 17 February 2015–11 March 2015. The area with low coherence corresponds to Area 6, namely the most densely-vegetated region circled by blue lines.

Compared with the phase-based InSAR methods, the pixel offset technique makes use of SAR amplitude information and can overcome the InSAR limitations in regions with low coherence. More contrasting features on the vegetated terrain surface (buildings or corner reflectors) can provide a better estimate in the application of the offset tracking technique. From the results of the pixel offset technique, Area 6, which provided sparse interferometric information during DInSAR processing, preserved more radar backscatter changes via the use of amplitude information. Thus, the pixel offset approach should be considered a more robust method to resolve landslide movements with decorrelation problems.

To conclude, landslides with more contrasting ground features (natural or man-made) could highly benefit from hybrid-SAR via the use of both phase and amplitude information; thus, more deformation signals could be detected in densely-vegetated areas.

5.4. Sliding Direction Sensitivity

In InSAR processing, a movement parallel to the LOS could be fully registered, while the movement orthogonal to LOS cannot be registered. Especially for landslides that have a strong horizontal component, InSAR-derived movements would significantly underestimate the real deformation vector. In most cases, the majority of the velocity is along the line of the maximum slope. Based on a coefficient (C-index) proposed by Notti [34], which indicates the ratio between the velocity projected along the slope (VSLOPE) and the velocity in the LOS (VLOS), we can estimate the percentage of movement detected along the slope using interferometric techniques. The values of C depend only on the radar LOS geometry and landslide topographical geometry. Using the expressions below, we can obtain the C-index map in Figure 11.

$$C = [\cos(s) * \sin(a - 90) * N] + \{[-1 * \cos(s) * \cos(a - 90)] * E\} + [\cos(s) * H];$$
$$H = \sin(\alpha);$$
$$N = \cos(90 - \alpha) * \cos(n);$$
$$E = \cos(90 - \alpha) * \cos(e);$$

(3)

where α is the LOS incident angle, n is the angle of the LOS with respect to north, e is the angle of the LOS with respect to east, s is the slope and a is the aspect.

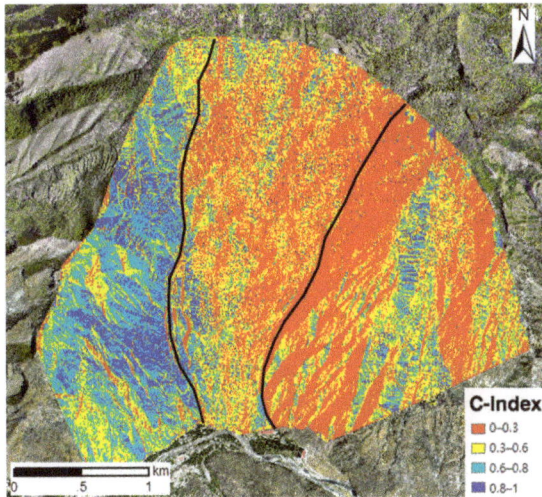

Figure 11. The C-index map of the Xishancun Landslide for the descending geometry.

Negative values of the C-index represent that the direction of movement is reversed in the LOS geometries. The values close to one mean that the movement of the landslide is mostly registered along the LOS direction. The Xishancun Landslide is a south-oriented slope, and the C coefficient map shows that the western slope is registered for approximately 50% and that the eastern slope is registered for only 20% in the LOS descending geometry.

Giant complex landslides usually have both slowly-moving planar slides and rotational landslides with vertical movement at the crown and horizontal movement at the toe. Hence, a single interferometric technique cannot derive comprehensive movement information for the whole landslide body. Unlike the InSAR technique, the pixel offset technique can provide two-dimensional displacement information. Both azimuth and range displacements derived are projections of the 3D displacement vector onto corresponding dimensions. The mathematical expressions are as follows:

$$D_N sin\alpha sin\theta - D_E cos\alpha sin\theta + D_V cos\theta = d_{rg}$$
$$D_N cos\alpha + D_E sin\alpha = d_{az}$$

(4)

where D_N, D_E and D_V represent displacements in the northing, easting and vertical directions, respectively. The parameters α and θ represent the heading angle and nominal incidence angle at the interest point, respectively, and d_{rg} and d_{az} are displacements measured in the range and azimuth directions, respectively. For the Xishancun Landslide, the equation can be derived as follows:

$$d_{rg} = [-0.076\ 0.539\ 0.839]\ [D_N\ D_E\ D_V]^T$$
$$d_{az} = [-0.990\ -0.139\ 0]\ [D_N\ D_E\ D_V]^T$$

(5)

Thus, the range displacements are sensitive to the easting and vertical directions, while azimuth displacements are sensitive to the northing direction. As the Xishancun Landslide is a south-facing slope, it is relatively safe to estimate that its deformation is mainly in the northing and vertical directions. As a result, to derive the two-dimensional movement of the landslide, the pixel offset technique should be involved to estimate the vertical and northward movements.

By incorporating both phase-based and amplitude-based techniques into the hybrid-SAR processing, various landslides with different sliding directions should be evaluated and distinguished.

5.5. Borehole Surface Displacements of Hybrid-SAR versus In Situ Measurements

To obtain a more robust analysis for the hybrid-SAR application, the displacement results produced by the phase-based and amplitude-based methods have been compared with the inclinometers of corresponding boreholes covering the period of the TerraSAR acquisitions. The actual locations of the four boreholes are shown in Figure 2. The measurements of inclinometers Bh1 and Bh2 (Figure 12) are only compared with the results of the offset tracking method, as the conventional DInSAR technique failed to obtain the displacement in this area because of the impact of dense vegetation cover in this area. The measurements of inclinometers Bh3 and Bh4 (Figure 12) are only compared with the results of the DInSAR processing, as the offset tracking technique failed to obtain a robust value because of low SNR values on the two borehole points. In the following figures, the x-axis represents the dates of acquisition, while on the y-axis are the displacement values of relative inclinometers. The positive value represents movement downwards along the slope, while the negative value represents movement upwards along the slope. In the following comparison, the accumulated displacements of inclinometers are converted to monthly displacement velocities, as almost all of the inclinometers show uniform deformation patterns during the monitoring period. Similarly, the displacements achieved from phase-based and amplitude-based processing are also converted to monthly displacement velocities.

From Figure 12, the borehole Bh1 shows a velocity along the maximum slope angle of 36.5 mm/month, while the velocity of the Bh2 instrument is −9.7 mm/month. The velocity values of borehole Bh3 and Bh4 are both 2.75 mm/month. Theoretically, the monthly inclinometer displacements of all four boreholes should be larger than the hybrid-SAR displacements [35], as the radar benchmark displacement is measured along the LOS direction of the satellite, which is only a component of the real movement vector. In contrast, the inclinometer measures the real displacement along the maximum slope angle direction.

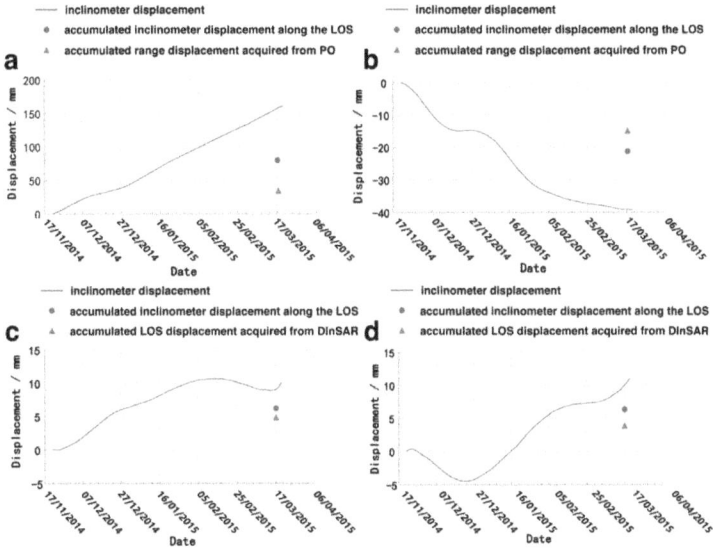

Figure 12. Comparisons of inclinometer displacements, accumulated inclinometer displacement along the LOS and accumulated hybrid-SAR LOS displacements of boreholes Bh1 (**a**); Bh2 (**b**); Bh3 (**c**); and Bh4 (**d**) measured from 21 November 2014–17 March 2015.

To compare the two datasets more robustly, we projected the displacement vector of the inclinometers along the radar LOS direction through a simple equation:

$$I_{LOS} = I_{SLOPE} * \sin \theta \tag{6}$$

where I_{LOS} is the inclinometer displacement along the LOS direction, I_{SLOPE} is the measured inclinometer displacement and θ is $33°$, the look angle of the TerraSAR satellite.

With the equations correcting the inclinometer measurements to the LOS direction, the results show that the velocity components along the LOS of Bh1 and Bh2 are 19.8 mm/month and −5.3 mm/month, respectively, while the range velocities achieved from the amplitude-based pixel offset technique are 7.8 mm/month and −3.8 mm/month, respectively. The velocity values along the LOS of Bh3 and Bh4 are both 1.5 mm/month, while the LOS velocities acquired from the phase-based DInSAR technique are 1.2 mm/month and 0.9 mm/month, respectively. In Figure 12, the finally accumulated inclinometer displacements along the LOS and hybrid-SAR LOS displacements of four boreholes until the date of 17 March 2015 are marked as red points and green triangles, respectively.

The above analysis reveals that the LOS displacements measured by hybrid-SAR may underestimate the real displacement determined via in situ monitoring, i.e., the borehole inclinometers. However, the displacement tendency of every single borehole confirms that the satellite monitoring is consistent with the in situ monitoring in displacement scale and magnitude.

5.6. Summary of Hybrid-SAR Applications and Discussion of Future Works

A summary of the application circumstances for the two independent techniques and the combined hybrid-SAR technique is listed below in Table 5. This list thoroughly reveals the advantages and disadvantages of these techniques.

Table 5. Application circumstances for the phase-based InSAR technique, amplitude-based pixel offset technique and hybrid-SAR technique.

Application Circumstances	Phase-Based InSAR Technique	Amplitude-Based Pixel Offset Technique	Hybrid-SAR Technique
Satellite acquisition geometry	Only suitable for good terrain visibility	Only suitable for good terrain visibility	Only suitable for good terrain visibility
Landslide deformation magnitude	Suitable for small deformation magnitudes	Valid for large deformation magnitudes	Valid for both small and large deformation magnitudes
Sliding mode	Suitable for creeping, extremely slow earthflow and slow vertical gravitational deformation	Suitable for rotational landslides and episodic movement	Suitable for various sliding modes, combining the advantages of two techniques
Densely vegetation cover impact	Affected, but may be effective if there is a high density of permanent scatterers	Affected but may be effective if there are highly contrasting ground features	Affected but may be effective if there are enough permanent scatterers and highly contrasting ground features
Sliding direction	Suitable for sliding in satellite LOS direction	Effective for sliding in both range and azimuth direction	Effective for sliding in both range and azimuth direction

Future improvements could be obtained when long-term datasets are available to generate time series displacement fields for this landslide. Time series hybrid-SAR, which is able to obtain a 3D deformation map with a time span of a year or more, can further improve the analysis of the landslide mechanism and improve the early warning capabilities. Then, the different temporal-spatial and scale-related characteristics of phase-based InSAR and amplitude-based offset tracking techniques could be analyzed and evaluated. The exploitation of a seamless handover scheme and also the definition of utilization criteria of the hybrid-SAR technique would be carried out step by step in giant complex landslide research. A validation analysis could be carried out in more detail by combining in situ monitoring data with time series deformation results. In this way, the integration of hybrid-SAR time series displacements and ground-based monitoring data can facilitate a better understanding of the landslide kinematics and the relationship to triggering factors.

6. Conclusions

Occurring on the eastern edge of the Tibetan Plateau in China, the Wenchuan Earthquake triggered a number of landslides that slid onto populated towns and villages. Instead of having a unitary characteristic and definite deformation mode, giant landslides normally have many different topographic features, complex deformation patterns and different orientations with respect to radar satellites. Thus, complex giant landslides require the use of different analysis methods to enhance the coverage of different types of measurements.

In giant landslide research, the good terrain visibility to the radar sensor is a prerequisite for the application of the hybrid-SAR technique. When considering the landslide deformation magnitude, high-resolution InSAR techniques can be used to derive reliable movement for slow-moving landslides with no or little decorrelation and with the displacement gradients not exceeding the measurable threshold. On the other hand, for fast-moving landslides with movements greater than the SAR image pixel size and even large enough to show significant change on the radar backscatter, the pixel offset techniques based on the amplitude information can achieve a better result for displacement monitoring. Moreover, dense vegetation cover hinders the exploitation of both techniques. Especially for the rotational component of a complex landslide, the vertical and horizontal measurements are both needed, which can be resolved by the measurement of the range and azimuth offsets using pixel-offset techniques. Hence, the hybrid-SAR technique, which makes joint use of both phase-based and amplitude-based methods, could be considered a robust methodological method to retrieve the deformation of giant complex landslides.

In the Xishancun Landslide, the phase-based DInSAR technique successfully provides a good preliminary interpretation of landslide-prone areas, even when the satellite data acquired are not abundant to conduct a robust time series study. In particular, the slow motions of the earthflow and creeping regions show quite obvious fringes on the interferograms, with a displacement rate of 1 cm/month. Unfortunately, the densely-vegetated terrain on the landslide body invalidated the use of the DInSAR approach in this specific region. However, due to the presence of contrasting features (buildings and corner reflectors), offset tracking processing was successfully carried out and retrieved the significant deformation tendency. Moreover, a typical rotational slide was also identified in the results of both the offset tracking method and InSAR method. This slide is moving 2–3 cm per month downwards and towards the riverbank. The combined analysis of the hybrid-SAR technique provides reliable identification and monitoring of the Xishancun Landslide at the preliminary research stage.

Our work confirms hybrid-SAR to be able to promptly identify diverse landslides with different deformation magnitudes, sliding modes and slope geometries, even when the available satellite data are limited. The application of the hybrid-SAR technique allows for a comprehensive and preliminary identification of areas with significant movement and provides reliable data support for the forecasting and monitoring of landslides. Moreover, the effective evaluations of phase-based and amplitude-based techniques in different application circumstances should be carried out in advance. Joint analysis and cross-validation are both needed to thoroughly enhance the measurements of the entire landslide.

Acknowledgments: This work was supported by the projects funded by the 973 National Basic Research Program (No. 2013CB733204 and No. 2013CB733203) and the Fundamental Research Funds for the Central Universities (Tongji University). The TerraSAR-X Stripmap datasets are copyrighted by DLR/Infoterra GmbH.

Author Contributions: Tengteng Qu carried out SAR data processing, interpreted the results and wrote the original manuscript. Chun Liu and Ping Lu supervised the research and contributed to the manuscript writing. Hangbin Wu, Xiaohang Shao, Hong Wan and Nan Li contributed to in situ measurement processing. Rongxing Li was involved in the data interpretation, analysis and edited the manuscript. All authors have read and approved the final manuscript.

Conflicts of Interest: The authors declare no conflict of interest.

References

1. Huang, R.; Li, W.L. Analysis of the geo-hazards triggered by the 12 May 2008 Wenchuan Earthquake, China. *Bull. Eng. Geol. Environ.* **2009**, *68*, 363–371. [CrossRef]
2. Liu, C.; Li, W.; Wu, H.; Lu, P.; Sang, K.; Sun, W.; Chen, W.; Hong, Y.; Li, R. Susceptibility evaluation and mapping of China's landslides based on multi-source data. *Nat. Hazards* **2013**, *69*, 1477–1495. [CrossRef]
3. Xu, C.; Xu, X.; Yao, X.; Dai, F. Three (nearly) complete inventories of landslides triggered by the 12 May 2008 Wenchuan Mw 7.9 Earthquake of china and their spatial distribution statistical analysis. *Landslides* **2014**, *11*, 441–461. [CrossRef]
4. Ye, X.; Kaufmann, H.; Guo, X. Landslide monitoring in the Three Gorges area using D-InSAR and corner reflectors. *Photogram. Eng. Remote Sens.* **2004**, *70*, 1167–1172. [CrossRef]
5. Zhao, C.; Lu, Z.; Zhang, Q.; de La Fuente, J. Large-area landslide detection and monitoring with ALOS/PALSAR imagery data over Northern California and Southern Oregon, USA. *Remote Sens. Environ.* **2012**, *124*, 348–359. [CrossRef]
6. Ferretti, A.; Prati, C.; Rocca, F. Permanent scatterers in SAR interferometry. *IEEE Trans. Geosci. Remote Sens.* **2001**, *39*, 8–20. [CrossRef]
7. Colesanti, C.; Ferretti, A.; Prati, C.; Rocca, F. Monitoring landslides and tectonic motions with the permanent scatterers technique. *Eng. Geol.* **2003**, *68*, 3–14. [CrossRef]
8. Ferretti, A.; Fumagalli, A.; Novali, F.; Prati, C.; Rocca, F.; Rucci, A. A new algorithm for processing interferometric data-stacks: Squeesar. *IEEE Trans. Geosci. Remote Sens.* **2011**, *49*, 3460–3470. [CrossRef]
9. Berardino, P.; Fornaro, G.; Lanari, R.; Sansosti, E. A new algorithm for surface deformation monitoring based on small baseline differential sar interferograms. *IEEE Trans. Geosci. Remote Sens.* **2002**, *40*, 2375–2383. [CrossRef]

10. Casu, F.; Manzo, M.; Lanari, R. A quantitative assessment of the SBAS algorithm performance for surface deformation retrieval from DInSAR data. *Remote Sens. Environ.* **2006**, *102*, 195–210. [CrossRef]

11. Hooper, A.; Zebker, H.; Segall, P.; Kampes, B. A new method for measuring deformation on volcanoes and other natural terrains using InSAR persistent scatterers. *Geophys. Res. Lett.* **2004**, *31*. [CrossRef]

12. Hooper, A.; Segall, P.; Zebker, H. Persistent scatterer interferometric synthetic aperture radar for crustal deformation analysis, with application to Volcán Alcedo, Galápagos. *J. Geophys. Res. Solid Earth* **2007**, *112*, B07407. [CrossRef]

13. Werner, C.; Wegmüller, U.; Strozzi, T.; Wiesmann, A. Interferometric point target analysis for deformation mapping. In Proceedings of the 2003 IEEE International Geoscience and Remote Sensing Symposium, IGARSS'03, Toulouse, France, 21–25 July 2003; pp. 4362–4364.

14. Strozzi, T.; Wegmüller, U.; Keusen, H.R.; Graf, K.; Wiesmann, A. Analysis of the terrain displacement along a funicular by SAR interferometry. *IEEE Geosci. Remote Sens. Lett.* **2006**, *3*, 15–18. [CrossRef]

15. Lu, P.; Catani, F.; Tofani, V.; Casagli, N. Quantitative hazard and risk assessment for slow-moving landslides from persistent scatterer interferometry. *Landslides* **2014**, *11*, 685–696. [CrossRef]

16. Li, X.F.; Peter, M.J.; Chen, F.; Zhao, Y.H. Measuring displacement field from TerraSAR-X amplitude images by subpixel correlation: An application to the landslide in shuping, Three Gorges Area. *Acta Petrol. Sin.* **2011**, *27*, 3843–3850.

17. Singleton, A.; Li, Z.; Hoey, T.; Muller, J.-P. Evaluating sub-pixel offset techniques as an alternative to D-InSAR for monitoring episodic landslide movements in vegetated terrain. *Remote Sens. Environ.* **2014**, *147*, 133–144. [CrossRef]

18. Shi, X.; Zhang, L.; Balz, T.; Liao, M. Landslide deformation monitoring using point-like target offset tracking with multi-mode high-resolution TerraSAR-X data. *ISPRS J. Photogram. Remote Sens.* **2015**, *105*, 128–140. [CrossRef]

19. Raspini, F.; Ciampalini, A.; Del Conte, S.; Lombardi, L.; Nocentini, M.; Gigli, G.; Ferretti, A.; Casagli, N. Exploitation of amplitude and phase of satellite SAR images for landslide mapping: The case of Montescaglioso (South Italy). *Remote Sens.* **2015**, *7*, 14576–14596. [CrossRef]

20. Bhattacharya, A.; Mukherjee, K.; Kuri, M.; Vöge, M.; Sharma, M.; Arora, M.; Bhasin, R.K. Potential of SAR intensity tracking technique to estimate displacement rate in a landslide-prone area in Haridwar region, India. *Nat. Hazards* **2015**, *79*, 2101–2121. [CrossRef]

21. Massonnet, D.; Feigl, K.L. Radar interferometry and its application to changes in the Earth's surface. *Rev. Geophys.* **1998**, *36*, 441–500. [CrossRef]

22. Jiang, M.; Li, Z.; Ding, X.; Zhu, J.-J.; Feng, G. Modeling minimum and maximum detectable deformation gradients of interferometric SAR measurements. *Int. J. Appl. Earth Obs. Geoinform.* **2011**, *13*, 766–777. [CrossRef]

23. Chen, M.; Tomás, R.; Li, Z.; Motagh, M.; Li, T.; Hu, L.; Gong, H.; Li, X.; Yu, J.; Gong, X. Imaging land subsidence induced by groundwater extraction in Beijing (China) using satellite radar interferometry. *Remote Sens.* **2016**, *8*, 468. [CrossRef]

24. Yun, S.H.; Zebker, H.; Segall, P.; Hooper, A.; Poland, M. Interferogram formation in the presence of complex and large deformation. *Geophys. Res. Lett.* **2007**, *34*, 237–254. [CrossRef]

25. De Zan, F. Accuracy of incoherent speckle tracking for circular gaussian signals. *IEEE Geosci. Remote Sens. Lett.* **2014**, *11*, 264–267. [CrossRef]

26. Lu, P.; Shang, Y. Active tectonics revealed by river profiles along the Puqu fault. *Water* **2015**, *7*, 1628–1648. [CrossRef]

27. Qi, S.; Xu, Q.; Lan, H.; Zhang, B.; Liu, J. Spatial distribution analysis of landslides triggered by 12 May 2008 Wenchuan Earthquake, China. *Eng. Geol.* **2010**, *116*, 95–108. [CrossRef]

28. Cui, P.; Chen, X.-Q.; Zhu, Y.-Y.; Su, F.-H.; Wei, F.-Q.; Han, Y.-S.; Liu, H.-J.; Zhuang, J.-Q. The Wenchuan Earthquake (12 May 2008), Sichuan Province, China, and resulting geohazards. *Nat. Hazards* **2011**, *56*, 19–36. [CrossRef]

29. Dai, F.; Xu, C.; Yao, X.; Xu, L.; Tu, X.; Gong, Q. Spatial distribution of landslides triggered by the 2008 Ms 8.0 Wenchuan Earthquake, China. *J. Asian Earth Sci.* **2011**, *40*, 883–895. [CrossRef]

30. Lu, P.; Wu, H.; Qiao, G.; Li, W.; Scaioni, M.; Feng, T.; Liu, S.; Chen, W.; Li, N.; Liu, C. Model test study on monitoring dynamic process of slope failure through spatial sensor network. *Environ. Earth Sci.* **2015**, *74*, 3315–3332. [CrossRef]

31. Notti, D.; Meisina, C.; Zucca, F.; Colombo, A. Models to predict persistent scatterers data distribution and their capacity to register movement along the slope. In Proceedings of the Fringe 2011 Workshop, ESRIN, Frascati, Italy, 19–23 September 2011; pp. 17–23.
32. Handwerger, A.L.; Roering, J.J.; Schmidt, D.A.; Rempel, A.W. Kinematics of Earthflows in the Northern California coast ranges using satellite interferometry. *Geomorphology* **2015**, *246*, 321–333. [CrossRef]
33. Lu, P.; Bai, S.; Casagli, N. Spatial relationships between landslide occurrences and land cover across the Arno river basin (Italy). *Environ. Earth Sci.* **2015**, *74*, 5541–5555. [CrossRef]
34. Notti, D.; Herrera, G.; Bianchini, S.; Meisina, C.; García-Davalillo, J.C.; Zucca, F. A methodology for improving landslide PSI data analysis. *Int. J. Remote Sens.* **2014**, *35*, 2186–2214.
35. Tofani, V.; Raspini, F.; Catani, F.; Casagli, N. Persistent Scatterer Interferometry (PSI) technique for landslide characterization and monitoring. *Remote Sens.* **2013**, *5*, 1045–1065. [CrossRef]

remote sensing

MDPI

Article

Landslide Deformation Analysis by Coupling Deformation Time Series from SAR Data with Hydrological Factors through Data Assimilation

Yanan Jiang [1], Mingsheng Liao [1,2,*], Zhiwei Zhou [3,†], Xuguo Shi [1,†], Lu Zhang [1,2,†] and Time Balz [1]

[1]   State Key Laboratory of Information Engineering in Surveying, Mapping and Remote Sensing, Wuhan University, 129 Luoyu Road, Wuhan 430079, China; Yananjiang@whu.edu.cn (Y.J.); xuguoshi@whu.edu.cn (X.S.); luzhang@whu.edu.cn (L.Z.); balz@whu.edu.cn (T.B.)
[2]   Collaborative Innovation Center for Geospatial Technology, 129 Luoyu Road, Wuhan 430079, China
[3]   Global Navigation Satellite System Research Centre, Wuhan University, 129 Luoyu Road, Wuhan 430079, China; zhiwei8848@gmail.com
*   Correspondence: liao@whu.edu.cn; Tel.: +86-27-6877-8070; Fax: +86-27-6877-8229
†   These authors contributed equally to this work.

Academic Editors: Zhenhong Li, Roberto Tomas, Zhong Lu and Prasad S. Thenkabail
Received: 2 December 2015; Accepted: 14 February 2016; Published: 25 February 2016

Abstract: Time-series SAR/InSAR techniques have proven to be effective tools for measuring landslide movements over large regions. Prior studies of these techniques, however, have focused primarily on technical innovation and applications, leaving coupling analysis of slope displacements and trigging factors as an unexplored area of research. Linking potential landslide inducing factors such as hydrology to SAR/InSAR derived displacements is of crucial importance for understanding landslide deformation mechanisms and could support the development of early-warning systems for disaster mitigation and management. In this study, a sequential data assimilation method named the Ensemble Kalman filter (EnKF), is adopted to explore the response mechanisms of the Shuping landslide movement in relation to hydrological factors. Previous research on the Shuping landslide area shows that the reservoir water level and rainfall are the two main triggering factors in slope failures. To extract the time-series deformations for the Shuping landslide area, Pixel Offset Tracking (POT) technique with corner reflectors was adopted to process the TerraSAR-X StripMap (SM) and High-resolution Spotlight (HS) images. Considering that these triggering factors are the primary causes of displacement fluctuations in periodic displacement, time-series decomposition was carried out to extract the periodic displacement from the POT measurements. The correlations between the periodic displacement and the inducing factors were qualitatively estimated through a grey relational analysis. Based on this analysis, the EnKF method was adopted to explore the response relationships between the displacements and triggering factors. Preliminary results demonstrate the effectiveness of EnKF in studying deformation response mechanisms and understanding landslide development processes.

Keywords: landslide; pixel offset tracking; corner reflectors; displacement; time-series; triggering factors; data assimilation; Ensemble Kalman filter

1. Introduction

Landslides are a serious problem that can cause great loss to life and property. In the last century, due to global climate change and human impact on the environment, landslides have occurred with increasing frequency with corresponding economic losses and fatalities [1–3]. The Qianjiangping landslide located near Shuping for example, moved rapidly on 14 July 2003 and killed 24 people with the direct economic losses of about seven million USD [4]. Deformation monitoring is therefore essential since it can help us understand the characteristics and trends in landslide evolution and is particularly important for the development of preventative measures such as disaster forecasting and early warning systems [5–7].

In recent years, time-series SAR/InSAR techniques have been successfully used for detecting earth surface deformation, including applications for landslide monitoring. These techniques include Persistent Scatterer SAR Interferometry (PSInSAR^{TM}) [8], Small Baselines (SBAS) [9], Stable Point Network (SPN) [10], Interferometric Point Target Analysis (IPTA) [11], Coherent Pixels Technique (CPT) [12], Advanced InSAR algorithm (SqueeSAR) [13], Pixel Offset Tracking (POT) [14] and other InSAR Time Series Analysis (TSA) techniques [15]. All these techniques permit observation of the deformation distribution over the aerial extent of a landslide, unlike ground measurement techniques such as GPS measurements that monitor a limited set of points [16]. To date, studies have focused primarily on technical innovation and applications, but few have considered the interpretation of landslide displacement measurements derived from SAR/InSAR images in relation to potential triggering factors, although this work is of crucial importance to understand deformation mechanisms and assist early-warning and disaster forecasting.

To extend this research, we propose a data assimilation methodology that enables a coupled analysis between time-series SAR measurements and hydrological data. Data assimilation makes use of all available information from measurements and models to estimate unknowns, thereby reducing predictive uncertainties [17–19]. This method has been examined and applied in a number of fields such as atmospheric and oceanic modelling for numerical prediction [19], regional- to global-scale hydrometeorology for reflecting the land surface-atmosphere interaction [20], and catchment scale hydrology for parameter estimation and hydrological analysis [21]. However, there are few applications of data assimilation in landslide disaster research. We investigated the feasibility of this methodology to study the deformation mechanism of the Shuping landslide area in relation to hydrological data, e.g., water level and rainfall.

The Shuping landslide area is near the Three Gorges Reservoir. Previous studies revealed that the Shuping slope stability was affected by the fluctuation of water level and rainfall, exhibiting a maximum accumulative deformation of more than 1 m during 2008 and 2010 [22]. In a previous study, we investigated the effectiveness of a modified POT method for fast-moving landslide monitoring at the Shuping landslide site [22]. The deformation results were consistent with historic displacement measurements acquired by GPS [23]. Coupling analysis, however, between the slope displacements as derived from the POT technique and potential landslide inducing factors such as hydrology at Shuping region is still an unexplored area of research.

In this paper, we present a new approach that applies EnKF to fill the gap in the research that relates SAR/InSAR derived displacements to deformation mechanisms. The geological and topographical conditions accounting for relatively stable long-term trends are not considered in this paper, only the seasonally changing triggering factors related to periodic displacement fluctuation [24–26]. Corner reflectors installed at/near the landslide zone served as point targets to measure slope movements using the modified POT technique. A time-series decomposition method was used to separate the POT displacement into periodic and long-term trend terms prior to analysis of the interactions between landslide movements and hydrological factors. We conducted a grey relational analysis and thus qualitatively evaluated the role of hydrological factors in the variation of periodic displacement. In order to understand how inducing factors affect the development processes of Shuping landslides, a sequential data assimilation method, the Ensemble Kalman filter (EnKF) was adopted as it incorporates the available observations sequentially in time to explore the relationship between the landslide periodic motion and hydrological factors. Predictions based on StripMap (SM) mode and High-resolution Spotlight (HS) mode TerraSAR-X imagery were compared and analyzed in relation to fluctuation changes in the water level and rainfall. Preliminary results demonstrate the feasibility of our proposed method for understanding the response relationship between POT derived landslide movements and hydrological factors.

2. Methodology

2.1. Measuring Landslide Deformation with SAR/InSAR Data

Advanced remote sensing techniques based on Synthetic Aperture Radar (SAR) data are powerful methods for detecting and monitoring gradual ground surface deformations at a reasonable cost. Several authors have applied Interferometric Synthetic Aperture Radar (InSAR) to update landslide inventories and the activity status of slope deformations at a regional scale [27–32], but some drawbacks exist with regards to temporal and spatial decorrelations and the atmospheric phase screen effects [30,33].

These limitations have been partially resolved by time-series InSAR techniques, such as persistent scatterer SAR interferometry (PSInSAR™) [8], small baselines (SBAS) [9] or SqueeSAR [13,34]. These techniques make use of large sets of SAR images acquired at different times over the same area, permitting observation of the temporal evolution of surface deformations. However, these advanced time-series InSAR method cannot effectively extract surface displacement in areas with dense vegetation coverage and mutation surface deformation due to loss of coherence [35–37].

The POT technology is based on amplitude cross-correlation between SAR images, and can make up for the limitations of InSAR technology [14]. If stable Point Targets (PT), such as corner reflectors, are used, displacement can be measured at a cm-level accuracy [22]. POT technology has been recognized as an effective and robust tool for capturing the fast movement of the land surface caused by events such as earthquakes, landslides, and glacier motions [22,38–40].

In our previous study, a modified POT technique was employed to detect the deformation at the Shuping landslide area [22]. Two stacks of TerraSAR-X datasets acquired from 2008 to 2010 were analyzed to characterize the historic evolution of the Shuping landslide. In this study, corner reflectors installed at/near the landslide zone were utilized as PTs to analyze the spatial-temporal pattern of landslide deformations. Consequently, displacement measurements at these PTs were reasonably considered as a surrogate for local landslide deformations and adopted in a response analysis. The process flow of our proposed POT is illustrated in Figure 1—the details of this technique can be found in [22].

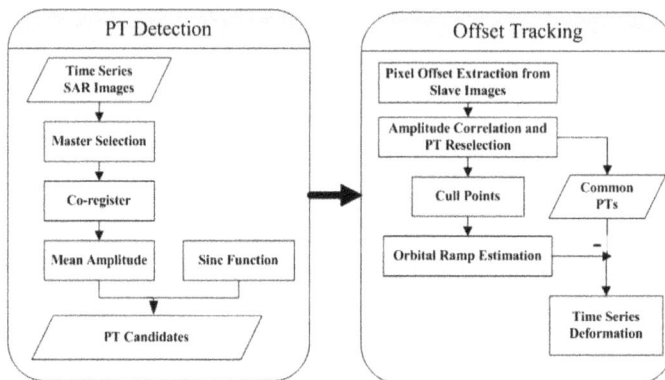

Figure 1. The process flow of PT offset tracking.

2.2. Methods for Time-Series Decomposition

In this study, we conducted a coupling analysis between the POT time-series measurements and the landslide destabilizing factors, e.g., the fluctuation of the reservoir water level and rainfall. Assuming that these hydrological factors with seasonal variations are the primary cause of displacement fluctuations [24–26], time-series decomposition is needed to separate displacement extracted by the POT technique into a trend term and a periodic term. The time-series decomposition requires input data with the same time intervals. Because the time-series POT measurements are incomplete, as some dates are missing, a cubic spline interpolation was thus adopted to generate data for these missing but expected revisit times.

The processed displacement in the study area is a nonlinear, nonstationary curve influenced by reservoir level fluctuations, seasonal precipitation and other random disturbances. To eliminate the effect caused by random noise, de-noising processing must be applied to measurements. Then, the displacement can be separated into trend and periodic components. The trend component was computed by a quadratic polynomial fitting method and the seasonal component was calculated using the difference between the displacement time-series and the previously estimated trend component. The processing procedure for time series decomposition is shown in Figure 2.

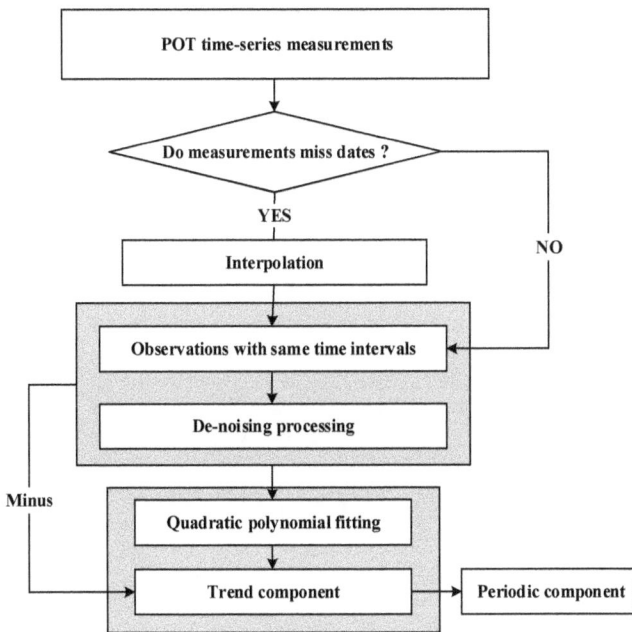

Figure 2. The flow chart for time series decomposition.

The moving average method is believed to be an effective way to reduce the short-term uncertainty caused by random disturbances and highlights longer-term cycles or trends [41,42]; thus, it was applied to the displacement time-series derived from the SM stack. The time intervals for the moving average method, were set at less than the cycle of the periodical factors to protect the seasonal signal from damage. In this study, 16 historical series of displacements before 15 February 2009 in the SM stack were set as the first moving window to conduct the moving average analysis, roughly one half-cycle of the hydrological factors (approximately one cycle occurs per year). However, the HS stack derived deformation time-series has only one complete cycle (one cycle per year) during the period from February 2009 to April 2010. Therefore, it was impossible to use 16 historical series observations to conduct the moving average analysis and at the same time obtain a complete period for the cyclical variation of displacement. An alternative de-noising method must be found.

In the grey forecasting model, the Accumulated Generating Operation (AGO) is adopted in time-series data processing since it can weaken the randomness of the raw data and remove the interference from noise [43,44]. In this study, the capacity of AGO to eliminate random noise in deformations derived from SAR images will be explored. The time series of displacements at the corner reflectors derived from SM dataset were employed to noise reduction using both the moving average method and the AGO method. To estimate the feasibility of the AGO method, the resulting seasonal deformations from the AGO method will be compared with that from the moving average method. These results are shown and discussed in Section 4.2.

2.3. Coupling Analysis Scheme Based on Data Assimilation

EnKF is a sequential data assimilation method that can incorporate available observations sequentially in time. It was first introduced by Envenson [45], and later clarified by Burgers *et al.* [46] and Van Leewen [47]. The EnKF based upon the Monte Carlo method, applies an ensemble of model states to represent model estimation and continuously updates error statistics over time [20]. An analysis scheme operates directly on the ensemble of model states when observations become available for data assimilation. The EnKF has been applied in a number of fields such as atmospheric and oceanic modelling for numerical prediction [18], regional- to global-scale hydrometeorology for reflecting the land surface-atmosphere interaction [20], and catchment scale hydrology for parameter estimation and hydrological analysis [21]. However, EnKF or indeed other data assimilation methods have rarely been applied in landslide research. Given its widespread use in many domains, we adopted the EnKF data assimilation method to study the deformation mechanism of a landslide in a reservoir with the influence of hydrological factors, water level and rainfall.

2.3.1. Dynamic Process of a Landslide Evolution

Landslide evolution can be considered as a differential function that temporally integrates dynamic states and model parameters under a driving force, such as rainfall or the reservoir level. The functions for a landslide process can be denoted as a nonlinear stochastic process,

$$S_{t+1} = f(S_t, U_{t+1}) + w_{t+1}, \tag{1}$$

where t denotes the time step, S is the state vector consisting of model parameters and variables, f is the nonlinear forecast operator, U is a set of externally time-dependent forcing variables (e.g., rainfall), and the noise term w accounts for model error.

Observations using different instruments or techniques can be directly or indirectly related to the model states and denoted as,

$$y_t = H(S_t) + \varepsilon_t, \tag{2}$$

where y denotes the observation vector, H is the observation operator that represents the deterministic relationship between observation y and the true state S. Before data assimilation, the POT technique is adopted to retrieve landslide displacement from original SAR signals/images; then, time-series decomposition is applied to get the periodic component from the POT measurements. Because the state vectors include periodic displacement, we designated a linear observation operator with only 0 s and 1 s as its entries. The observations are perturbed by white noise ε that represents the possible measurement errors. This noise is mutually uncorrelated in space and time, with means equaling zero and standard deviations scaling to the current values of variables and dependent on the measurement accuracy of the monitoring tools and methods. For example, the achievable accuracy of displacement measured by POT can be theoretically expressed as [48]:

$$\sigma = \sqrt{\frac{3}{10N}} \frac{\sqrt{2 + 5\gamma^2 - 7\gamma^4}}{\pi \gamma^2}, \tag{3}$$

where γ is the coherence value and N is the number of pixels within the estimation window, equal to 11×11 in our study. The coherence of the corner reflectors usually ranges from 0.8 to 0.9. Selecting 0.8 as the coherence value in our study yielded a calculated accuracy of approximately 0.0364 m. Consequently, we applied a scaler factor of 0.4 to the values for periodic term displacement based on the POT results.

In order to get optimal estimates for the parameters and states of interest, we combined complementary information from Equations (1) and (2), which is the basic idea of data assimilation. The EnKF sequential data assimilation method uses an ensemble of state vectors (*a priori* S_t) to represent the true probability distribution of the state vectors (*a posteriori* S_{t+1}) by integrating the *priori* forward in time as observations (y_t) become available. The sequential data assimilation process is illustrated in Figure 3.

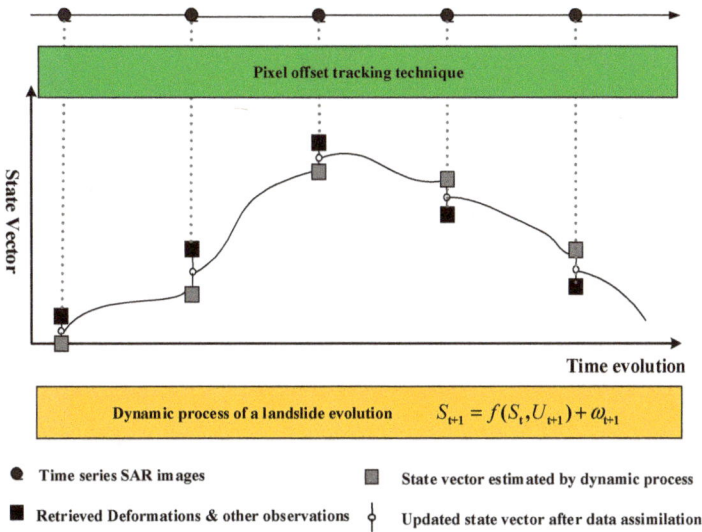

Figure 3. The sequential data assimilation process for this study.

2.3.2. Framework of Ensemble Kalman Filter

The EnKF algorithm consists of three processes for each assimilation step: a forecast based on current state variables (*i.e.*, solve landslide evolution equations with current static and dynamic parameters), data assimilation (computation of a Kalman gain), and updating the state variables and parameters.

Similar to Equation (1), the model forecast is executed in the EnKF for each ensemble member as follows:

$$S_{t+1}^{i-} = f(S_t^{i+}, U_{t+1}^i) + \omega_{t+1}^i, \qquad \omega_{t+1}^i \sim N(0, Q_{t+1}), \qquad i = 1, 2, \ldots, N_e \ , \tag{4}$$

where N_e is the number of ensemble members, S_{t+1}^{i-} is the component of the ith ensemble member forecast at time $t+1$, S_t^{i+} is the ith updated ensemble member at time t, U_{t+1}^i is the ith perturbed forcing variables, and ω_{t+1}^i is independent white noise for the forecast mode, drawn from multi-normal distributions with zero mean and specified covariance Q_{t+1}. At time $t+1$, the observation ensemble member can be written as,

$$y_{t+1}^i = HS_{t+1}^{i+} + \varepsilon_{t+1}^i, \qquad \varepsilon_{t+1}^i \sim N(0, R_{t+1}), \qquad i = 1, 2, \ldots, N_e \ , \tag{5}$$

where y_{t+1} is observation vector at time $t+1$; HS_{t+1} is the observation data calculated from the true state S_{t+1}, and ε^i_{t+1} is the noise term with zero mean and specified covariance R_{t+1}.

In the analysis step, the observations are perturbed by adding random perturbations and used for updating the state variation of each member. The update process can be expressed as,

$$S^{i+}_{t+1} = S^{i-}_{t+1} + K_{t+1}(y^i_{t+1} - HS^{i-}_{t+1}), \tag{6}$$

$$K_{t+1} = P^-_{t+1}H^T(HP^-_{t+1}H^T + R_{t+1})^{-1}, \tag{7}$$

$$P_{t+1} \approx \frac{1}{N_e - 1} \sum_{i=1}^{N_e} \left\{ \left[S^{i-}_{t+1} - \overline{S^-_{t+1}} \right] \left[S^{i-}_{t+1} - \overline{S^-_{t+1}} \right]^{-1} \right\}, \tag{8}$$

$$\overline{S^-_{t+1}} = \frac{1}{N_e} \sum_{i=1}^{N_e} S^{i-}_{t+1}, \tag{9}$$

where S^+_{t+1} is the updated state vector after assimilation at time $t+1$, K_{t+1} is the Kalman gain matrix determining the weight between the modeling and observation states in the assimilation process, P^-_{t+1} is the forecasted background covariance matrix at time $t+1$; $\overline{S^-_{t+1}}$ is the mean of N_e forecasted state vectors at time $t+1$.

Finally, the updated ensemble S^{i+}_{t+1} is figured out, of which we compute the mean to get the updated state vector S^+_{t+1} which completes the process of data assimilation at time $t+1$. Then, the updated ensemble S^{i+}_{t+1} is iterated forward along with S^{i+}_t to repeat the entire process for the next time step.

3. Study Area and Test Datasets

3.1. Study Area

The Three Gorges Dam (TGD) in China is the largest hydropower project in the world. Since the Three Gorges Reservoir (TGR) went into operation in 2003, water level fluctuations have changed the physical and mechanical properties of rock and soil around the reservoir. The resulting ground deformation has re-activated previous landslides and created new landslide hazards [35]. Preconstruction landslide investigations for the TGD identified the Shuping landslide area as an old landslide that was re-activated during the first impoundment of the TGR [49].

The landslide is located in the Shuping Village of Zigui County on the south bank of the Yangtze River (Figure 1), approximately 49 km upstream of the Three Gorges Dam site [22,50]. The landslide extends into the Yangtze River and lies at an elevation between 65 and 400 m with a width of about 650 m. It is a south–north oriented landslide with a slope gradient varying from 22° in the upper part to 35° in the lower part. The landslide has an approximately 40–70 m thick sliding mass consisting of about 20 million m³ of sandy mudstone and muddy sandstone belonging to the Triassic Badong formation (T2b). Geological surveys show that landslides are likely to occur within this stratigraphic unit in the Three Gorges area [51,52].

From Figure 4 we can see that in the past, a valley has divided the Shuping landslide into eastern Block 1 and western Block 2 portions. Since 2005, 18 corner reflectors have been installed sparsely at Shuping landslide to help understand the landslide evolution [53,54], as presented in Figure 5. Out of the 18 corner reflectors, four (CR8, CR13, CR17, CR18) showed up only on images acquired after 24 January 2009. Among them, corner reflector CR12 is located on a stable area outside but adjoined to the landslide and thus was used as a reference point [50].

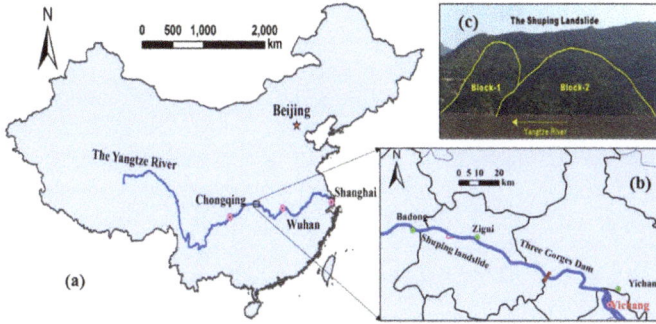

Figure 4. Location of the Shuping landslide in China (**a**,**b**) and a photo of Shuping landslide taken from opposite bank facing south (**c**) [22].

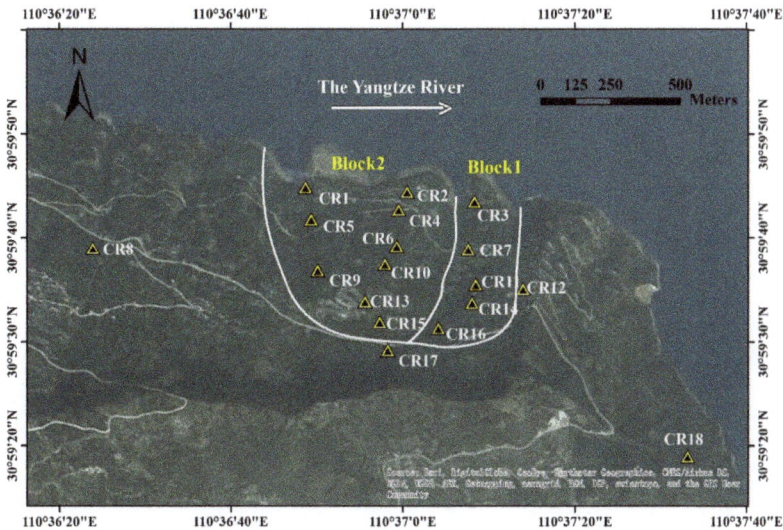

Figure 5. The distribution of 18 corner reflectors at Shuping landslide.

The Three Gorges area belongs to a subtropical zone with a humid monsoon climate [55]. The average annual rainfall exceeds 1200 millimeters, concentrated between June and October. The Shuping landslide area is located in a precipitation area characterized by heavy rain and storms. Historic displacement measurements at Shuping using GPS and an extensometer indicated that the rate of the ground deformation increased with the drawdown of the reservoir level and during the period of increased rainfall in the wet season [4,35].

3.2. Datasets

In this study, we focus on Shuping landslide aiming to illustrate the suitability of data assimilation for understanding the interaction between landslide movements derived from POT and the influence of hydrological factors (*i.e.*, reservoir water level and rainfall). For this purpose, four time-series datasets have been considered and analyzed. Two of them correspond to the displacement time-series of the landslide, which were measured by POT with TerraSAR-X StripMap (SM) and High Spotlight (HS) mode imageries. In addition, the other two correspond to the triggering factors, which are supposed to be related with the landslide deformation mechanism.

The basic parameters of the SAR images are shown in Table 1. The SAR monitoring results were obtained by processing 34 SM-mode and 36 HS-mode TerraSAR-X images through the POT approach. A detailed description of POT technology and data processing can be found in [22]. The measurements derived from the SM and HS dataset are two-dimensional in both azimuth and range direction.

Table 1. Basic parameters about the TerraSAR-X images.

Parameter	SM Mode Data	HS Mode Data
Orbit direction	Descending	Descending
Look angle	24°	39°
Heading	190.7°	189.6°
Antenna polarization mode	VV	HH
Azimuth spacing	1.96 m	0.87 m
Range spacing	0.91 m	0.45 m
Date of master image	17 November 2009	4 July 2009
First acquisition	21 July 2008	21 February 2009
Last acquisition	1 May 2010	15 April 2010

The daily reservoir water level was available for the period of June 2008 to May 2010. In November 2008, the TGR had become fully operational when its highest water level reached 175 m. Since then, reservoir water level fluctuates between 145 m and 175 m over the course of a year. These seasonal changes in water level exhibit clear seasonal changes due to artificial flood control [55,56].

Daily rainfall data are also available from the meteorological station at Badong in 2008 and Zigui between 2009 and 2010. The rainy season of Shuping landslide area lasts from June to October each year. The rainfall data in Figure 6 displays clear seasonal variations due to monsoon influences.

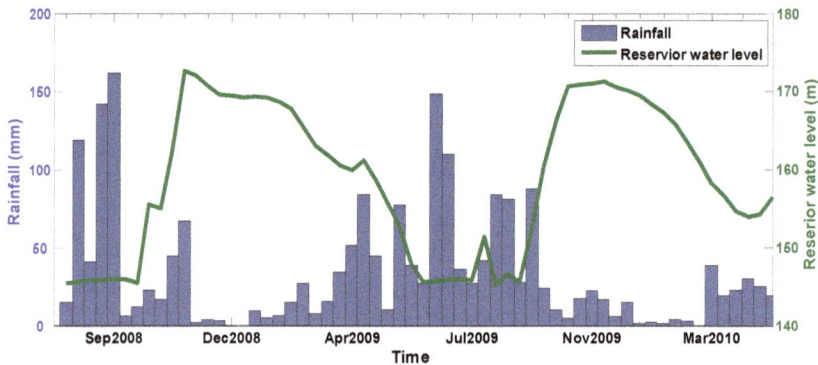

Figure 6. The fluctuations of the synchronous reservoir water level and rainfall.

As can be seen from Figure 6, the water level fluctuated in a cycle opposite to the natural precipitation conditions. The reservoir began impounding at the end of the wet season in October and quickly reached the maximum water level and maintained this from November to February for power generation and navigation. Then, the reservoir water level was gradually decreased to the minimum previous level during spring, for flood control.

4. Experimental Results and Analyses

4.1. Deformation Characteristics of Shuping Landslide Area

As can be seen in Figures 7 and 8 three CRs, namely CR8, CR17 and CR18 are located on a stable region outside the landslide area, and remained stable over a long period of nearly two years. The cumulative displacements derived from both HS and SM stacks are not more than 0.1 m. These results are reasonable and agree with the actual situation, which confirms the reliability of POT in deformation monitoring. The distribution of displacement derived from HS and SM data are almost identical in both azimuth and range at same locations, although the HS mode starting acquisition date was seven months later with a 15° difference in look angle compared to the SM mode.

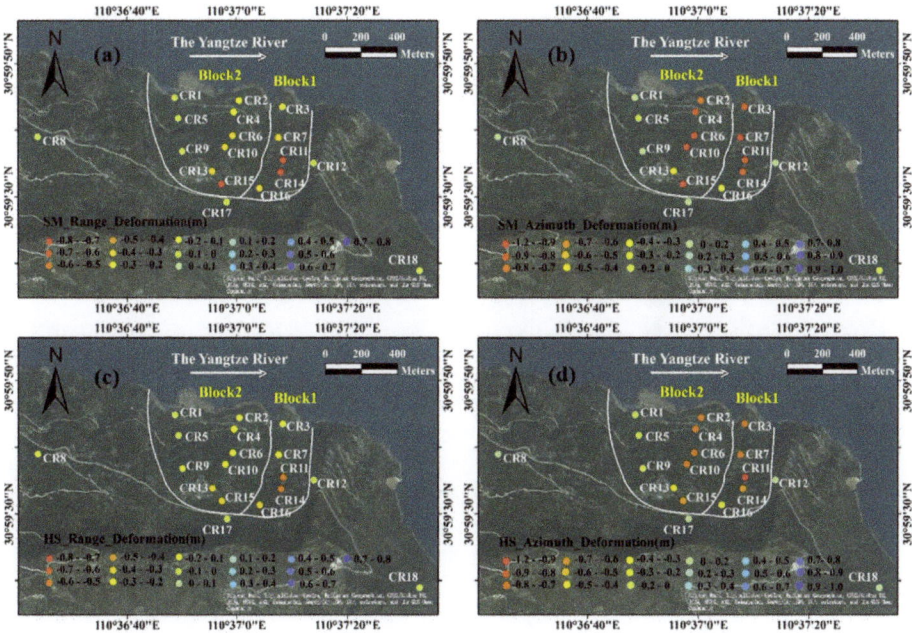

Figure 7. Monitoring displacements from 21 July 2008 to 1 May 2010 using SM data in (**a**) range and (**b**) azimuth; Monitoring Displacements from 21 February 2009 to 15 April 2010 using HS data in (**c**) range and (**d**) azimuth.

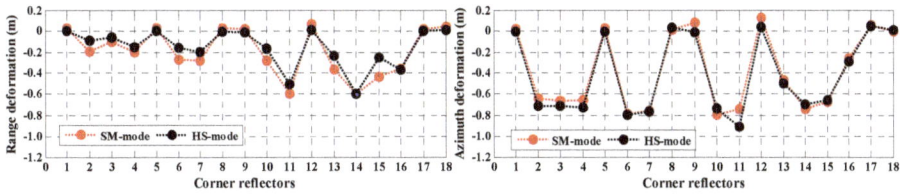

Figure 8. Displacements of CRs at Shuping landslide in range (**left**) and azimuth (**right**) during February 2009 to April 2010. The black dot represents the SM-mode measurements, and the red dot represents the HS-mode measurements.

Also shown in Figures 7 and 8 the spatial pattern of the measurements revealed that the deformation of the Shuping landslide varies with location. The deformation distribution of Block-1 was quite different with that of Block-2. The cumulative displacements at PTs within Block-1 are almost uniformly distributed with approximate meter-level displacements moving toward Yangtze River in azimuth. An approximate elevation-dependent displacement shows up as a large deformation near the upper part of Block-1, while an almost stable state near the lower part of Block-1 in range. In contrast, large displacements only show up in the eastern part of Block-2 neighboring Block-1, while there is a nearly stable state present in the western part of Block-2. The cumulative azimuth and range displacements at PTs within the eastern part of Block-2 are similar in spatial pattern to that of Block-1, implying that these two areas may share a similar deformation mechanism.

As shown in Table 2, the other fourteen corner reflectors installed at the slide are artificially categorized into three groups, considering their geographic locations and evolution features. The CR1, CR5 and CR9 corner reflectors located in the western part of Block-2 were assigned in group 1. As can be seen in Figures 7 and 8 displacements in this group were 0.1 m at the most, from which it can be inferred that the western part of Block-2 was in a rather stable state from July 2008 to May 2010.

Table 2. Groups of 18 corner reflectors.

Groups	No. of Corner Reflectors
Reference point	CR12
Outside landslide area	CR8, CR17 and CR18
Group1	CR1, CR5 and CR9
Group2	CR2, CR3, CR4, CR6, CR7 and CR10
Group3	CR11, CR13, CR14, CR15 and CR16

The CR2, CR3, CR4, CR6, CR7 and CR10 corner reflectors near the upper part and the profile center of the landslide were denoted as Group 2; the CR11, CR13, CR14, CR15 and CR16 corner reflectors near the lower part of the landslide were marked as Group 3. These two groups experienced relatively significant displacement (Figures 7 and 8) and exhibited similar temporal evolution features in both the azimuth and range during the study period, with similar abrupt displacements during April to June in 2009 (Figure 9). However, difference displacements still existed in the range direction between HS and SM measurements during the period February 2009 to June 2009. Such discrepancies are primarily caused by the big difference in look angle [22], since the range measurements has a cosine of the look angle relation with the vertical displacement. Consequently, when vertical displacement stays constant, a larger look angle (HS-mode) corresponds to a larger range deformation.

The CRs in groups 2 and 3 show similar evolution features, two of the corner reflectors in Groups 2 and 3 were selected as the objects for analysis. One was CR2 located in the eastern part of Block-2, belonging to the upper part of the landslide, the other was CR14 installed in Block-1, belonging to the lower part of the landslide.

4.2. Time-Series Deformation Decomposition

As discussed in Section 2.2, the moving average method is the ideal method for eliminating the short-term uncertainty caused by random disturbances; however, it is a calculation from a moving window with 16 historical series observations, and infeasible for measurements from a HS stack, since it has only one complete cycle (approximately one cycle per year) from February 2009 to April 2010; otherwise, we could not obtain one complete period of the cyclical variation of displacement. Thus, AGO was adopted as an alternative to weaken the random disturbances from SAR derived displacements.

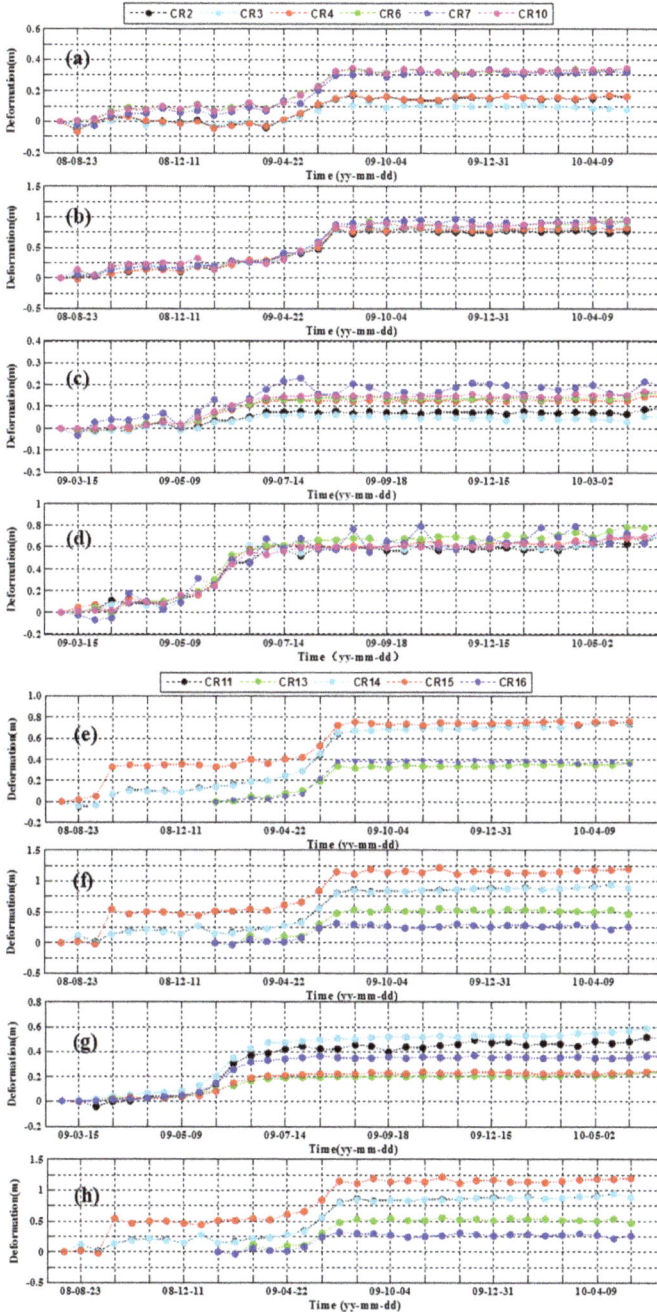

Figure 9. Time-series displacement of corner reflectors in Groups 2 and 3. (**a–d**) are displacement at CRs in Group 2; (**a,b**) are displacement in range and azimuth using SM stack; (**c,d**) are displacement in range and azimuth using HS stack; (**e–h**) are displacement at CRs in Group 3; (**e,f**) are displacement in range and azimuth using SM stack; (**g,h**) are displacement in range and azimuth using HS stack.

Displacement time-series at CR2 (in Block-2) and CR14 (in Block-1) derived from the SM dataset were used to estimate the feasibility of the AGO method in eliminating the randomness of the raw data. The de-noised displacements were separated into a trend and periodic components. The trend component was computed by a quadratic polynomial fitting method, while the seasonal component was calculated using the difference between the displacement time-series and the previously estimated trend component. The resulting deformation components were rendered in Figures 10 and 11.

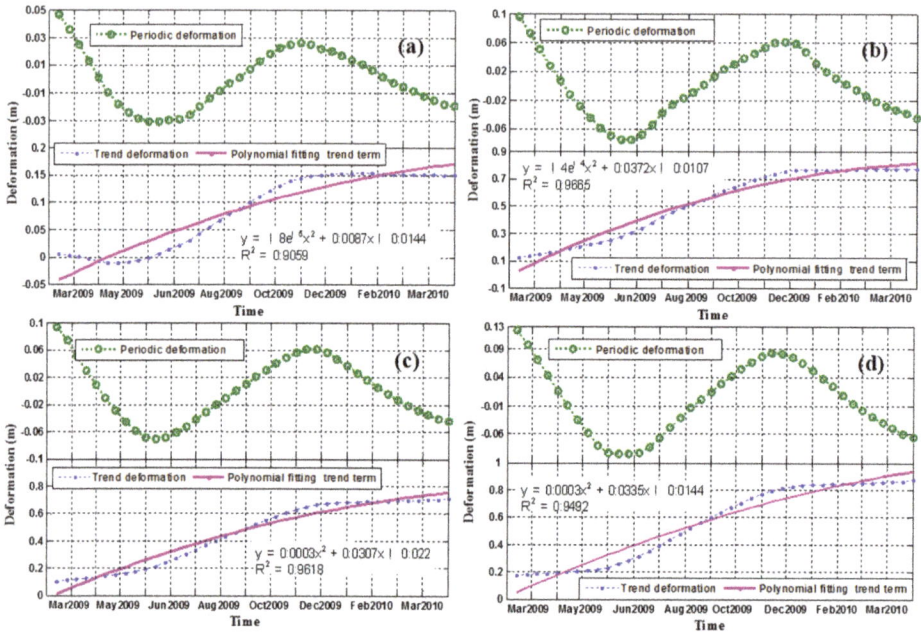

Figure 10. The decomposition results of CR2 and CR14 using the moving average method; (**a**,**b**) are range and azimuth results of CR2; (**c**,**d**) are range and azimuth results of CR14.

As can been in Figure 10a–d, the results from moving average method at CR2 and CR14 shared similar trends of both components' evolution; as was the case using AGO method, shown in Figure 11a'–d'. These results are reasonable and confirm the stability of both methods in reducing short-term random disturbances whilst the calculated seasonal deformations at CR2 and CR14 using both methods had similar trends (also shown in Figures 10 and 11) with minor discrepancies (Figure 12). However, because of the numerical sum of the AGO method, the derived results in Figure 11 are cumulative deformations, not the actual deformations at corner reflectors. As a result, the numerical values of both trend and periodic deformations were amplified many times as compared to those of the moving average method. For the trend component results from the AGO method, the magnitude depended on the time-series deformations of the corresponding corner reflectors, as shown in Figure 9. As for the periodic component, since it was controlled by the same seasonal changes of water level and rainfall, the amplitude of the fluctuation was uniform and approximately ten times the amplitude calculated using the moving average method, as can be seen in Figure 12.

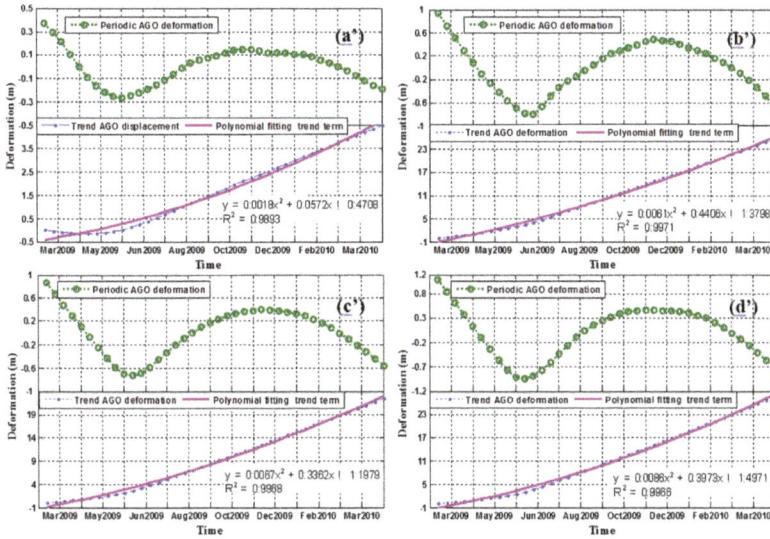

Figure 11. The decomposition results of CR2 and CR14 using accumulated generating operation; (a',b') are range and azimuth results of CR2; (c',d') are range and azimuth results of CR14.

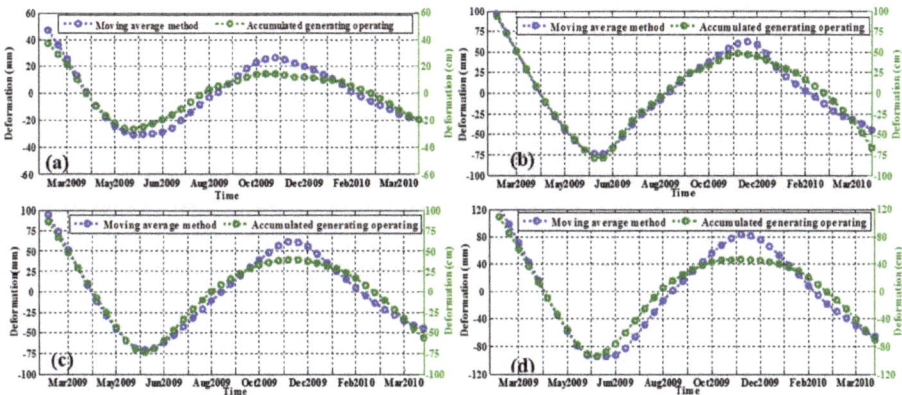

Figure 12. A comparison of the calculated seasonal deformations between the moving average method and the accumulated generating operation method. Deformation time series in both range and azimuth were plot in (a,b) for CR2 and (c,d) for R14.

However, this different amplitude did not involve obstructions in subsequent processing, because we took the computation from the moving average method as a reference unit in the follow-up experiments. As a result, the AGO method could be used as an alternative and we compared it to the moving average method to make sure that AGO can be used to reduce short-term random disturbances. Even though AGO can be used as an alternative method, it still has disadvantages since it may smooth the peaks of the data, as shown in Figure 12. The moving average method can counteract the over-smoothing effect of AGO. Consequently, if sufficient data is available, the moving average method is recommended. Other methods, such as wavelet analysis in frequency domain, can certainly be considered to get rid of the short-term disturbance.

4.3. Impact Factors for Landslide Deformation

In Section 4.2, time-series decomposition was applied on two corner reflectors (CR2 and CR14) to extract the periodic displacement from the POT measurements. In this section, the periodic displacement of the CR14 corner reflector with severe fluctuation was chosen to analyze the relationship between the deformation and the causes.

We applied a grey relational analysis to determine the correlation between periodic deformation, water level and rainfall. The grey system theory proposed by Deng in 1982 [44] has been shown to be useful when dealing with incomplete and uncertain information. The grey relational analysis based on grey system theory can be effectively used to solve the complicated interrelationships among multiple performance characteristics [57]. Through grey relational analysis, one can obtain the grey relational degree to evaluate the correlation of different measurement data. The range of the relational degree value is −1 to 1; the closer the value to 1, the higher the correlation of two sequences. The calculated grey relational degrees are list in Table 3.

Table 3. The correlation between periodic deformation and reservoir level as well as rainfall.

Data Source		Reservoir Level	Rainfall
SM mode	range	0.878	0.776
	azimuth	0.878	0.776
HS mode	range	0.864	0.807
	azimuth	0.863	0.807

As can be seen in Table 3, the periodic component of the SAR time-series measurements was strongly correlated with reservoir level changes and seasonal precipitation. As shown in Figure 13, the periodic displacement fluctuated along with the variation of reservoir level and rainfall. It increased during the wet season with the water level remaining stable or rising, and decreased gradually with the reservoir level failing at the beginning of rainy season, indicating that the fluctuation of the reservoir water level was the major factor influencing deformation in the Shuping landslide area.

In our previous study [22], we demonstrated that the total deformation was affected by the drawdown of the water level from April to June, but not by the sharp rise of the water level in October. From Figure 13, the water level rose rapidly from 145 m to 170 m during September to November both in 2008 and 2009. The change in deformation was very small, however, especially after excluding the effects of precipitation. Conversely, a very large deformation gradient occurred when the reservoir water fell quite rapidly from 170 m to 145 m in the period of November to April in both 2008 and 2009.

Since the rainy season of the Three Gorges area lasts from June to October, as can be seen in Figure 13, impacts of the rainfall on displacement appeared before the reservoir impoundment during the period of July 2008 to September 2008 and the period of June 2009 to September 2009. Larger periodic displacement occurred at the end of the rainy season when the reservoir water level had reached its highest in November 2008 and 2009. Combining these analyses, it is apparent that the fluctuation of reservoir water levels and rainfall were two significant triggering factors influencing the deformation in the Shuping landslide area.

4.4. Data Assimilation Results

Although the relationships among the rainfall, the water level, and landslide displacement were qualitatively analyzed in Section 4.3, a quantitative representation of this relationship needs further confirmation. The EnKF was adopted to validate the quantitative interactions of the Shuping landslide movement under the influence of hydrological factors.

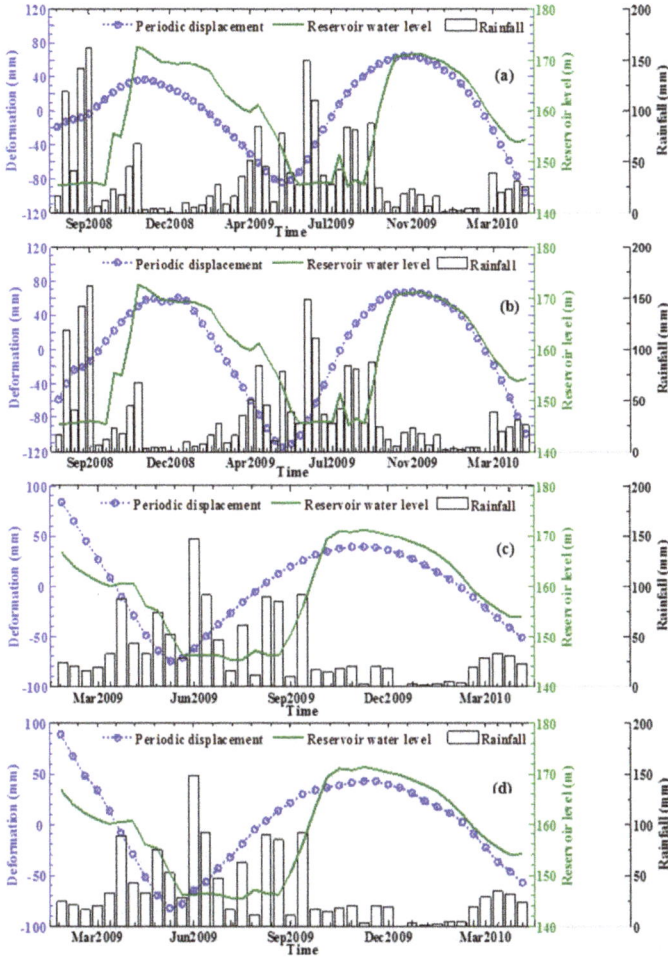

Figure 13. The periodic component of the SAR time-series derived from SM mode data in range (**a**) and azimuth (**b**) and that from HS mode data in range (**c**) and azimuth (**d**).

The measured deformations are available at an interval of 11 days, during this short time-step, the variation of the periodic displacement was very small. Thus, the fluctuation of the periodic motion can be perceived as smooth, as shown in Figure 13. With this in mind, we considered periodic displacement to be a function of the water level and rainfall. Thus, a Taylor series expansion could be applied. The dynamic equation of periodical deformation can be expressed as:

$$d(w_{t+1}^i, r_{t+1}^i) = d(w_t^i, r_t^i) + \left(\frac{\partial d}{\partial w}\right)_{w_t}(w_{t+1}^i - w_t^i) + \frac{1}{2}\left(\frac{\partial^2 d}{\partial w^2}\right)_{w_t}(w_{t+1}^i - w_t^i)^2 + \left(\frac{\partial d}{\partial r}\right)_{r_t}(r_{t+1}^i - r_t^i)$$
$$+ \frac{1}{2}\left(\frac{\partial^2 d}{\partial r^2}\right)_{r_t}(r_{t+1}^i - r_t^i)^2 + g_t^i \qquad g_t^i \sim N(0, \Omega_t), \ i = 1, 2, \ldots, N_e \tag{10}$$

where d is the periodic motion, w_t and r_t is the reservoir water level and rainfall at time t respectively, $\frac{\partial d}{\partial w}, \frac{\partial d}{\partial r}, \frac{\partial^2 d}{\partial w^2}$, and $\frac{\partial^2 d}{\partial r^2}$ are the first and second partial derivatives of periodic displacement *versus* water level w and rainfall r, g is the third order or higher remainder term of the Taylor expansion with a small value. In quantitative terms, the effects of the trigging factors are not comparable because of different units of measurement and dimensions. To eliminate the effect of the index dimension, Min-Max Normalization was carried out [58].

In this study, the ensemble number Ne was set to 1000; the standard deviation g was scaled with a scaling factor of 10% multiplied to the current forecast values. Since, $\frac{\partial d}{\partial w}$, $\frac{\partial d}{\partial r}$, $\frac{\partial^2 d}{\partial w^2}$, and $\frac{\partial^2 d}{\partial r^2}$ are partial derivatives, we could not accurately estimate certain values. Thus, we take these four partial values as model parameters in the state vector and updated them synchronously during assimilation. Based on these parameter settings and those discussed in Section 2.2, the model iteratively ran through the time-steps in sequence while being assimilated to the water level and rainfall observations.

The assimilation and simulation results after the parameters were optimized for the periodic motion of the CR14 corner reflector in both range and azimuth from February 2009 to April 2010 are shown in Figure 14. These figures show that, as the assimilation is progressing, the data assimilation estimates provided good representation of the SAR derived displacements and the optimized parameter simulation results become more reliable than the testing results after running a few steps. The mean absolute error of the assimilation results were shown in Figure 14. We can see that at the beginning of data assimilation, the estimation error of EnKF was relatively high with maximum error at approximately 25 mm, but, after a few steps forward, it decreased gradually. However, it could not reach zero due to either the incomplete parameters or the observation errors already incorporated in the dynamic process.

Since the four parameters used in the dynamic equation were synchronously optimized by assimilation observations during the data assimilation process. This allowed the predictive uncertainties of the periodical deformation to be reduced. In this study, data assimilation progressed until finishing assimilation the observations acquired on 20 April 2009 and 15 April 2009 for SM- and HS-mode series. Then, we adopted the updated model parameters to make a short term forecast for the periodic term displacement. As can been seen in Table 4, the short term forecasts are consistent with our analysis of the relationship between the deformation and the impact factors in Section 4.3. In concrete terms, the hydrologic conditions of the Shuping landslide area are as follows: the water level of TGR remains low during the period from May to September and the rainy season in the Three Gorges area does not begin until June. Consequently, in May, the corresponding displacement variation affected by the fluctuation of water level and rainfall reduces to the lowest value; that is to say, the periodic motion of CR14 falls to the lowest level in May. Conversely, with the commencement of the rainy season in June, a gradual increase of the periodic displacement occurred. As can be seen in Table 4, although the results from the two stacks do not strictly align in terms of timing, these results are consistent with the developing trend, which suggests the feasibility of data assimilation for studying the combined effects of presumed triggering factors on landslide deformation. Considering that we placed the uncertain parameters of the model into the state vector to synchronize updates, this approach relies on state-parameter dependencies and is not able to provide good estimates if the dependencies are weak. To address this challenge, novel variants of the EnKF, such as dual estimate approaches [59], may be useful remedies.

Results confirm that EnKF is suitable for landside research. It can continually update landside states and model parameters when new data become available. Recently, the EnKF has become popular in many areas of the earth sciences because it is easy to use, is flexible, and has relatively few restrictive assumptions [19,21]. However, the potential of EnKF in landside disaster research needs further exploration to address ongoing issues in the coupling mode between the landslide observations (including the *in situ* measurements and remote sensing retrieval data) and other mechanism models (e.g., Transient Rainfall Infiltration and Grid-based Regional Slope-Stability Model (TRIGRS) and in the Fast Lagrangian Analysis of Continua model (FLAC3D)). If these issues are settled, then this method could assist in disaster early-warning systems and forecasting.

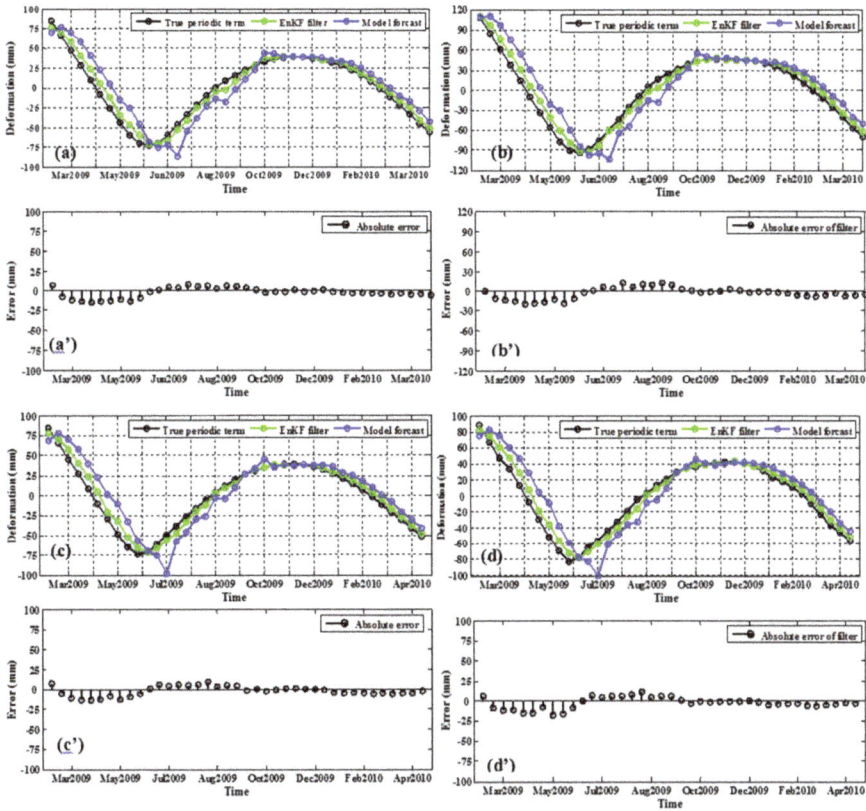

Figure 14. The assimilation results and their corresponding mean absolute error of periodic motion at point CR14 from February 2009 to April 2010; among them (**a,b**) were assimilation and simulation results from SM mode stack in range and azimuth separately with (**a′,b′**) as the corresponding mean absolute error; (**c,d**) were assimilation and simulation results from HS mode stack in range and azimuth separately with (**c′,d′**) as the corresponding mean absolute error.

Table 4. A short term forecast of the periodic term displacement.

Date	Range (mm)	Azimuth (mm)
26 April 2010 (HS)	−46.9	−54.3
1 May 2010 (SM)	−49.4	−59.1
7 May 2010 (HS)	−43.6	−50.1
12 May 2010 (SM)	−42.9	−51.4

In specific applications, the success of data assimilation is quite dependent on the error specification in the model, error in observations, and the ensemble size. When designing the data assimilation scheme, EnKF requires error estimations of the dynamic process and error estimation for the observations to properly couple model predictions with observations. However, it is very difficult to quantify these errors because the sources and statistical structure of each error are often unknown. In this study, we assumed that larger value variables could introduce larger errors; thus, the error noises were represented by adding a scaling factor to the then-current values of the variable. In actual applications, it is preferable to overestimate rather than underestimate errors as the underestimation may result in filter divergences. However, a much larger bias would make the system unstable and propagate improper information to the next time step, thus spoiling assimilation performance. The ensemble is a set of realizations in the EnKF and used to approximate the probability distribution of the state variables. Generally, enlarging the ensemble size enables the EnKF to propagate the error information more accurately; however, at the same time, it increases the computational burden. Consequently, if there is no calculation burden, we prefer a larger ensemble size (e.g., 1000) to get a more accurate result; or if a large complex model is adopted in the assimilation scheme, we have to make a compromise between estimation accuracy and computational feasibility.

5. Conclusions

In recent years, time-series SAR techniques have been successfully applied to detect landslide deformation. However, most of these studies focused primarily on technical innovation and applications. Few studies have considered the further interpretation of SAR/InSAR derived landslide displacements in relation to potential triggering factors, although this work is of crucial importance to understand deformation mechanisms and assist disaster early-warning and forecasting.

In this study, we executed a coupled analysis based on data assimilation methodology to explore the displacement response in the Shuping landslide area under the influence of hydrological factors. Corner reflectors installed at/near the landslide zone were utilized in the Pixel Offset Tracking (POT) technique to analyze the spatial-temporal evolution of landslide deformation. The measurements with SM- and HS-mode TerraSAR-X imageries showed consistency in their spatial-temporal patterns. Because the corner reflectors usually maintain very high coherence (>0.8), the achievable accuracy can reach cm-level.

Since the reservoir water level and rainfall are the primary causes of displacement fluctuations, they can be regarded as periodic displacement. Therefore, we performed time-series decomposition on the POT monitoring results to obtain the periodic displacement. Our seasonal deformation results showed that the AGO can be used as an alternative method to weaken the random disturbance from SAR derived displacements. However, if sufficient data is available, the moving average method is preferable. Then, a grey relational analysis was conducted to estimate the grey relational degree between the periodic displacement and the inducing factors.

We adopted a sequential data assimilation method named EnKF to study the quantitative interactions of the landslide movement under the influence of hydrological factors. By incorporating available observations sequentially in time, the variables and parameters used in dynamic equations were synchronously optimized, the resulting assimilated estimates matched well with the SAR derived displacements while the predictive uncertainties were reduced over time. The predictions of periodic displacement for SM- and HS-mode time-series were compared and analyzed with the seasonal variation of the trigging factors. These results showed similarities with developing trends and were consistent with variation of the trigging factors.

Preliminary results confirmed that the EnKF is feasible for studying the deformation response mechanisms of landslide areas near a reservoir under the influence of hydrological factors. To facilitate the success of data assimilation processing, error specification must be consistent with the model capability and the measurement flexibility. We propose the EnKF sequential data assimilation method as a possible approach to get an accurate estimation of a current landslide state and generate prognostic variables for landslide evolution, which are of particular importance for disaster forecasting and developing early warning systems.

As the triggering factors of rainfall episodes and reservoir water levels continue to change, the variation in the Shuping landslide kinematics will also continue over time. Though the current deformation may not have a devastating impact, significant deformation showed up in Block-1 and the eastern part of Block-2. The maximum cumulative deformation exceeded 1 m during the period of July 2008 to May 2009. In contrast, a rather stable state was identified in the western part of Block-2. Therefore, the landslide deformation monitoring with mechanism analysis is a vital tool for understanding the further progress of the landslides and subsequently developing early hazard warning systems. Further studies will be followed by combining multi-track SAR data to derive the three-dimensional displacement fields for the Shuping landslide area, since they are favored by geologists when investigating earth surface deformation mechanisms.

Acknowledgments: This work is financially supported by the National Key Basic Research Program of China (Grant No. 2013CB733205 and Grant No. 2013CB733204). The authors would like to thank the Astrium Satellite Company and the German Aerospace Center (DLR) for providing the TerraSAR-X datasets through the General AO project (GEO0606) and TSX-Archive-2012 AO project (GEO1856).

Author Contributions: Yanan Jiang implemented the methodology and finished the manuscript organization. Mingsheng Liao supervised and guided the research. Zhiwei Zhou and Lu Zhang provided valuable suggestions for the revision. Xuguo Shi and Lu Zhang modified the Pixel Offset Tracking algorithm for this study. All authors contributed extensively with writing this paper.

Conflicts of Interest: The authors declare no conflict of interest.

References

1. Au, S.W.C. Rain-induced slope instability in Hong Kong. *Eng. Geol.* **1998**, *51*, 1–36. [CrossRef]
2. Schuster, R.L.; Highland, L. *Socioeconomic and Environmental Impacts of Landslides in the Western Hemisphere*; U.S. Geological Survey (Citeseer): Denver, CO, USA, 2001.
3. Huang, R.Q. Some catastrophic landslides since the twentieth century in the southwest of China. *Landslides* **2009**, *6*, 69–81.
4. Wang, F.W.; Zhang, Y.M.; Huo, Z.T.; Peng, X.M.; Araiba, K.; Wang, G.H. Movement of the shuping landslide in the first four years after the initial impoundment of the Three Gorges Dam reservoir, China. *Landslides* **2008**, *5*, 321–329. [CrossRef]
5. Crosta, G.; Frattini, P.; Agliardi, F. Deep seated gravitational slope deformations in the European Alps. *Tectonophysics* **2013**, *605*, 13–33. [CrossRef]
6. Michoud, C.; Baumann, V.; Lauknes, T.; Penna, I.; Derron, M.H.; Jaboyedoff, M. Large slope deformations detection and monitoring along shores of the Potrerillos Dam reservoir, argentina, based on a small-baseline InSR approach. *Landslides* **2015**, *4*. [CrossRef]
7. Liu, C.Z.; Liu, Y.H.; Wen, M.S.; Li, T.F.; Lian, J.F.; Qin, S.W. Geo-hazard initiation and assessment in the Three Gorges reservoir. In *Landslide Disaster Mitigation in Three Gorges Reservoir, China*; Springer: Berlin, Germany, 2009; pp. 3–40.
8. Ferretti, A.; Prati, C.; Rocca, F. Permanent scatterers in SAR interferometry. *IEEE Trans. Geosci. Remote. Sens.* **2001**, *39*, 8–20. [CrossRef]
9. Berardino, P.; Fornaro, G.; Lanari, R.; Sansosti, E. A new algorithm for surface deformation monitoring based on small baseline differential SAR interferograms. *IEEE Trans. Geosci. Remote. Sens.* **2002**, *40*, 2375–2383. [CrossRef]
10. Arnaud, A.; Adam, N.; Hanssen, R.; Inglada, J.; Duro, J.; Closa, J.; Eineder, M. ASAR-ERS interferometric phase continuity. In Proceedings of the IEEE International Geoscience and Remote Sensing Symposium, Toulouse, France, 21–25 July 2003; pp. 1133–1135.
11. Werner, C.; Wegmüller, U.; Strozzi, T.; Wiesmann, A. Interferometric point target analysis for deformation mapping. In Proceedings of the IEEE International Geoscience and Remote Sensing Symposium, Toulouse, France, 21–25 July 2003; pp. 4362–4364.
12. Blanco-Sanchez, P.; Mallorquí, J.J.; Duque, S.; Monells, D. The Coherent Pixels Technique (CPT): An advanced DInSAR technique for nonlinear deformation monitoring. *Pure Appl. Geophys.* **2008**, *165*, 1167–1193. [CrossRef]
13. Ferretti, A.; Fumagalli, A.; Novali, F.; Prati, C.; Rocca, F.; Rucci, A. A new algorithm for processing interferometric data-stacks: SqueeSAR. *IEEE Trans. Geosci. Remote. Sens.* **2011**, *49*, 3460–3470. [CrossRef]
14. Strozzi, T.; Luckman, A.; Murray, T.; Wegmüller, U.; Werner, C.L. Glacier motion estimation using SAR offset-tracking procedures. *IEEE Trans. Geosci. Remote. Sens.* **2002**, *40*, 2384–2391. [CrossRef]
15. Notti, D.; Calò, F.; Cigna, F.; Manunta, M.; Herrera, G.; Berti, M.; Meisina, C.; Tapete, D.; Zucca, F. A user-oriented methodology for DInSAR time series analysis and interpretation: Landslides and subsidence case studies. *Pure Appl. Geophys.* **2015**, *172*, 3081–3105. [CrossRef]

16. Tomás, R.; Li, Z.; Liu, P.; Singleton, A.; Hoey, T.; Cheng, X. Spatiotemporal characteristics of the Huangtupo landslide in the Three Gorges region (China) constrained by radar interferometry. *Geophys. J. Int.* **2014**, *197*, 213–232. [CrossRef]

17. Daley, R. *Atmospheric Data Analysis, Cambridge Atmospheric and Space Science Series*; Cambridge University Press: Cambridge, UK, 1991.

18. Talagrand, O. Assimilation of observations, an introduction. *J. Meteorol. Soc. Jpn.* **1997**, *75*, 81–99.

19. Evensen, G. *Data Assimilation—The Ensemble Kalman Filter*, 2nd ed.; Springer: Berlin, Germany, 2009.

20. Huang, C.L.; Li, X.; Lu, L.; Gu, J. Experiments of one-dimensional soil moisture assimilation system based on Ensemble Kalman filter. *Remote Sens. Environ.* **2008**, *112*, 888–900. [CrossRef]

21. Xie, X.H.; Zhang, D.X. Data assimilation for distributed hydrological catchment modeling via ensemble kalman filter. *Adv. Water Resour.* **2010**, *33*, 678–690. [CrossRef]

22. Shi, X.G.; Zhang, L.; Balz, T.; Liao, M.S. Landslide deformation monitoring using point-like target offset tracking with multi-mode high-resolution TerraSAR-X data. *ISPRS J. Photogramm.* **2015**, *105*, 128–140. [CrossRef]

23. Wang, L. Research of Recurrence Mechanism and Prediction of Shuping Landslide under Water Level Variation and Rainfall in Three Gorges Reservoir Area. Master's Thesis, China Three Gorges University, Yichang, China, 15 May 2014.

24. Ren, F.; Wu, X.L.; Zhang, K.X.; Niu, R.Q. Application of wavelet analysis and a particle swarm-optimized support vector machine to predict the displacement of the shuping landslide in the Three Gorges, China. *Environ. Earth Sci.* **2015**, *73*, 4791–4804. [CrossRef]

25. Peng, L.; Niu, R.Q.; Wu, T. Time series analysis and support vector machine for landslide displacement prediction. *J. Zhejiang Univ.* **2013**, *47*, 1672–1679.

26. Zhang, J.; Yin, K.L.; Wang, J.J.; Huang, F.M. Diaplacement prediction of baishuihe landslide based on time series and pso-svr model. *Chin. J. Rock Mech. Eng.* **2015**, *34*, 382–391.

27. Carnec, C.; Massonnet, D.; King, C. Two examples of the application of SAR interferomety to sites of small extent. *Geophys. Res. Lett.* **1996**, *23*, 3579–3582. [CrossRef]

28. Berardino, P.; Costantini, M.; Franceschetti, G.; Iodice, A.; Pietranera, L.; Rizzo, V. Use of differential SAR interferometry in monitoring and modelling large slope instability at maratea (Basilicata, Italy). *Eng. Geol.* **2003**, *68*, 31–51. [CrossRef]

29. Catani, F.; Farina, P.; Moretti, S.; Nico, G.; Strozzi, T. On the application of SAR interferometry to geomorphological studies: Estimation of landform attributes and mass movements. *Geomorphology* **2005**, *66*, 119–131. [CrossRef]

30. Lauknes, T.R.; Piyush Shanker, A.; Dehls, J.F.; Zebker, H.A.; Henderson, I.H.C.; Larsen, Y. Detailed rockslide mapping in northern norway with small baseline and persistent scatterer interferometric SAR time series methods. *Remote Sens. Environ.* **2010**, *114*, 2097–2109. [CrossRef]

31. Massonnet, D.; Feigl, K.L. Radar interferometry and its application to changes in the earth's surface. *Rev. Geophys.* **1998**, *36*, 441–500. [CrossRef]

32. Bianchini, S.; Herrera, G.; Mateos, R.; Notti, D.; Garcia, I.; Mora, O.; Moretti, S. Landslide activity maps generation by means of Persistent Scatterer interferometry. *Remote Sens.* **2013**, *5*, 6198–6222. [CrossRef]

33. Liu, P.; Li, Z.H.; Hoey, T.; Kincal, C.; Zhang, J.F.; Zeng, Q.M.; Muller, J.-P. Using advanced InSAR time series techniques to monitor landslide movements in badong of the three gorges region, China. *Int. J. Appl. Earth Obs.* **2013**, *21*, 253–264. [CrossRef]

34. Raspini, F.; Moretti, S.; Casagli, N. Landslide mapping using SqueeSAR data: Giampilieri (Ialy) case study. In *Landslide Science and Practice*; Springer: Berlin, Germany, 2013; Volume 1, pp. 147–154.

35. Singleton, A.; Li, Z.; Hoey, T.; Muller, J.-P. Evaluating sub-pixel offset techniques as an alternative to D-InSAR for monitoring episodic landslide movements in vegetated terrain. *Remote Sens. Environ.* **2014**, *147*, 133–144. [CrossRef]

36. Fan, J.H.; Lin, H.; Xia, Y.; Zhao, H.L.; Guo, X.F.; Li, M. Mapping the deformation of shuping landslide using DInSAR and offset tracking methods. In *Landslide Science for a Safer Geoenvironment*; Sassa, K., Canuti, P., Yin, Y., Eds.; Springer: Berlin, Germany, 2014; pp. 319–324.

37. Raspini, F.; Ciampalini, A.; del Conte, S.; Lombardi, L.; Nocentini, M.; Gigli, G.; Ferretti, A.; Casagli, N. Exploitation of amplitude and phase of satellite SAR images for landslide mapping: The case of montescaglioso (south Italy). *Remote Sens.* **2015**, *7*, 14576–14596. [CrossRef]

38. Sansosti, E.; Berardino, P.; Bonano, M.; Calò, F.; Castaldo, R.; Casu, F.; Manunta, M.; Manzo, M.; Pepe, A.; Pepe, S. How second generation SAR systems are impacting the analysis of ground deformation. *Int. Appl. Earth Obs.* **2014**, *28*, 1–11. [CrossRef]

39. Hu, X.; Wang, T.; Liao, M.S. Measuring coseismic displacements with point-like targets offset tracking. *IEEE Geosci. Remote Sens.* **2014**, *11*, 283–287. [CrossRef]

40. Li, X.F.; Muller, J.P.; Fang, C.; Zhao, Y.H. Measuring displacement field from TerraSAR-X amplitude images by subpixel correlation: An application to the landslide in Shuping, Tree Grges area. *Acta Petrol. Sin.* **2011**, *27*, 3843–3850.

41. Seng, H.S. A new approach of moving average method in time series analysis. In Proceedings of the New Media Studies (CoNMedia), Tangerang, Indonesia, 27–28 November 2013; pp. 1–4.

42. Xu, F.; Wang, Y.; Du, J.; Ye, J. Study of displacement prediction model of landslide based on time series analysis. *Chin. J. Rock Mech. Eng.* **2011**, *30*, 746–751.

43. Kayacan, E.; Ulutas, B.; Kaynak, O. Grey system theory-based models in time series prediction. *Expert Syst. Appl.* **2010**, *37*, 1784–1789. [CrossRef]

44. Deng, J.L. Control problems of grey systems. *Syst. Control Lett.* **1982**, *1*, 288–294.

45. Evensen, G. Sequential data assimilation with a nonlinear quasi-geostrophic model using Monte Carlo methods to forecast error statistics. *J. Geophys. Res.* **1994**, *99*, 10143. [CrossRef]

46. Burgers, G.; van Jan Leeuwen, P.; Evensen, G. Analysis scheme in the ensemble Kalman filter. *Mon. Weather Rev.* **1998**, *126*, 1719–1724. [CrossRef]

47. Van Leeuwen, P.J. Comment on data assimilation using an ensemble kalm$an filter technique. *Mon. Weather Rev.* **1999**, *127*, 1374–1377. [CrossRef]

48. De Zan, F. Accuracy of Incoherent Speckle Tracking for Circular Gaussian Signals. *IEEE Geosci. Remote Sens.* **2014**, *11*, 264–267. [CrossRef]

49. Chen, D.; Xue, G.; Xu, F. *Study on the Engineering Geology Properties in Three Gorges*; Hubei Science and Technology Publisher: Wuhan, China, 1997.

50. Fu, W.X.; Guo, H.D.; Tian, Q.J.; Guo, X.F. Landslide monitoring by corner reflectors differential interferometry SAR. *Int. J. Remote Sens.* **2010**, *31*, 6387–6400. [CrossRef]

51. Wen, B.P.; Wang, S.J.; Wang, E.Z.; Zhang, J.M. Characteristics of rapid giant landslides in China. *Landslides* **2004**, *1*, 247–261. [CrossRef]

52. Wang, F.W.; Zhang, Y.M.; Huo, Z.T.; Peng, X.M. Monitoring on shuping landslide in the Three Gorges Dam Reservoir, China. In *Landslide Disaster Mitigation in Three Gorges Reservoir, China*; Springer: Berlin, Germany, 2009; pp. 257–273.

53. Xia, Y. Synthetic aperture radar interferometry. In *Sciences of Geodesy-I*; Springer: Berlin, Germany, 2010; pp. 415–474.

54. Xia, Y.; Kaufmann, H.; Guo, X.F. Landslide monitoring in the Three Gorges area using D-InSAR and corner reflectors. *Photogramm. Eng. Remote Sens.* **2004**, *70*, 1167–1172.

55. Tomás, R.; Li, Z.; Lopez-Sanchez, J.M.; Liu, P.; Singleton, A. Using wavelet tools to analyse seasonal variations from InSAR time-series data: A case study of the Huangtupo landslide. *Landslides* **2015**, *6*. [CrossRef]

56. Tullos, D. Assessing the influence of environmental impact assessments on science and policy: An analysis of the Three Gorges project. *J. Environ. Manag.* **2009**, *90*, S208–S223. [CrossRef] [PubMed]

57. Lin, J.L.; Lin, C.L. The use of the orthogonal array with grey relational analysis to optimize the electrical discharge machining process with multiple performance characteristics. *Int. J. Mach. Tools Manuf.* **2002**, *42*, 237–244. [CrossRef]

58. Li, X.Z.; Kong, J.M. Application of GA-SVM method with parameter optimization for landslide development prediction. *Nat. Hazards Earth Syst.* **2014**, *14*, 525–533. [CrossRef]

59. Yang, K.; Koike, T.; Kaihotsu, I.; Qin, J. Validation of a dual-pass microwave land data assimilation system for estimating surface soil moisture in semiarid regions. *J. Hydrometeorol.* **2009**, *10*, 780–793. [CrossRef]

remote sensing

MDPI

Article

Space-Borne and Ground-Based InSAR Data Integration: The Åknes Test Site

Federica Bardi [1,*], Federico Raspini [1], Andrea Ciampalini [1], Lene Kristensen [2], Line Rouyet [3], Tom Rune Lauknes [3], Regula Frauenfelder [4] and Nicola Casagli [1]

[1] Department of Earth Sciences, Università di Firenze, via La Pira 4, 50121 Firenze, Italy;
    federico.raspini@unifi.it (F.R.); andrea.ciampalini@unifi.it (A.C.); nicola.casagli@unifi.it (N.C.)
[2] Norwegian Water Resources and Energy Directorate (NVE), Ødegårdsvegen 176, 6200 Stranda,
    Norway; lkr@nve.no
[3] Norut Northern Research Institute, 9294 Tromsø, Norway; line.rouyet@norut.no (L.R.);
    tom.rune.lauknes@norut.no (T.R.L.)
[4] Norwegian Geotechnical Institute (NGI), Sognsveien 72, 0806 Oslo, Norway; Regula.Frauenfelder@ngi.no
*   Correspondence: federica.bardi@unifi.it; Tel.: +39-055-27-7777

Academic Editors: Zhenhong Li, Roberto Tomas, Heiko Balzter, Josef Kellndorfer and Prasad S. Thenkabail
Received: 10 December 2015; Accepted: 4 March 2016; Published: 12 March 2016

Abstract: This work concerns a proposal of the integration of InSAR (Interferometric Synthetic Aperture Radar) data acquired by ground-based (GB) and satellite platforms. The selected test site is the Åknes rockslide, which affects the western Norwegian coast. The availability of GB-InSAR and satellite InSAR data and the accessibility of a wide literature make the landslide suitable for testing the proposed procedure. The first step consists of the organization of a geodatabase, performed in the GIS environment, containing all of the available data. The second step concerns the analysis of satellite and GB-InSAR data, separately. Two datasets, acquired by RADARSAT-2 (related to a period between October 2008 and August 2013) and by a combination of TerraSAR-X and TanDEM-X (acquired between July 2010 and October 2012), both of them in ascending orbit, processed applying SBAS (Small BAseline Subset) method, are available. GB-InSAR data related to five different campaigns of measurements, referred to the summer seasons of 2006, 2008, 2009, 2010 and 2012, are available, as well. The third step relies on data integration, performed firstly from a qualitative point of view and later from a semi-quantitative point of view. The results of the proposed procedure have been validated by comparing them to GPS (Global Positioning System) data. The proposed procedure allowed us to better define landslide sectors in terms of different ranges of displacements. From a qualitative point of view, stable and unstable areas have been distinguished. In the sector concerning movement, two different sectors have been defined thanks to the results of the semi-quantitative integration step: the first sector, concerning displacement values higher than 10 mm, and the 2nd sector, where the displacements did not exceed a 10-mm value of displacement in the analyzed period.

Keywords: satellite interferometry; ground-based radar; radar data integration; rockslide monitoring

1. Introduction

On the basis of recent evaluations, landslides represent the most frequent geo-hazard, occurring worldwide more frequently than any other natural disaster, including earthquakes and volcanic eruptions [1]. Landslides pose great threats to human lives, causing thousands of deaths and injured people every year (e.g., [2–4]). Moreover, every year, landslides cause billions of dollars (e.g., [5–7]) of direct and indirect socio-economic losses, in terms of property and infrastructure damage and environmental degradation.

In addition, landslide disasters show a documented increasing trend, mainly owing to the over exploitation of natural resources, improper land use planning and growing urbanization, which determines an increase in the population exposed to the landslide risk [8]. In this sense, the effort of the scientific community is focused on determining every possible measure of risk mitigation.

At the basis of the risk mitigation strategies, there is the deep knowledge of the phenomena, which, in the landslide field, is related to monitoring activities. Dealing with landslides, monitoring activities rely on the measurement of displacement fields in order to assess the temporal evolution and spatial distribution of moving areas.

This type of information represents key parameters to geometry and kinematics assessment of a mass movement; it is of great value especially in those urbanized areas endangered by movement and where the investigated phenomenon is going to threat valuable elements at risk.

Remote sensing and Earth observation (EO) data have a major role to play for studying geohazard-related events at different stages, such as detection, mapping, hazard zonation, modeling, prediction and monitoring.

During the last decade, different monitoring and remote sensing techniques, devoted to landslide analysis, have undergone rapid development. Among them, Interferometric Synthetic Aperture Radar (InSAR) techniques have seen an increasingly greater spread. Firstly conceived of and developed for data acquired from space-borne platforms, InSAR methods were later applied also on ground-based platforms (GB-InSAR). Especially as regards landslide and unstable slope monitoring activities, GB-InSAR systems have become more and more popular over the last few years [9–11].

InSAR techniques belong to the family of active remote sensing techniques, and thanks to their intrinsic characteristics, they present many advantages in the field of landslide monitoring and management with respect to conventional, geodetic techniques.

Among the several advantages that could be counted, the possibility to collect systematic and easily updatable acquisitions and to produce time displacement maps of several square kilometer wide areas can be considered the crucial benefits of these techniques. Moreover, they are able to observe the investigated instable areas under any light and weather conditions, obtaining displacement measurements with high precision.

PS-InSAR (Permanent Scatterer InSAR) [12,13] was the first technique, developed by TRE (TeleRilevamento Europa), specifically implemented for the processing of several (at least 15 or more) co-registered, multi-temporal space-borne SAR images of the same target area.

This kind of technique is useful in order to obtain the deformation time series and the deformation velocity of stable reflective point-wise targets, called PS, with respect to a reference point considered as stable. These targets are represented by hand-made artifacts (e.g., buildings, railways) and/or natural targets, such as rocky outcrops. The measurement of the PS displacement occurs along the satellite line of sight (LOS).

Specifically, the precision on the deformation rate is about 0.1–1 mm/y [12–17] by using satellite InSAR techniques.

Several other approaches have been proposed for the processing of multi-interferometric long series of SAR images; most of them have been satisfactorily compared by [16] and by [18].

Among the several approaches, the Small BAseline Subset (SBAS) technique uses small baselines, multilook data and a coherent-based selection criterion [18,19].

The ground-based SAR interferometer (GB-InSAR) is a terrestrial system that emits and receives microwaves moving along a rail track, multiple times [9–11,20,21]. Its cross-range resolution is directly proportional to the length of the rail. This kind of sensor measures both the amplitude and the phase of the radar signal. The phase can be profitably used in order to monitor ground deformation. The GB-InSAR can acquire an image every few minutes, allowing the monitoring of faster movements with respect to the satellite sensors. As regards GB-InSAR techniques, they are able to acquire sub-millimetric deformation rates [11]. Finally, the possibility to retrieve the temporal evolution of a single landslide(s) system without physical access to the unstable slope or the necessity of positioning any targets on the ground is a great advantage when the observed area is a steep, mountainous slope [22–24].

In this paper, in order to improve the applicability of InSAR techniques in the field of landslide monitoring, a proposal of the integration between ground-based and satellite InSAR datasets is presented. The integration is possible thanks to the intrinsic features of the techniques, which can be considered partially complementary, in terms of spatial and temporal resolution.

The integration procedure is based on three main steps: a qualitative integration, with the implementation of a geodatabase to differentiate stable from unstable areas; a semi-quantitative integration, which is based on data homogenization and evaluation of macro-areas with different displacement values; and a quantitative integration, where data can also be analyzed in terms of time series, which can be used to apply forecasting algorithms. This third step is possible only if high precision long time series data are available.

In this work, the Åknes test site has been selected to apply the first and second steps of the proposed procedure. The Åknes rockslide is located on the western coast of Norway, a country highly susceptible to large rockslides due to its numerous fjords, steep topography and high relief [25]. The Åknes rockslide is an unstable mass rock of about 50 million m^3 [26]. The unstable area represents a threat, in case of collapse, for the several communities located on the same fjord (Sunnylvsfjorden), mainly in terms of a possible induced tsunami. The availability of GB-InSAR and satellite InSAR data, with a period of overlapping measurements, makes the rockslide suitable to test the proposed integration procedure, in its first and second steps. Thanks to the implementation of this new approach, more precise information on the ground displacement pattern was obtained, together with an implementation in data coverage on the observed scenario. These improvements could be helpful in risk mitigation strategies.

2. Materials and Methods

2.1. Data Integration

The intrinsic features of ground-based and satellite InSAR techniques, in terms of both their advantages and limitations, make them particularly suitable to be applied together in the field of landslide mapping, monitoring and risk management [27–29]. These techniques are indeed in a way that is complementary and suitable to be used in a synergistic way. On the one hand, satellite InSAR techniques are useful for monitoring unstable areas under specific conditions: the main limitations related to its applicability regard the satellite revisiting time, the slope exposure with respect to the sensor LOS and the velocity of the investigated movements with respect to the wavelength and the repeat-pass interval [30]. Due to the inherent limitations of current space observation systems and relevant data processing technique, satellite InSAR techniques are currently applicable only to two classes of the Cruden and Varnes (1996) [31] classification: extremely slow and very slow movements (velocity < 16 mm/y and 16 mm/y ⩽ velocity < 1.6 m/y, respectively). Nevertheless, the satellite InSAR technique ability of measuring very slow and smooth ground displacements represents a valuable support to landslide hazard prevention activities over wide areas, giving the opportunity to detect extremely slow, precursor movements that usually occur several weeks or months before the catastrophic failure, preceding major landslide disasters [32,33]; in the case of moving rockslides, the technique can also be useful to detect acceleration phases (tertiary creep) [34]. On the other hand, GB-InSAR allows a continuous monitoring of the displacements ranging from a few millimeters per day up to 1 or more meters per day, on a more local scale [11]. Furthermore, the instrument flexibility enables the investigation of shadow areas of hillsides not illuminated by the beam of the satellite radar sensor, and it allows to choose, during the installation phase, the best line of sight (LOS). These characteristics make this technique particularly useful for emergency phases [35].

Because of the above-mentioned characteristics and differences, the integration of these techniques enables us to obtain useful information on ground displacement patterns, with high precision and improved spatio-temporal resolution and coverage, compared to the stand-alone use of each technique.

In this paper, an attempt to define a procedure to integrate ground-based and satellite InSAR datasets is described, in order to use the results of both techniques, in those areas where both data are available, as in the case of the Åknes test site.

A schematic description of the proposed integration methodology is shown in Figure 1.

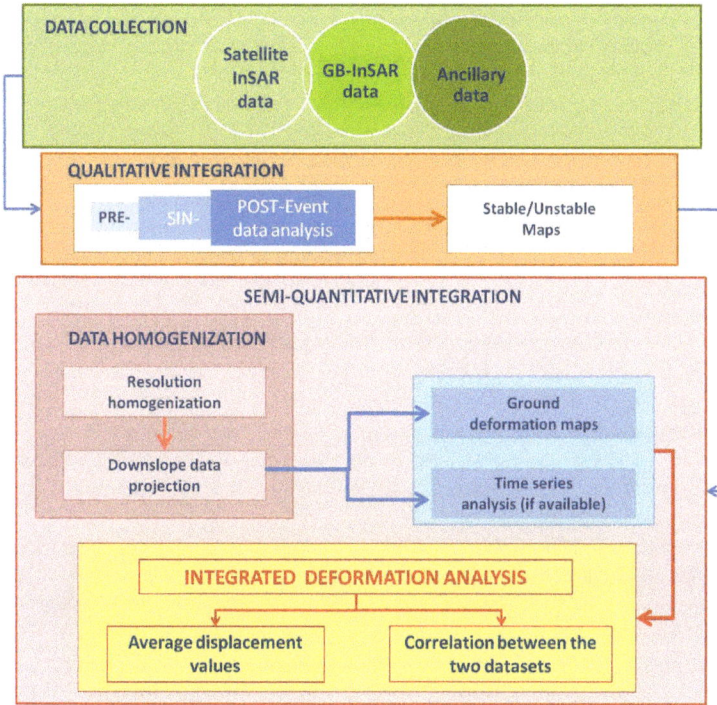

Figure 1. Schematic representation of the proposed integration methodology.

Data integration is performed during the post-processing phase, and it can be achieved by using a qualitative and/or a semi-quantitative approach.

2.1.1. Qualitative Integration

The qualitative approach is preparatory, allowing the collection of more information about displacements, also in those areas concerning shadowing if observed using only a single dataset.

Therefore, a preliminary phase is characterized by data collection, including the acquisition of ancillary data (*i.e.*, orthophotos of the study area, pre-existing landslide inventory maps, the results of geological and geomorphological field surveys, *etc.*) and the available SAR data, acquired both from space-borne and ground-based platforms (Figure 1).

In case of landslide collapse, datasets are distinguished, on the basis of the acquisition time, in pre-, during and post-event datasets, and they are compared with each other considering the time span to which they refer. Generally, SAR satellite data cover longer periods than GB-InSAR data, thanks to the existence of historical archives resulting from the spatial missions carried out over recent years, starting from 1992 with the launch of the ERS-1 satellite.

On the other hand, ground-based systems generally provide post- or at least during event data, because they are usually installed in emergency phases, when the landslide has already collapsed. An exception is represented by rockslides, which are often monitored also during the pre-collapse phase, to detect acceleration phases, as in the case of the Åknes rockslide.

By using GIS (Geographical Information System) platforms, InSAR data can be easily observed in a two-dimensional geo-referred space. Integrated satellite and ground-based displacement maps can be performed to distinguish the areas characterized by displacements from the areas without evidence of movements, following a binary approach (stable/unstable). The combined use of satellite and ground-based SAR data can provide ground displacement measurements with higher spatio-temporal resolution and coverage than the stand-alone use of the techniques, which can be satisfactorily applied for mapping and monitoring landslide phenomena [27,36].

In order to validate the integration methodology, the results can be qualitatively compared to the available ancillary data or to the results of scheduled field surveys.

2.1.2. Semi-Quantitative Integration

Moving to the semi-quantitative approach, as the first step, it is necessary to take into account the different temporal and spatial resolutions of compared datasets, in terms of time and space. Therefore, firstly, data have to be homogenized. The level of homogenization that is possible to obtain defines the integration accuracy.

Generally, datasets with finer temporal resolution are re-sampled, reducing their resolution to values comparable to the ones of the other datasets. For example, with respect to the satellites (in the best conditions, new generation sensors are able to acquire every day [37]), ground-based systems have a better resolution; they are indeed able to acquire several times per day; however, ground-based datasets are rescaled to become comparable to data acquired by space-borne sensors. The same happens for the spatial resolution, where data are re-sampled referring to the pixel size of data with coarser resolution (generally satellite data).

Another important limitation on the quantitative integration is related to the intrinsic feature of the technique to be capable of acquiring only the LOS component of displacements. With more than 2 viewing geometries, the real displacement vector can be obtained as proposed by Wright *et al.*, 2004 [38]. With a range of approximation, 2 geometries could be enough to compute the real displacement vector components [38]. In the best conditions, satellite data are available in two different acquisition geometries: ascending and descending. Combining the information obtained by these different geometries, it is possible to extract the vertical and horizontal (in the east-west direction) components of the movement [39]. Therefore, two of the three components of the real vector displacement are identified; these two components allow the detection with a good approximation of the real displacement vector. Once the "almost real" displacement vector is identified, satellite data are compared to ground-based datasets, in order to calculate the displacement percentage also detected by GB-InSAR systems. Generally, because of the different acquisition geometries, space-borne sensors are able to estimate a small component of the 3D real motions; on the contrary, ground-based platforms are able to detect a big part of the real vector of displacement by using an LOS as parallel as possible to the displacement direction.

Usually, satellite datasets are often available in only one of the possible configurations (ascending or descending). In these very frequent cases, satellite and ground-based data are projected on a common direction, considered as the most probable direction, along which the real displacement develops. Generally, this direction is approximately considered the downslope direction; it is almost true if the investigated phenomena have a surface rupture as parallel as possible to the topography, which is typical of translational slides. The methodology used for the projection is based on a simplified formula with respect to the equation introduced by [38]; the proposed method is described in [40] and later in [27,41,42]. The input data consist of the angular values of aspect and slope (derivable from the digital terrain model) of the area affected by the investigated landslide phenomena and the azimuth angle and the incidence angle of the LOS, both from the satellite and the GB-InSAR.

After calculating the direction cosines of LOS and slope (respectively functions of azimuth and incidence angles and aspect and slope angles) in the directions of zenith (Zlos, Zslope), north (Nlos, Nslope) and east (Elos, Eslope), the coefficient C is defined as follows (Equation (1)):

$$C = Z_{los} \times Z_{slope} + N_{los} \times N_{slope} + E_{los} \times E_{slope} \qquad (1)$$

C gives information about the portion (percentage) of "real" displacement detected by spaceborne/ground-based sensors.

The "real" displacement (Dreal, Equation (2)) is defined as the ratio between the displacement recorded along the LOS (Dlos) and the C value:

$$D_{real} = D_{los}/C \qquad (2)$$

By using the MATLAB interface, an automation of the projection procedure is obtained.

In some cases, the projection algorithm could be applied to a set of angular conditions in which the LOS and the downslope angle directions are nearly perpendicular: in these situations, the projection could lead to large estimate errors, especially when small movements in the LOS are processed. These particular values, for simplicity called "outliers" (Figure 2), can be highlighted by applying a filter on the MATLAB script [43]. This problem is generally more common in satellite datasets, where the LOS is likely to be very different from the downslope direction; it is less common in the case of ground-based projected data, because of the similarity between the GB-InSAR LOS and the downslope direction of the investigated slope.

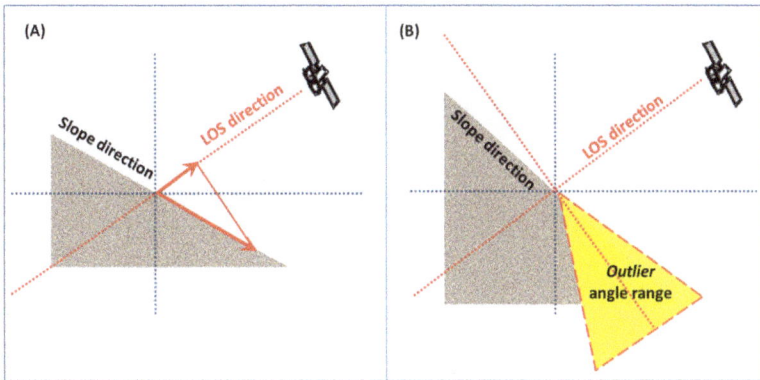

Figure 2. Projection geometry of LOS movement on slope direction (**A**) and the definition of outlier angle range (**B**) (modified from [39]).

Finally, the result of data integration by using a semi-quantitative approach consists of the production of ground deformation maps or ground velocity maps in relation to the projected input data (displacement or velocity values). These maps, containing the information related to both satellite and ground-based InSAR data, can also be superimposed on ancillary data. An additional step in the quantitative approach to the integration methodology is represented by the time series analysis, in case of the availability of long time series; comparing displacement/velocity time series of selected points on the ground deformation/velocity maps allows the quantitative information deriving from projected data to be increased. The points, selected for extracting time series, are generally chosen in areas where the radar signal is characterized by high stability, a high signal/noise ratio, high power and coherence parameters.

For these high quality points, both the satellite- and ground-based time series can be analyzed and compared.

2.2. Åknes Test Site

2.2.1. Location

The Åknes rockslide is located in the northwestern side of Sunnylvsfjorden, a branch of the Storfjorden, on the western Norwegian coast (Figure 3). It is considered one of the most hazardous rockslide areas in Norway [44], including about 50 million m^3 of rock [26], characterized by continuous creep. The location of the landslide body, which develops above the fjord and near several communities, makes the surrounding area exposed to a high level of risk, mainly in terms of a possible tsunami, induced by the collapse of rock material into the fjord [45,46]. The area also represents one of Norway's most visited tourist attractions, thanks to the natural beauty of the mentioned fjord (the nearby Geirangerfjord is listed on UNESCO's World heritage list). The importance of the study area makes Åknes one of the most investigated and monitored rockslides in the world.

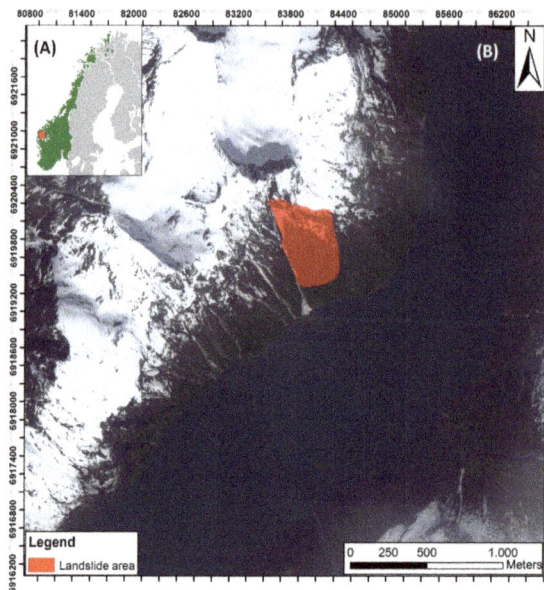

Figure 3. Åknes rockslide location on the Norwegian coast (**A**) and zoom on the area affected by the rockslide, whose location is emphasized by a red polygon (**B**).

2.2.2. Geological and Geomorphological Setting

From a geological point of view, the Åknes rockslide is located in the Western Gneiss Region of Norway. Gneisses of Proterozoic age represent the bedrock's dominant lithology of the area, and they exhibit the effects of alteration and reshuffle suffered during the Caledonian orogeny [47].

The geological setting of the area, especially as regards gneisses lithologies, is characterized by intense foliation, which strongly contributes to determine slope instability; the main cracks are indeed developed along foliation plans [48].

Various sub-domains can be distinguished in the rockslide body [44].

Kristensen *et al.* [49] identified two main sectors based on different deformation patterns and consequently different risk scenarios. The first scenario corresponds to the upper portion of the slope, and it is characterized by higher displacement values with respect to the second scenario, which corresponds to the middle sector of the slope; nowadays, the lower part of the slope is not affected by significant deformation [44]; therefore, it is not taken into account in this work (Figure 4).

As regards the deformation pattern, the upper part of the landslide is mainly concerning extension forces, in contrast to the lower part, which is characterized by generalized compression. The upper zone is delimited by a back scarp, which is controlled by pre-existing foliation planes or pre-existing fractures; the major upper crack is about 800 m long. The basal sliding surface seems to be controlled by foliation, as well; it dips 30°–35° to S-SSE, and it extends almost parallel to the topography. A steeply-dipping, NNW-SSE-trending fault defines the western boundary, and the eastern boundary is defined by a gently-dipping, NNE-SSW-trending fault [44]. The toe location is not so clear, but it can be approximately located at 75–100 m a.s.l. [50] (Figure 4).

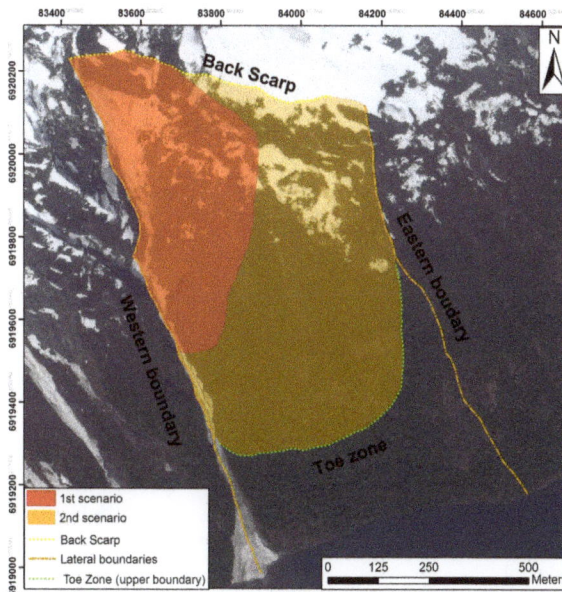

Figure 4. Rockslide distinction in two different risk scenarios and its delimitation taking into account the main detected geomorphological features [44,50].

More information about the rock slope geological setting can be found in [51–53] and in [54], where the results of extensive field and non-field work activities are summarized, such as geological and geophysical investigations, numerical modelling, *etc.*

2.3. Available Datasets

2.3.1. GB-InSAR Monitoring Activity

Åknes is one of the most studied rockslides in the world.

It takes advantage of one of the best organized early warning systems, which, in turn, can exploit a really complete monitoring network [55,56].

Moreover, starting from 2006, GB-InSAR campaigns have also been performed, in the context of the Åknes/Tafjord project (now part of the Norwegian Water Resources and Energy Directorate, in Norwegian Norges vassdrags- og energidirektorat (NVE) [57]). The instrument was provided by the Italian Society Ellegi LiSALab s.r.l. [58]. Five GB-InSAR campaigns have been achieved, specifically in the summer seasons of 2006, 2008, 2009, 2010 and 2012 (Table 1).

The system was installed on the opposite side of the fjord with respect to the rockslide, in a location named Oaldsbygda (Figure 5), equipped with an Internet network and a power supply. Very strong atmospheric effects concern the images acquired from this location, because of the sudden and very fast changes of local atmospheric conditions on the fjord, which is crossed by the radar signal during the acquisitions. Because of these technical problems, the nominal sub-millimetric accuracy of GB-InSAR systems is here reduced to millimetric accuracy. The stability threshold value has been fixed at 2 mm. Technical features of the GB-InSAR system are listed in Table 2.

Table 1. Dates of ground-based (GB)-InSAR campaigns.

GB-InSAR Campaigns
21 July 2006–25 October 2006
17 July 2008–13 October 2008
1 July 2009–17 October 2009
9 July 2010–31 October 2010
12 July 2012–24 October 2012

Table 2. Parameters of the employed radar system [49].

	GB-InSAR
Rail length	3 m
Central frequency	17.2 GHz
Bandwidth	60 MHz
Number of frequencies	2501
Steps along the rail	601
Image acquisition time	8 min
Processed image range	1800–4200 m
Processed image azimuth	±1200 m
Distance to the back scarp	3000 m

Despite the atmospheric disturbance, whose impact was reduced during the processing phase, using proper filters, the availability of several campaigns makes the Oaldsbygda data useful to analyze the evolution of the deformation pattern of the rockslide.

Data recorded during the campaigns with the Oaldsbygda instrument, performed in the spring-summer seasons of 2009, 2010 and 2012, have been selected for the analysis; they have been processed using LiSALab s.r.l. software.

Data have been analyzed in the GIS environment, upon the application of georeferencing transformations on the SAR images.

All campaigns started in July and stopped in October; their duration differs only by a few days. In Figure 6, cumulated displacement maps related to the total period of each acquisition campaign are displayed. Lower in the slope, the correlation is much lower (weak coherence) due to dense vegetation.

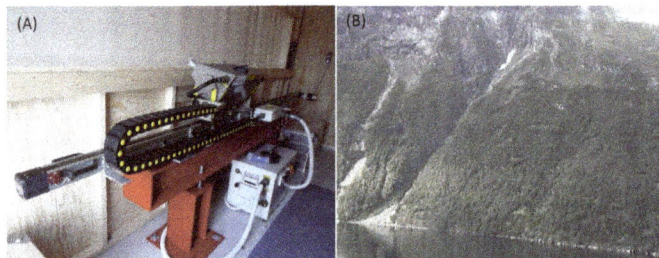

Figure 5. GB-InSAR system employed at Oaldsbygda (**A**) and its view of the Åknes slope (**B**).

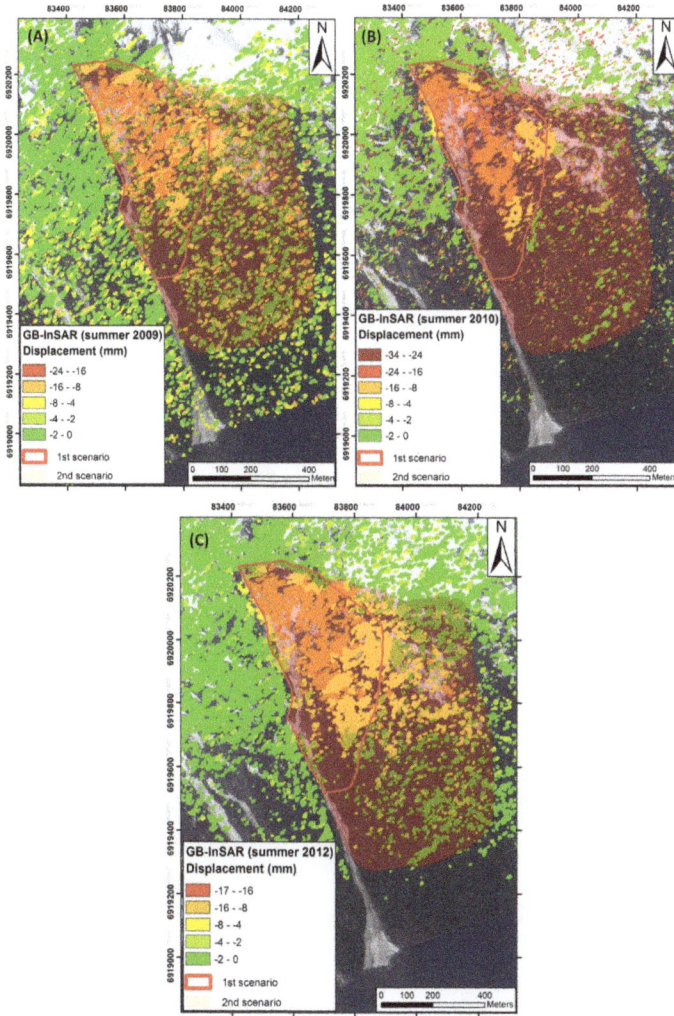

Figure 6. Cumulated displacement maps of GB-InSAR campaigns from Oaldsbygda. (**A**) Time interval of 108 days, between July and October 2009; (**B**) time interval of 114 days, between July and October 2010; (**C**) time interval of 104 days, between July and October 2012. Negative displacements represent movements approaching the sensor.

The maps in Figure 6 show a quite consistent deformation pattern, measured from year to year with a maximum of 34 mm in about a 4-month interval from July–October 2010. In greater detail, movements mainly concern the upper portion of the rockslide (1st scenario in Figure 4), reaching values of about 24 mm in the period between July and October 2009, about 34 mm during July–October 2010 and about 17 mm in 2012, during the same months. The GB-InSAR ability to detect only the LOS (line of sight) component of the displacement vector [11] has been partially overcome by comparing GB-InSAR data with GPS datasets [49,54]: this comparison allowed estimating the percentage of real displacement detected by the Oaldsbygda instrument, thanks to the possibility to record a 3D displacement vector with GPS instruments (Table 3). GPS locations are displayed by arrows in Figure 7.

Comparison results show that, from Oaldsbygda, the central part of the rockslide is highly detectable by GB-InSAR, which is able to identify about 70% of the real displacement, as recorded in Table 3 (bigger arrows in Figure 7). The upper sector of the landslide is mainly affected by vertical movements; therefore, in this area the instrument of Oaldsbygda is able to detect only 30% of the displacement (Table 3; smaller arrows in Figure 7).

Table 3. Displacement registered by GPSs (from [54]) compared to average displacement registered by GB-InSAR (mm/y) at GPS locations (GPS locations are as shown in Figure 7; G1 represents the base station location).

GPS Stations	GPS 3D Movement	Displacement Registered by GB-InSAR at GPS Location	LOS % of True Vector
G2	85.1 mm/y	30 mm/y	35%
G3	81.4 mm/y	24 mm/y	29%
G4	2.8 mm/y	1 mm/y	30%
G5	30.6 mm/y	20 mm/y	66%
G6	17.6 mm/y	13 mm/y	73%
G7	25.8 mm/y	19 mm/y	75%
G8	14.7 mm/y	10 mm/y	71%
G9	4.9 mm/y	3 mm/y	63%

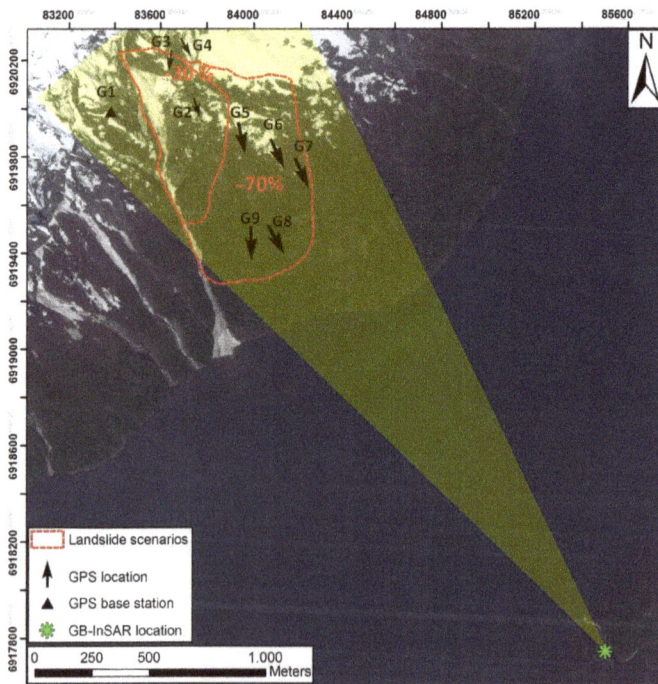

Figure 7. Location of GB-InSAR instrument and relative field of view. Black arrows refer to *in situ* instruments (GPS); their size is proportional to the percentage of displacement, registered by GPS instruments, and detected by GB-InSAR. The upper portion of the slope is less detectable (about 30%) than the lower sector (about 70%) from the GB-InSAR location. Background ortophoto is obtained from Virtual Earth imagery; GPS data from [49].

2.3.2. Satellite InSAR Monitoring Activity

In the case of satellite InSAR acquisitions, as well as in the case of GB-InSAR monitoring activities, the vegetation cover that affects the lower part of the landslide causes a loss of coherence in the radar signal. Moreover, the available satellite InSAR datasets present non-optimal LOS for measuring the horizontal component of the movement. Generally, Earth observing satellites, flying in polar orbits, have a direction close to NS; in this condition, horizontal displacements are almost undetectable because of their deep incident angles; moreover if snow or vegetation cover concerns the observed area or if there is high variability in soil moisture, the loss in coherence of the radar signal negatively influences radar images. These unfavorable conditions widely concern mountainous Norwegian regions.

In the specific case of Åknes, two datasets are available, acquired by RADARSAT-2 (14 images) and by a combination of TerraSAR-X and TanDEM-X (13 images), both in ascending orbit, which is the best configuration to avoid layer effects in east-facing slopes' observations. The LOS of both the satellites ranges between 76° and 77° in azimuth and between 25° and 28° in look angle. The RADARSAT-2 dataset includes images acquired between October 2008 and August 2013, with a revisiting time of 24 days; whereas TerraSAR-X/Tandem-X acquisitions are related to a shorter period, between July 2010 and October 2012, with a repeat time of 11 days. Only summer-early autumn periods are meaningful, because in other periods, the snow cover causes interferometric decorrelation (*i.e.*, different scattering properties from one scene to another).

RADARSAT-2 and TerraSAR-X data have been processed [59], applying the SBAS (Small BAseline Subset; [19,60,61]) method.

As for GB-InSAR data, satellite datasets have been compared to GPS data [59]. Considering the small component of displacements detectable by space-borne platforms, the comparison has been implemented projecting GPS data onto the LOS direction of the satellites. Satellite LOSs have indeed angles of about 62° for RADARSAT-2 and 65° for TerraSAR-X/Tandem-X, with respect to the horizontal direction, and an ENE orientation. This acquisition geometry makes the LOS near to perpendicular to the downslope direction.

The comparison provided a good correspondence between mean velocity values registered by satellite and mean velocities registered by GPS instruments, which have been estimated to reach approximately 20 mm/y. The same stability threshold value (± 2 mm/y) has been used both for C-band and X-band data, in order to make all of the used satellite data, acquired by different satellite sensors, as comparable as possible. Moreover, this value is acceptable, as it does not exceed the precision of the satellite InSAR technique. The results also show that RADARSAT-2 datasets are more relevant than TerraSAR-X/Tandem-X datasets, which registered maximum velocity values around 16 mm/y (Figure 8). This discrepancy is partially explained, besides the difference in the covered time periods, considering that, despite the improvement of spatial resolution, the use of X-band sensors' atmospheric effects may be severe. It is shown that radar with a wavelength shorter than 4 cm is more vulnerable to atmospheric effects. Therefore, the 3 cm-long X band radar is more influenced by rain and cloud than the 5.6 cm-long C band [18,62]. As a consequence, the extraction of phase variations related to displacement for each scatterer can be inaccurate: during phase unwrapping, the component of deformation becomes indistinguishable from terms related to atmospheric effects, leading to the underestimation of the actual deformation patterns. This problem affects more those landslides located in areas affected by persistent rainfall and cloud cover; this is the case of the Åknes rockslide.

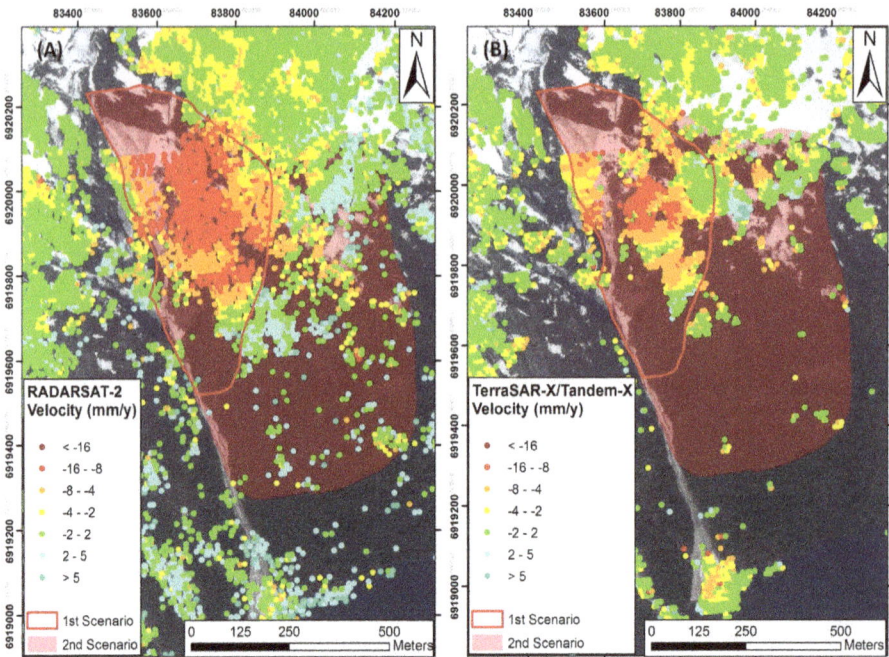

Figure 8. Mean velocities registered by RADARSAT-2 (**A**) and TerraSAR-X/Tandem-X (**B**) between 2010 and 2012 (values detectable on the LOS of the satellites). The red line corresponds to the landslide limit in the most dangerous scenario.

3. Results

The availability of both GB-InSAR and satellite InSAR data, for the overlapping periods of measurements, makes Åknes a suitable case study to test the suggested integration procedure, firstly from a qualitative point of view and later from a semi-quantitative point of view, as proposed in Figure 1.

3.1. Qualitative Integration Results

During the first step of integration, a qualitative analysis of different available data is required. For the Åknes test site, an orthophoto of the study area is available, together with InSAR datasets. Both satellite and GB-InSAR data refer to a pre-event phase, as the Åknes rockslide has not collapsed yet.

Qualitative data integration has been performed in the GIS environment, overlapping all of the available datasets (shown separately in Figures 6–8) on the available orthophoto. Data have been classified into two categories, to distinguish stable areas from unstable areas, independent of data type and relative reference period. The stability threshold has been fixed at 2 mm, according to the accuracy of both the GB-InSAR and satellite InSAR technique (Figure 9).

The main advantage of this integration step is in the improved density of measurements, which determines an almost complete coverage of the study area. Moreover, the distinction between stable and unstable areas is strongly emphasized: Figure 9 clearly shows that the main movements occur in the upper portion of the rockslide (first scenario in Figure 4). This area has been selected to apply the proposed integration methodology, in a more quantitative way.

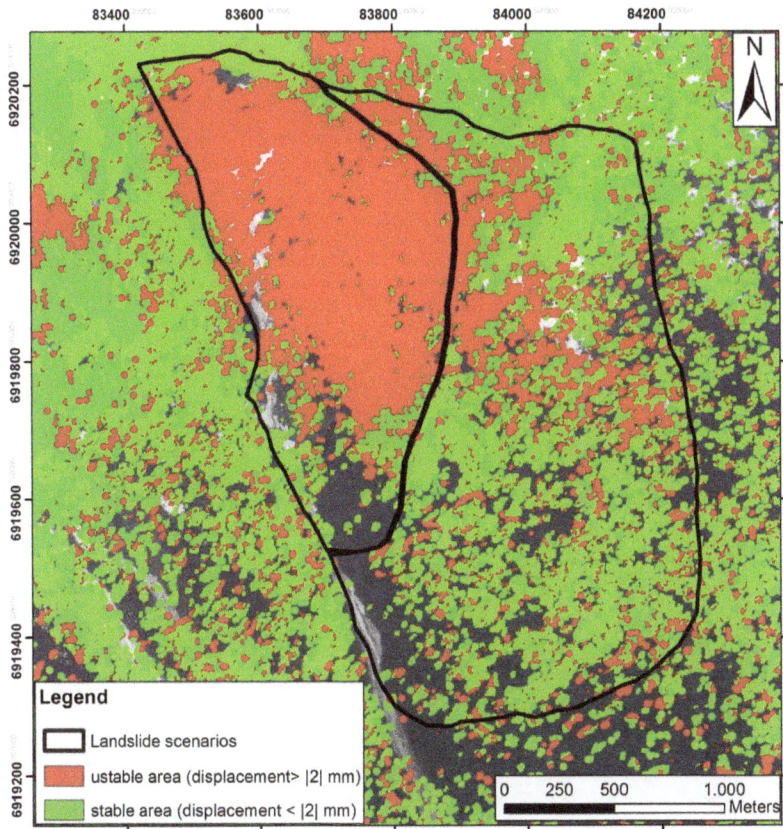

Figure 9. Qualitative integration between GB-InSAR available datasets (Figures 6 and 7) and RADARSAT-2 and TerraSAR-X/Tandem-X datasets (Figure 8). Data have been overlapped on an orthophoto of the study area. Unstable areas are displayed in red, whereas stable areas are shown in green.

3.2. Semi-Quantitative Integration Results

Among the available datasets, data referring to a period between July and October 2010 have been selected: this period indeed corresponds to the best overlap between GB-InSAR and satellite datasets.

Firstly, data have been homogenized in terms of LOS. Assuming that the most probable displacement direction is the downslope direction, both GB-InSAR and satellite data have been projected along this chosen direction. The downslope direction, especially for the middle part of the landslide, is indeed considered approximately similar to the real displacement direction, as the rockslide movement can be considered parallel to the topographic surface. In the upper part of the landslide, on the contrary, the main component of movement is vertical.

Starting from LOS displacement values, the objective was to evaluate the percentage of displacement detected by the instrument with respect to the downslope direction and to compare these percentages with the displacement values obtained by GPS campaigns (real displacement values) (Table 3; Figure 7).

The results can be summarized in a map showing the percentage of "real" displacement values (considering as "real" the displacements that happen in the downslope direction) detected in each sector of the landslide. Figure 10 shows the percentages of "real" displacement values detected by the Oaldsbygda SAR system. The upper portion of the landslide (emphasized in black in Figure 10) represents an area where the displacement direction is almost vertical and, for this reason, barely detectable along the GB-InSAR LOS (detectable displacement: about 30% of total). In any case, except for the upper sector, it is considered acceptable to assume the downslope direction as the direction of real displacement. The remaining portion of the analyzed rockslide sector seems to be highly detectable by the Oaldsbygda instrument, which can observe, along its LOS, a percentage between 60% and 90% of the "real" displacement vector. The map clearly shows a strong correspondence between the percentage of displacement detectable by GB-InSAR, considering as the "real" displacement direction the downslope direction, and the percentage of displacement detectable by GB-InSAR if compared with real displacement values registered by GPSs (Table 3; Figure 7).

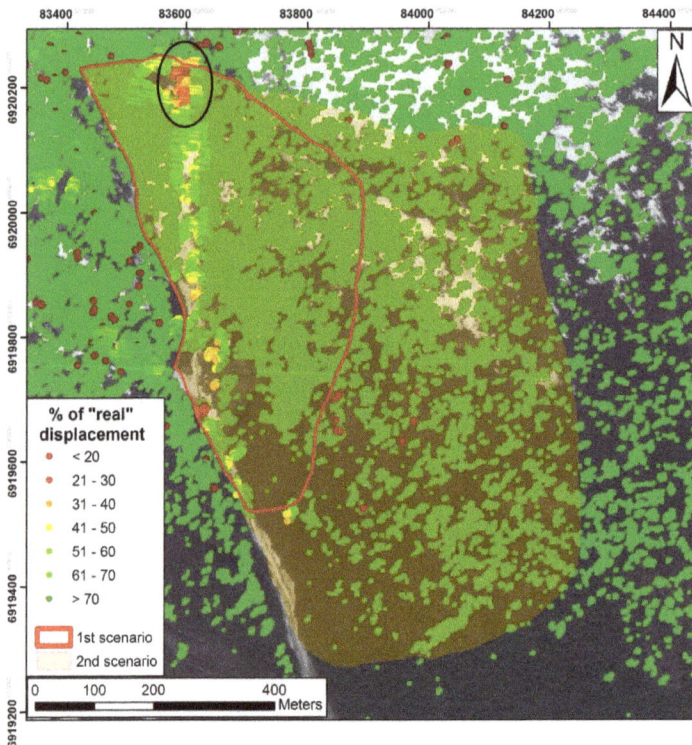

Figure 10. Map related to the % of "real" displacements detectable along the GB-InSAR LOS, considering as "real" the downslope direction. The sector emphasized by a black oval represents the upper part of the landslide, mainly concerning vertical movements.

Satellite InSAR data have also been projected downslope, following the same procedure applied for GB-InSAR datasets.

Because of the two employed satellite platforms being characterized by similar LOS directions (about 76° and 77° in azimuth and about 25° and 28° in look angle), the results of the projections are almost similar, as well (Figure 11).

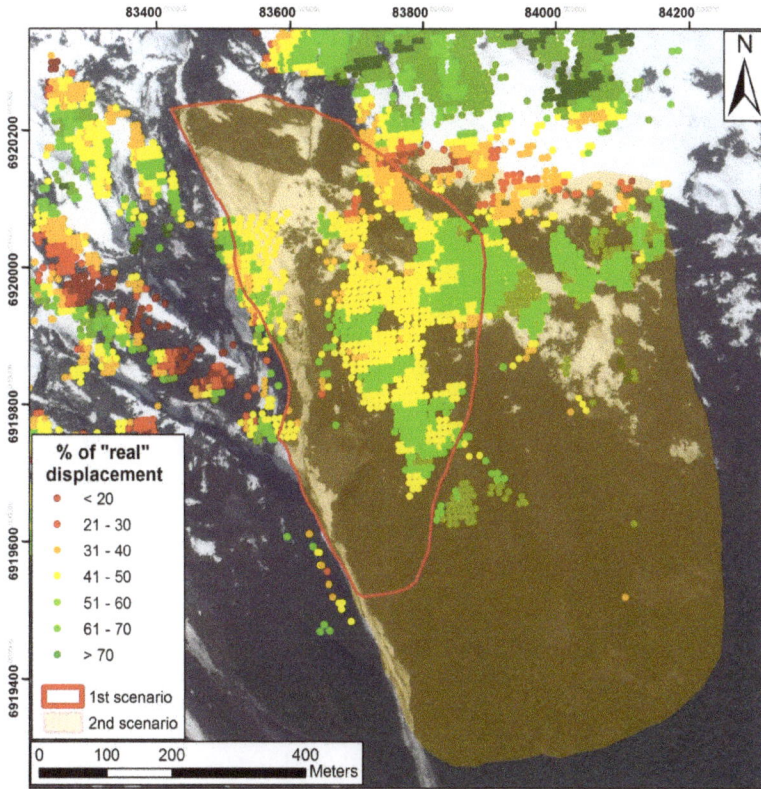

Figure 11. The % of "real" displacements detected by RADARSAT-2 and TerraSAR-X/Tandem-X satellites. The red line corresponds to the landslide limit in the most dangerous scenario (first scenario in Figure 4).

Along satellites' LOS, about 40%–50% on average of the "real" displacement is detectable. Instead, the maximum observable displacement corresponds to 60% of its "real" value. The upper part of the landslide, corresponding to the upper crack, is completely not visible from space-borne platforms (shadow area).

After the projection, the GB-InSAR dataset has been resampled in order to make it comparable to satellite data, in terms of temporal and spatial resolution. Finally, slope displacement values have been calculated (Figure 12A).

Some observations can be pointed out: first of all, the detectable area from the different sensors is not the same, specifically GB-InSAR data coverage is lower in the middle and lower parts of the slope than the satellite data coverage, which in turn is lacking in the upper part of the landslide, near the major crack. Concerning the displacements, RADARSAT-2 shows very similar patterns to the GB-InSAR datasets (Figure 12A–C). To better compare the datasets, a restricted period of about three months (from July–October 2010) has been selected in the RADARSAT-2 acquisitions, in order to make it comparable to the available data of the 2010 GB-InSAR campaign.

Data projection allowed displaying the two datasets on the same map (Figure 12C). Displacement values have been compared; a stability threshold of 5 mm has been fixed. This threshold value is almost similar to the displacement accuracy of the employed satellite datasets.

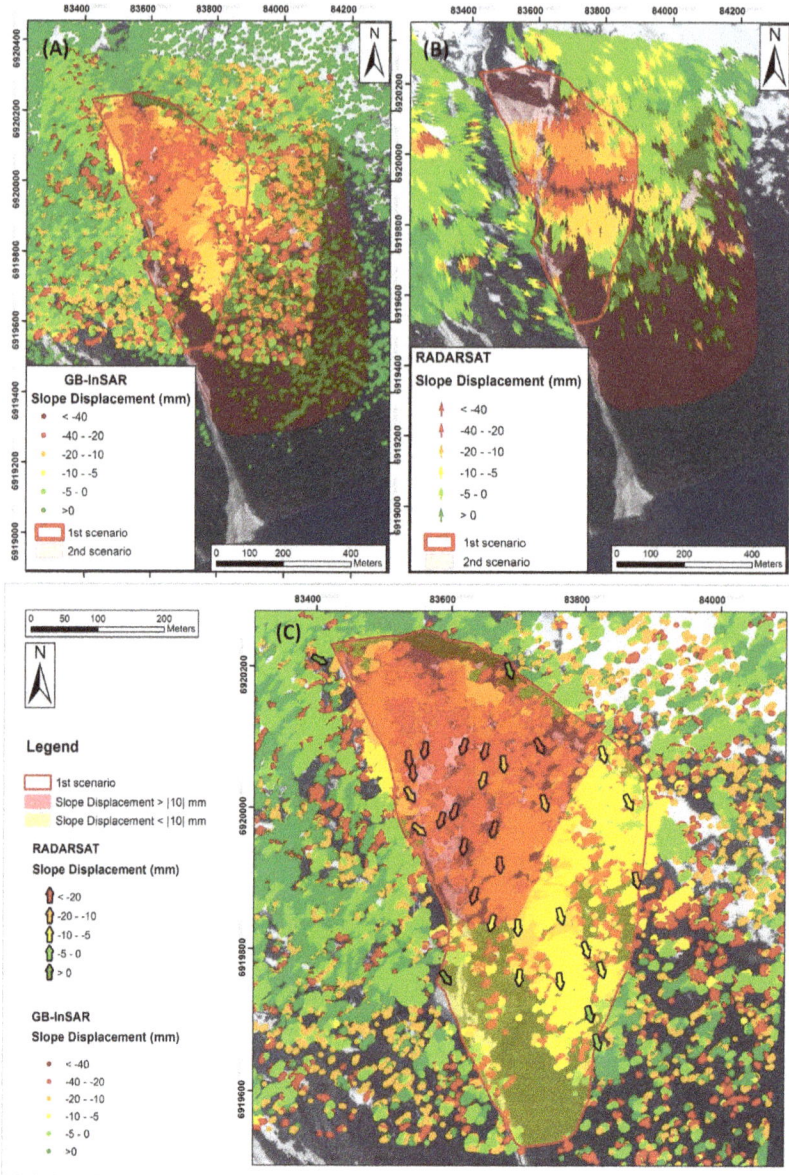

Figure 12. (**A**) Slope displacement (mm) registered by GB-InSAR in the 2010 campaign; (**B**) slope displacement (mm) registered by RADARSAT-2 in 2010; (**C**) GB-InSAR and RADARSAT-2 datasets' integration. The map in (**C**) shows "real" displacements, detected by the two different platforms; data refer to the period between July and October 2010; the red line corresponds to the landslide limit in the most dangerous scenario (first scenario in Figure 4).

Displacement values, detected by RADARSAT-2 and the GB-InSAR, agree to define the "middle" part of the landslide (defining as "landslide" only the portion of the rockslide, which is included in the limits that define the most dangerous possible scenario (first scenario in Figure 4) as concerns the major movement, reaching more than 40 mm of displacement in about three months). The upper part of the landslide, mainly affected by vertical displacements, shows displacement values around 5 mm in the downslope direction: this is probably an underestimation related to the differences between real displacement directions and the downslope direction. A further distinction in two sectors of the first landslide scenario is obtained and presented in Figure 12C. The upper part of the scenario (in red in Figure 12) represents the area concerning higher displacements, with respect to the lower part (in yellow in Figure 12). In Figure 13, the box plots and the histograms of data referring to the two different identified sectors are displayed. In Table 4, basic statistical parameters of the four datasets are displayed. Datasets show high dispersion, with a high number of outliers (very high standard deviations, as shown in Table 4); median values are lower than mean values, indicating negative skewness distributions (Figure 13B–D). Anyway, statistical analysis supports the identification of the two sectors. As regards the first sector, 50% of the GB-InSAR data distribution ranges between |11| mm and |22| mm; it does not differ so much from the RADARSAT-2 dataset, where 50% of the distribution ranges between |9| mm and |21| mm. Mean and median values detected by GB-InSAR for the first sector are respectively |17| mm and |16| mm; whereas mean and median values detected by RADARSAT-2 in the same sector are |18| mm and |13| mm. Generally, these distributions show higher values than datasets related to the second sector, where 50% of the values registered by GB-InSAR range between |6| mm and |11| mm and 50% of the values detected by RADARSAT-2 range between |4| mm and |9| mm. Mean and median values detected by GB-InSAR for the second sector are respectively |10| mm and |9| mm; instead, RADARSAT-2 distribution mean and median values, in the second sector, are |9| mm and |7| mm (Figures 12C and 13, Table 4). Considering these results, in spite of the high dispersion of the distributions, it is possible to assume that the second sector is affected by lower displacement values than the first sector and that the limit between the two sectors can be fixed at |10| mm.

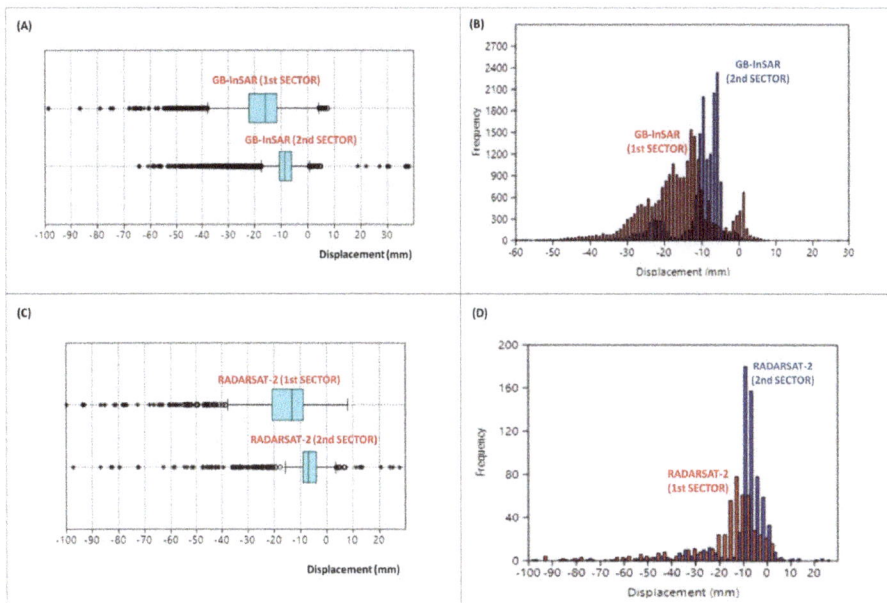

Figure 13. Box plots (**A**) and histograms (**B**) of GB-InSAR datasets, referring to the first and second sectors identified in Figure 12; box plots (**C**) and histograms (**D**) of RADARSAT-2 datasets, referring to the first and second sectors identified in Figure 12.

Table 4. Basic statistic parameters of the analyzed datasets (GB-InSAR data in the 1st and 2nd sectors and RADARSAT-2 data in the 1st and 2nd sectors).

	GB-InSAR (1st Sector)	GB-InSAR (2nd Sector)	RADARSAT-2 (1st Sector)	RADARSAT-2 (2nd Sector)
Mean	−17 mm	−10 mm	−18 mm	−9 mm
Median	−16 mm	−9 mm	−13 mm	−7 mm
SD	10	7	17	13
25 Percentile	−22 mm	−11 mm	−21 mm	−9 mm
75 Percentile	−12 mm	−6 mm	−9 mm	−4 mm

TerraSAR-X/Tandem-X data, referring to the same period (July–October 2010) have also been projected in the downslope direction and organized in order to be comparable to the other datasets (Figure 14).

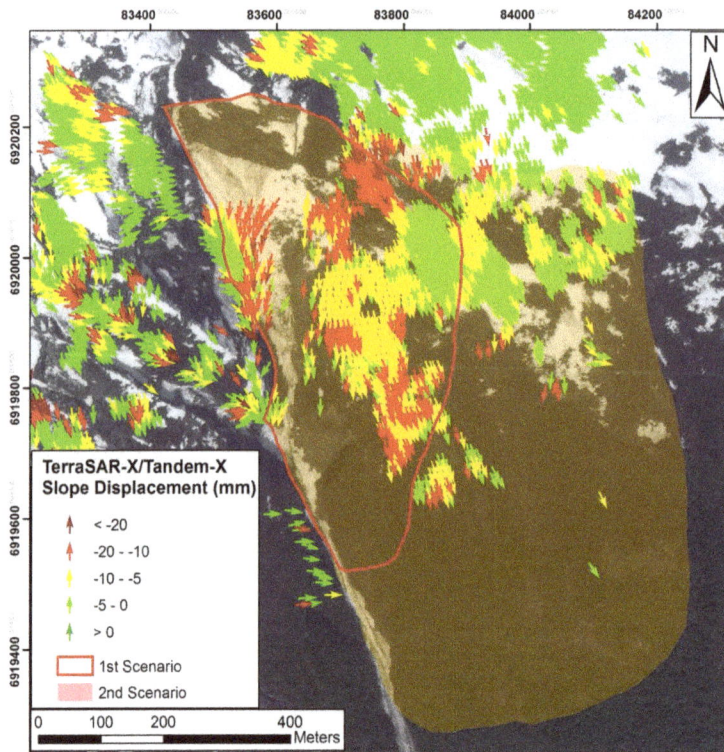

Figure 14. Slope displacements registered by TerraSAR-X/Tandem-X in 2010. Arrows are oriented along the downslope direction. The red line corresponds to the landslide limit in the most dangerous scenario.

TerraSAR-X/Tandem-X datasets show displacement values lower than values obtained by the RADARSAT-2 platform and GB-InSAR system. This underestimation, as mentioned before, can be mainly related to the higher atmospheric disturbance, concerning new-generation X-band satellites. Anyway, the TerraSAR-X/Tandem-X data comparison with GB-InSAR and RADARSAT-2 datasets can be considered satisfactory from the qualitative point of view.

4. Discussion

The measurement of the superficial displacement of a sliding mass often represents the most effective method for defining its kinematic behavior, allowing one to observe the relationship to triggering factors and to assess the effectiveness of the mitigation measures. Hence, once we suspect an area to be sliding or when dealing with a known landslide, it is mandatory to retrieve accurate and timely updated information on the rate and extent of the occurring deformation. It may be necessary to perform several measurement campaigns to confirm suspected movements, several months to characterize the type of deformation and years of continuous monitoring to fully understand the kinematics of a sliding mass.

Repeat surveys of benchmarks allow the periodic estimation of the extent and rates of deformation. These techniques provide punctual information and are time consuming and resource intensive, since a great deal of time and economic resources are required for timely updates. In most of the cases, these methods produce scattered measurements with an uneven temporal distribution. Indeed, due to the high cost for the establishment and maintenance of an observation network, sparse measurement points are materialized and are infrequently surveyed due to logistics. Considering the characteristics and logistics of the Åknes rockslide, Earth observation and remote sensing have an important role to play for studying landslide-related deformation, as they can regularly measure surface stability over large areas.

It is worth remembering that there is no monitoring system valid for all cases; in fact, every system must be designed purposely for a specific site, because the precursors and monitored parameters may largely vary depending on the type of landslide. Whatever the type of landslide, InSAR-based techniques (both ground based and space borne), thanks to their wide spatial coverage and their millimeter accuracy, are ideally suited to measure the spatial extent and magnitude of landslide-related surface deformation. Outputs from interferometric analysis ensures an almost spatially continuous coverage of information on surface deformation and related hazards. This definitely improves confidence on the spatial pattern of the examined phenomenon.

A further benefit of satellite SAR techniques is the generation of time series of the relative LOS position for each target in correspondence with each SAR acquisition, allowing the analysis of the temporal evolution of displacement and a look back at displacement that already has taken place. The possibility to retrieve a retrospective view of displacement is a unique opportunity for studying the evolution of the uninstrumented sector of a phenomena.

Unlike the conventional ground-based technologies (which record the displacement of targets, specific points or individual reflectors), InSAR, through the generation of interferograms, can provide 2D maps of changes in the satellite-to-target path between the acquisition times of the two SAR scenes. Hence, InSAR provided a significantly increased coverage of information, leading to a better overall understanding of movements of a sliding mass. This aspect is of paramount importance in the about 50 million m^3 Åknes rockslide, where a single moving benchmark can be related to localized displacements of an individual block and not to the general instability of the whole landslide body. Moreover, the information coverage allows one to accurately map the extension of the threatened area, its rate of deformation and to define, accordingly, risk scenarios.

The Åknes landslide is a large phenomenon that cannot be stabilized and may accelerate suddenly. The monitoring of its surface displacement is thus crucial for the prevention and forecast of collapse. In the case of the Åknes landslide, the synergic use of multiple SAR sensors (ground based and space borne) can lead to redundant measurements, allowing a more advanced and realistic mapping and classification of the phenomenon and a better understand of the deformation pattern. The redundancy of monitoring data is very important, being the basic condition for the design and implementation of any early warning system. Integrated use of multi-source monitoring data reduces the possibility of missing events or generating false alarms and the consequent loss of confidence and reliability of the system.

In recent years, the improved capabilities of new generation X-band satellites in terms of flexibility and time performance (revisiting time and timeliness of delivery) contributed to the use of satellite SAR sensors as operational monitoring tools [63,64]. Remote monitoring represents more and more a tool for surveying and/or early warning.

5. Conclusions

In this work, an attempt to integrate ground-based and satellite InSAR datasets is proposed. The main objective is to improve the knowledge obtainable from InSAR techniques, in the field of landslide mapping and monitoring.

The proposed procedure suggests three main steps to be performed: a qualitative phase, the result of which follows a binary approach, with the distinction, in the observed scenario, of stable and unstable areas; a semi-quantitative phase, where the distinction of macro-areas concerning different displacement ranges is possible; a quantitative integration, during which time series analysis is performed, and forecasting algorithms for the evaluation of the future behavior of the landslide can be applied.

The Åknes Norwegian rockslide has been selected to test the proposed procedure.

Firstly, data acquired by satellite and GB-InSAR platforms have been analyzed separately. Data analysis allows one to define landslide sectors concerning higher displacements, which correspond to the upper portion of the slope. Satellite data have been acquired by RADARSAT-2 and TerraSAR-X/Tandem-X satellite platforms. As in the GB-InSAR acquisitions, satellite data have been acquired in summer seasons. For the application of the integration algorithm, a defined period in the available datasets has been selected. The period between July and October 2010 has been chosen, as it is characterized by the best overlap between ground-based and satellite acquisitions.

The different datasets have been homogenized in terms of spatial and temporal resolution and also as regarding their different LOS. Homogenized data have been integrated and analyzed on the same map, in the GIS environment.

Data integration allowed increasing the data coverage on the observed scenario, which becomes widely detectable. Data projection allowed better defining of the real value of the displacement vector in the observed scenario: projection reliability has been tested comparing its result with data acquired by GPS campaigns, available in the literature. The projection also allowed defining the upper portion of the landslide as the main vertical displacements of concern.

Data integration from a semi-quantitative point of view has also been performed, allowing proposing the distinction of the upper sector of the landslide, defined as concerning higher displacements, into two sub-sections: the upper portion concerning displacements higher than 10 mm in the period between July and October 2010; and the lower portion concerning displacements lower than 10 mm in the same period.

Unfortunately, the unavailability of long time series of the observed datasets made it impossible to obtain also a quantitative integration for the selected test site.

Acknowledgments: The authors are grateful to LiSALab Company for the GB-InSAR data processing. The TerraSAR-X data have been provided by the German Aerospace Center (DLR) under the TSX-AO (TerraSAR-X Announcement of Opportunity) project Contract #GEO0764. RADARSAT-2 data have been provided by Kongsberg Satellite Services (KSAT) through an agreement with the Norwegian Space Centre. The authors also appreciate the help from staff of the Norwegian Geotechnical Institute, which contributed to this research production: especially, they would like to thank Bjorn Kalsnes, Vidar Kveldsvik and Jose Cepeda for their support. The authors are also grateful to the NVE (Norwegian Water Resources and Energy Directorate) staff, especially they would like to thank Lars Harald Blikra, for supporting this research. Michele Crosetto, Michel Jaboyedoff, Tazio Strozzi and an anonymous reviewer are thanked for their suggestions and comments, which significantly improved the original manuscript. The authors are also grateful to the Guest Editors, Zhenhong Li and Roberto Tomas, for their suggestions to improve this work.

Remote Sens. **2016**, *8*, 237

Author Contributions: All of the authors participated in editing and reviewing the manuscript. Federica Bardi, Federico Raspini and Andrea Ciampalini developed the integration procedure of GB-InSAR and satellite InSAR data. Lene Kristensen contributed to GB-InSAR data processing and post-processing. Line Rouyet and Tom Rune Lauknes contributed to satellite InSAR data processing. Regula Frauenfelder and Nicola Casagli provided a general review. All contributed to the writing of this manuscript.

Conflicts of Interest: The authors declare no conflict of interest.

References

1. IGOS (Integrated Global Observing Strategy). GEOHAZARDS Theme Report: For the Monitoring of Our Environment from Space and from Earth. European Space Agency publication, 2004. Available online: http://unesdoc.unesco.org/images/0014/001405/140532eo.pdf (accessed on 9 October 2015).

2. Columbia University. Global Landslide Total Economic loss Risk Deciles. 2000, *vol.1*. Available online: http://sedac.ciesin.columbia.edu/data/set/ndh-landslide-total-economic-loss-risk-deciles (accessed on 9 October 2015).

3. Guzzetti, F. Landslide fatalities and evaluation of landslide risk in Italy. *Eng. Geol.* **2000**, *58*, 89–107. [CrossRef]

4. Petley, D.N. The global occurrence of fatal landslides in 2007. *Geophys. Res. Abstr.* **2008**, *10*, EGU2008-A-10487.

5. Schuster, R.L.; Highland, L. *Socioeconomic and Environmental Impacts of Landslides in the Western Hemisphere*; US Department of the Interior, US Geological Survey: Cartagena, Colombia, 2001.

6. Canuti, P.; Casagli, N.; Ermini, L.; Fanti, R.; Farina, P. Landslide activity as a geoindicator in Italy: Significance and new perspectives from remote sensing. *Environ. Geol.* **2004**, *45*, 907–919. [CrossRef]

7. Kjekstad, O.; Highland, L. Economic and social impacts of landslides. In *Landslides–Disaster Risk Reduction*; Springer: Berlin, Germany; Heidelberg, Germany, 2009; pp. 573–587.

8. Nadim, F.; Kjekstad, O.; Peduzzi, P.; Herold, C.; Jaedicke, C. Global landslide and avalanche hotspots. *Landslides* **2006**, *3*, 159–173. [CrossRef]

9. Luzi, G.; Pieraccini, M.; Mecatti, D.; Noferini, L.; Guidi, G.; Moia, F.; Atzeni, C. Ground-based radar interferometry for landslides monitoring: Atmospheric and instrumental decorrelation sources on experimental data. *IEEE Trans. Geosci. Remote Sens.* **2004**, *42*, 2454–2466. [CrossRef]

10. Luzi, G. Ground based SAR interferometry: A novel tool for geoscience. In *Geoscience and Remote Sensing*; New Achievements, InTech; Imperatore, P., Riccio, D., Eds.; 2010; pp. 1–26. Available online: http://www.intechopen.com/articles/show/title/ground-based-sar-interferometry-a-novel-toolfor-geoscience (accessed on 20 May 2015).

11. Monserrat, O.; Crosetto, M.; Luzi, G. A review of ground-based SAR interferometry for deformation measurement. *ISPRS J. Photogramm. Remote Sens.* **2014**, *93*, 40–48. [CrossRef]

12. Ferretti, A.; Prati, C.; Rocca, F. Non linear subsidence rate estimation using Permanent Scatterers in differential SAR interferometry. *IEEE Trans. Geosci. Remote Sens.* **2000**, *38*, 2202–2212. [CrossRef]

13. Ferretti, A.; Prati, C.; Rocca, F. Permanent scatterers in SAR interferometry. *IEEE Trans. Geosci. Remote Sens.* **2001**, *39*, 8–20. [CrossRef]

14. Ferretti, A.; Bianchi, M.; Prati, C.; Rocca, F. Higher-order permanent scatterers analysis. *Eurasip J. Appl. Signal Process.* **2005**, *20*, 3231–3242. [CrossRef]

15. Hanssen, R.S. Satellite radar interferometry for deformation monitoring: A priori assessment of feasibility and accuracy. *Int. J. Appl. Earth Obs. Geoinform.* **2005**, *6*, 253–260.

16. Raucoules, D.; Bourgine, B.; Michele, M.; Le Gozannet, G.; Closset, L.; Bremmer, C.; Veldkamp, H.; Tragheim, D.; Bateson, L.; Crosetto, M.; *et al.* Validation and intercomparison of persistent scatterers interferometry: PSIC4 project results. *J. Appl. Geophys.* **2009**, *68*, 335–347. [CrossRef]

17. Crosetto, M.; Monserrat, O.; Iglesias, R.; Crippa, B. Persistent scatterer interferometry: Potential, limits and initial C- and X-band comparison. *Photogramm. Eng. Remote Sens.* **2010**, *76*, 1061–1069. [CrossRef]

18. Crosetto, M.; Monserrat, O.; Cuevas-Gonzàlez, M.; Devanthèry, N.; Crippa, B. Persistent scatterer interferometry: A review. *ISPRS J. Photogramm. Remote Sens.* **2015**. in press.

19. Berardino, P.; Fornaro, G.; Lanari, R.; Sansosti, E. A new algorithm for surface deformation monitoring based on small baseline differential SAR interferograms. *IEEE Trans. Geosci. Remote Sens.* **2002**, *40*, 2375–2383.

20. Rudolf, H.; Leva, D.; Tarchi, D.; Sieber, A.J. Mobile and versatile SAR system. In Proceedings of the IEEE 1999 International Geoscience and Remote Sensing Symposium, Hamburg, Germany, 28 June–2 July 1999; pp. 592–594.

21. Tarchi, D.; Rudolf, H.; Luzi, G.; Chiarantini, L.; Coppo, P.; Sieber, A.J. SAR interferometry for structural changes detection: A demonstration test on a dam. In Proceedings of the IEEE 1999 International on Geoscience and Remote Sensing Symposium, Hamburg, Germany, 28 June–2 July 1999; pp. 1522–1524.

22. Massonnet, D.; Feigl, K.L. Radar interferometry and its application to changes in the Earth's surface. *Rev. Geophys.* **1998**, *36*. [CrossRef]

23. Singhroy, V.; Mattar, K.E.; Gray, A.L. Landslide characterisation in Canada using interferometric SAR and combined SAR and TM images. *Adv. Space Res.* **1998**, *21*, 465–476. [CrossRef]

24. Crosetto, M.; Monserrat, O.; Cuevas, M.; Crippa, B. Spaceborne differential SAR interferometry: Data analysis tools for deformation measurement. *Remote Sens.* **2011**, *3*, 305–318. [CrossRef]

25. Lauknes, T.R.; Shanker, A.P.; Dehls, J.F.; Zebker, H.A.; Henderson, I.H.C.; Larsen, Y. Detailed rockslide mapping in northern Norway with small baseline and persistent scatterer interferometric SAR time series methods. *Remote Sens. Environ.* **2010**, *114*, 2097–2109.

26. Blikra, L.H. The Åknes rockslide, Norway. In *Landslides: Types, Mechanisms and Modeling*; Clague, J.J., Stead, D., Eds.; Cambridge University Press: Cambridge, UK, 2012; pp. 323–335.

27. Bardi, F.; Frodella, W.; Ciampalini, A.; Bianchini, S.; Del Ventisette, C.; Gigli, G.; Fanti, R.; Moretti, S.; Basile, G.; Casagli, N. Integration between ground based and satellite SAR data in landslide mapping: The San Fratello case study. *Geomorphology* **2014**, *223*, 45–60. [CrossRef]

28. Tofani, V.; Del Ventisette, C.; Moretti, S.; Casagli, N. Integration of remote sensing techniques for intensity zonation within a landslide area: A case study in the northern Apennines, Italy. *Remote Sens.* **2014**, *6*, 907–924. [CrossRef]

29. Eriksen, H.Ø.; Lauknes, T.R.; Larsen, Y.; Dehls, J.F.; Grydeland, T.; Bunkholt, H. Satellite and Ground-Based Interferometric Radar Observations of an Active Rockslide in Northern Norway. In *Engineering Geology for Society and Territory*; Springer International Publishing: Cham, Switzerland, 2015; pp. 167–170.

30. Zebker, H.A.; Villasenor, J. Decorrelation in interferometric radar echoes. *IEEE Trans. Geosci. Remote Sens.* **1992**, *30*, 950–959. [CrossRef]

31. Cruden, D.M.; Varnes, D.J. Landslide types and processes. In *Landslides: Investigation and Mitigation*; Transportation Research Board, Special Report, 247; Turner, A.K., Schuster, R.L., Eds.; Transportation Research Board, National Research Council, National Academy Press: Washington, DC, USA, 1996; pp. 36–75.

32. Raspini, F.; Ciampalini, A.; Del Conte, S.; Lombardi, L.; Nocentini, M.; Gigli, G.; Ferretti, A.; Casagli, C. Exploitation of amplitude and phase of satellite SAR images for landslide mapping: the case of Montescaglioso (South Italy). *Remote Sens.* **2015**, *7*, 14576–14596. [CrossRef]

33. Bianchini, S.; Cigna, F.; Righini, G.; Proietti, C.; Casagli, N. Landslide hotspot mapping by means of persistent scatterer interferometry. *Environ. Earth. Sci.* **2012**, *67*, 1155–1172. [CrossRef]

34. Lauknes, T.R. Rockslide Mapping in Norway by Means of Interferometric SAR Time Series Analysis. Ph.D. Thesis, University of Trømso (UIT), Trømso, Norway, 2010.

35. Intrieri, E.; Gigli, G.; Mugnai, F.; Fanti, R.; Casagli, N. Design and implementation of a landslide early warning system. *Eng. Geol.* **2012**, *147–148*, 124–136. [CrossRef]

36. Ciampalini, A.; Bardi, F.; Bianchini, S.; Frodella, W.; Del Ventisette, C.; Moretti, S.; Casagli, N. Analysis of building deformation in landslide area using multisensor PSInSAR technique. *Int. J. Appl. Earth Obs. Geoinform.* **2014**, *33*, 166–180. [CrossRef]

37. Bovenga, F.; Wasowski, J.; Nitti, D.O.; Nutricato, R.; Chiaradia, M.T. Using COSMO SkyMed X-band and ENVISAT C-band SAR interferometry for landslides analysis. *Remote Sens. Environ.* **2012**, *119*, 272–285. [CrossRef]

38. Wright, T.J.; Parsons, B.E.; Lu, Z. Toward mapping surface deformation in three dimensions using InSAR. *Geophys. Res. Lett.* **2004**, *31*, L010607. [CrossRef]

39. Tofani, V.; Raspini, F.; Catani, F.; Casagli, N. Persistent Scatterer Interferometry (PSI) technique for landslide characterization and monitoring. *Remote Sens.* **2013**, *5*, 1045–1065. [CrossRef]

40. Colesanti, C.; Wasowski, J. Investigating landslides with space-borne Synthetic Aperture Radar (SAR) interferometry. *Eng. Geol.* **2006**, *88*, 173–199. [CrossRef]

41. Cascini, L.; Fornaro, G.; Peduto, D. Advanced low- and full-resolution DInSAR map generation for slow-moving landslide analysis at different scales. *Eng. Geol.* **2010**, *112*, 29–42. [CrossRef]

42. Herrera, G.; Gutiérrez, F.; Garcìa-Davalillo, J.C.; Guerrer, J.; Notti, D.; Galve, J.P.; Fernàndez-Merodo, J.A.; Cooksley, G. Multi-sensor advanced DInSAR monitoring of very slow landslides: The Tena Valley case study (Central Spanish Pyrenees). *Remote Sens. Environ.* **2013**, *128*, 31–43. [CrossRef]

43. Barbieri, M.; Corsini, A.; Casagli, N.; Farina, P.; Coren, F.; Sterzai, P.; Leva, D.; Tarchi, D. *Space-Borne and Ground-Based SAR Interferometry for Landslide Activity Analysis and Monitoring in the Appennines of Emilia Romagna (Italy): Review of Methods and Preliminary Results*; European Space Agency, (Special Publication): Noordwijk, The Netherlands, 2004; pp. 463–470.

44. Ganerød, G. V.; Grøneng, G.; Rønning, J. S.; Dalsegg, E.; Elvebakk, H.; Tønnesen, J. F.; Kveldsvik, V.; Eiken, T.; Blikra, L. H.; Braathen, A. Geological model of the Åknes rockslide, western Norway. *Eng. Geol.* **2008**, *102*, 1–18. [CrossRef]

45. Blikra, L.H.; Braathen, A.; Derron, M.H.; Eiken, T.; Kveldsvik, V.; Grøneng, G.; Dalsegg, E.; Elvebakk, H.; Roth, M. The Åkerneset slope failure—A potential catastrophic rockslide in western Norway? *Abstr. Proc. Geol. Soc. Nor.* **2005**, *1*, 15–16.

46. Eidsvik, U.M.; Medina-Cetina, Z.; Kveldsvik, V.; Glimsdal, S.; Harbitz, C.B.; Sandersen, F. Risk assessment of a tsunamigenic rockslide at Åknes. *Nat. Hazards* **2011**, *56*, 529–545. [CrossRef]

47. Tveten, E.; Lutro, O.; Thorsnes, T. *Geologisk kart over Norge, bergrunnskart Ålesund, 1:250,000, (Ålesund, Western Norway)*; Geological Survey of Norway: Trondheim, Norway, 1988 (In Norwegian)

48. Heincke, B.; Gunther, T.; Dalsegg, E.; Ronning, J.S.; Ganerød, G.V.; Elvebakk, H. Combined three-dimensional electric and seismic tomography study on the Aknes rockslide in western Norway. *J. Appl. Geophys.* **2010**, *70*, 292–306. [CrossRef]

49. Kristensen, L.; Rivolta, C.; Dehls, J.; Blikra, L.H. GB-InSAR measurement at the Åknes rockslide, Norway. In Proceedings of the International Conference Vajont 1963–2013. Thoughts and Analyses after 50 years since the Catastrophic Landslide, Padua, Italy, 8–10 October 2013.

50. Frei, C. H.; Loew, S.; Leuenberger-West, F. First results of a large-scale multi-tracer test within an unstable rockslide area (Åknes, Norway). *Geophys. Res. Abstr.* **2008**, *10*, SRef-ID:1607-7962/gra/EGU2008-A-08930.

51. Kveldsvik, V.; Nilsen, B.; Einstein, H.H.; Nadim, F. Alternative approaches for analyses of a 100,000 m^3 rock slide based on Barton-Bandis shear strength criterion. *Landslides* **2007**, *5*, 161–176. [CrossRef]

52. Kveldsvik, V.; Kanya, A.M.; Nadim, F.; Bhasin, R.; Nilsen, B.; Einstein, H.H. Dynamic distinct-element analysis of the 800 m high Aknes rock slope. *Int. J. Rock Mech. Min. Sci.* **2009**, *46*, 686–698. [CrossRef]

53. Kveldsvik, V.; Einstein, H.H.; Nilsen, B.; Blikra, L.H. Numerical analysis of the 650,000 m^2 Aknes rock slope based on measured displacements and geotechnical data. *Rock. Mech. Rock. Eng.* **2009**, *42*, 689–728. [CrossRef]

54. Nordvik, T.; Nyrnes, E. Statistical analysis of surface displacements—An example from the Åknes rockslide, western Norway. *Nat. Hazards Earth. Syst. Sci.* **2009**, *9*, 713–724. [CrossRef]

55. Blikra, L.H. The Åknes rockslide: Monitoring, threshold values and early-warning. In Proceedings of the 10th International Symposium on Landslides and Engineered Slopes, Xian, China, 30 June–4 July 2008; pp. 1089–1094.

56. Lacasse, S.; Eidsvig, U.; Nadim, F.; Hoeg, K.; Blikra, L.H. Evaluation of Åknes rockslide hazard using event trees. In Proceedings of the 42nd U.S. Rock Mechanics Symposium (USRMS), ARMA-08–340, San Francisco, CA, USA, 29 June–2 July 2008.

57. The Norwegian Water Resources and Energy Directorate (NVE). Available online: http://www.nve.no/ english/ (accessed on 12 March 2015).

58. Tarchi, D.; Casagli, N.; Fanti, R.; Leva, D.; Luzi, G.; Pasuto, A.; Pieraccini, M.; Silvano, S. Landslide monitoring by using ground-based SAR interferometry: An example of application to the Tessina landslide in Italy. *Eng. Geol.* **2003**, *1*, 15–30. [CrossRef]

59. Northern Research Institute of Trømso (NORUT) (Trømso, Norway). Unpublished work, 2013.

60. Hooper, A.; Zebker, H.A.; Segall, P.; Kampes, B. A new method for measuring deformation on volcanoes and other natural terrains using InSAR persistent scatterers. *Geophys. Res. Lett.* **2004**, *31*, 1–5. [CrossRef]

61. Lauknes, T.R.; Zebker, H.A.; Larsen, L. InSAR deformation time series using an L1-Norm small-baseline approach. *IEEE Trans. Geosci. Remote Sens.* **2011**, *49*, 536–546. [CrossRef]

62. Lillesand, T.M.; Kiefer, R.W.; Chipman, J.W. *Remote sensing and Image Interpretation*, 5th ed.; Wiley, Cop.: New York, NY, USA, 2004.

63. Raspini, F.; Moretti, S.; Fumagalli, A.; Rucci, A.; Novali, F.; Ferretti, A.; Prati, C.; Casagli, N. The COSMO-SkyMed constellation monitors the costa concordia wreck. *Remote Sens.* **2014**, *6*, 3988–4002. [CrossRef]

64. Ciampalini, A.; Raspini, F.; Bianchini, S.; Tarchi, D.; Vespe, M.; Moretti, S.; Casagli, N. The costa concordia last cruise: The first application of high frequency monitoring based on COSMO-SkyMed constellation for wreck removal. *J. Photogramm. Remote Sens.* **2016**. [CrossRef]

remote sensing

MDPI

Article

Evaluation of the Use of Sub-Pixel Offset Tracking Techniques to Monitor Landslides in Densely Vegetated Steeply Sloped Areas

Luyi Sun * and Jan-Peter Muller

University College London, Mullard Space Science Laboratory, Holmbury St. Mary, Surrey RH5 6NT, UK; muller@ucl.ac.uk

* Correspondence: luyi.sun.12@ucl.ac.uk; Tel.: +44-1483-204-167; Fax: +44-1483-278-312

Academic Editors: Zhenhong Li, Roberto Tomas, Zhong Lu, Richard Gloaguen and Prasad S. Thenkabail
Received: 1 June 2016; Accepted: 10 August 2016; Published: 17 August 2016

Abstract: Sub-Pixel Offset Tracking (sPOT) is applied to derive high-resolution centimetre-level landslide rates in the Three Gorges Region of China using TerraSAR-X Hi-resolution Spotlight (TSX HS) space-borne SAR images. These results contrast sharply with previous use of conventional differential Interferometric Synthetic Aperture Radar (DInSAR) techniques in areas with steep slopes, dense vegetation and large variability in water vapour which indicated around 12% phase coherent coverage. By contrast, sPOT is capable of measuring two dimensional deformation of large gradient over steeply sloped areas covered in dense vegetation. Previous applications of sPOT in this region relies on corner reflectors (CRs), (high coherence features) to obtain reliable measurements. However, CRs are expensive and difficult to install, especially in remote areas; and other potential high coherence features comparable with CRs are very few and outside the landslide boundary. The resultant sub-pixel level deformation field can be statistically analysed to yield multi-modal maps of deformation regions. This approach is shown to have a significant impact when compared with previous offset tracking measurements of landslide deformation, as it is demonstrated that sPOT can be applied even in densely vegetated terrain without relying on high-contrast surface features or requiring any de-noising process.

Keywords: landslide monitoring; sub-Pixel Offset Tracking (sPOT); TerraSAR-X High-resolution Spotlight data; Corner Reflectors vs. natural scatterers; densely vegetated terrain

1. Introduction

Remote sensing, especially in the microwave region, has become the most convenient and feasible tool widely applied in deformation mapping. In the Three Gorges Region (TGR), due to the often limited access to Global Positioning System (GPS) measurements, the high costs of skilled labour and instrumentation required, it is difficult to obtain sufficient local geodetic measurements [1]. The usage of satellite remote sensing data for landslide studies in the TGR can be traced back to the 1980s [2]. Due to the high humidity caused by the monsoon climate of this region, optical sensors are often limited in obtaining an effective time series of measurements. A Synthetic Aperture Radar (SAR), which is able to work both day and night during all weather conditions and which repeatedly acquires time series of high-resolution images covering large areas, has been recognized as an effective and powerful sensor for landslide monitoring [1,3].

The differential Interferometric SAR (DInSAR), which is capable of detecting surface deformation over a large area in the direction of the satellite Line of Sight (LOS) with a centimetre-to-millimetre precision, has been extensively applied to monitor volcanic activities, earthquakes, mining deformations, glacier movement, subsidence and landslides [4–13]. Time series algorithms have been developed to extend the use of DInSAR for temporal evolution of ground deformation, which can be essentially divided into two broad categories: the Persistent Scatterer (PS) InSAR [14–16] and Small Baseline Subset (SBAS) [17,18].

However, the applications of DInSAR/time series DInSAR in the Three Gorges Region are limited by the difficulties arising from steep slopes, dense vegetation cover and high humidity. The experiment using PS-InSAR with ENVISAT data to measure deformation in the TGR area did not find sufficient PS points [19], which lead to the failure of phase unwrapping [3]. The attempt of using PS-InSAR with ASAR images to monitor the Shuping landslide failed for the same reason [1]. Experiments applying the SBAS method on TerraSAR-X (TSX) data in the Three Gorge Region did not find significant dependence upon the perpendicular baseline or dramatic increase of reliable scatterers over time, suggesting the use of SBAS method has limited benefit in this case [20].

In addition, DInSAR measurements on the Shuping landslide yielded varied results in previous studies. Fu et al. obtained DInSAR measurements on 12 corner reflectors (CRs) (−1–11 cm in 140 days) of the Shuping landslide using five ENVISAT ASAR images spanning from September 2005 to March 2006. The investigated period missed the most active period of the Shuping landslide [21]. Xia et al. used the same 12 CRs to derive linear displacement rates of 1–11 cm/year from September 2005 to June 2007 [22], in contradiction with the results of [21] with a different deformation rate but an overlapping observation period between September 2005 and March 2006. For the period between September 2005 and June 2007, extensometer measurements show 50–70 cm/year displacement with a dramatic increase from May to August 2007 [23], which is different from the linear trend monitored by PS-InSAR.

It should be noted there is a limitation of DInSAR with regard to the maximum detectable displacement. If no prior knowledge of the deformation is provided, which is usually the case, the implementation of phase unwrapping relies on an assumption that the phase difference between any two neighbouring pixels does not exceed $\pm\pi$. This implies the maximum detectable deformation per pixel is half wavelength. In addition, phase gradients larger than 0.5 fringes may cause large-scale unwrapping errors, which means the displacement gradient between two neighbouring pixels is limited to 1/4 wavelength [24]. Thus, the maximum detectable displacement gradient (DDG) of InSAR measurements is

$$D = \lambda/4\mu \tag{1}$$

where D denotes the maximum DDG, λ is the wavelength of the SAR sensor and μ is the pixel size of the SAR images for classical interferometry, or distance between persistent scatterers for PS techniques. The value of D depends on the satellite. For example, in the case of TerraSAR-X Hi-resolution Spotlight (with wavelength 0.031 m, pixel size 0.456 m) data, the maximum DDG is 0.0059 using a small multi-looking factor of 2. This means that over a ground distance of 1 m (about 1 pixel in the case of TSX Hi Res data), a displacement of 0.59 cm in one revisit cycle (11 days) will be underestimated even when given very high phase coherence. In a real scenario, the coherence is usually lower, especially in densely vegetated areas. The theoretical limit will drop with the coherence leading to further underestimation, which is the case in our study. Many slow-moving landslides (~1.6 m/year as defined in [25,26] and cases reported in [27,28]) can exceed this threshold of displacement gradient, especially near the landslide boundary.

The sub-pixel Offset Tracking (sPOT) technique (sometimes referred to as Pixel Offset Mapping) has previously been applied to monitor glacier movements, volcanic activities and co-seismic tears in the solid earth resulting from severe earthquakes to address the technical defects and limitations of conventional DInSAR techniques, particularly their sparse coverage and the impact of dense vegetative cover [29]. In the past, studies on offset tracking techniques to measure slope movements are dominated by using optically sensed imagery from spaceborne or airborne platforms [30–35].

For SAR sensors, initially medium resolution SAR imagery were employed in offset tracking for measurements of very large deformation (metres to tens of metres) [36–38]. Intensity Tracking (based on Normalized Cross Correlation) was proposed and implemented on a set of ERS-1/2 SAR data acquired from March 1992 to February 1996 in order to estimate the motion of Monacobreen in Northern Svalbard [36]. It indicated that in the case of various SAR missions (RADARSAT, ERS-2, ENVISAT, ALOS) with a more than 24 day revisit interval, intensity based tracking is the only technique able to correctly measure glacier movement. The work by [37] proposed a PO-SBAS approach using ENVISAT data to measure large displacements (several metres) occurring in the inner part of the Sierra Negra caldera due to the October 2005 eruption. This PO-SBAS approach attempted to minimize the perpendicular baseline via small baseline combinations of offset pairs. However, the TSX data employed in our study consistently has short baselines ranging from 12 m to 220 m, so the benefit of creating a SBAS network is limited. With the availability of higher resolution SAR data, Manconi et al. obtained post-event deformation maps for emergency evaluation of a large, rapidly-moving (10–20 m) landslide [39–41]. The same PO-SBAS approach was applied to ascending and descending pairs of COSMO-SkyMed images to retrieve 3D deformation of the Montescaglioso landslide (Italy) of which the main movement occurred over 15–20 min with an average velocity of about 0.5–1 m per minute.

Sub-Pixel Offset Tracking has recently been employed to derive centimetre-level landslide rates in the Three Gorges Region using 1–3 m resolution space-borne SAR images. Li et al. [42] used four pairs of TSX HS images to derive 2D (azimuth and slant range) landslide displacement in the Three Gorges Region. The results indicate May–August 2009 was the most active period of the Shuping landslide. However, due to the impact of dense seasonal vegetation cover, some results still show false deformation features in the slant range direction. Singleton et al. [20] conducted further analysis focusing on the 540,000 m^2 area centred on the landslide blocks. This work focused on the use of previously installed Corner Reflectors (CRs) with offset tracking, to derive a deformation magnitude for each CR in order to plot time series deformation curves, confirming a dramatic increase in landslide rates from May to August in 2009. However, it was pointed out that the errors associated with the corner reflector measurements are an order of magnitude lower than those calculated from densely vegetated terrain. Also, it was pointed out in [43] using ground-based SAR (GB-SAR) data to measure the displacement from artificial CRss, the main constraint of the offset tracking technique is the need of CRs. This raises a large question for the vast majority of regions where no CRs are available especially in densely vegetated terrain. The question arises: are sPOT techniques able to correctly measure landslide rates? This is the starting point of this study.

As the main objective of this study, the potential of using natural scatterers is assessed on deformation measurements using an offset tracking approach by combining sub-pixel cross-correlation with a time series statistical analysis, which makes a significant difference in that it does not rely on high contrast surface features (e.g., Corner Reflectors). Unlike the scenario of a very large deformation [36–38,44], this study aims to exploit the use of offset tracking with time series high-resolution SAR data covering two years, to derive the temporal evolution and spatial distribution of a slow-moving landslide with an active period of months and accumulative displacement of up to 1 m per year. The study area is characterized by dense vegetation cover on steep slopes, which causes rapid decrease of temporal correlation/low coherence of DInSAR on natural scatterers. Given the deformation velocity, the offsets caused by seasonal changes of vegetation cannot be ignored, which increases the challenge of the use of natural scatterers.

In this paper, sub-Pixel Offset Tracking is applied in monitoring ground deformation in densely vegetated terrain and concentrating on the evaluation of its general application in the vast majority of regions where CRs are not available. Firstly, the landslide displacement rates in the field site, Shuping landslide area, were measured from artificial CRs using the fully available 2 year time series TerraSAR-X (TSX) Hi-resolution Spotlight data acquired from February 2009–April 2010 and January 2012–February 2013. Secondly, the correlation between the landslide displacements and water level variations of the Three Gorges Reservoir were then analysed to infer a possible failure mechanism for the Shuping landslide. Finally, as a key part of this study, the capability of sPOT techniques for measuring ground displacements in densely vegetated areas was assessed by a statistical analysis of deformation magnitudes derived from natural scatterers on the whole landslide body. Based on the above analysis, an approach is proposed to extend the applications of sPOT to densely vegetated terrain without requiring artificial CRs.

2. Study Site

The Three Gorges Region (TGR) of China, which is located between latitude 28°32′N–31°44′N and longitude 105°44′E–111°39′E, is the region directly or indirectly involved in the submersion of the water storage of the Three Gorges Project (TGP). It stretches along the Yangtze River including 16 county-level divisions of the Chongqing municipality and 4 divisions of the Hubei province. The Three Gorges Dam (TGD), located at Sandouping Town to the west of the city of Yichang, China, is one of the world's largest civil engineering structures, which blocks water to form a 660 km long and ≈1.1 km wide reservoir. The water level of the Yangtze River in the TGD rose from 66 m to 135 m, 156 m and eventually 175 m above sea level during the three impoundments in 2003, 2006 and 2009. The Three Gorges Project (TGP) does a remarkable job of generating a huge amount of electric power as well as controlling floods and improving the shipping capacity of the Yangtze River. However, the construction and operation of Three Gorges Dam resulted in a significant land use change, which altered energy and water budgets, affected the regional weather and climate patterns, and is linked to the dramatically increased geological hazards dominated by landslide activities in the Three Gorges Region [45–47]. Numerous landslide activities have occurred in residential areas with high population density, causing a lot of wasted resources and loss of property.

As the construction and operation of the Three Gorges Dam raised major concerns about its environmental impacts, a number of studies have been carried out on several topics, including terrestrial ecosystems, sedimentation, pollution, river discharges, regional climate and induced geological hazards dominated by landslides [48–53].

Most of the landslides which occurred in the Three Gorges Region are identified as being triggered by water, with the variations of reservoir water level and seasonal heavy rainfall being the two main factors [23,54].

The field site, Shuping landslide area, is located on the south bank of the Yangtze River near Shazhenxi Town, Zigui County with centre coordinates of 30.996°N, 110.609°E as indicated in Figure 1a. The Shuping landslide was identified as an ancient landslide during the field investigations before the construction of Three Gorges Dam [55]. This area is underlaid by muddy sandstone and sandy mudstone of the Triassic Badong formation. The landslide is composed of two blocks as marked in Figure 1b facing the North, with a width of about 650 m, elevation ranging from 65 to 400 m, thickness of 40–70 m, volume of about 20 million m^3 and average slope varying from 22° on the upper part to 35° on the lower part [54]. The landslide area is characterised by terraced slopes densely covered with orange trees. The landscape photos of the Shuping landslide area in Figure 2 show cracks on the local infrastructure and one photo of one of the CRs is shown.

The Shuping landslide is a typical slope accumulation landslide where deformation has increased since the water impoundment of the Three Gorges Reservoir in 2003. In June 2003, significant deformation appeared on the slope and it acute from 8 February 2004 on. This serious deformation posed a significant danger to 580 inhabitants and 163 houses directly in its path and most of the inhabitants had moved of the landslide area by May 2004. According to GPS measurements, from January 2004–October 2006, when the reservoir water level varied between 135 and 145 m, the ground deformation of Shuping landslide area was predominantly a combination of squirm and uniform deformation. The accumulative displacement reached 300 mm from August 2004 to August 2006, 250 mm from August 2006 to July 2007, 500 mm from August 2007 to February 2009, and 700 mm from February 2009 to February 2010 according to extensometer measurements along the centre line of eastern block [23,54]. Following this for every single year, the deformation magnitude periodically fluctuates with variations in the reservoir water level which also coincides with rainfall periodicity [56].

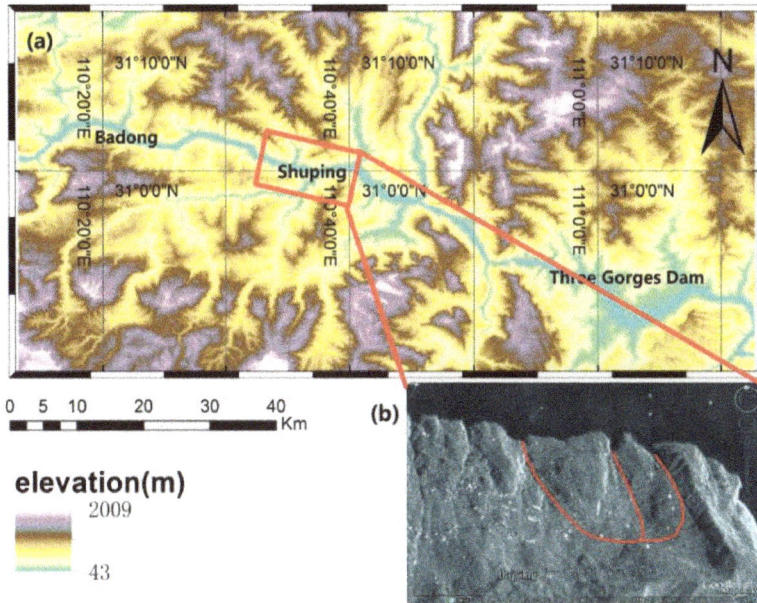

Figure 1. (**a**) Location of the Shuping landslide area; (**b**) Perspective view of landslide body shown in TerraSAR-X Hi-resolution Spotlight amplitude image superimposed in Google Earth with landslide blocks marked in red. Data source: TerraSAR-X © DLR <2009>.

Figure 2. (**a**) Landscape of the landslide area; (**b**) Cracks on local infrastructure caused by the landslide; (**c**) one of the Corner Reflectors installed in the landslide area. Photos were taken during a field campaign in May 2014.

3. Data and Methods

3.1. Data

The data employed in this research uses the TerraSAR-X Hi-resolution Spotlight (TSX HS) data. Fifty-seven archived TSX High-resolution Spotlight (HS) images were acquired from 21 February 2009–15 April 2009 and 2 January 2012–23 February 2013 over the Shuping landslide area in the Three Gorges Region. The extent of Spotlight data coverage is shown by the rectangle frame in Figure 1a. The metadata of the two annual time series of TSX HS data is listed in Table 1.

Table 1. Metadata of the two stacks of SAR data using the parameters from the first image of each stack, as the values remain very close for all subsequent acquisitions.

	TerraSAR-X High-Resolution Spotlight Data	
Annual time series	2009–2010	2012–2013
First acquisition	21 February 2009	2 January 2012
Last acquisition	15 April 2010	23 February 2013
Satellite orbit heading (°)	190.552	189.617
Wavelength (m)	0.031	0.031
Incidence angle (°)	43.690	43.602
Range pixel spacing (m)	0.456	0.455
Azimuth pixel spacing (m)	0.862	0.873
Range resolution (m)	0.851	0.852
Azimuth resolution (m)	1.100	1.100

3.2. Methods: Sub-Pixel Offset Tracking (sPOT) Techniques

An alternative to the use of SAR interferometry is to measure sub-pixel offsets between the SAR images. This can be achieved by FFT-based correlation (sometimes referred to as phase correlation) or Normalized Cross Correlation [57]. Due to the high noise level of SAR images, cross-correlation is more robust (found out in experiments) and therefore chosen for this study. One of the first offset tracking applications to the Three Gorges Area is shown by Li et al. [58] and more recently in [20]. We refer to these as sub-Pixel offset tracking (sPOT) techniques.

The Normalized Cross Correlation (NCC) derives a set of 2-dimensional (2D) offsets between pre-event and post-event images. NCC is a traditional method for image registration. It is applied to the intensity bands of cross event images to detect ground deformation through a measure of similarity between window pairs extracted from pre-event and post-event images. The similarity, which is defined as the correlation coefficient, is computed as follows:

$$NCC = \frac{\sum_{m=1}^{N_x} \sum_{n=1}^{N_y} \left[\left(i_1\left(m,n\right) - \overline{i_1} \right) \cdot \left(i_2\left(m,n\right) - \overline{i_2} \right) \right]}{\sqrt{\sum_{m=1}^{N_x} \sum_{n=1}^{N_y} \left(i_1\left(m,n\right) - \overline{i_1} \right)^2} \sqrt{\sum_{m=1}^{N_x} \sum_{n=1}^{N_y} \left(i_2\left(m,n\right) - \overline{i_2} \right)^2}} \tag{2}$$

where i_1 and i_2 denote pre-event and post-event images with a two-dimensional offset (a, b), which can be described as $i_2\left(x,y\right) = i_1\left(x - a, y - b\right)$. $N_x \times N_y$ is the correlation window size which can be modified by the application requirements. $\overline{i_1}$ and $\overline{i_2}$ are the mathematical expectation values of the cross-event image pair:

$$\overline{i_1} = \frac{1}{N_x \times N_y} \sum_{m=1}^{N_x} \sum_{n=1}^{N_y} i_1\left(m,n\right) \tag{3}$$

$$\overline{i_2} = \frac{1}{N_x \times N_y} \sum_{m=1}^{N_x} \sum_{n=1}^{N_y} i_2\left(m,n\right) \tag{4}$$

The NCC method searches for maximum correlation (i.e., maximum similarity) between window pairs formed by the pre-event and post-event images. Those window pairs for which a maximum correlation detected, are considered as corresponding pairs. After locating the corresponding pixels in the master and slave images, the 2D offsets of the slave image w.r.t. the master image can be obtained. To achieve a sub-pixel accuracy of correlation, two categories of approaches are usually used: (1) image intensities are oversampled prior to cross-correlation; (2) Without oversampling of the intensity bands, cross-correlation is done in the original image resolution, correlation peaks are located by polynomial fitting.

In this paper, all data are processed using the following step-by-step approach:

- For each data stack (2009–2010 and 2012–2013), the first acquisition was used as the master image. All the slave images were co-registered with respect to the same master to sub-pixel accuracy. Topographic distortions were modeled using a reference DEM (SRTM 1 arc-second global DEM) and precise orbital data and subtracted before the cross-correlation.
- Images are cropped to the landslide sub-area as inputs to the cross-correlation included within COSI_Corr [59–61]. At this point, the azimuth and range deformation fields are derived.
- Time series histograms of the range/azimuth deformation fields are plotted for the measurements derived on the landslide blocks and the measurements on the stable ground respectively.
- To correct the centroid shifts (mainly caused by the impact of vegetation) on every histogram, the time series histograms of the measurements from stable ground were all fitted by Gaussian functions. The centroid location of every Gaussian peak was taken as a reference to correct the centroid offsets for the corresponding histograms of range/azimuth offsets of the landslide area.
- From the change in histograms, the temporal evolution of the landslide is shown and the active period of the landslide can be identified, as well as the deformation scale.
- Using a correlation coefficient of 0.25 as the threshold, all pixels with correlation above this value are plotted to show the spatial distribution of azimuth and slant range offsets occurred in February 2009–April 2010 and January 2012–February 2013. The two maps can be plotted for each salve acquisition date in the data stack.

4. Results

4.1. Time Series Landslide Rates Derived from Corner Reflectors (CRs) Using Sub-Pixel Offset Tracking

Subsets of landslide sub-areas were cropped from 35 pairs of TerraSAR-X Hi-resolution Spotlight (TSX HS) images acquired from 21 February 2009–15 April 2010 and 20 pairs from 2 January 2012–23 February 2013. The sPOT method was applied to every co-registered subset pair using 20090221 and 20120102 images, respectively, as the common master image for each annual time series (i.e., 2009–2010 and 2012–2013, respectively), to retrieve deformations along the range (satellite line-of-sight) and the azimuth (along-track) direction. The acquisition dates and estimated baselines of employed data (in brackets) are listed in Tables 2 and 3 with each image named after the acquisition date in the format "yyyymmdd".

Table 2. TSX HS data employed in 2009–2010 time series analysis.

Common Master	Slave (Perpendicular Baseline)		
20090304 (192 m)	20090315 (125 m)	20090326 (040 m)
20090406 (028 m)	20090417 (080 m)	20090428 (050 m)
20090509 (040 m)	20090520 (041 m)	20090531 (051 m)
20090611 (052 m)	20090622 (125 m)	20090703 (074 m)
20090714 (072 m)	20090725 (137 m)	20090805 (071 m)
20090816 (120 m)	20090827 (074 m)	20090907 (040 m)
20090918 (156 m)	20090929 (180 m)	20091010 (046 m)
20091112 (085 m)	20091123 (017 m)	20091204 (089 m)
20091215 (043 m)	20091226 (066 m)	20100106 (105 m)
20100117 (145 m)	20100128 (033 m)	20100219 (150 m)
20100304 (097 m)	20100313 (220 m)	20100324 (102 m)
20100404 (111 m)	20100415 (123 m)	

Table 3. TSX HS data employed in 2012–2013 time series analysis.

Common Master	Slave (Perpendicular Baseline)		
20120113 (035 m)	20120124 (012 m)	20120204 (094 m)
20120215 (074 m)	20120226 (055 m)	20120308 (021 m)
20120319 (081 m)	20120330 (029 m)	20120421 (058 m)
20120524 (064 m)	20120615 (191 m)	20120820 (183 m)
20120922 (083 m)	20121025 (002 m)	20121127 (082 m)
20130110 (025 m)	20130121 (040 m)	20130201 (160 m)
20130212 (017 m)	20130223 (029 m)	

There have been artificial CRs installed in the Three Gorges Region since 2000 [62]. Seventeen CRs are identified in the Shuping landslide area from the TSX Hi-resolution Spotlight image as shown in Figure 3.

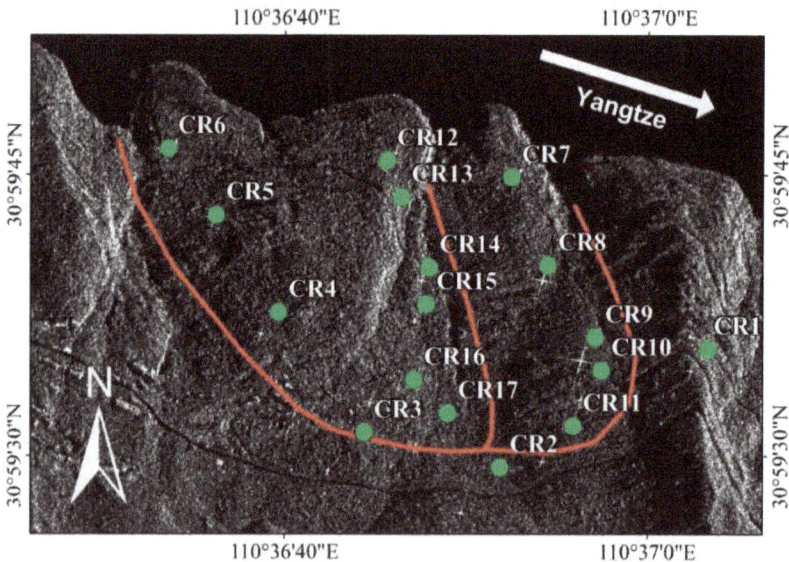

Figure 3. Location of Corner Reflectors in the Shuping landslide area, shown in TerraSAR-X amplitude, with landslide boundaries corresponding to the two landslide blocks. Data source: TerraSAR-X © DLR <2009>.

The correlation coefficients of all CRs are examined prior to the time series analysis. As shown in Figure 4, the correlation coefficient of CR3 is very low (around 0.2) throughout the 2012–2013 stack, which will lead to inconclusive cross-correlation. CR5 shows inconsistencies in the correlation coefficient, because it is missing from the SAR amplitude images during 15 June 2012–10 January 2013 possibly due to reinstallation or orientation adjustment. Thus, CR3 and CR5 were both excluded from the analysis of the two annual time series.

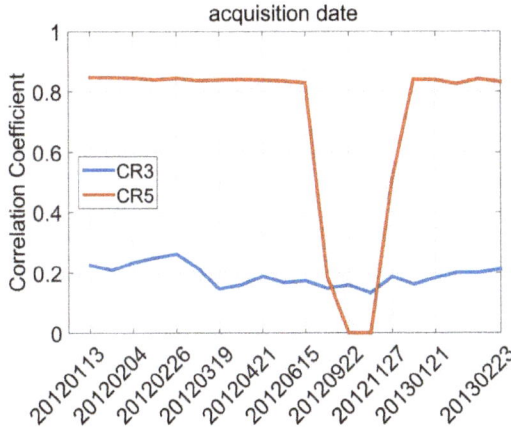

Figure 4. Correlation coefficients of CRs derived by sub-pixel offset tracking in the Shuping landslide area. Acquisition date is displayed in the format of 'yyyymmdd'.

The deformation magnitudes of the remaining 14 CRs (as shown in Figure 3) were extracted to plot time series landslide rates. No de-noising or filtering steps were applied. As all data was acquired with right looking SAR in the descending mode, the negative magnitude of the azimuth deformation corresponds to the reverse along-track direction (predominantly to the North) and the positive magnitude of range deformation represents the movement away from the sensor.

The topographic distortions of the range offsets were modeled by using a reference DEM and precise orbital data and subtracted before cross-correlation. To reduce the background noise, CR1 was taken as a reference point for all the other CRs as it is identified as located on the stable ground.

The two annual time series of landslide rates derived from CRs are shown in Figures 5 and 6.

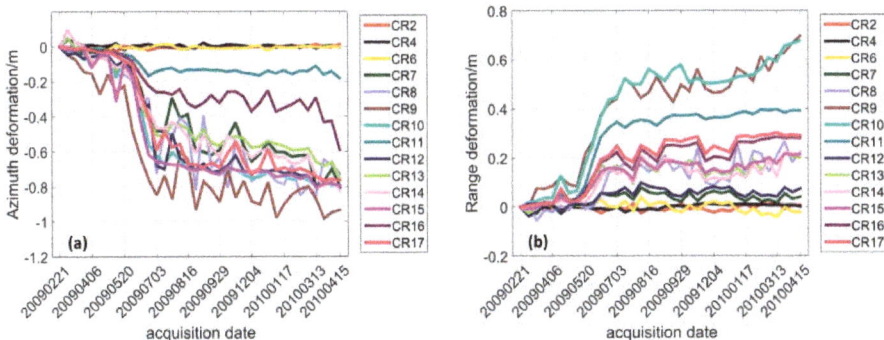

Figure 5. A 2009–2010 time series deformation measured from Corner Reflectors: (a) Azimuth deformation; (b) Slant range deformation. Acquisition date is displayed in the format of 'yyyymmdd'.

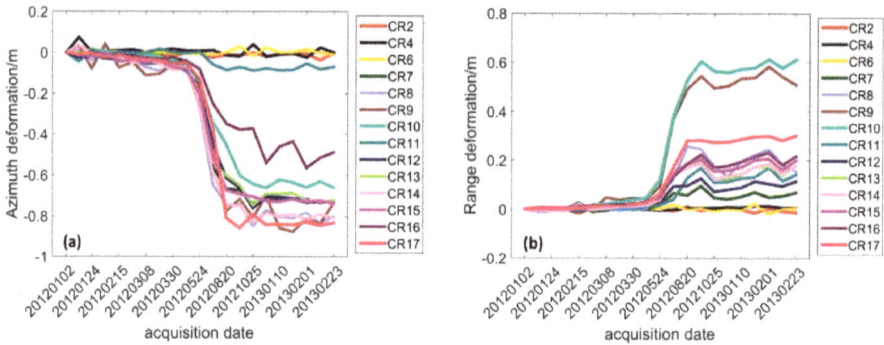

Figure 6. A 2012–2013 time series deformation measured from Corner Reflectors: (**a**) Azimuth deformation; (**b**) Slant range deformation. Acquisition date is displayed in the format of 'yyyymmdd'.

4.2. The Correlation Between the Landslide Deformation and Water Level Variations

To study the relationship between the landslide displacements and the operation of the Three Gorges Dam, the derived landslide rates of CR9 and CR15 (taken as examples due to the typical deformation patterns) were plotted against water level measurements of the Three Gorges Reservoir for the same time periods, from February 2009–April 2010 and January 2012–February 2013. The water level measurements can be accessed from the Three Gorges Corporation website: http://www.ctg.com.cn/inc/sqsk.php.

As shown in Figures 7 and 8, the water level measurements over the same time period show a consistent seasonal pattern with a lower level in the flood season and normal levels in other seasons. This is strongly correlated with the active period of the Shuping landslide. There is no correlation between the displacements and the big rise in water levels in September, this will be addressed in the Discussion Section 5.4.

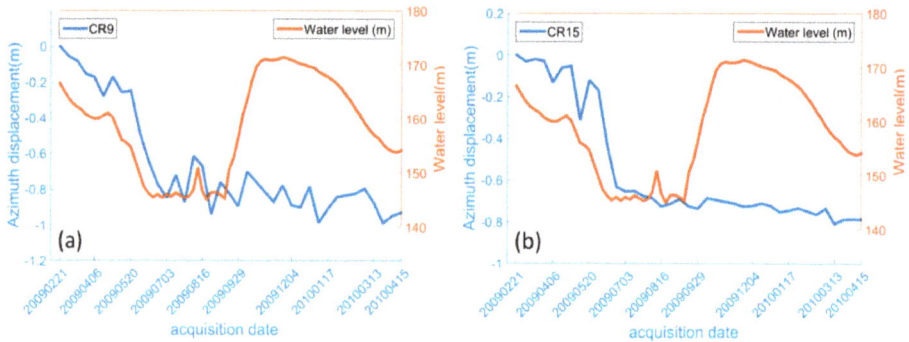

Figure 7. (**a**) Azimuth displacements of CR9 versus water level measurements of Three Gorges Reservoir from February 2009 to April 2010; (**b**) Azimuth displacements of CR15 versus water level measurements of Three Gorges Reservoir from February 2009 to April 2010. Acquisition date is displayed in the format of 'yyyymmdd'.

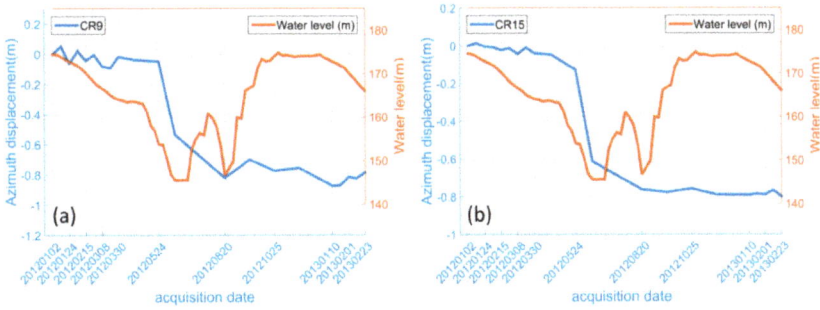

Figure 8. (**a**) Azimuth displacements of CR9 versus water level measurements of Three Gorges Reservoir from January 2012 to February 2013; (**b**) Azimuth displacements of CR15 versus water level measurements of Three Gorges Reservoir from January 2012 to February 2013. Acquisition date is displayed in the format of 'yyyymmdd'.

4.3. Assessment of Using Natural Scatterers with sPOT Techniques to Monitor Landslide Movement in Densely Vegetated Terrain

Artificial CRs are not widely available along the banks of Yangtze River due to the large number and scale of landslides in the Three Gorges Area and the high costs of the associated building works as well as the huge difficulties in physical access [63–65]. In order to assess the use of sPOT techniques in densely vegetated terrain without relying on CRs, statistics of deformation measurements derived from natural scatterers were compared to those derived from CRs. The analysis was conducted in the 540,000 m^2 area covering the two landslide blocks. All contributions from CRs were masked out from the original azimuth/range deformation output. No de-noising or filtering steps were applied. Results are shown in Figure 9.

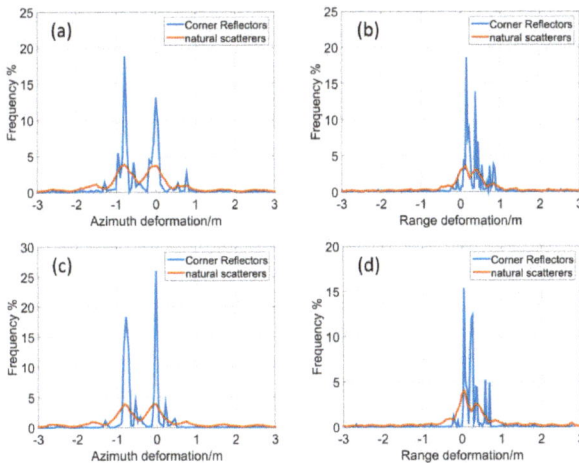

Figure 9. (**a**) Histograms of azimuth deformation derived from 21 February 2009–15 April 2010 image pair from natural scatterers vs. CRs inside the landslide boundary; (**b**) Histograms of range deformation derived from the 21 February 2009–15 April 2010 image pair from natural scatterers vs. CRs inside the landslide boundary; (**c**) Histograms of azimuth deformation derived from 2 January 2012–23 February 2013 image pair from natural scatterers vs. CRs inside the landslide boundary; (**d**) Histograms of range deformation derived from the 2 January 2012–23 February 2013 image pair from natural scatterers vs. CRs inside the landslide boundary. Modified from [66].

From the comparisons of displacement histograms shown in Figure 9, we can observe that measurements from natural scatterers show the same range of deformation magnitudes as those derived from CRs.

4.4. Statistical Analysis Combined with sPOT for General Use in Landslide Monitoring in Densely Vegetated Areas

The histograms of azimuth/range deformation measured from the natural scatterers of the landslide blocks and adjacent stable ground are compared in Figure 10.

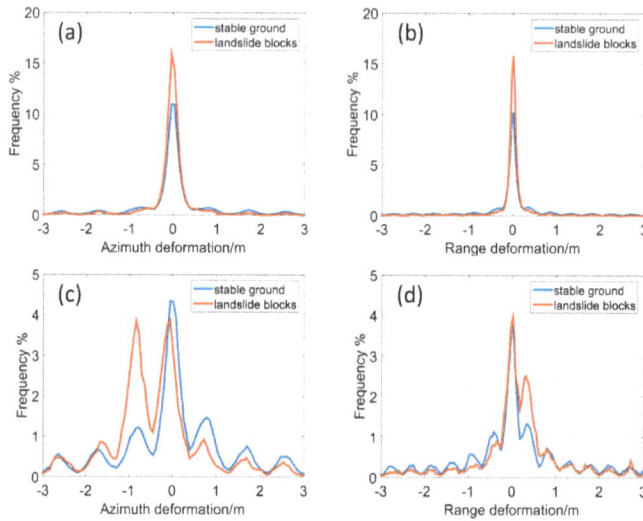

Figure 10. Comparison of azimuth/range deformation histograms of natural scatterers derived from stable ground and landslide blocks; (**a**) azimuth displacement derived from 2 January 2012–13 January 2012 image pair; (**b**) range displacement derived from 2 January 2012–13 January 2012 image pair; (**c**) azimuth displacement derived from 2 January 2012–23 February 2013 image pair; (**d**) range displacement derived from 2 January 2012–23 February 2013 image pair.

In Figure 10a,b, before the occurrence of the landslide the histograms of the 2D deformation measured from landslide blocks and stable ground have very similar distributions with only one main lobe centred on a zero offset. After the displacements as shown in Figure 10c,d the histogram of the stable ground still retains a single, main lobe centred on the zero offset with increased side lobes (probably due to the vegetation impacts during the over one year interval). The histogram of landslide blocks has a dramatic impact in changing the distribution, with a secondary lobe centred on positive value for range displacement and negative value for azimuth displacement, in addition to very similar side lobes found in the histogram of the stable ground.

Following on from the above analysis, a new approach combining sub-pixel cross-correlation and statistical processing is proposed to monitor landslides in such challenging areas for general use when high-contrast surface features are very few or not available.

The processing flow of the new sPOT approach is shown in Figure 11. Firstly, using the annual time series data acquired from 21 February 2009–15 April 2010 with the image of 21 February 2009 as the master image and all the others as slave images; and the other annual time series data acquired from 2 January 2012–23 February 2013 with the image of 2 January 2012 as the master image and all the others as slave images, co-registration was carefully applied to achieve 1/100–1/10 pixel accuracy. Secondly, sub-pixel Cross Correlation was applied to the co-registered time series to derive range/azimuth deformation fields of the Shuping landslide.

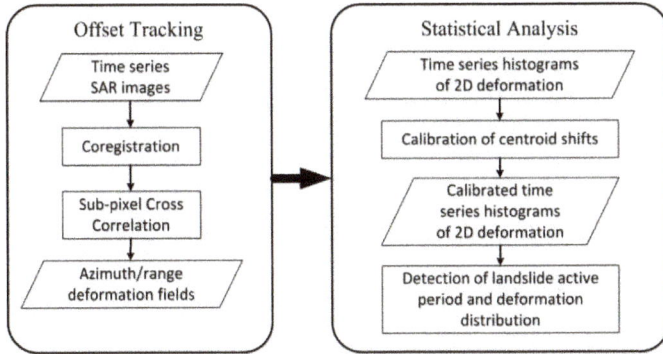

Figure 11. Workflow of the approach combining sPOT and statistical analysis.

To distinguish the measurements of CRs from natural scatterers and demonstrate the effectiveness of the proposed approach, all contributions from CRs in the range/azimuth deformation fields were masked out beforehand.

This approach applies statistical analysis to the derived deformation fields, by plotting histograms of range/azimuth time series offsets derived from landslide blocks. The constant offsets, showing as centroid shifts in the histograms of the stable area, were corrected by Gaussian fitting to the time series histograms of the range/azimuth offsets derived from the stable ground.

The Gaussian model to fit is given by

$$y = \sum_{i=1}^{n} a_i \exp\left[-\left(\frac{x - b_i}{c_i} \right)^2 \right], \quad 1 \le n \le 8 \tag{5}$$

The number of Gaussian functions was increased one by one until the fit computation converged or reached the maximum number of fitting functions. Examples of the fitted Gaussian functions are plotted against the original histograms in Figure 12, showing the main lobes and secondary lobes are all fitted.

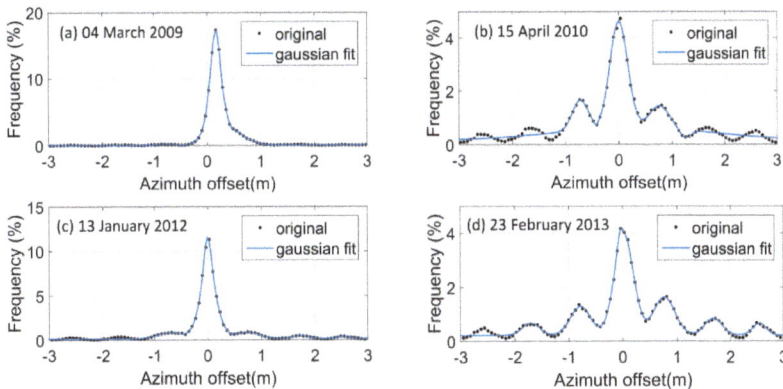

Figure 12. Gaussian fitted histogram vs. the original histogram of azimuth offsets derived from stable area (**a**) 4 March 2009; (**b**) 15 April 2010; (**c**) 13 January 2012; (**d**) 23 February 2013.

The centroid offsets of the Gaussian fitted deformation histograms derived from the 2009–2010 and 2012–2013 annual time series on the stable ground are shown in Figure 13.

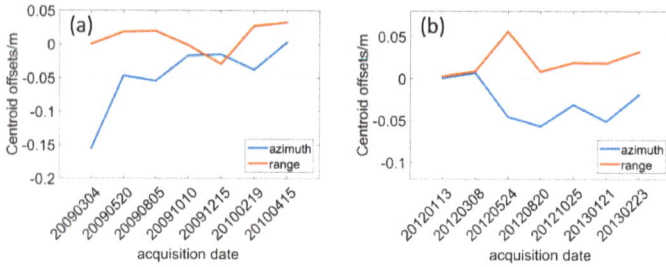

Figure 13. Centroid offsets of deformation histograms derived from natural scatterers on the stable ground. (**a**) Results of 2009–2010 time series; (**b**) Results of 2012–2013 time series. Acquisition date is displayed in the format of 'yyyymmdd'.

The centroid location of every Gaussian peak was taken as a reference to correct the centroid offsets for corresponding histograms of range/azimuth offsets derived from the landslide blocks. After correction, the histograms (referred to as "calibrated histograms" in this paper) of the 2D deformation fields derived from the stable ground adjacent to the landslide area are shown in Figure 14.

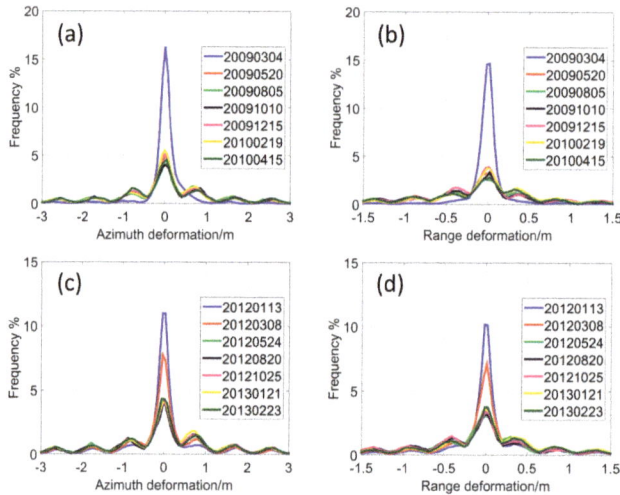

Figure 14. Calibrated time series histograms of 2D offsets derived from natural scatterers on the stable ground: (**a**) azimuth offsets derived from 2009–2010 time series of TSX HS data; (**b**) range offsets derived from 2009–2010 time series of TSX HS data; (**c**) azimuth offsets derived from 2012–2013 time series of TSX HS data; (**d**) range offsets derived from 2012–2013 time series of TSX HS data. Acquisition date is displayed in the format of 'yyyymmdd'.

In Figure 14, the calibrated histograms have the main lobe centered on the coordinate origin point with symmetrical small side lobes, indicating the calibration was successful.

The calibrated time series histograms of the azimuth/range displacement derived from the Shuping landslide blocks are shown in Figure 15.

In Figure 15, we can observe that the envelope of histograms slowly moves backwards (azimuth deformation) and forwards (range deformation) with time and the distribution of offsets gradually spread out indicating that different scatters have different landslide rates.

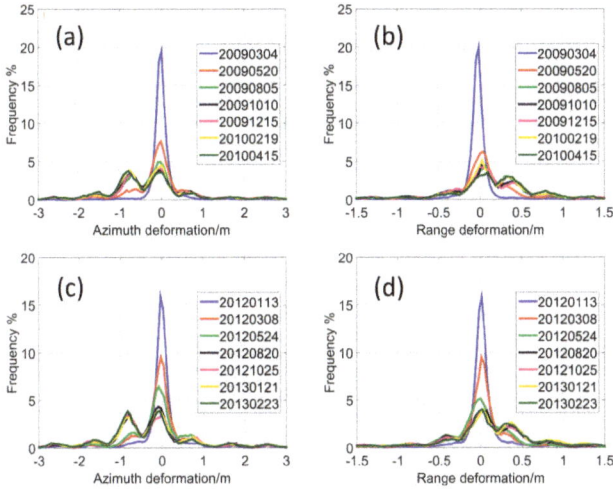

Figure 15. Calibrated time series histograms of 2D displacements derived from natural scatterers in the landslide area: (**a**) Azimuth deformation histograms of 2009–2010 time series; (**b**) Range deformation histograms of 2009–2010 time series; (**c**) Azimuth deformation histograms of 2012–2013 time series; (**d**) Range deformation histograms of 2012–2013 time series. Acquisition date is displayed in the format of 'yyyymmdd'. Modified from [66].

Using the correlation coefficient of 0.25 as the threshold, all pixels with correlation above this value are plotted to present the spatial distribution of azimuth and slant range displacements which occurred in February 2009–April 2010 and January 2012–February 2013, as shown in Figures 16 and 17. Range offsets beyond −1–+1 m and azimuth offsets beyond −2–+2 m are removed for better visualization. The two displacement intervals are identified on the histograms.

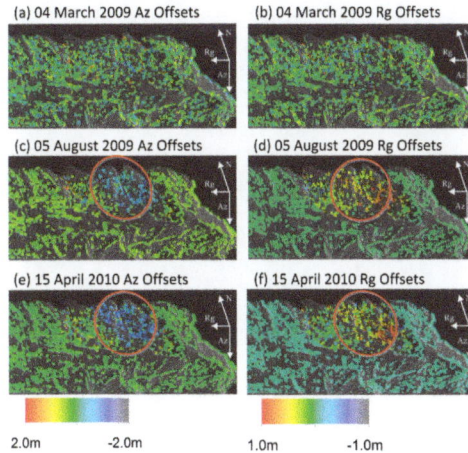

Figure 16. Spatial distribution of the 2D displacement measured from the whole area. (**a,b**) Azimuth/range offsets of the 21 February 2009–4 March 2009 pair; (**c,d**) Azimuth/range offsets of the 21 February 2009–5 August 2009 pair; (**e,f**) Azimuth/range offsets of the 21 February 2009–15 April 2010 pair. The arrows marked as "N, Rg, Az" refer to the North, slant range and azimuth directions. All these scatterers have correlation coefficient no less than 0.25.

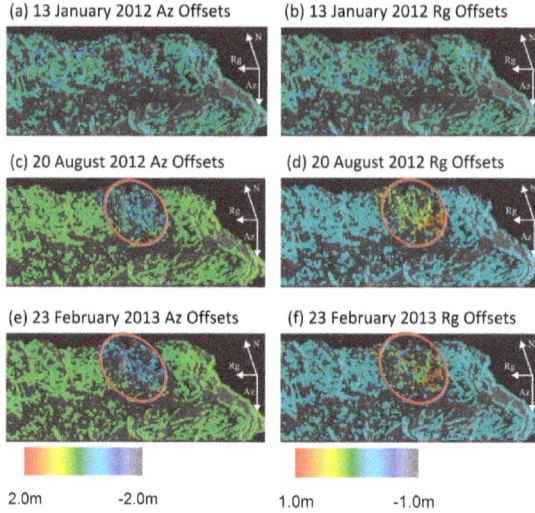

(a) 13 January 2012 Az Offsets   (b) 13 January 2012 Rg Offsets

(c) 20 August 2012 Az Offsets   (d) 20 August 2012 Rg Offsets

(e) 23 February 2013 Az Offsets   (f) 23 February 2013 Rg Offsets

2.0m   −2.0m   1.0m   −1.0m

Figure 17. Spatial distribution of the 2D displacement measured from the whole area. (**a,b**) Azimuth/range offsets of the 2 January 2012–13 January 2012 pair; (**c,d**) Azimuth/range offsets of the 2 January 2012–20 August 2012 pair; (**e,f**) Azimuth/range offsets of the 2 January 2012–23 February 2013 pair. The arrows marked as "N, Rg, Az" refer to the North, slant range and azimuth directions. All these scatterers have correlation coefficient no less than 0.25.

5. Discussion

5.1. Performance Assessment of sPOT on Vegetated Surface

The performance of sPOT in the vegetated areas is assessed by cumulative histograms of azimuth and range offsets [20,67] derived from a rectangular area (242,035 m^2) on the stable ground adjacent to the landslide body. In COSI_Corr, the sub-pixel accuracy is achieved by a quadratic polynomial interpolation of the correlation peak instead of oversampling the SAR intensities. Therefore, we only alter the correlation window size in the tests.

Cumulative Distribution Functions (CDFs) of azimuth/range displacements are plotted for different correlation window sizes, as shown in Figure 18.

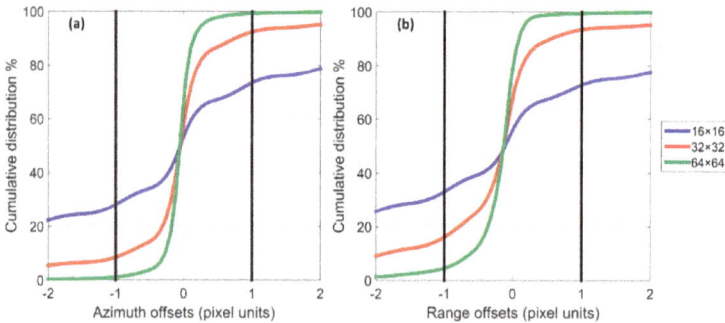

Figure 18. (**a**) Cumulative histograms of azimuth offsets derived from the vegetated surface over a stable area adjacent to the landslide body; (**b**) Cumulative histograms of range offsets derived from the vegetated surface over a stable area adjacent to the landslide body. This is plotted for different correlation window sizes of 16 × 16, 32 × 32 and 64 × 64.

The elapsed time of each parameter setting in a 64 bit Windows 7 system (processor speed: 2.3 GHz, RAM: 8 GB) is listed in Table 4.

Table 4. Processing Time Corresponding to Different Window Sizes of Cross-Correlation, Taking Into Account the Time Consumption of Image Co-Registration.

Correlation Window Size	Processing Time
16 × 16	30 min
32 × 32	39 min
64 × 64	78 min

We can see that using a correlation window size of 32 × 32, in the CDFs of both azimuth and range displacements, over 75% of pixels are characterised by displacements around zero and within ±1.0 pixel offset range. A larger window size improves the accuracy but dramatically increases the processing time (detailed in Table 4). Larger window sizes also increase artifacts and reduce the resolution of the deformation fields [67]. In the above experiments, it is found that a 32 × 32 correlation window size fulfils the research objectives with a reasonable time consumption and is therefore chosen in the processing for both CRs and vegetated surface.

Slight offsets of the centroid are observed from the CDFs of azimuth/range displacements (Figure 18), which very likely results from the impact of dense vegetative cover. This is also pointed out in Section 4.3 via the time series histograms of 2D deformation derived from stable ground, which can be corrected by the proposed calibration technique. As long as the majority of the pixels are characterised with deformation around a certain magnitude within a reasonable offset range, the parameters are considered satisfactory to provide sufficient robustness for sPOT method in the vegetated terrain.

5.2. Accuracy Assessment of Sub-Pixel Offset Tracking (sPOT)

The applicability of sPOT techniques to monitor landslides is determined by their accuracy, which consists of image co-registration errors and the uncertainty associated with Cross Correlation. In theory, the uncertainty of Cross Correlation can be calculated as the standard deviation error of the determination of the correlation peak [68], expressed as follows:

$$\sigma = \sqrt{\frac{3}{10N}} \frac{\sqrt{2 + 5\gamma^2 - 7\gamma^4}}{\pi\gamma^2} \tag{6}$$

where γ is the cross-correlation coefficient; N is the number of independent samples involved in the Cross Correlation, referring to the original image resolution element. The correlation peak is then interpolated using a quadratic polynomial for 1/4 pixel accuracy.

Thus, with a correlation window size of 32 × 32 and a correlation coefficient no less than 0.783 for all CRs, the Cross Correlation has an uncertainty of 0.02 pixels. This is validated by a simulation of cross-correlation using the same parameters with the image acquired on 21 February 2009 as the master and the same image shifted by 5 pixels in slant range direction and 8 pixels in inverse azimuth direction as the slave. The 2D offsets derived by the cross-correlation are analysed and shown in Table 5. These results are obtained with the correlation coefficients ranging from 0.806 to 0.999, almost the same correlation coefficients measured from CRs.

From Table 5, we can see that with the correlation coefficients of no less than 0.8, the cross-correlation measures a mean offset of 5 pixels in the range direction and −8 pixels in the azimuth direction, exactly the same offsets as the image shifted prior to the simulation. The standard deviation errors are 0.022 pixels and 0.021 pixels respectively in the range and azimuth directions. This is in alignment with the theoretical uncertainty calculated for CRs (with a correlation coefficient no less than 0.783) using Equation (6). Thus, the calculated uncertainty is believed to be a good estimate of practical errors of cross-correlation.

Table 5. Statistics of range and azimuth offsets derived from the 20090221 image and the same image shifted by 5 pixels in range direction and 8 pixels in inverse azimuth direction.

	Mean (Pixel)	Std (Pixel)	Max (Pixel)	Min (Pixel)
Range offset	5.000	0.022	5.497	4.503
Azimuth offset	−8.000	0.021	−7.503	−8.496

It is worth noting that the coregistration is not perfect. The residual errors from the coregistration step may lead to uncompensated image offsets which can be mixed in with the investigated displacements [37]. Thus, the overall errors of offset tracking should consider the coregistration errors with the standard deviation errors of cross-correlation. Taking into account both of the cross-correlation uncertainty and the co-registration errors up to 1/10 pixels, the theoretical accuracy of sPOT comes to 0.12 pixels. Substituting the range pixel spacing of 45.6 cm and azimuth pixel spacing of 86.2 cm of TSX Spotlight data into Equation (6), the accuracy of offsets measured from CRs is 5.5 cm in the range direction and 10.3 cm in the azimuth direction. Thus, the offset tracking technique has sufficient accuracy on CRs to monitor Shuping landslides with regard to the annual displacement rate up to 1 m in the azimuth direction and up to 0.7 m in the range direction.

The corner reflector measurements were compared with the results of the same period presented in [20], the differences between slant range/azimuth offsets are shown in Table 6.

Table 6. Comparison between the corner reflector measurements derived in this study and the results presented in a previous study [20].

	Mean Difference (m)	Standard Deviation (m)	RMS Errors (m)
Range offset	0.006	0.031	0.032
Azimuth offset	0.025	0.084	0.088

As shown in Table 6, the root mean square error (RMSE) of offset measurements is 0.088 m in azimuth direction and 0.032 m in range direction, both within the expected accuracy of corner reflector measurements, which reaches a good agreement from a statistical standpoint.

The accuracy of the offsets derived from natural scatterers in the vegetated terrain is assessed by simulation using the correlation coefficients of 21 February 2009–15 April 2010 image pair as inputs. The histogram of the correlation coefficients of natural scatterers is plotted in Figure 19. All contributions from artificial CRs were masked out before analysis.

The accuracy consists of the simulated uncertainties using Equation (6) and co-registration errors of 1/10 pixel size. The cumulative distributions of the 2D accuracy are shown in Figure 20.

From Figure 20, we can see that over 75% of natural scatterers have improved accuracy rates of 34 cm in the azimuth direction and 18 cm in the range direction. For a typical correlation coefficient of 0.25, the lowest accuracy is 24 cm in the azimuth direction and 13 cm in the range direction. Hence, the accuracy of the natural scatterers is statistically significant in measuring the Shuping landslides with regard to the annual displacement rate up to 1 m in the azimuth direction and 0.7 m in the range direction.

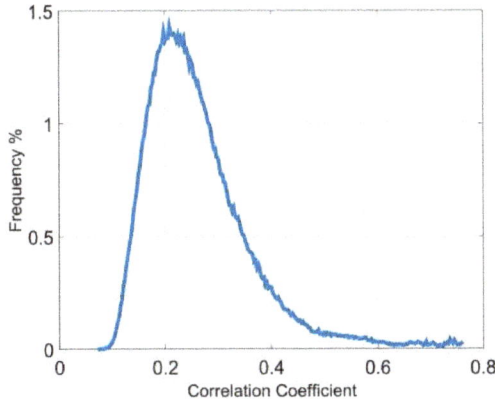

Figure 19. Histogram of the correlation coefficients of 21 February 2009–15 April 2010 image pair processed by sPOT method, the contributions from corner reflectors have been masked out before the assessment.

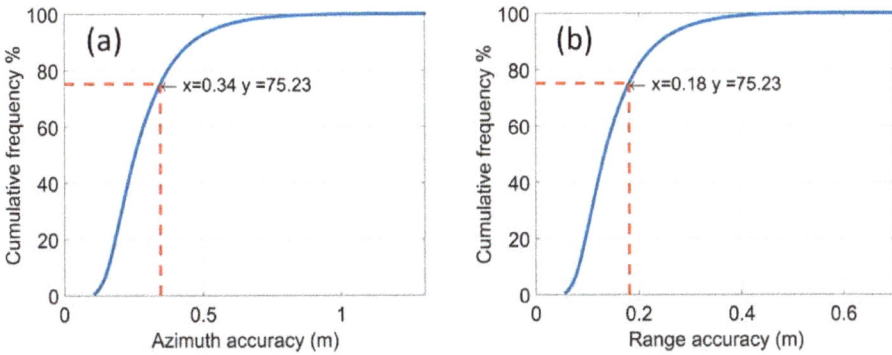

Figure 20. (**a**) Cumulative distributions of the accuracy of azimuth displacements derived from natural scatterers; (**b**) Cumulative distributions of the accuracy of range displacements derived from natural scatterers. Both of the Cross Correlation uncertainty and co-registration errors are considered.

5.3. Validation of Derived Shuping Landslide Rates

Due to the lack of availability of in-situ measurements of CRs in the Shuping landslide area, the offset tracking results are verified by the extensometer measurements presented by Wang et al. [23] on the eastern block of the Shuping landslide (where CR7-11 is located).

The measured range and azimuth deformation represent the components of the actual displacement projected on the slant range and azimuth directions. The displacement vector measured by extensometers is along the slope surface and centreline of the eastern block [23], approximately along the gradient of the elevation model. Slope degrees are derived using 1 arc-second SRTM DEM for each elevation value. By the decomposition of the extensometer measurements on to the North, East and Up directions, the azimuth and slant range offsets dr and da can be resolved using the following Equations [69]:

$$
\begin{aligned}
d_r &= -d_u\cos\theta_{inc} + \sin\theta_{inc}\left[d_n\cos\left(\alpha_h - \tfrac{3\pi}{2}\right) + d_e\sin\left(\alpha_h - \tfrac{3\pi}{2}\right)\right] \\
&= -d_u\cos\theta_{inc} - d_n\sin\theta_{inc}\sin\alpha_h + d_e\sin\theta_{inc}\cos\alpha_h
\end{aligned}
\tag{7}
$$

233

$$da = d_n\sin\left(\alpha_h - \frac{3\pi}{2}\right) - d_e\cos\left(\alpha_h - \frac{3\pi}{2}\right)$$
$$= d_n\cos\alpha_h + d_e\sin\alpha_h$$

(8)

where d_u, d_n, d_e are the Upward, Northward and Eastward displacement, d_r and d_a denote the range and azimuth displacement derived from SAR images, θ_{inc} is the antenna incidence angle, α_h represents the satellite orbit heading (the angle formed between North and azimuth direction).

CR9 in the eastern block is selected for the comparison because its elevation of 277 m is closest to one of the extensometers "SP1-9" crossing the 275 m contour as presented in [23]. The SP1-9 shows 70 cm increase of displacement from February–August 2009.

With an orientation of 355° (measured in Google Earth using the centreline of the eastern block), slope degrees of 22° (derived from the DEM), accumulative displacement of 0.7 m is decomposed and inversed to the radar geometry using Equations (7) and (8). The inversion obtained −0.62 m of azimuth displacement and 0.33 m of slant range displacement for the period of 1 February–01 August 2009. The accumulative displacements derived by offset tracking on CR7 for 21 February–5 August 2009 are −0.6 m in the azimuth direction and 0.44 m in the range direction. We can see that the azimuth displacement derived by offset tracking is very close to the extensometer measurement with only a difference of 2 cm, while the range displacement has a difference of 11 cm. This is probably because the extensometer only measures the displacement projection along the orientation of the block (355°) approximately to the North and the decomposed range displacement is only part of the projection of actual displacement along the slant range. Overall, the extensometer measurements revealed a consistent pattern of all extensometers with a dramatic increase of the displacement from May to August 2009, which is in alignment with the pattern detected by offset tracking.

5.4. Landslide Mechanism Inferred From This Study

The landslide rates measured from the two annual time series show a consistent seasonal pattern of the deformation magnitude of all CRs located in the landslide blocks, whilst the displacements of CRs on the stable ground fluctuate around zero. The corner reflector measurements inside the landslide boundary show a dramatic increase over the same time period over both observation years (i.e., May to August in 2009 and 2012), which implies that the landslide is likely caused by the same driving factors.

In Figures 7 and 8, the measurements of CR9 and CR15 both show a strong correlation with the water level variations of Three Gorges Reservoir in 2009–2010 and 2012–2013 time series observation. It is evident that the landslide active period coincides with the sharp drawdown of the water level of the Three Gorges Dam Reservoir. This suggests a strong connection between the landslide displacements and the operation of the Three Gorges Dam. As there is no deformation observed after the big rise of water level in September, the failure mechanism can be considered as follows: When the reservoir water level increases, the voids of the soil are gradually filled with water during this process. Then the water level reaches its highest level (175 m) and remains in this level for a period. The ground water table gradually reaches a higher elevation and remains in this state. The pressure inside the landslide body balances with the pressure formed by the 175 m water level, which maintains the slope stability. When the water level sharply drops, the ground water content within the landslide body does not drop at the same rate. This results in greater water content inside the landslide body, which forms an outward pressure, leading to the loss of slope stability. However, as the water level drawdown happens synchronously with the seasonal rainfall variations (due to the drawdown being enacted to mitigate against flooding effects form the high seasonal rainfall), a further study is required to differentiate the impacts of these two factors to fully understand the mechanism of the Shuping landslide.

5.5. Potential and Limitations of sPOT to Monitor Landslides in Densely Vegetated Areas

The sub-Pixel Offset Tracking (sPOT) approach only utilizes intensity bands of the satellite imagery to retrieve 2D ground deformation. It is less sensitive to Atmospheric phase screening (APS) and low phase coherence, not requiring phase unwrapping which leads to most of the failures in DInSAR time series approaches due to the low density of Persistent Scatter points. Thus, sPOT techniques potentially has the capability and advantage of measuring the deformation of slope movements with the speed exceeding the maximum detectable displacement of DInSAR or map mass movements in challenging areas such as densely vegetated steep terrain.

The sPOT techniques can be applied in measuring surface deformation when the expected accuracy is sufficient. This is jointly determined by the deformation rates, availability of high-resolution imagery and surface features in the targeted area. Using hi-resolution SAR imagery, sPOT can be used for the monitoring of slow-moving landslides and complement other applications of DInSAR, since sPOT has no limitation on the maximum detectable deformation gradient (DDG) and is still applicable to deformation measurements in the direction perpendicular to satellite LOS.

The deformation rates of the Shuping landslide show that the maximum displacement assumption of DInSAR is not valid in this region, suggesting offset tracking is the only viable alternative method when high resolution imagery is available in order to achieve sufficient accuracy with regard to the displacement rates. Similar to the PSI and SBAS methods, the displacement derived by offset tracking is relative to the stable ground. The constant offsets were removed in the step of centroid shift correction. There can be several sources of the constant offsets, such as the seasonal changes of vegetation cover, the general change of backscatters due to different view angles between passes, etc.

In addition, when there is a lack of ground truth measurements, the results derived from another independent dataset can be utilized in the validation. Thus, for potentially unstable slopes, more than one dataset acquired over the same time period is important for deformation monitoring.

The proposed sPOT approach has been shown of being capable of measuring landslide rates in densely vegetated terrain. Instead of only measuring the deformation magnitude of sparsely distributed 17 CRs, this approach provides a synoptic overview of the deformation fields of all pixels in the landslide area by statistical analysis. From the centroid location changes of every annual time series histograms, the active period of Shuping landslide can be detected. In Figure 15, for both azimuth and range displacements, we can see the histogram centroid stays the same during from 21 February 2009–20 April 2009 and 2 January 2012–24 May 2012 and then shows a sudden change from May of each annual time series as well as a wider and split distribution of histograms. In the periods during 5 August 2009–15 April 2010 and 20 August 2012–23 February 2013 the centroid rarely moves with slight changes in the envelope shapes. From the statistical analysis, the active period of the Shuping landslide is identified as May to August annually, whilst after August, the slope tends to be relatively stable, which shows an accurate correlation with the corner reflector measurements. Offset maps (Figures 16 and 17) show the spatial distribution of the deformation and a distinguishable pattern representing the ongoing landslide.

6. Conclusions

Monitoring of landslides using DInSAR in the Three Gorges Region has received extensive attention over recent years due to the challenges posed in this region. Sub-Pixel Offset Tracking (sPOT) is here shown as an alternative method to address several issues that DInSAR encountered in previous research.

In this study, we demonstrated the capability of sub-Pixel Offset Tracking (sPOT) techniques to monitor relatively fast slope movements in densely vegetated areas with and without the presence of artificial CRs. Although lower accuracy is expected by using sPOT, as long as the accuracy is sufficient with regard to the deformation rates in the study area, sPOT has the advantage of measuring ground displacement perpendicular to the satellite line-of-sight. In addition to DInSAR techniques, sPOT should also be applied to assess if the assumption of DInSAR can be fulfilled. As only SAR amplitude is employed in the processing, sPOT is less sensitive to changes of Atmospheric Phase Screen and low phase coherence. It is not limited by the maximum detectable displacement (MMD) of DInSAR or time series DInSAR as it is not based on any assumption required by phase unwrapping.

Artificial CRs can help to achieve higher accuracy for ground deformation measurements made from sPOT measurements in densely vegetated terrain, whist in the vast majority of regions where such high-contrast features are not available, the proposed approach is able to independently measure ground deformation in terms of the detection of the landslide scale, active period and distribution of deformation magnitude of all the scatterers for the whole landslide area.

In this paper, the statistical analysis of landslide rates derived over vegetated surfaces shows a dramatic increase of landslide displacement rates in the time period approximately from May to August in 2009–2010 and 2012–2013. In each annual time series, the landslide active period coincided with a large drawdown of the reservoir water level in the flood season, suggesting that in Shuping there is a strong connection between the formation of landslides and the operation of the Three Gorges Dam.

Acknowledgments: This research is linked to the ESA-MOST DRAGON-3 Project #10665: Monitoring ground surface displacements in China from EO through case studies of landslides in the Three Gorges Area, crustal tectonic movement in Tibet and subsidence in South China. The work is supported by the UCL Dean's Prize and China Scholarship Council Scholarship. The TerraSAR-X data employed in this study was provided by the German Aerospace Centre (DLR) under data grant GEO0112. We thank J. Zhang, Q. Jiao and T. Xue from China Earthquake Administration for support on fieldwork, and A. Singleton from the University of Glasgow for providing CR measurements for comparison. COSI_Corr was downloaded from http://www.tectonics.caltech.edu/slip_history/spot_coseis/download_software.html for non-commercial research.

Author Contributions: First draft by L. Sun; Supervision and selection of areas by J.-P. Muller as well as editing of original and revised manuscripts prior to submission; Joint field work by L. Sun and J.-P. Muller in the area in May 2014.

Conflicts of Interest: The authors declare no conflict of interest. The founding sponsors had no role in the design of the study; in the collection, analyses, or interpretation of data; in the writing of the manuscript, and in the decision to publish the results.

References

1. Liao, M.; Tang, J.; Wang, T.; Balz, T.; Zhang, L. Landslide monitoring with high-resolution sar data in the Three Gorges Region. *Sci. China Earth Sci.* **2012**, *55*, 590–601. [CrossRef]
2. Wang, Z.H. Remote sensing in the preparatory work of China's hydropower construction. *Remote Sens. Land Resour.* **1995**, *7*, 1–8.
3. Perski, Z.; Wojciechowski, T.; Borkowski, A. Persistent scatterer sar interferometry applications on landslides in carpathians (Southern Poland). *Acta Geodyn. Geomater.* **2010**, *7*, 1–7.
4. Gabriel, A.K.; Goldstein, R.M.; Zebker, H.A. Mapping small elevation changes over large areas: Differential radar interferometry. *J. Geophys. Res.* **1989**, *94*, 9183–9191. [CrossRef]
5. Goldstein, R.M.; Engelhardt, H.; Kamb, B.; Frolich, R.M. Satellite radar interferometry for monitoring ice sheet motion: Application to an antarctic ice stream. *Science* **1993**, *262*, 1525–1530. [CrossRef] [PubMed]
6. Massonnet, D.; Rossi, M.; Carmona, C.; Adragna, F.; Peltzer, G.; Feigl, K.; Rabaute, T. The displacement field of the landers earthquake mapped by radar interferometry. *Nature* **1993**, *364*, 138–142. [CrossRef]
7. Zebker, H.A.; Rosen, P. On the derivation of coseismic displacement fields using differential radar interferometry: The landers earthquake. In Proceedings of the IEEE International Geoscience and Remote Sensing Symposium, Surface and Atmospheric Remote Sensing: Technologies, Data Analysis and Interpretation, Pasadena, CA, USA, 8–12 August 1994; pp. 286–288.
8. Shan, X.; Ma, J.; Wang, C.; Liu, J.; Song, X.; Zhang, G. Co-seismic ground deformation and source parameters of Mani M7.9 earthquake inferred from spaceborne D-InSAR observation data. *Sci. China Earth Sci.* **2004**, *47*, 481–488. [CrossRef]
9. Colesanti, C.; Wasowski, J. Investigating landslides with space-borne synthetic aperture radar (SAR) interferometry. *Eng. Geol.* **2006**, *88*, 173–199. [CrossRef]
10. Tomas, R.; Herrera, G.; Lopez-Sanchez, J.M.; Vicente, F.; Cuenca, A.; Mallorquí, J.J. Study of the land subsidence in Orihuela city (SE Spain) using PSI data: Distribution, evolution and correlation with conditioning and triggering factors. *Eng. Geol.* **2010**, *115*, 105–121. [CrossRef]

11. Milillo, P.; Fielding, E.J.; Shulz, W.H.; Delbridge, B.; Burgmann, R. Cosmo-skymed spotlight interferometry over rural areas: The slumgullion landslide in colorado, USA. *IEEE J. Sel. Top. Appl. Earth Obs. Remote Sens.* **2014**, *7*, 2919–2926. [CrossRef]
12. Liu, D.; Shao, Y.; Liu, Z.; Riedel, B.; Sowter, A.; Niemeier, W.; Bian, Z. Evaluation of InSAR and TomoSAR for monitoring deformations caused by mining in a mountainous area with high resolution satellite-based SAR. *Remote Sens.* **2014**, *6*, 1476–1495. [CrossRef]
13. Fernández, J.; Romero, R.; Carrasco, D.; Luzón, F.; Araña, V. InSAR volcano and seismic monitoring in Spain. Results for the period 1992–2000 and possible interpretations. *Opt. Lasers Eng.* **2002**, *37*, 285–297. [CrossRef]
14. Ferretti, A.; Prati, C.; Rocca, F. Permanent scatterers in SAR interferometry. *IEEE Trans. Geosci. Remote Sens.* **2001**, *39*, 8–20. [CrossRef]
15. Hooper, A.; Zebker, H.; Segall, P.; Kampes, B. A new method for measuring deformation on volcanoes and other natural terrains using InSAR persistent scatterers. *Geophys. Res. Lett.* **2004**. [CrossRef]
16. Ferretti, A.; Savio, G.; Barzaghi, R.; Borghi, A.; Musazzi, S.; Novali, F.; Prati, C.; Rocca, F. Submillimeter accuracy of InSAR time series: Experimental validation. *IEEE Trans. Geosci. Remote Sensi.* **2007**, *45*, 1142–1153. [CrossRef]
17. Lundgren, P.; Usai, S.; Sansosti, E.; Lanari, R.; Tesauro, M.; Fornaro, G.; Berardino, P. Modeling surface deformation observed with synthetic aperture radar interferometry at Campi Flegrei Caldera. *J. Geophys. Res. Solid Earth* **2001**, *106*, 19355–19366. [CrossRef]
18. Berardino, P.; Fornaro, G.; Lanari, R.; Sansosti, E. A new algorithm for surface deformation monitoring based on small baseline differential SAR interferograms. *IEEE Trans. Geosci. Remote Sens.* **2002**, *40*, 2375–2383. [CrossRef]
19. Muller, J.; Zeng, Q.; Li, Z.; Liu, J.; Austin, N.; Brown, D.; Nightingale, M.; Zhang, J.; Gong, L.; Ouyang, Z. Dragon Project 2558: Exploitation of SAR and optical imagery for monitoring the environmental impacts of the Three Gorges Dam. In Proceedings of the 2008 Dragon Symposium, Beijing, China, 21–25 April 2008.
20. Singleton, A.; Li, Z.; Hoey, T.; Muller, J.P. Evaluating sub-pixel offset techniques as an alternative to D-InSAR for monitoring episodic landslide movements in vegetated terrain. *Remote Sens. Environ.* **2014**, *147*, 133–144. [CrossRef]
21. Fu, W.; Guo, H.; Tian, Q.; Guo, X. Landslide monitoring by corner reflectors differential interferometry SAR. *Int. J. Remote Sens.* **2010**, *31*, 6387–6400. [CrossRef]
22. Xia, Y. Synthetic aperture radar interferometry. In *Sciences of Geodesy I: Advances and Future Directions*; Xu, G., Ed.; Springer: Berlin/Heidelberg, Germany, 2010; pp. 415–474.
23. Wang, F.; Yin, Y.; Huo, Z.; Zhang, Y.; Wang, G.; Ding, R. Slope deformation caused by water-level variation in the Three Gorges Reservoir, China. In *Landslides: Global Risk Preparedness*; Sassa, K., Rouhban, B., Briceño, S., McSaveney, M., He, B., Eds.; Springer: Berlin/Heidelberg, Germany, 2013; pp. 227–237.
24. Chen, C.W.; Zebker, H.A. Network approaches to two-dimensional phase unwrapping: Intractability and two new algorithms. *J. Opt. Soc. Am. A* **2000**, *17*, 401–414. [CrossRef]
25. Cruden, D.M.; Varnes, D.J. Landslide types and processes. In *Landslides: Investigation and Mitigation*; Transportation Research Board: Washington, DC, USA, 1996; pp. 36–75.
26. Hungr, O.; Leroueil, S.; Picarelli, L. The varnes classification of landslide types, an update. *Landslides* **2014**, *11*, 167–194. [CrossRef]
27. Longoni, L.; Papini, M.; Brambilla, D.; Arosio, D.; Zanzi, L. The role of the spatial scale and data accuracy on deep-seated gravitational slope deformation modeling: The Ronco Landslide, Italy. *Geomorphology* **2016**, *253*, 74–82. [CrossRef]
28. Brückl, E.; Brunner, F.; Lang, E.; Mertl, S.; Müller, M.; Stary, U. The gradenbach observatory—Monitoring deep-seated gravitational slope deformation by geodetic, hydrological, and seismological methods. *Landslides* **2013**, *10*, 815–829. [CrossRef]
29. Michel, R.; Avouac, J.P.; Taboury, J. Measuring ground displacements from SAR amplitude images: Application to the landers earthquake. *Geophys. Res. Lett.* **1999**, *26*, 875–878. [CrossRef]
30. Kääb, A. Monitoring high-mountain terrain deformation from repeated air and spaceborne optical data: Examples using digital aerial imagery and ASTER data. *ISPRS J. Photogramm. Remote Sens.* **2002**, *57*, 39–52. [CrossRef]

31. Yamaguchi, Y.; Tanaka, S.; Odajima, T.; Kamai, T.; Tsuchida, S. Detection of a landslide movement as geometric misregistration in image matching of SPOT HRV data of two different dates. *Int. J. Remote Sens.* **2003**, *24*, 3523–3534. [CrossRef]

32. Delacourt, C.; Allemand, P.; Casson, B.; Vadon, H. Velocity field of the "la clapière" landslide measured by the correlation of aerial and quickbird satellite images. *Geophys. Res. Lett.* **2004**. [CrossRef]

33. Wangensteen, B.; Guðmundsson, Á.; Eiken, T.; Kääb, A.; Farbrot, H.; Etzelmüller, B. Surface displacements and surface age estimates for creeping slope landforms in Northern and Eastern Iceland using digital photogrammetry. *Geomorphology* **2006**, *80*, 59–79. [CrossRef]

34. Debella-Gilo, M.; Kääb, A. Sub-pixel precision image matching for measuring surface displacements on mass movements using normalized cross-correlation. *Remote Sens. Environ.* **2011**, *115*, 130–142. [CrossRef]

35. Bennett, G.; Roering, J.; Mackey, B.; Handwerger, A.; Schmidt, D.; Guillod, B. Historic drought puts the brakes on earthflows in Northern California. *Geophys. Res. Lett.* **2016**. [CrossRef]

36. Strozzi, T.; Luckman, A.; Murray, T.; Wegmuller, U.; Werner, C.L. Glacier motion estimation using SAR offset-tracking procedures. *IEEE Trans. Geosci. Remote Sens.* **2002**, *40*, 2384–2391. [CrossRef]

37. Casu, F.; Manconi, A.; Pepe, A.; Lanari, R. Deformation time-series generation in areas characterized by large displacement dynamics: The SAR amplitude pixel-offset SBAS technique. *IEEE Trans. Geosci. Remote Sens.* **2011**, *49*, 2752–2763. [CrossRef]

38. Casu, F.; Manconi, A. Four-dimensional surface evolution of active rifting from spaceborne SAR data. *Geosphere* **2016**, *12*, 697–705. [CrossRef]

39. Manconi, A.; Casu, F.; Ardizzone, F.; Bonano, M.; Cardinali, M.; De Luca, C.; Gueguen, E.; Marchesini, I.; Parise, M.; Vennari, C.; et al. Brief communication: Rapid mapping of landslide events: The 3 December 2013 montescaglioso landslide, Italy. *Nat. Hazards Earth Syst. Sci.* **2014**, *14*, 1835–1841. [CrossRef]

40. Elefante, S.; Manconi, A.; Bonano, M.; De Luca, C.; Casu, F. Three-dimensional ground displacements retrieved from SAR data in a landslide emergency scenario. In Proceedings of the 2014 IEEE Geoscience and Remote Sensing Symposium, Quebec City, QC, USA, 13–18 July 2014; pp. 2400–2403.

41. Raspini, F.; Ciampalini, A.; Del Conte, S.; Lombardi, L.; Nocentini, M.; Gigli, G.; Ferretti, A.; Casagli, N. Exploitation of amplitude and phase of satellite SAR images for landslide mapping: The case of Montescaglioso (South Italy). *Remote Sens.* **2015**, *7*, 14576–14596. [CrossRef]

42. Li, X. The Application of Sub-Pixel Correlation on the Measurement of Landslide Deformation—Taking the Shuping Landslide as an Example. Ph.D. Thesis, Peking University, Beijing, China, 2011.

43. Monserrat, O.; Moya, J.; Luzi, G.; Crosetto, M.; Gili, J.A.; Corominas, J. Non-interferometric GB-SAR measurement: Application to the vallcebre landslide (Eastern Pyrenees, Spain). *Nat. Hazards Earth Syst. Sci.* **2013**, *13*, 1873–1887. [CrossRef]

44. Raucoules, D.; De Michele, M.; Malet, J.P.; Ulrich, P. Time-variable 3D ground displacements from high-resolution synthetic aperture radar (SAR). Application to la valette landslide (South French Alps). *Remote Sens. Environ.* **2013**, *139*, 198–204. [CrossRef]

45. Jiao, M.; Song, L.; Wang, J.; Ke, Y.; Zhang, C.; Zhou, T.; Xu, Y.; Jiang, T.; Zhu, C.; Chen, X.; et al. Addressing the potential climate effects of China's Three Gorges project. *Water Energy Int.* **2013**, *70*, 59–60.

46. Wang, T.; Perissin, D.; Liao, M.; Rocca, F. Deformation monitoring by long term D-InSAR analysis in Three Gorges Area, China. In Proceedings of the IGARSS 2008—2008 IEEE International Geoscience and Remote Sensing Symposium, Boston, MA, USA, 7–11 July 2008; pp. IV-5–IV-8.

47. Wang, T.; Perissin, D.; Rocca, F.; Liao, M.-S. Three Gorges Dam stability monitoring with time-series InSAR image analysis. *Sci. China Earth Sci.* **2011**, *54*, 720–732. [CrossRef]

48. Müller, B.; Berg, M.; Yao, Z.P.; Zhang, X.F.; Wang, D.; Pfluger, A. How polluted is the Yangtze River? Water quality downstream from the Three Gorges Dam. *Sci. Total Environ.* **2008**, *402*, 232–247. [CrossRef] [PubMed]

49. Wu, J.; Gao, X.; Giorgi, F.; Chen, Z.; Yu, D. Climate effects of the Three Gorges Reservoir as simulated by a high resolution double nested regional climate model. *Quat. Int.* **2012**, *282*, 27–36. [CrossRef]

50. Xu, K.; Milliman, J.D. Seasonal variations of sediment discharge from the Yangtze River before and after impoundment of the Three Gorges Dam. *Geomorphology* **2009**, *104*, 276–283. [CrossRef]

51. Xu, X.; Tan, Y.; Yang, G.; Li, H.; Su, W. Impacts of China's Three Gorges Dam project on net primary productivity in the reservoir area. *Sci. Total Environ.* **2011**, *409*, 4656–4662. [CrossRef] [PubMed]

52. Zhao, F.; Shepherd, M. Precipitation changes near Three Gorges Dam, China. Part I: A spatiotemporal validation analysis. *J. Hydrol.* **2011**, *13*, 735–745. [CrossRef]

53. Guo, H.; Hu, Q.; Zhang, Q.; Feng, S. Effects of the Three Gorges Dam on Yangtze River flow and river interaction with Poyang Lake, China: 2003–2008. *J. Hydrol.* **2012**, *416–417*, 19–27. [CrossRef]

54. Wang, F.; Zhang, Y.; Huo, Z.; Peng, X.; Araiba, K.; Wang, G. Movement of the shuping landslide in the first four years after the initial impoundment of the Three Gorges Dam Reservoir, China. *Landslides* **2008**, *5*, 321–329. [CrossRef]

55. Chen, D.; Xue, G.; Xu, F. *Study on the Engineering Geology Properties in Three Gorges*; Hubei Science and Technology Publisher: Wuhan, China, 1997.

56. Sassa, K.; Fukuoka, H.; Wang, F.; Wang, G. *Landslides: Risk Analysis and Sustainable Disaster Management*; Springer Science & Business Media: Berlin/Heidelberg, Germany, 2006.

57. Zitova, B.; Flusser, J. Image registration methods: A survey. *Image Vis. Comput.* **2003**, *21*, 977–1000. [CrossRef]

58. Li, X.; Muller, J.-P.; Fang, C.; Zhao, Y. Measuring displacement field from TerraSAR-X amplitude images by sub-pixel correlation: An application to the landslide in shuping, Three Gorges Area. *Acta Petrol. Sin.* **2011**, *27*, 3843–3850.

59. Leprince, S.; Ayoub, F.; Klingert, Y.; Avouac, J.P. Co-registration of optically sensed images and correlation (COSI-Corr): An operational methodology for ground deformation measurements. In Proceedings of the Geoscience and Remote Sensing Symposium, Barcelona, Spain, 23–28 July 2007; pp. 1943–1946.

60. Leprince, S.; Barbot, S.; Ayoub, F.; Avouac, J.P. Automatic and precise orthorectification, coregistration, and subpixel correlation of satellite images, application to ground deformation measurements. *IEEE Trans. Geosci. Remote Sens.* **2007**, *45*, 1529–1558. [CrossRef]

61. Ayoub, F.; Leprince, S.; Avouac, J.-P. Co-registration and correlation of aerial photographs for ground deformation measurements. *ISPRS J. Photogramm. Remote Sens.* **2009**, *64*, 551–560. [CrossRef]

62. Xia, Y.; Kaufmann, H.; Guo, X.F. Landslide monitoring in the Three Gorges Area using D-InSAR and corner reflectors. *Photogramm. Eng. Remote Sens.* **2004**, *70*, 1167–1172.

63. Liu, J.G.; Mason, P.J.; Clerici, N.; Chen, S.; Davis, A.; Miao, F.; Deng, H.; Liang, L. Landslide hazard assessment in the Three Gorges Area of the Yangtze river using ASTER imagery: Zigui–Badong. *Geomorphology* **2004**, *61*, 171–187. [CrossRef]

64. Bai, S.-B.; Wang, J.; Lü, G.-N.; Zhou, P.-G.; Hou, S.-S.; Xu, S.-N. Gis-based logistic regression for landslide susceptibility mapping of the Zhongxian segment in the Three Gorges Area, China. *Geomorphology* **2010**, *115*, 23–31. [CrossRef]

65. Xu, X.; Tan, Y.; Yang, G. Environmental impact assessments of the Three Gorges project in China: Issues and interventions. *Earth-Sci. Rev.* **2013**, *124*, 115–125. [CrossRef]

66. Sun, L.; Muller, J.-P.; Singleton, A.; Li, Z.; Liu, D.; Riedel, B.; Niemeier, W.; Liang, C.; Zeng, Q.; Jiao, J. Monitoring ground surface displacements in the Three Gorges area, crustal tectonic movement in Tibet and subsidence in South China. In Proceedings of the Dragon 3Mid Term Results, Chengdu, China, 26–29 May 2015.

67. Yun, S.H.; Zebker, H.; Segall, P.; Hooper, A.; Poland, M. Interferogram formation in the presence of complex and large deformation. *Geophys. Res. Lett.* **2007**. [CrossRef]

68. De Zan, F. Accuracy of incoherent speckle tracking for circular gaussian signals. *IEEE Geosci. Remote Sens. Lett.* **2014**, *11*, 264–267. [CrossRef]

69. Hanssen, R.F. *Radar Interferometry: Data Interpretation and Error Analysis*; Springer: New York, NY, USA, 2001.

remote sensing

MDPI

Article

Landslide Displacement Monitoring by a Fully Polarimetric SAR Offset Tracking Method

Changcheng Wang [1], Xiaokang Mao [1,2] and Qijie Wang [1,*]

[1] School of Geosciences and Info-Physics, Central South University, Changsha 410083, China;
    wangchangcheng@csu.edu.cn (C.W.); 135011077@csu.edu.cn (X.M.)
[2] China Railway Siyuan Survey and Design Group Company, Wuhan 430063, China
* Correspondence: qjwang@csu.edu.cn; Tel.: +86-731-8883-0573

Academic Editors: Roberto Tomas, Zhenhong Li, Richard Gloaguen and Prasad S. Thenkabail
Received: 2 May 2016; Accepted: 21 July 2016; Published: 28 July 2016

Abstract: Landslide monitoring is important for geological disaster prevention, where Synthetic Aperture Radar (SAR) images have been widely used. Compared with the Interferometric SAR (InSAR) technique, intensity-based offset tracking methods (e.g., Normalized Cross-Correlation method) can overcome the limitation of InSAR's maximum detectable displacement. The normalized cross-correlation (NCC) method, based on single-channel SAR images, estimates azimuth and range displacement by using statistical correlation between the matching windows of two SAR images. However, the matching windows—especially for the boundary area of landslide—always contain pixels with different moving characteristics, affecting the precision of displacement estimation. Based on the advantages of polarimetric scattering properties, this paper proposes a fully polarimetric SAR (PolSAR) offset tracking method for improvement of the precision of landslide displacement estimation. The proposed method uses the normalized inner product (NIP) of the two temporal PolSAR Pauli scattering vectors to evaluate their similarity, then retrieve the surface displacement of the Slumgullion landslide located in southwestern Colorado, USA. A pair of L-band fully polarimetric SAR images acquired by the Jet Propulsion Laboratory's Uninhabited Aerial Vehicle Synthetic Aperture Radar (UAVSAR) system are selected for experiment. The results show that the Slumgullion landslide's moving velocity during the monitoring time ranges between 1.6–10.9 mm/d, with an average velocity of 6.3 mm/d. Compared with the classical NCC method, results of the proposed method present better performance in the sub-pixel estimation. Furthermore, it performs better when estimating displacement in the area around the landslide boundaries.

Keywords: landslide monitoring; displacement; offset tracking; PolSAR

1. Introduction

As a kind of natural geological disaster, landslide has caused enormous property damage and a large number of casualties all over the world [1]. Landslides occur more often and cause more serious loss in developing countries [2]. To prevent or reduce damages caused by landslides, it is necessary to monitor them. Common landslide monitoring methods are in-situ observations, GPS, and remote sensing techniques [3]. Among them, remote sensing can automatically monitor the displacement or changes of the whole landslide, not only of a few observation points. This technique has been widely applied in the three aspects of landslide investigations, namely landslide recognition, landslide monitoring, landslide hazards forecasting [4]. For optical remote sensing, cloudy or rainy areas always pose difficulties to regular landslide monitoring [5]. However, synthetic aperture radar (SAR) systems can monitor landslides in all weather and with all illumination conditions. With these advantages, SAR images have been used for many studies to monitor landslides in recent decades [6,7].

Differential SAR interferometry (D-InSAR) is one of the two main approaches used to estimate the landslide displacement using SAR data. It uses SAR images' phase difference information, and has been widely used in land deformation monitoring [8–13]. Furthermore, as polarimetric information can help to improve the coherence, some researchers have combined polarimetric SAR (PolSAR) data with D-InSAR to improve the results [14–16]. Although the D-InSAR technique has been successfully applied in retrieving highly accurate landslide displacement in some case studies [17], it also has some problems. One issue is that it only measures the landslide displacement in the line-of-sight (LOS) direction [18,19]. Therefore, its result has neither the deformation along the azimuth direction, nor the displacement information when the moving direction of the landslide is perpendicular to the LOS direction. Second, the D-InSAR method cannot work when the landslide displacement exceeds the maximum detectable displacement. Last but not least, its successful application is often limited by de-correlation effects (phase noise). The other method is the offset tracking method, which uses SAR images' intensity information to estimate the deformation along the azimuth and range direction [20–23]. As SAR intensity information is more stable than phase information, it is easier to compute landslide displacement in both the azimuth and range direction by intensity-based offset tracking methods [24]. The precision of this method is related to the spatial resolution of SAR images, which is lower than that of the D-InSAR method [19]. However, with the new generation of high resolution sensors (e.g., TerraSAR-X, Radarsat-2, and high resolution airborne SAR systems), the precision of this method has been improved [25]. The normalized cross-correlation (NCC) offset tracking is a classical intensity-based displacement estimation approach [26] which has been widely used for glacier velocity estimation [23], earthquake displacement [27–30], landslide monitoring, and so on. In addition, as polarimetric SAR images provide more valuable information than that of single channel SAR images, they may help to improve the precision of intensity-based deformation estimation. However, few studies have taken into account the polarimetric information for landslide displacement estimation. Most studies use single-channel SAR intensity images to estimate landslide deformation. Therefore, we attempt to use PolSAR images to improve the accuracy of the intensity-based deformation estimation. Erten, et al. [31,32] proposed two PolSAR tracking methods for glacier velocity estimation, one based on the mutual information of temporal polarimetric covariance matrices and the other based on maximum likelihood estimation. These two methods improved the precision of the offset tracking method compared with single SAR. These methods measure the second-order statistical dependence between two temporal polarimetric covariance matrices. In this paper, taking the advantages of polarimetric scattering properties, we propose a new polarimetric SAR tracking method to improve the accuracy of landslide displacement estimation.

The rest of the paper is arranged as follows. Section 2 simply introduces the classical NCC method and presents the principle of the proposed polarimetric SAR offset tracking method for landslide monitoring. Section 3 illustrates the experimental results. Then, we make a performance analysis of the proposed method in Section 4. Conclusions are drawn in Section 5.

2. Methodology

2.1. The Classical Normalized Cross-Correlation Tracking Algorithm

The normalized cross-correlation (NCC) tracking method is a classical intensity-based matching method, which finds the best match by maximizing a similarity measurement of candidate local blocks of two images. Then, the local offset of each pixel is determined by the peak value of the corresponding candidate correlation plane [33]. For two SAR images, I_M and I_S, the correlation (ρ) between two image templates is calculated with Equation (1).

241

$$\rho = \frac{\left| \sum\limits_{i=1}^{M} \sum\limits_{j=1}^{N} (m_{i,j} - \mu_m)(s_{i,j} - \mu_s) \right|}{\sqrt{\sum\limits_{i=1}^{M} \sum\limits_{j=1}^{N} (m_{i,j} - \mu_m)^2 \sum\limits_{i=1}^{M} \sum\limits_{j=1}^{N} (s_{i,j} - \mu_s)^2}} \tag{1}$$

where the template size is $M \times N$, $i \in \{1, 2, \ldots, M\}$ and $j \in \{1, 2, \ldots, N\}$. $m_{i,j}$ and $s_{i,j}$ are intensity values at pixel (i, j) within the template of I_M and I_S, respectively. μ_m and μ_s are the mean of the template of I_M and I_S, respectively.

After defining a searching window in image I_S, the NCC correlation plane can be calculated by shifting pixel by pixel. Then, we can find the peak value of the correlation plane. To achieve a sub-pixel accuracy, correlations around the peak location should be interpolated. Finally, the sub-pixel offsets along the azimuth and range directions are determined by the peak of the interpolated correlation plane.

2.2. The Proposed Polarimetric Tracking Algorithm

A fully PolSAR system acquires four-channel images in the horizontal (H) and vertical (V) polarization basis. Then, the observed scattering matrix can be expressed as a 4-D polarimetric vector $\vec{K} = (S_{HH}, S_{HV}, S_{VH}, S_{VV})^T$ [34]. For a monostatic backscattering system and a reciprocal target, the scattering matrix is symmetric; that is, $S_{HV} = S_{VH}$. Then, the 4-D scattering vector can be reduced to a 3-D polarimetric vector $\vec{K} = \left[S_{HH}, \sqrt{2}S_{HV}, S_{VV} \right]^T$ [35].

Matching algorithms are based upon an assumption that the matched point and the reference point have similar characteristics. However, traditional single channel SAR images only have very limited information for reference in each pixel, so matching algorithms will use the statistical properties of the neighboring pixels to select tie-points. Compared with the single polarization SAR image, the fully polarimetric SAR image has more information contained in one pixel, which can better represent the scattering mechanisms of the targets. Thus, for PolSAR image registration, we can assume that the matched pixel and the reference pixel have the same polarimetric scattering mechanisms. As PolSAR is sensitive to the structure and orientation of the targets, many different types of target decomposition methods have been widely used to analyze the polarimetric scattering mechanisms of the targets [36–40]. For example, the Pauli decomposition—a widely used coherent target decomposition method for the PolSAR images—decomposes the target into three different basic scattering mechanisms: the surface or odd-bounce contributions ($(S_{HH} + S_{VV})/\sqrt{2}$), the double-bounce or even-bounce contributions ($(S_{HH} - S_{VV})/\sqrt{2}$), and the diffuse (volume) contributions ($\sqrt{2}S_{HV}$, in the monostatic case, where $S_{HV} = S_{VH}$). Surface contributions are mainly from point scatterers, flat surface, or trihedral corner reflectors. The double-bounce contributions are mainly from dihedral structures like ground–wall or ground–trunk. The diffuse (volume) contributions correspond to the statistical distributed scatterers like the forest canopy. After vectorization, the target's Pauli decomposition vector can be written as in Equation (2).

$$\vec{K}_p = \left[\ (S_{HH} + S_{VV})/\sqrt{2}, \quad (S_{HH} - S_{VV})/\sqrt{2}, \quad \sqrt{2}S_{HV} \ \right]^T \tag{2}$$

According to the basic assumption mentioned above, we use the normalized inner product (NIP) of the two temporal PolSAR Pauli scattering vectors $\vec{K}_p^1$ and $\vec{K}_p^2$ for each image pixel as the criteria for matching.

$$\delta = \frac{\left| \vec{K}_p^1 \times \vec{K}_p^2 \right|}{\left\| \vec{K}_p^1 \right\| \times \left\| \vec{K}_p^2 \right\|} \tag{3}$$

The normalized inner product $\delta \in [0,1]$ measures how well the vectors $\overrightarrow{K}_p^1$ and $\overrightarrow{K}_p^2$ align linearly. When $\delta = 1$, the two scattering vectors are perfectly aligned or are linearly correlated. Conversely, when $\delta = 0$, the two scattering vectors are orthogonal, which means they are totally uncorrelated. For the two calibrated PolSAR data with the same acquiring conditions, amplitudes of each polarization channel for the same target are thought to be the same or similar. In this case, the NIP can be used to evaluate the similarity between two scattering vectors [41]. As PolSAR images are also affected by noise, we use the mean value $\langle \delta \rangle$ of the polarimetric normalized inner product (PolNIP) of the neighboring pixels as the final correlation value. Like the traditional NCC method, offsets between the master and slave polarimetric SAR images are estimated by the location of the peak in the PolNIP matrix. Finally, the interpolation around the peak of the PolNIP matrix is used to achieve sub-pixel offsets.

3. Study Area and Experimental Results

In order to validate the proposed method, two L-band fully PolSAR images acquired by the Jet Propulsion Laboratory's Uninhabited Aerial Vehicle Synthetic Aperture Radar (UAVSAR) system are selected. These images cover the Slumgullion landslide (37°59′30″ N, 107°15′25″ W) located in the San Juan Mountains of Colorado, USA. It is reported that the Slumgullion landslide has been moving continually for over 300 years. It is one of the most famous landslides and is an Area of Critical Environmental Concern (ACEC) of USA. Its active part is about 3.9 km long [42], with widths ranging from 100 m to 500 m. The ground surface elevation along the landslide is between 2750 m and 3650 m [43,44]. A number of studies and experiments have been carried out relating to this area. Many scientists have made efforts to analyze its motion characteristics and inner mechanism. Milillo, etc. 2014 [45] used the COSMO-SkyMed Spotlight interferometry with a data acquisition interval of 24 h to inverse the Slumgullion landslide motion. In recent years, more and more fully polarimetric SAR data have been acquired by spaceborne and airborne SAR systems. The increasing polarimetric information provides opportunities to improve the accuracy of displacement estimation.

The two UAVSAR PolSAR images were acquired on 19 August 2011 and 9 May 2012, respectively, with an interval of 264 days. Table 1 lists their detailed information. According to the former GPS surveying results, the maximum displacement in the neck area of the Slumgullion landslide can reach about three meters during the time interval of data acquisition. In this case, it is impossible for the traditional D-InSAR method to estimate the landslide displacement, because the displacement exceeds the maximum detectable deformation gradient of the D-InSAR method.

Table 1. Detailed information of the UAVSAR polarimetric synthetic aperture radar (PolSAR) datasets.

Acquired Time	Polarization	Pixel Spacing (Azimuth)	Pixel Spacing (Range)	Wavelength
2011-08-19	HH, HV, VH, and VV	0.60 m	1.67 m	23.8 cm
2012-05-09	HH, HV, VH, and VV	0.60 m	1.67 m	23.8 cm

The azimuth displacement in Figure 1a shows that there is an obvious movement along the Slumgullion landslide, especially in its active part. The Slumgullion landslide can be simply divided into three main parts: the upper part, the hopper and neck area, and the lower part. The upper part indicates normal faults and tension cracks [44]. For the hopper and neck area, the elevation decreases sharply. The lower part indicates thrust faults [44]. Due to the complex fault movement inside the Slumgullion landslide, the range displacement (Figure 1b) shows more complex characteristics. Figure 1c shows the 2D displacement vectors composed of the azimuth and range displacement in Figure 1a,b. The displacement vector image presents the detailed information of the moving direction and scale. Based on Figure 1, we produce a simplified interpretive map of the Slumgullion landslide, as shown in Figure 2.

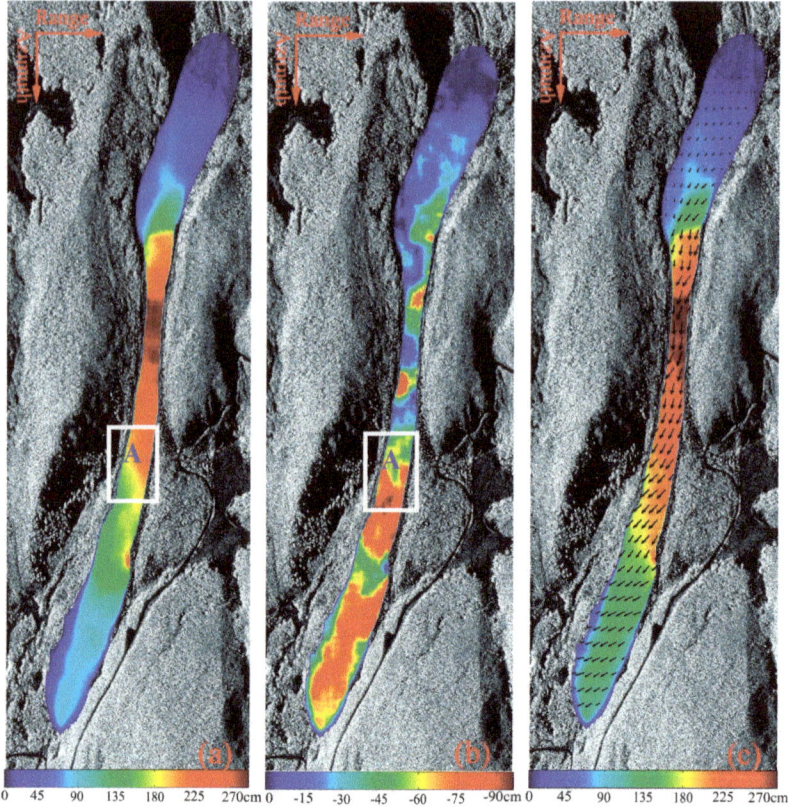

Figure 1. Displacement of the Slumgullion landslide. The background is the SAR intensity image. (**a**) the azimuth displacement; (**b**) the range displacement; (**c**) the 2D displacement vector.

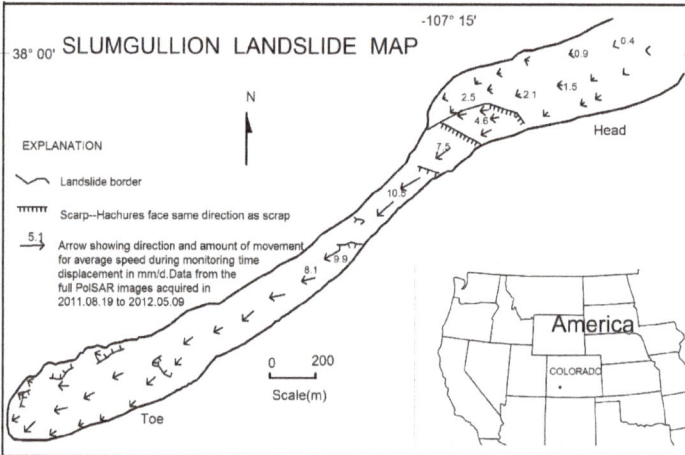

Figure 2. Simplified interpretive map of the Slumgullion landslide map, produced based on the the 2D displacement vector image Figure 1c.

During the time interval between these two PolSAR images, the head of the Slumgullion landslide slid with an average rate of 1.6–2.1 mm/d. For the hopper and neck area—the narrowest part of the landslide, which looks like a funnel-shaped stretch [44]—the sliding rate reaches the maximum 10.1–10.9 mm/d. As described in reference [42], the monitoring site (located in the hopper and neck area) had an average velocity of 10 mm/d during the period of 12 August 2004–4 June 2006. The landslide then grows wider and slides more slowly downward. The average displacement rate for this area is 4.0–5.0 mm/d. More detailed landslide geological interpretations can be found in [42–44].

4. Discussion

4.1. Performance of Sub-Pixel Estimation

To analyze the sub-pixel accuracy of the proposed method, we select the confidence measure Q introduced by Erten [31] as the criterion. The confidence measure Q is defined in Equation (4).

$$Q = \frac{max(PolNIP) - mean(PolNIP)}{mean(PolNIP) - min(PolNIP)} \tag{4}$$

The larger the Q value, the steeper the peak of PolNIP. In the sub-pixel matching method, the best matching position is found by maximizing the interpolation of the matching template's correlation surface. Thus, the precision of the sub-pixel matching position is closely correlated to the steepness of the peak area around the estimated pixel. A larger Q value means more precise offset estimation. We select a pixel from the active area and the stable area to compare the matching performance of the NCC and the proposed PolNIP methods. Figure 3 shows the contour map of the interpolated correlation surface around the estimated pixel produced by the NCC and PolNIP methods.

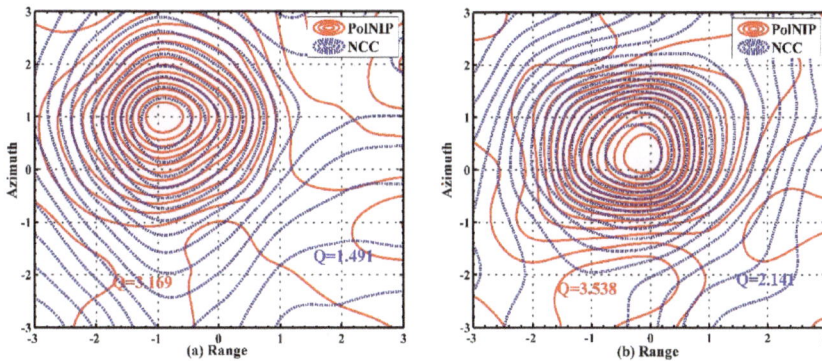

Figure 3. Contour map of the correlation (**a**) active area; (**b**) stable area.

In Figure 3, the contour maps of both the active area and the stable area produced by the PolNIP method are more concentrated around the peak than that of the NCC method. As we know, the denser the contour line, the steeper the peak, and the smaller the error of sub-pixel estimation. Therefore, compared with the traditional NCC method, the PolNIP method can improve the accuracy of sub-pixel matching with polarimetric information. Figure 4 shows the confidence measure (Q value) images of the whole experimental area obtained by the NCC and the PolNIP. We can find that, in either the moving area (along the Slumgullion landslide) or the stable area, the Q values of the PolNIP method are larger than that of the NCC method. Furthermore, we make two statistical histograms (Figure 5) of the Q values of these two methods. The histograms indicate that the PolNIP method has higher confidence.

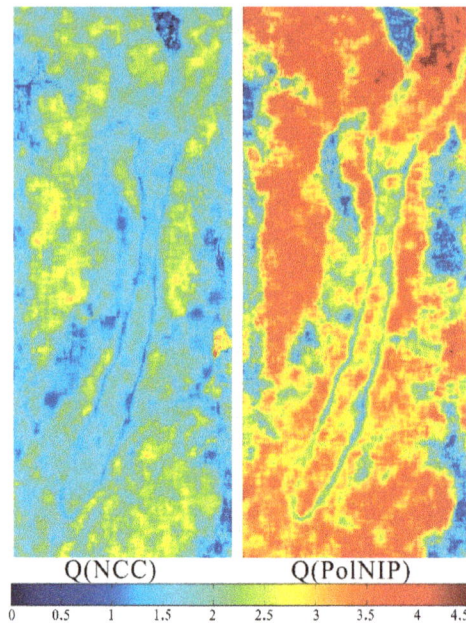

Figure 4. Quality measure Q of the normalized cross-correlation (NCC) and the polarimetric normalized inner product (PolNIP) methods. In reference to the SAR intensity image in Figure 1, the quality maps show that the confidence in the shadow or water areas (the dark area in Figure 1) is much lower than that of the other areas.

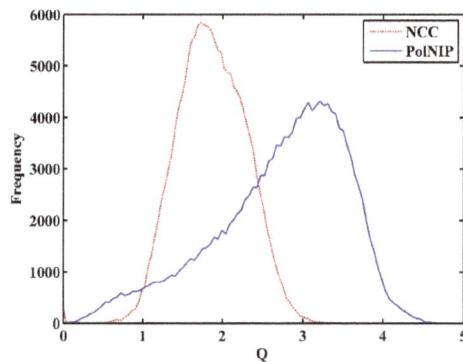

Figure 5. Histograms of the quality measure Q of the NCC and the PolNIP methods.

4.2. Performance of Displacement Estimation on Landslide Boundaries

The Slumgullion landslide has irregular boundaries, so the accurate estimation of the boundaries' displacement is very important for the analysis of the structural features of the landslides. As the NCC uses a matching window (e.g., 32 × 32 pixels) to compute the normalized cross-correlation between two images, the matching window always contains both the moving area and the stable area around the landslide boundaries, affecting the estimation accuracy. Additionally, the size of the matching window significantly affects the correlation value and the precision of the displacement estimation. Figure 6 shows the estimated azimuth displacement maps produced by the NCC with HH channel data (top) and the PolNIP (bottom) method for the sub-area *A* in Figure 1 under different matching windows. We can see that as the matching window size increases, the estimated displacements become more accurate. The results of small matching windows contain more errors. However, results of large matching windows for the NCC method become blurred along the boundaries of the landslide, and the larger the matching window, the more blurred the displacement map. On the contrary, displacements at the boundaries obtained by the proposed PolNIP method only vary a little with increasing matching window size. Figure 7 shows the results of the range direction. Although the range resolution (1.67 m, as shown in Table 1) is much coarser than the azimuth resolution (0.60 m), we can see that the estimated range displacement of the NCC method is still more blurred than that of the PolNIP method. In addition, with the increasing window size, the estimated displacement variation tendency around the boundaries in the range direction agrees with that of the azimuth direction.

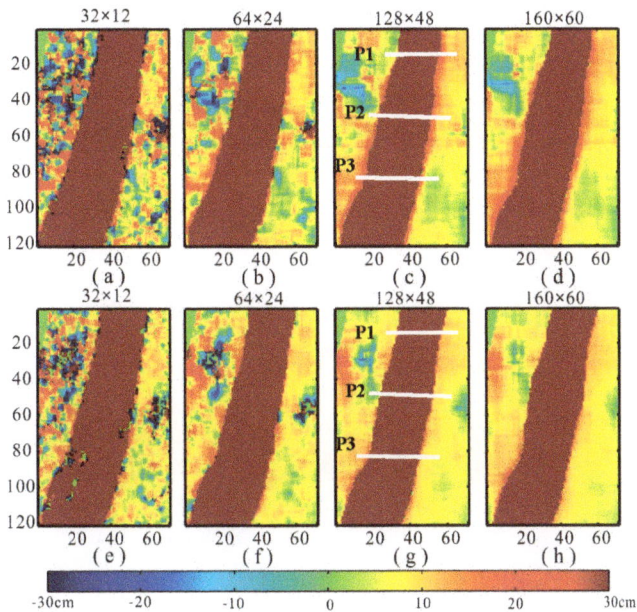

Figure 6. Estimated azimuth displacement maps of the white rectangle *A* in Figure 1. The displacement maps were computed by the NCC (**a–d**) and the proposed PolNIP (**e–h**) methods under different sizes of windows, such as 32 × 12, 64 × 24, 128 × 48, and 160 × 60 pixels. To highlight the difference of offset estimation with these two methods, we readjust the display range to [−30, 30] cm.

Figure 8 shows the azimuth (top) and range (bottom) displacement profiles of the NCC and the PolNIP methods with a matching window of size 128 × 48 pixels. Figure 8a indicates that the NCC and the PolNIP methods get relatively smooth curves of the azimuth displacement in the sliding area. However, around the landslide boundaries, the curve of the PolNIP method indicates more significant inflexion points than that of the NCC method. These inflexion points are the transition points between the stable area and the moving area. On the contrary, the curves of the range displacement present slight fluctuations in the sliding area. In addition, curves of the PolNIP method have significant inflexion points.

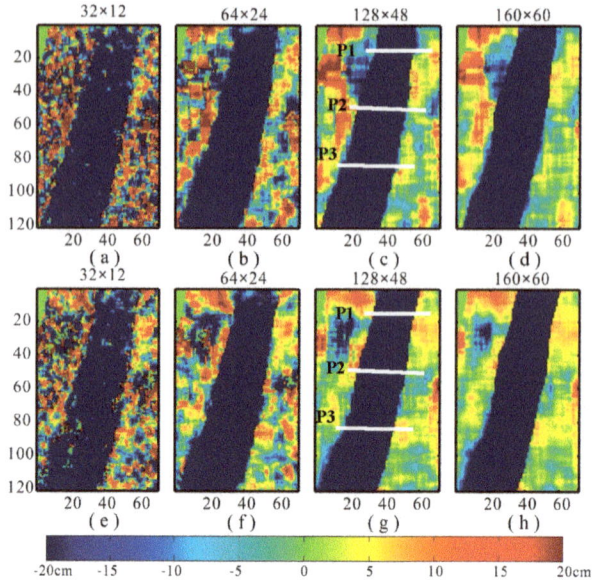

Figure 7. Estimated range displacement maps of the white rectangle *A* in Figure 1. The displacement maps were computed by NCC (**a–d**) and the proposed PolNIP (**e–h**) methods under different sizes of windows, such as 32 × 12, 64 × 24, 128 × 48, and 160 × 60 pixels. We also readjust the display range to [−20, 20] cm to highlight the difference.

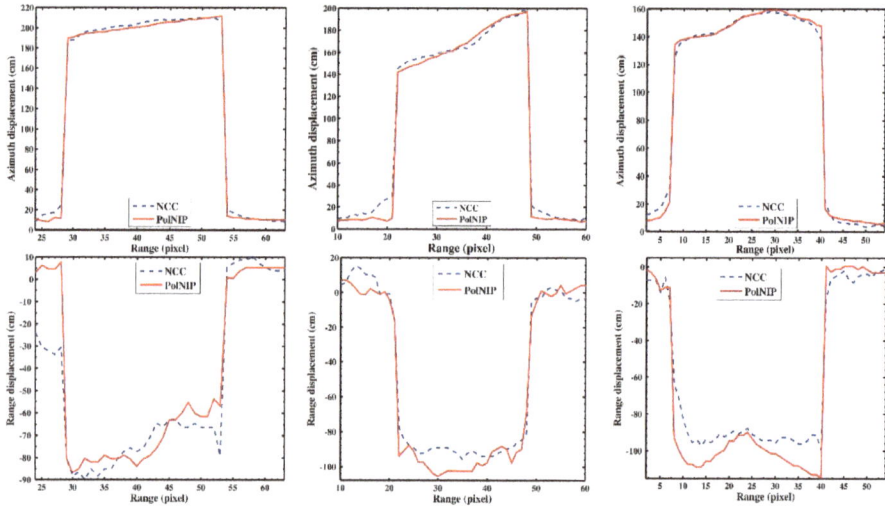

Figure 8. Azimuth (**top**) and range (**bottom**) displacement profiles of the white lines (P1, P2, and P3) in Figure 6c,g and Figure 7c,g with a estimation window of size 128 × 48 pixels. Blue dashed lines and red solid lines indicate the displacement profiles of the NCC and the PolNIP methods, respectively.

To figure out why the estimation results along the landslide boundaries obtained by the NCC method are more blurred, we plot a schematic diagram of the offsets estimation over an area containing stable and moving pixels, as shown in Figure 9. We assume the motion area in the slave image shifts two pixels along the azimuth direction relative to the master image. When the matching window in the slave image slides along the azimuth direction from $s(x, -4)$ to $s(x, 4)$ (relative to the center pixel in the master block), the fitted curve of the correlation value should have two peaks at $s(x, 0)$ and $s(x, 2)$, respectively. However, as shown in Figure 10, the normalized cross-correlation value of the NCC method is ambiguous for the peak of $s(x, 2)$, which affects the sub-pixel accuracy offsets estimation. On the contrary, the proposed PolNIP method can get two peaks (as shown in Figure 10)—$s(x, 0)$ and $s(x, 2)$, respectively. The reason is that the adjacent pixels' NCC value vary slightly due to similar intensity within the matching windows. In this case, it is not beneficial for offsets estimation of landslide boundary area. However, when the matching window slides one pixel for the PolNIP method, the polarimetric similarity for all pairs of temporal pixels' Pauli scattering vectors may dramatically change. Then, the adjacent pixels' PolNIP values have larger difference, which is beneficial for the location of a local peak for the fitted curve.

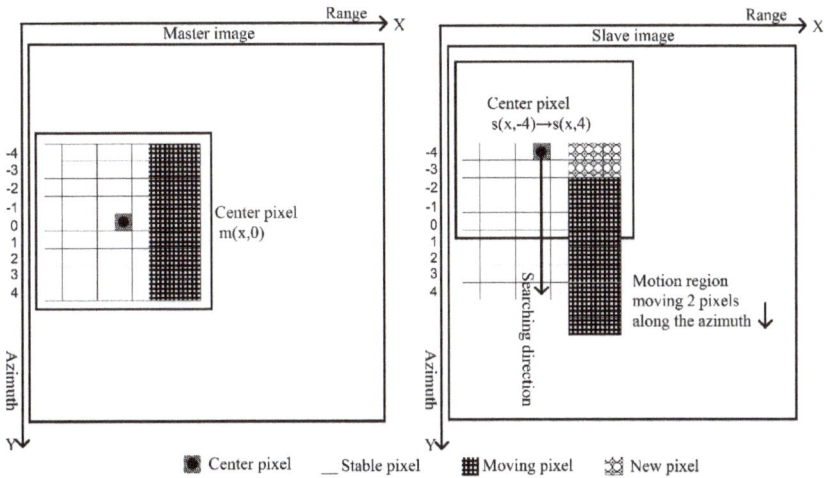

Figure 9. Schematic diagram for the offsets estimation of the NCC method over an area containing stable and moving pixels.

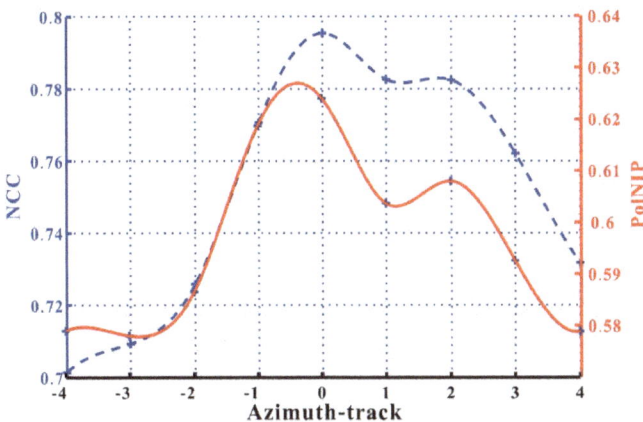

Figure 10. Fitted curve of the correlation value of the NCC method (the blue dashed line) and the PolNIP method (the red solid line).

5. Conclusions

Based on the assumption that different ground objects can be distinguished by their polarimetric scattering properties, a new correlation measure based on fully PolSAR images for offset tracking method has been proposed for landslide displacement estimation. We apply it to the Slumgullion landslide displacement monitoring. The results show that from 19 August 2011 to 9 May 2012, the sliding velocity in different parts of the Slumgullion landslide ranged between 1.6–10.9 mm/d, which is consistent with former publications [42–44,46]. Compared with *in-situ* and GPS observations in the former publications, the landslide displacement maps produced by the proposed method have higher spatial resolution, which is beneficial for geologists in interpreting the formation activity of the Slumgullion landslide.

The proposed method has been evaluated on two aspects. One is the performance analysis on sub-pixel displacement estimation. A confidence measure (Q) was used to estimate the quality of the sub-pixel estimation. The value of Q calculated by the proposed method is about twice that of the NCC method. The other aspect is the performance analysis of displacement estimation on landslide boundaries. The matching window of the NCC method always contains both moving area and stable area around the landslide boundaries, which affects the displacement estimation accuracy. In addition, with the increasing size of matching windows, the azimuth and range displacements at the landslide boundaries estimated by the NCC method become more blurred. This means that the matching windows contain some areas with different moving characteristics, especially for the landslide boundaries. However, the proposed PolNIP method uses each pixel's polarimetric information to partly reduce the effect of hybrid moving characteristics within the matching windows. The results of the proposed PolNIP method contain more detailed displacement information about the landslide boundaries. However, like the NCC method, the performance of the PolNIP method also depends on image resolution. In our experiments, we use high resolution PolSAR data, which makes it possible to use offset tracking method for monitoring of the landslide displacement. For medium resolution SAR data (e.g., Sentinel-1), the method will not deliver comparable results.

Acknowledgments: The work was supported by the National Natural Science Foundation of China (No. 41371335), the National Basic Research Program of China (No. 2013CB733303) and the Natural Science Foundation of Hunan Province, China (No. 2016JJ2141). The authors would like to thank Schulz William H. from USGS for providing GPS data and his helpful suggestions. The authors would like to thank NASA JPL for providing UAVSAR PolSAR images.

Author Contributions: Changcheng Wang conceived the idea, designed the experiments, wrote and revised the paper; Xiaokang Mao performed the experiments and revised the paper; Qijie Wang analyzed the experimental results and revised the paper.

Conflicts of Interest: The authors declare no conflict of interest.

References

1. Dai, F.C.; Lee, C.F.; Ngai, Y.Y. Landslide risk assessment and management: An overview. *Eng. Geol.* **2002**, *64*, 65–87. [CrossRef]
2. Guzzetti, F.; Carrara, A.; Cardinali, M.; Reichenbach, P. Landslide hazard evaluation: A review of current techniques and their application in a multi-scale study, Central Italy. *Geomorphology* **1999**, *31*, 181–216. [CrossRef]
3. Varnes, D.J. *Landslide Hazard Zonation: A Review of Principles and Practice*; United Nations Educational Scientific And cultural Organization (UNESCO) Natural Hazards Press: Paris, France, 1984.
4. Scaioni, M.; Longoni, L.; Melillo, V.; Papini, M. Remote sensing for landslide investigations: An overview of recent achievements and perspectives. *Remote Sens.* **2014**, *6*, 9600–9652. [CrossRef]
5. Delacourt, C.; Allemand, P.; Casson, B.; Vadon, H. Velocity field of the "La Clapière" landslide measured by the correlation of aerial and QuickBird satellite images. *Geophys. Res. Lett.* **2004**, *31*, L15619. [CrossRef]
6. Rosen, P.A.; Hensley, S.; Joughin, I.R.; Madsen, S.N.; Rodriguez, E.; Goldstein, R.M. Synthetic aperture radar interferometry. *Proc. IEEE* **2000**, *88*, 333–382. [CrossRef]

7. Crosetto, M.; Gili, J.A.; Monserrat, O.; Cuevas-González, M.; Corominas, J.; Serral, D. Interferometric SAR monitoring of the Vallcebre landslide (Spain) using corner reflectors. *Natl. Hazards Earth Syst. Sci.* **2013**, *13*, 923–933. [CrossRef]

8. Ao, M.; Wang, C.; Xie, R.; Zhang, X.; Hu, J.; Du, Y.; Li, Z.; Zhu, J.; Dai, W.; Kuang, C. Monitoring the land subsidence with persistent scatterer interferometry in Nansha District, Guangdong, China. *Natl. Hazards* **2015**, *75*, 2947–2964. [CrossRef]

9. Berardino, P.; Fornaro, G.; Lanari, R.; Sansosti, E. A new algorithm for surface deformation monitoring based on small baseline differential SAR interferograms. *IEEE Trans. Geosci. Remote Sens.* **2002**, *40*, 2375–2383. [CrossRef]

10. Crosetto, M.; Castillo, M.; Arbiol, R. Urban subsidence monitoring using radar interferometry: Algorithms and validation. *Photogramm. Eng. Remote Sens.* **2003**, *69*, 775–783. [CrossRef]

11. Ferretti, A.; Prati, C.; Rocca, F. Nonlinear subsidence rate estimation using permanent scatterers in differential SAR interferometry. *IEEE Trans. Geosci. Remote Sens.* **2000**, *38*, 2202–2212. [CrossRef]

12. Sun, Q.; Zhang, L.; Ding, X.L.; Hu, J.; Li, Z.W.; Zhu, J.J. Slope deformation prior to Zhouqu, China landslide from InSAR time series analysis. *Remote Sens. Environ.* **2015**, *156*, 45–57. [CrossRef]

13. Hu, J.; Wang, Q.; Li, Z.; Zhao, R.; Sun, Q. Investigating the Ground Deformation and Source Model of the Yangbajing Geothermal Field in Tibet, China with the WLS InSAR Technique. *Remote Sens.* **2016**, *8*, 191. [CrossRef]

14. Pipia, L.; Fabregas, X.; Aguasca, A.; Lopez-Martinez, C.; Duque, S.; Mallorqui, J.J.; Marturia, J. Polarimetric differential SAR interferometry: First results with ground-based measurements. *IEEE Geosci. Remote Sens. Lett.* **2009**, *6*, 167–171. [CrossRef]

15. Navarro-Sanchez, V.D.; Lopez-Sanchez, J.M.; Vicente-Guijalba, F. A contribution of polarimetry to satellite differential SAR interferometry: Increasing the number of pixel candidates. *IEEE Geosci. Remote Sens. Lett.* **2010**, *7*, 276. [CrossRef]

16. Navarro-Sanchez, V.D.; Lopez-Sanchez, J.M.; Ferro-Famil, L. Polarimetric approaches for persistent scatterers interferometry. *IEEE Trans. Geosci. Remote Sens.* **2014**, *52*, 1667–1676. [CrossRef]

17. Colesanti, C.; Wasowski, J. Investigating landslides with space-borne Synthetic Aperture Radar (SAR) interferometry. *Eng. Geol.* **2006**, *88*, 173–199. [CrossRef]

18. Gabriel, A.K.; Goldstein, R.M.; Zebker, H.A. Mapping small elevation changes over large areas: Differential radar interferometry. *J. Geophys. Res. Solid Earth* **1989**, *94*, 9183–9191. [CrossRef]

19. Bamler, R.; Eineder, M. Accuracy of differential shift estimation by correlation and split-bandwidth interferometry for wideband and delta-k sar systems. *IEEE Geosci. Remote Sens. Lett.* **2005**, *2*, 151–155. [CrossRef]

20. Casu, F.; Manconi, A.; Pepe, A.; Lanari, R. Deformation time-series generation in areas characterized by large displacement dynamics: The SAR amplitude pixel-offset SBAS technique. *IEEE Trans. Geosci. Remote Sens.* **2011**, *49*, 2752–2763. [CrossRef]

21. R. Scheiber, M.; Jager, P.; Prats-Iraola, F.; De Zan, D. Geudtner, Speckle tracking and interferometric processing of TerraSAR-X TOPS data for mapping nonstationary scenarios. *IEEE J. Sel. Topics Appl. Earth Observ. Remote Sens.* **2015**, *8*, 1709–1720. [CrossRef]

22. Wegmuller, U.; Werner, C.; Strozzi, T.; Wiesmann, A. Ionospheric electron concentration effects on SAR and INSAR. In Proceedings of the 2006 IEEE International Symposium on Geoscience and Remote Sensing, Denver, CO, USA, 31 July 2006–4 August 2006; pp. 3731–3734.

23. Li, J.; Li, Z.; Wang, C.; Zhu, J.; Ding, X. Using SAR offset-tracking approach to estimate surface motion of the South Inylchek Glacier in Tianshan. *Chin. J. Geophys. Chin. Ed.* **2013**, *56*, 1226–1236. (In Chinese)

24. Debella-Gilo, M.; Kääb, A. Sub-pixel precision image matching for measuring surface displacements on mass movements using normalized cross-correlation. *Remote Sens. Environ.* **2011**, *115*, 130–142. [CrossRef]

25. Lowry, B.; Gomez, F.; Zhou, W.; Mooney, M.A.; Held, B.; Grasmick, J. High resolution displacement monitoring of a slow velocity landslide using ground based radar interferometry. *Eng. Geol.* **2013**, *166*, 160–169. [CrossRef]

26. Zitova, B.; Flusser, J. Image registration methods: A survey. *Image Vis. Comput.* **2003**, *21*, 977–1000. [CrossRef]

27. Fielding, E.J.; Lundgren, P.R.; Taymaz, T.; Yolsal-Çevikbilen, S.; Owen, S.E. Fault-Slip source models for the 2011 M 7.1 Van Earthquake in Turkey from SAR Interferometry, Pixel Offset Tracking, GPS, and Seismic Waveform Analysis. *Seismol. Res. Lett.* **2013**, *84*, 579–593. [CrossRef]

28. Feng, G.; Li, Z.; Shan, X.; Zhang, L.; Zhang, G.; Zhu, J. Geodetic model of the 2015 April 25 Mw 7.8 Gorkha Nepal Earthquake and Mw 7.3 aftershock estimated from InSAR and GPS data. *Geophys. J. Int.* **2015**, *203*, 896–900. [CrossRef]
29. Wang, T.; Wei, S.; Jónsson, S. Coseismic displacements from SAR image offsets between different satellite sensors: Application to the 2001 Bhuj (India) earthquake. *Geophys. Res. Lett.* **2015**, *42*, 7022–7030. [CrossRef]
30. Wang, T.; Jónsson, S. Improved SAR amplitude image offset measurements for deriving three-dimensional coseismic displacements. *IEEE J. Sel. Topics Appl. Earth Observ. Remote Sens.* **2015**, *8*, 3271–3278. [CrossRef]
31. Erten, E. Glacier velocity estimation by means of a polarimetric similarity measure. *IEEE Trans. Geosci. Remote Sens.* **2013**, *51*, 3319–3327. [CrossRef]
32. Erten, E.; Reigber, A.; Hellwich, O.; Prats, P. Glacier velocity monitoring by maximum likelihood texture tracking. *IEEE Trans. Geosci. Remote Sens.* **2009**, *47*, 394–405. [CrossRef]
33. Hii, A.J.H.; Hann, C.E.; Chase, J.G.; Van Houten, E.E.W. Fast normalized cross-correlation for motion tracking using basis functions. *Comput. Methods Programs Biomed.* **2006**, *82*, 144–156. [CrossRef] [PubMed]
34. Lee, J.S.; Pottier, E. *Polarimetric Radar Imaging: From Basics to Applications*; Francis Group Raton (CRC) Press: Boca Raton, FL, USA, 2009.
35. Cloude, S.R. *Polarisation: Applications in Remote Sensing*; Oxford University Press: Oxford, UK, 2009.
36. Cameron, W.L.; Leung, L.K. Feature motivated polarization scattering matrix decomposition. In Proceedings of the IEEE 1990 International Radar Conference, Arlington, VA, USA, 7–10 May 1990; pp. 549–557.
37. Krogager, E. New decomposition of the radar target scattering matrix. *Electr. Lett.* **1990**, *26*, 1525–1527. [CrossRef]
38. Freeman, A.; Durden, S.L. A three-component scattering model for polarimetric SAR data. *IEEE Trans. Geosci. Remote Sens.* **1998**, *36*, 963–973. [CrossRef]
39. Cloude, S.R.; Pottier, E. An entropy based classification scheme for land applications of polarimetric SAR. *IEEE Trans. Geosci. Remote Sens.* **1997**, *35*, 68–78. [CrossRef]
40. Yamaguchi, Y.; Moriyama, T.; Ishido, M.; Yamada, H. Four-component scattering model for polarimetric SAR image decomposition. *IEEE Trans. Geosci. Remote Sens.* **2005**, *43*, 1699–1706. [CrossRef]
41. Yang, J.; Peng, Y.N.; Lin, S.M. Similarity between two scattering matrices. *Electr. Lett.* **2001**, *37*, 1. [CrossRef]
42. Schulz, W.H.; McKenna, J.P.; Biavati, G.; Kibler, J.D. Characteristics of Slumgullion landslide inferred from subsurface exploration, in-situ and laboratory testing, and monitoring. In Proceedings of the 1st North American Landslide Conference, Vail, CO, USA, 3–10 June 2007; pp. 3–8.
43. Coe, J.A.; Ellis, W.L.; Godt, J.W.; Savage, W.Z.; Savage, J.E.; Michael, J.A.; Debray, S. Seasonal movement of the Slumgullion landslide determined from Global Positioning System surveys and field instrumentation, July 1998–March 2002. *Eng. Geol.* **2003**, *68*, 67–101. [CrossRef]
44. Fleming, R.W.; Baum, R.L.; Giardino, M. *Map and Description of the Active Part of the Slumgullion Landslide, Hinsdale County, Colorado*; US Department of the Interior, US Geological Survey: Denver, CO, USA, 1999.
45. Milillo, P.; Fielding, E.J.; Shulz, W.H.; Delbridge, B.; Burgmann, R. COSMO-SkyMed spotlight interferometry over rural areas: The Slumgullion landslide in Colorado, USA. *IEEE J. Sel. Top. Appl. Earth Observ. Remote Sens.* **2014**, *7*, 2919–2926. [CrossRef]
46. Parise, M.A.; Coe, J.A.; Savage, W.Z.; Varnes, D.J. The Slumgullion landslide (southwestern Colorado, USA): Investigation and monitoring. In Proceedings of the International Conference FLOWS, Sorrento, Italy, 11–13 May 2003.

remote sensing
MDPI

Article

Pi-SAR-L2 Observation of the Landslide Caused by Typhoon Wipha on Izu Oshima Island

Manabu Watanabe [1,*], Rajesh Bahadur Thapa [1] and Masanobu Shimada [1,2]

[1] Earth Observation Research Center, Japan Aerospace Exploration Agency, 2-1-1 Sengen, Tsukuba, Ibaraki 305-8505, Japan; rajesh.thapa@jaxa.jp (R.B.T.); shimada@g.dendai.ac.jp (M.S.)
[2] Tokyo Denki University, Ishizaka, Hatoyama-machi, Hiki-gun, Saitama 350-0394, Japan
* Correspondence: watanabe.manabu@jaxa.jp; Tel.: +81-50-3362-3849; Fax: +81-29-868-2961

Academic Editors: Zhenhong Li, Roberto Tomas, Ioannis Gitas and Prasad S. Thenkabail
Received: 24 December 2015; Accepted: 22 March 2016; Published: 29 March 2016

Abstract: Pi-SAR-L2 full polarimetic data observed in four different observational directions over a landslide area on Izu Oshima Island, induced by Typhoon Wipha on 16 October 2013, were analyzed to clarify the most appropriate L-band full polarimetric parameters and observational direction to detect a landslide area. Japanese airborne Pi-SAR-L2 and PiSAR-L data were used in this analysis. Several L-band full polarimetric parameters, including backscattering coefficient (σ°), coherence between two polarimetric states, four-component decomposition parameters (double-bounce/volume/surface/helix scattering), and eigenvalue decomposition parameters (entropy/α/anisotropy), were calculated to determine the most appropriate parameters for detecting landslide areas. The change in land cover from forest before the disaster to bare soil after the disaster was detected well by α, and coherence between HH and VV. Observational data from the bottom to the top of the landslide detected the landslide well, whereas observations from the opposite sides were not as useful, indicating that a smaller local incident angle is better to distinguish landslide and forested areas. Soil from the landslide intruded into the urban areas; however, none of the full polarimetric parameters showed any significant differences between the landslide-affected urban areas after the disaster and unaffected areas before the disaster.

Keywords: polarimetry; disaster; L-band SAR

1. Introduction

Full polarimetric SAR data are capable of identifying radar scattering mechanisms on the ground, and they have been used to estimate land cover class by connecting the radar backscattering mechanism to the land cover condition both by day and by night in all weather conditions. Such characteristics make these data applicable to the detection of a disaster area, especially for emergency observations made soon after a disaster happens. Watanabe *et al.* [1] used Japanese L-band satellite SAR (PALSAR; Phased Array type L-band Synthetic Aperture Radar) full polarimetric data to detect landslide areas induced by the Iwate-Miyagi Nairuku earthquake of 2008, using the surface scattering component of a three-component decomposition model. Furthermore, σ°_{HV} has also been used to distinguish landslide areas with rough surfaces from other surface scattering areas, such as pastures and vacant pieces of land with smooth surfaces.

Polarimetric decomposition analysis was conducted on the data before and after a landslide event with ALOS PALSAR data [2]. For the detection of landslides areas, 30-m resolution full polarimetric data using unsupervised classification based on the Entropy-α plane are more useful than 10-m resolution single-polarization data. Czuchlewski *et al.* [3] use L-band airborne SAR polarimetry data, and identify the extent of the landslide, using scattering entropy, anisotropy, and pedestal height. They also pointed out that one post-event single polarized SAR image is insufficient for distinguishing and mapping landslides. Rodriguez *et al.* [4] use L-band airborne SAR polarimetry data, and show the landslide scar areas are dominated by single-bounce scattering and the surrounding forested regions are dominated by volume scattering. Radar vegetation index, pedestal height, and entropy are used to identify forest, to separate the landslide area. Shimada *et al.* [5] used Japanese L-band airborne SAR (Pi-SAR-L2; Polarimetric and Interferometric Airborne Synthetic Aperture Radar L2) data to show that the change of land cover from forest before a disaster to bare soil after a disaster was detected well by the polarimetric coherence between HH and VV ($\gamma_{(HH)-(VV)}$). Shibayama *et al.* [6] confirm the usefulness of $\gamma_{(HH)-(VV)}$ for detecting a landslide. They also pointed out that in landslide areas, the polarimetric indices of normalized surface scattering power (*ps*), normalized volume scattering power (*pv*), and $\gamma_{(HH)-(VV)}$ change drastically with the local incidence angle, whereas in forested areas, these indices are stable, regardless of the change in the local incidence angle. Several full polarimetric parameters have been suggested to detect a landslide area since now. In this study, airborne full polarimetric L-band SAR data, obtained both before and after a landslide event, were used to determine the most appropriate full polarimetric parameters and observation direction for identifying an area affected by a landslide induced by heavy rain. The data used in our analysis are unique for two reasons:

(1) hey comprise full polarimetric data observed just after the disaster (landslide).
(2) They were observed from four different observational directions at the same time after the disaster. One of the directions was also observed before the disaster.

To generalize our method, simple radar scattering models from rough surface were applied, as discussed in Section 4. The models were evaluated for two sites using three different local incident angles with two polarizations, simultaneously, which has rarely been undertaken for a landslide area.

2. Pi-SAR-L2 Data and Field Experiment

On 16 October 2013, Typhoon Wipha struck Izu Oshima Island, which is located 100 km south of Tokyo (Figure 1), generating a rainfall rate that was recorded at 122.5 mm/h. This heavy rain induced a large-scale landslide that affected an area of 1.14 million m^2 and led to 39 people being killed or missing. The Geospatial Information Authority of Japan (GSI) used aerial photographs taken after the disaster to produce a landslide map [7], and the main landslide areas are identified in Figure 2. The locations of many landslides can be observed in the mountain area, and some material displaced by the landslides intruded into residential areas. These data were used as the validation data.

The study area was observed before and after the disaster using Japanese airborne SAR (Pi-SAR-L and Pi-SAR-L2) (Table 1). The Pi-SAR-L2 observations were acquired in four different observational directions (L203201–L203204, Figure 1) six days after the disaster. The time required for the four flights was about one hour. Before the disaster, one Pi-SAR-L observation (L03801) had been made on 30 August 2000, in the same observational direction as L203201. Three of the four data (L203201, L203202, and L203203) were used to determine the parameters and directions most appropriate for detecting landslide areas. L203204 was not used, because its configuration (incident and azimuth angles to the landslide area) is almost same as L203202 data. These parameters for detecting landslide areas included backscattering coefficient ($\sigma°$), polarimetric coherence (γ), eigenvalue decomposition [8], and four-component decomposition parameters [9]. The γ is calculated from the correlation between two polarimetric states (HH-HV-VV base, (HH+VV)-(HH-VV)-(HV) base). The eigenvalue decomposition parameters consist of entropy/α/anisotropy, and they were obtained using PolSARPro [10]. Entropy represents the randomness of a scatterer, α represents the scattering mechanism (0° for surface scattering, 45° for dipole scattering or single scattering by a cloud of anisotropic particles, and 90° for double-bounce scattering), and anisotropy represents the relative importance of the second and the third eigenvalues. The four-component decomposition parameters (double-bounce/volume/surface/helix scattering) are related to surface, volume, double-bounce, and helix scattering components on the earth's surface, and they were obtained using a program of our own making. The processing window size for calculating the parameters was 7×7 pixels.

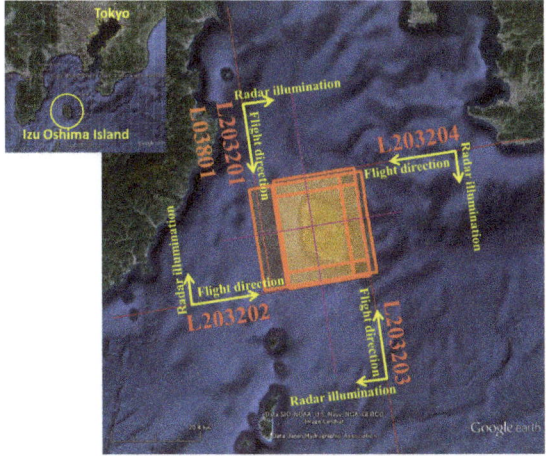

Figure 1. After the landslide: configuration of Pi-SAR-L2 observations performed using four different observational directions (L203201–L203204) on 22 October 2013. Before the landslide: Pi-SAR-L observation (L03801) performed using same flight course as L203201 on 30 August 2000.

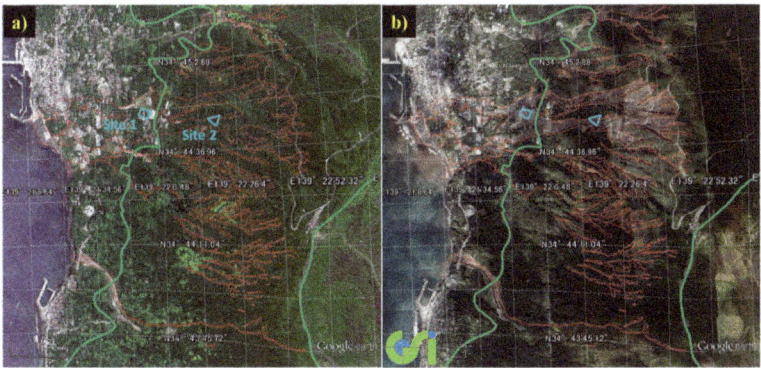

Figure 2. Optical image of the disaster area. Red polygon represents the landslide map, produced by GSI [8]. (**a**) Before the disaster (1 June 2010); (**b**) after the disaster (17 October 2013). The green line indicates the border between forest and other areas obtained from GSI, and the light blue polygon represents the field experiment sites.

Table 1. Specification of Pi-SAR and Pi-SAR-L2.

Items	Pi-SAR	Pi-SAR-L2
Band width	50 MHz	85 MHz
Sampling frequency	61.275 MHz	100 MHz
Operation height	6–12 km	6–12 km
Spatial resolution (slant)	2.5 m	1.76 m
Spatial resolution (azimuth, 4look *)	3.2 m	3.2 m
Noise equivalent sigma zero	−30 dB	−35 dB
Incidence angle	10~60 deg.	10–62 deg.
Polarimetry	full	full
Power	3.5 KW	3.5 KW

* Number of multi-look.

A forest area map, obtained from the national land numerical information download service managed by GSI (Figure 2), was used to identify the forest area before the disaster [11]. Field experiments were performed on 23 and 24 March 2015. The value of σ° obtained from bare soil is determined from the surface roughness, dielectric constant of the soil (equivalent to the volumetric soil moisture, Mv), and local incident angle. The parameter Mv was not measured in the field experiments because its value was different from that when the Pi-SAR-L2 observation was performed. However, surface roughness does not change much, if there is no disturbance. It was measured using a needle profiler at two sites (Sites 1 and 2, Figure 2) within the landslide area to evaluate the radar backscattering from bare soil (Figure 3). The red points shown in the photos were used to evaluate the surface roughness (s) and correlation length (l) normalized by wave number k (e.g., ks and kl). Site 1 was a slightly rough surface covered by soil with a slope of 5°, for which ks and kl were evaluated as 0.38 and 6.17, respectively. Site 2 was a rough surface covered by soil and volcanic rocks with a slope of 20°, for which the values of ks and kl were determined as 1.85 and 5.03, respectively.

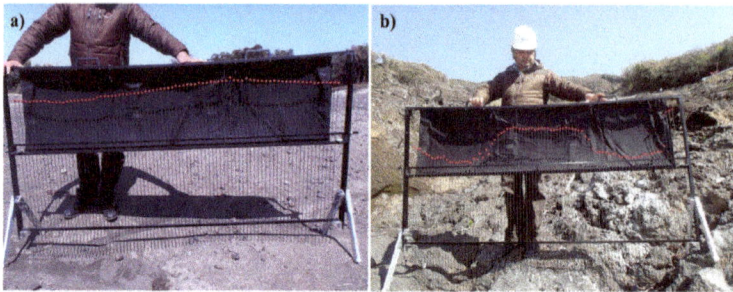

Figure 3. Site photos with a needle profiler. (**a**) Site 1; (**b**) Site 2.

There are three well known and simple surface scattering models. The first of these is the small perturbation model (SPM) [12], which is valid for smooth surfaces (*ks* < 0.3). The second, the physical optics model (POM) [13], is valid for slightly rough surfaces within the parameter ranges Mv < 0.25, $l^2 > 2.76 \cdot s\lambda$, and *kl* > 6. The third, the geometric optics model (GOM) [13], is valid for rough surfaces and predicts that $\sigma^\circ_{HH} = \sigma^\circ_{VV}$ at all incidence angles; this model is valid within the parameter ranges $kl > 6$, $l^2 > 2.76 \cdot s\lambda$, and $ks \cdot \cos\theta > 1.5$. The POM [13] is valid for Site 1, and is described by:

$$\sigma^0{}_{pp}(\theta) = k^2 \cos^2\theta \Gamma_p(\theta) \cdot \exp\left[-(2ks\cos\theta)^2\right] \sum_{n=1}^{\infty} \frac{(2ks\cos\theta)^{2n}}{n!} I \tag{1}$$

$$I_{exponential} = \frac{nl^2}{\left[n^2 + 2(kl\sin\theta)^2\right]^{3/2}} \tag{2}$$

In these equations, Γ(θ,ε) represents the Fresnel reflection coefficient, p represents the polarization (h or v), pp represents any combination of h and v, such as hh, hv, vh, vv, l is the correlation length, θ is the local incident angle, and ε is the relative dielectric constant, which is related to the soil moisture. The GOM [13] is almost valid for Site 2, and described by:

$$\sigma^0{}_{pp}(\theta) = \frac{\Gamma(0)\exp^2\left(-\tan^2\theta/2m^2\right)}{2m^2\cos^4\theta} \tag{3}$$

$$\Gamma(0) = \frac{1 - \sqrt{\varepsilon}}{1 + \sqrt{\varepsilon}} \tag{4}$$

where m is the surface slope and represented by $\sqrt{2}s/l$.

These models and parameters obtained in the field experiment were used to evaluate the observed σ° in Section 4.

3. Results

3.1. Landslide Area Detection with Full Polarimetric Parameters

The four-component decomposition image obtained by Pi-SAR-L2 (ID: L203201) is presented in Figure 4. Maximum and minimum digital number values for each four-component decomposition parameter were assigned to values of 0 and 255 in the RGB scale. The actual landslide is represented by the blue color, indicating that surface scattering is dominant. However, similar surface scattering can also be observed near the top of Mt. Mihara, the volcanic mountain located in the center of the island. The same situation was also observed for the other parameters (entropy, α, polarimetric coherence between (HH+VV) and (HH−VV) ($\gamma_{(HH+VV)-(HH-VV)}$, γ_{HH-VV}), and two of the representative parameters, $\gamma_{(HH+VV)-(HH-VV)}$, and α are shown in Figure 5a,c.

The differences in these parameters before and after the disaster ($\Delta\gamma_{(HH+VV)-(HH-VV)}$, $\Delta\alpha$) can be established by visual inspection (Figure 5b,d). It can be seen that the $\Delta\gamma_{(HH+VV)-(HH-VV)}$ shows differences for both the landslide area and the top of the mountain, where no landslide occurred, whereas $\Delta\alpha$ shows differences in the landslide area only. This indicates that α is better than $\gamma_{(HH+VV)-(HH-VV)}$ for detecting landslide areas in large scale images. Two other parameters ($\Delta\gamma_{(HH)-(VV)}$ and Δentropy) also showed the same characteristics as $\Delta\alpha$; however, four-component decomposition parameters ($\Delta\sigma^\circ_{Double}$, $\Delta\sigma^\circ_{Volume}$, $\Delta\sigma^\circ_{Surface}$, $\Delta\sigma^\circ_{Helix}$ in digital numbers), $\Delta\sigma^\circ$ (in digital number, and Δanisotropy, showed differences not only for the landslide area, but also the bare soil area, where no landslide occurred. The visual interpretation revealed that entropy, α, and γ_{HH-VV} were the prospective parameters for the detection of landslide areas from large scale images.

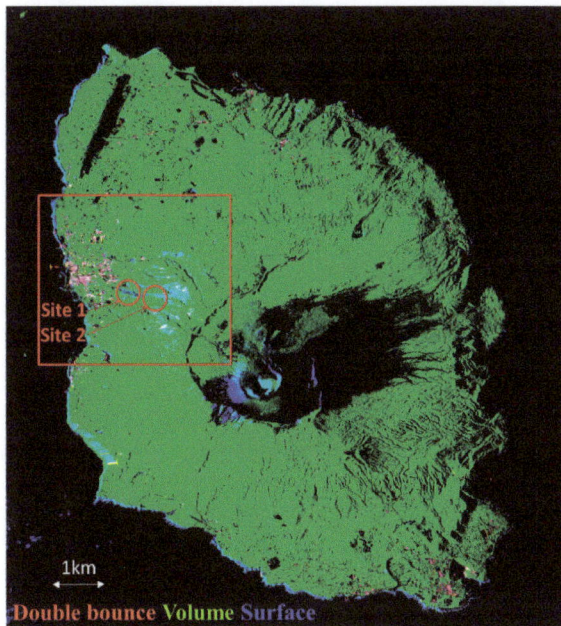

Figure 4. Four-component decomposition image obtained by Pi-SAR-L2 (ID: L203201, R: Double-bounce scattering G: Volume scattering B: Surface scattering). The main landslide area is surrounded by the red rectangle. The field experiment sites are represented by red circles.

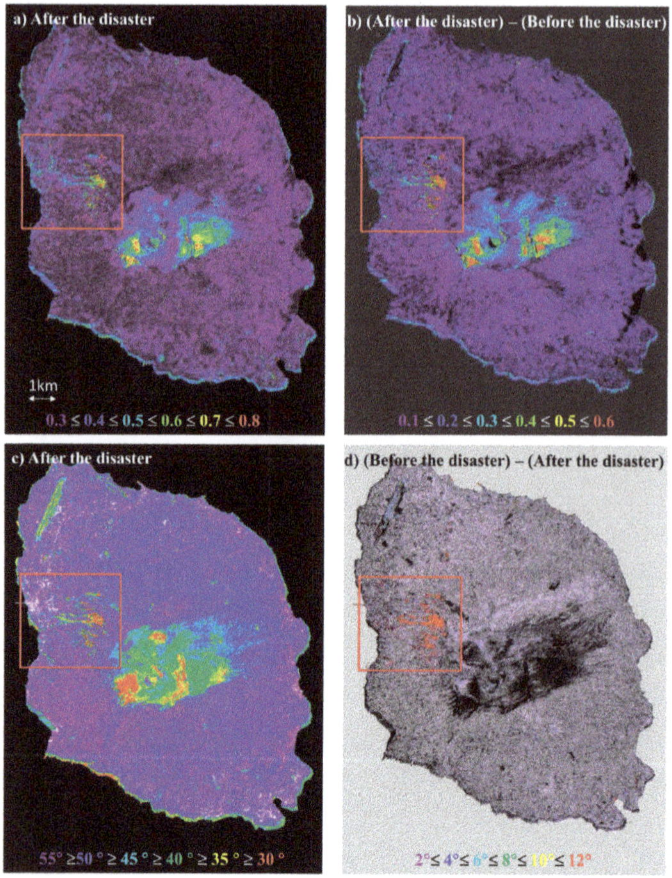

Figure 5. Full polarimetric parameters after the disaster and the difference between before and after the disaster. (**a**) $\gamma_{(HH+VV)-(HH-VV)}$; (**b**) $\Delta\gamma_{(HH+VV)-(HH-VV)}$; (**c**) α; and (**d**) $\Delta\alpha$.

The user's and producer's accuracies (Story and Congalton, [14]) for entropy, α, and γ_{HH-VV} were evaluated from the error matrices to estimate the classification accuracy for detecting the landslide areas shown in the red rectangle in Figure 5. Threshold levels were estimated from a cross-over point obtained from the histogram of the landslide and non-landslide areas for each parameter. The classification was conducted based on the threshold level. Two classes (landslide and no landslide) have been set, and classification accuracy for the case using parameters obtained after the disaster, and the case using the difference in the parameter values before and after the disaster are evaluated. The results are summarized in Table 2. User accuracy is a measure indicating the probability that a pixel is grouped into Class A, given that the classifier has labeled the pixel as belonging to Class A. The producer accuracy is a measure indicating the probability that the classifier has labeled an image pixel as belonging to Class A, given that the ground truth is Class A. The accuracies for $\gamma_{(HH+VV)-(HH-VV)}$ are also shown in the same table for reference. When the data before and after the disaster were used, the values of $\Delta\alpha$ and $\Delta\gamma_{HH-VV}$ show user's accuracies of about 58.7%–60.9% and producer's accuracies of about 33.8%–35.8%. The producer's accuracies are better than those of 25.9%–26.8% for Δentropy and $\Delta\gamma_{(HH+VV)-(HH-VV)}$. The accuracies are almost the same as those achieved when using parameters obtained after the disaster, *i.e.*, the values of α and γ_{HH-VV} show user's accuracies of about 59.5%–61.8% and producer's accuracies of about 34.0%–38.4%. The producer's accuracies of about 11.7%–11.8% for entropy and $\gamma_{(HH+VV)-(HH-VV)}$ obtained after the disaster are 14.2%–15% lower than Δentropy and $\Delta\gamma_{(HH+VV)-(HH-VV)}$. This may be due to the poor classification accuracy between a forest and landslide areas in these parameters.

The landslide area detection was performed for the forest area before the landslide using the forest map and the results are also presented in Table 2. Producer's accuracies are especially improved, and the accuracy changes from 35.8% to 52.2% for the α parameter (improved by 16.4%), and from 33.8% to 49.5% for the $\gamma_{(HH)-(VV)}$ parameters (improved by 15.7%).

Table 2. User's and producer's accuracies for detecting landslide areas using full polarimetric parameters.

Area	Full Polarimetric Parameters	Using Parameters Obtained after the Disaster		Using the Difference in the Parameter Values before and after the Disaster	
		User's Accuracy (%)	Producer's Accuracy (%)	User's Accuracy (%)	Producer's Accuracy (%)
All areas near the landslide	α	61.8	38.4	60.9	35.8
	$\gamma_{(HH)-(VV)}$	59.5	34.0	58.7	33.8
	Entropy	47.7	11.8	58.0	26.8
	$\gamma_{(HH+VV)-(HH-VV)}$	50.5	11.7	57.3	25.9
Forest areas before the landslide identified.	α	66.7	52.1	64.8	52.2
	γ_{HH-VV}	65.0	54.6	64.3	49.5

3.2. Landslide Area Detection in Three Different Observational Directions

The entropy, α, and anisotropy obtained from the three different observational directions (L203201, L203202, and L203203) with the forest mask, where only the forest area is picked up, are presented in Figure 6. The main landslide area is delineated by the red rectangle. The forest area is indicated as the yellow area and the blue area represents the bare soil area, which also indicates the landslide area. A visual inspection reveals that the landslide area is detected well in L203201, wherein observations are made from the bottom to the top of the landslide, whereas the clarity is lower in L203203, with observations made from the top to the bottom of the landslide.

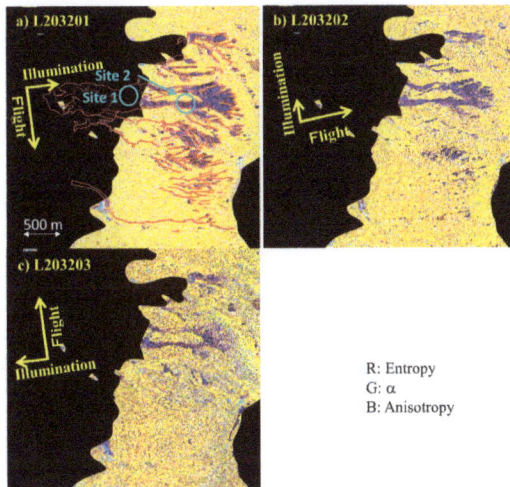

Figure 6. (**a**–**c**) entropy, α, and anisotropy obtained from three different observational directions. The landslide area used for validation [7] is shown by red lines in (**a**). The field experiment sites are represented by light blue circles.

The α and $\gamma_{(HH)-(VV)}$ values derived from L203201 and L203203 are presented in Table 3 for Site 1, Site 2, and a forest near Site 2. The values obtained from L203202 lie between those for L203201 and L203203 and have been omitted from the table. Small directional dependency can be observed for Site 1, because the slope is 5°. Small directional dependency is also observed for the forest area near Site 2, because random scattering in the forest canopy induces less directional dependency. However, relatively larger directional dependency can be observed for Site 2, because the slope is 20°.

Table 3. α and $\gamma_{(HH)-(VV)}$ values derived from L203201 and 203203 after the disaster.

		α			$\gamma_{(HH)-(VV)}$			$\gamma_{(HH+VV)-(HH-VV)}$		
		Mean	St. dev.	Diff.	Mean	St. dev.	Diff.	Mean	St. dev.	Diff.
Site 1	L203201	33.1	2.6	−0.1	0.70	0.05	0.06	0.50	0.09	0.11
	L203203	33.2	5.4		0.64	0.11		0.39	0.10	
Site 2	L203201	31.7	2.2	−12.3	0.77	0.05	0.32	0.61	0.07	0.27
	L203203	44.0	4.1		0.45	0.14		0.33	0.13	
Forest near Site 2	L203201	49.7	2.1	1.6	0.34	0.08	−0.01	0.26	0.08	0.03
	L203203	48.1	2.5		0.35	0.08		0.23	0.08	

Histograms of α for Site 2 and the forest near Site 2 are compared to examine the detectability of the landslide area in a forested region; the results are presented in Figure 7. The histograms for the forested area shows same for L203201 and L203203, indicating there is no local incident angle dependency. On the other hand, the histograms for the landslide area shows difference, indicating the local incident angle dependency. The results are consistent with the one obtained by Shibayama *et al.* [6]. The α value from landslide area in L203201 is clearly different from the value obtained from the forest, whereas the α value from landslide area in L203203 overlaps that from the forest. Some of the landslide area is misclassified as forest, which reduces the classification accuracy in L203203. The same pattern is observed for the $\gamma_{(HH)-(VV)}$ case. This indicates that a smaller local incident angle is better to distinguish landslide and forested areas.

The user's and producer's accuracies for detecting the landslide areas based on the different observational directions by α are summarized in Table 4. L203201 shows good accuracy with user's and producer's accuracies of 66.7% and 52.1%, respectively. L203203 shows the worst accuracy with user's and producer's accuracies of 59.1% and 16.4%, respectively, as was expected from the visual inspection in Figure 6.

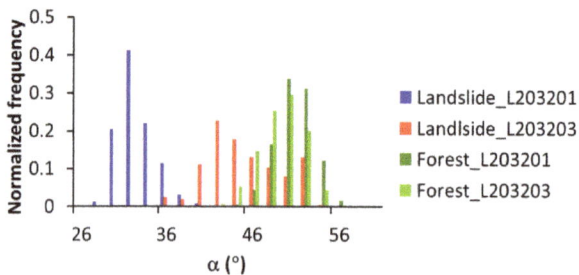

Figure 7. Histogram of α for Site 2 and a forest near Site 2.

Table 4. User's and producer's accuracies for detecting landslide areas based on different observational directions.

		L203201	L203202	L203203
Accuracy (%)	User's	66.7	60.3	59.1
	Producer's	52.1	27.3	16.4

4. Discussion

The accuracy of detection of the landslide area was improved when the forest area was picked up. The landslide area is picked up and shows the difference of entropy/α/anisotropy only before and after the disaster in Figure 8. Maximum and minimum values for each parameter are assigned to values of 0 and 255 in the RGB scale. Yellow areas seen in the upper right of Figure 8 indicate the significant change in entropy and α before and after the disaster. To a large extent, this area changed from forest to bare soil after the landslide. However, significant parameter change is not observed in the upper-left area of Figure 8. This is the residential area, which changed from a normal residential area before the disaster to a residential area with mud induced by the landslide after the disaster. No significant change could be detected by the full polarimetric parameters in this area. There are other areas without significant parameter change, such as in the lower right of Figure 8. Here, there are many narrow valleys and layover/shadowing might prevent any significant change in the full polarimetric parameters in this region. An outline of the detection accuracy by using $\Delta\gamma_{HH-VV}$ and $\Delta\alpha$ is presented in Figure 9. The misidentification of the residential area and the narrow valleys result in a low producer's accuracy of 33.8%–35.8%. On the other hand, the higher user's accuracy of 58.7%–60.9% is due to the detection of the change from forested area to landslide area.

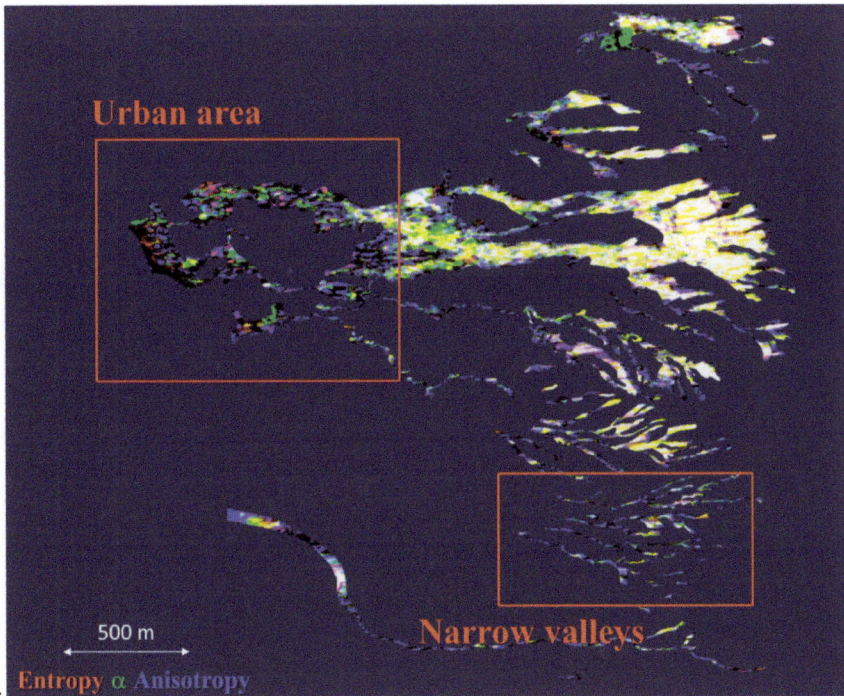

Figure 8. The difference in entropy/α/anisotropy before and after the disaster for the landslide area. Red: Δentropy Green: $\Delta\alpha$ Blue: Δanisotropy.

$\gamma_{(HH+VV)-(HH-VV)}$ for the landslide area is lower than $\gamma_{(HH)-(VV)}$ (Table 3). Very little double-bounce scattering, indicated by HH$-$VV component, is expected from the landslide area, and this led to the smaller $\gamma_{(HH+VV)-(HH-VV)}$. Since a smaller $\gamma_{(HH+VV)-(HH-VV)}$ indicates a smaller $\Delta\gamma_{(HH+VV)-(HH-VV)}$ between the landslide and forested area, detection accuracy for the $\gamma_{(HH+VV)-(HH-VV)}$ parameter is lower than that for the $\gamma_{(HH)-(VV)}$ parameter. The $\gamma_{(HH+VV)-(HH-VV)}$ parameter highlights changes at the top of Mt. Mihara (Figure 5b). Small changes in vegetation and soil moisture are expected between the August and October observations; $\gamma_{(HH+VV)-(HH-VV)}$ may be sensitive to such changes, unlike $\gamma_{(HH)-(VV)}$ and α.

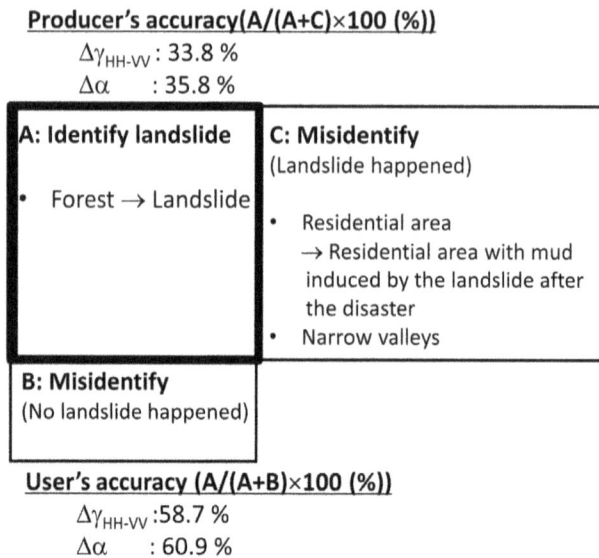

Producer's accuracy(A/(A+C)×100 (%))

$\Delta\gamma_{HH-VV}$: 33.8 %

$\Delta\alpha$: 35.8 %

A: Identify landslide	C: Misidentify

• Forest → Landslide	(Landslide happened)
	
	• Residential area
	    → Residential area with mud
	    induced by the landslide after
	    the disaster
	• Narrow valleys

B: Misidentify (No landslide happened)	

User's accuracy (A/(A+B)×100 (%))

$\Delta\gamma_{HH-VV}$:58.7 %

$\Delta\alpha$: 60.9 %

Figure 9. An outline of the detection accuracy using $\Delta\gamma_{HH-VV}$ and $\Delta\alpha$.

The accuracy is almost the same when using α and $\gamma_{(HH)-(VV)}$ after the disaster, and using the difference between α and $\gamma_{(HH)-(VV)}$ before and after the disaster. This indicates that α and $\gamma_{(HH)-(VV)}$ are good parameters to distinguish forest and bare soil areas. This is also supported by the accuracy improvement after the forest area is identified. The α and $\gamma_{(HH)-(VV)}$ obtained before the disaster are essential for distinguishing between areas of bare soil before the disaster and landslides induced by the disaster, and between bare soil and forest.

The values of $\sigma°$ obtained from Pi-SAR-L2 and estimated from theoretical models are plotted against the local incident angle in Figure 10. The L203201, L203202, and L203203 observations correspond to the minimum, median, and maximum local incident angle plots in the figure. Pi-SAR-L2 observations were performed six days after the disaster, and the value of Mv is assumed as 100% for Site 1. The values of $\sigma°$ obtained from Pi-SAR-L2 show the same pattern against the local incident angle, but with an offset of a few dB from the POM. If the value of ks was not 0.38 (measured on site), it was assumed as 0.5, and the model describes the data obtained from the three different local incident angles well. There was a 1.5-year interval between the Pi-SAR-L2 observations and surface roughness measurements and, thus, changes of 30% in the smoothness of the surface during that period could account for the difference of a few dB in the $\sigma°$ values. The GOM works well for Site 2 if *Mv* = 25% is assumed, except for the L203202 data. The dielectric constant of volcanic rock, which partly covers the surface of Site 2, is generally 5–10, and this corresponds to an Mv of 8%–18.8%. The assumption of *Mv* = 25% could be explained by the mixture of the rock and soil. At Site 2, a few larger rocks could not be characterized by the surface roughness measurements and the shadowing of these rocks might unexpectedly have affected the values of $\sigma°$ for the L203202 observations. However, the values of $\sigma°$ observed for the landslide areas are represented moderately by the simple surface scattering models.

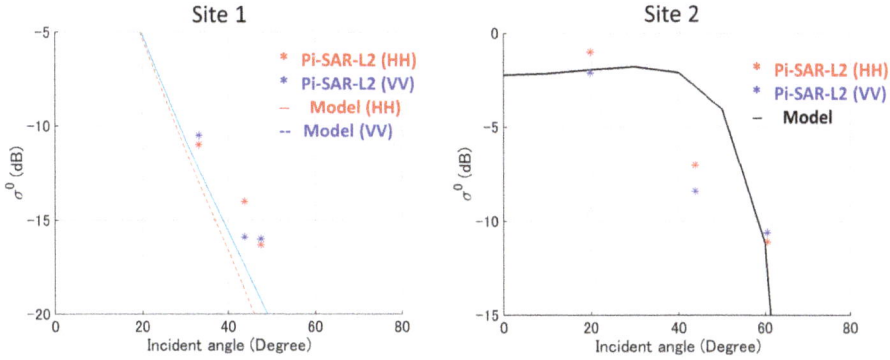

Figure 10. σ° obtained from Pi-SAR-L2 and estimated from theoretical models.

5. Conclusions

Pi-SAR-L2 full polarimetric data observed in four different observational directions over a landslide area were analyzed to clarify the most appropriate L-band full polarimetric parameters and observational direction to detect a landslide area. Data from one Pi-SAR-L observation performed before the disaster occurred were also used in this analysis. A summary of the preferable parameters to detect the landslide area was added in Table 5.

When the data before and after the disaster were used, the $\Delta\alpha$ and $\Delta\gamma_{HH\text{-}VV}$ showed user's accuracies of about 58.7%–60.9% and producer's accuracies of 33.8%–35.8%, indicating better performance than the other parameters, such as the four-component decomposition parameters ($\Delta\sigma^\circ_{Double}$, $\Delta\sigma^\circ_{Volume}$, $\Delta\sigma^\circ_{Surface}$, and $\Delta\sigma^\circ_{Helix}$), $\Delta\sigma^\circ$, $\Delta\gamma_{(HH+VV)\text{-}(HH-VV)}$, Δentropy, and Δanisotropy. Other two knowledge obtained from our analysis are:

✓ The detection accuracy is almost the same when using the parameters after the disaster, and using the difference between the parameters before and after the disaster.
✓ Producer's accuracies are improved, and the accuracy changes from 35.8% to 52.2% for the α parameter (improved by 16.4%), and from 33.8% to 49.5% for the $\gamma_{(HH)\text{-}(VV)}$ parameters (improved by 15.7%), when evaluated by the α and $\gamma_{(HH)\text{-}(VV)}$ parameters, if the forested area before the disaster is identified.

Table 5. Summary for the preferable parameters to detect landslide area.

Parameters	Local Incident Angle	Land Cover Change	
		Forest→Landslide	Residential area→Residential Area with Mud Induced by the Landslide
α, $\gamma_{HH\text{-}VV}$	Low ↓ High	Good ↓ Moderate	Not good.
Entropy, $\sigma^\circ_{Surface}$, $\gamma_{(HH+VV)\text{-}(HH-VV)}$	All	Moderate	
σ°_{Double}, σ°_{Volume}, σ°_{Helix}, σ°, Anisotropy	All	Not good	

However, the land cover change from the residential area before the disaster to the residential area with mud induced by the landslide after the disaster could not be detected by the full polarimetric parameters.

The landslide area was clearly identifiable using data observed from the bottom of the landslide to the top. The clarity was degraded when using data observed from the top of the landslide to the bottom, indicating that smaller local incident angle is better to distinguish landslide and forested area. The observed σ° for the landslide areas was moderately represented using two simple models: the POM for slightly rough surfaces, and the GOM for rough surfaces.

The α and $\gamma_{HH\text{-}VV}$ obtained from full polarimetric L-band SAR data are capable of identifying landslides, which is especially useful for emergency observations taken just after a disaster occurs; however, the parameters only detect the change from forest cover to bare soil. None of the representative full polarimetric parameters showed any significant differences between the landslide-affected urban areas after the disaster and unaffected areas before the disaster. The α and $\gamma_{(HH)\text{-}(VV)}$ obtained before the disaster are essential for distinguishing between areas of bare soil before the disaster and landslides induced by the disaster.

Acknowledgments: Pi-SAR-L2 and Pi-SAR data are copyrighted by the Japan Aerospace Exploration Agency (JAXA).

Author Contributions: M. Watanabe conceived, designed, and performed the experiments, analyzed the data, wrote the paper. R. B. Thapa performed the experiments, and R. B. Thapa and M. Shimada contributed to discussion and interpretation of the results.

Conflicts of Interest: The authors declare no conflict of interest.

References

1. Watanabe, M.; Yonezawa, C.; Iisaka, J.; Sato, M. ALOS/PALSAR full polarimetric observations of the Iwate-Miyagi Nairiku earthquake of 2008. *Int. J. Remote Sens.* **2012**, *33*, 1234–1245. [CrossRef]
2. Yonezawa, C.; Watanabe, M.; Saito, G. Polarimetric decomposition analysis of ALOS-PALSAR observation data before and after a landslide event. *Remote Sens.* **2012**, *4*, 2314–2328. [CrossRef]
3. Czuchlewski, K.R.; Weissel, J.K.; Kim, Y. Polarimetric synthetic aperture radar study of the Tsaoling landslide generated by the 1999 Chi-Chi earthquake, Taiwan. *J. Geophys. Res.* **2003**, *108*, 1–10. [CrossRef]
4. Rodriguez, K.M.; Weissel, J.K.; Kim, Y. Classification of landslide surfaces using fully polarimetric SAR: Examples from Taiwan. *IEEE Proc. Int. Geosci. Remote Sens. Symp.* **2002**, *2002*, 2918–2920.
5. Shimada, M.; Watanabe, M.; Kawano, N.; Ohki, M.; Motooka, T.; Wada, W. Detecting mountainous landslides by SAR polarimetry: A comparative study using Pi-SAR-L2 and X band SARs. *Trans. Jpn. Soc. Aeronaut. Space Sci., Aerosp. Technol. Jpn.* **2014**, *12*, 9–15. [CrossRef]
6. Shibayama, T.; Yamaguchi, Y.; Yamada, H. Polarimetric scattering properties of landslides in forested areas and the dependence on the local incidence Angle. *Remote Sens.* **2015**, *7*, 15424–15442. [CrossRef]
7. Geospatial Information Authority of Japan. Available online: http://www.gsi.go.jp/BOUSAI/h25-taihu26-index.html (accessed on 18 March 2015).
8. Cloude, S.R.; Pottier, E. A review of target decomposition theorems in radar polarimetry. *IEEE Trans. Geoscie. Remote Sens.* **1996**, *34*, 498–518. [CrossRef]
9. Yamaguchi, Y.; Yajima, Y.; Yamada, H. Four-component scattering model for polarimetric SAR image decomposition. *IEEE Trans. Geosci. Remote Sens.* **2005**, *43*, 1699–1706. [CrossRef]
10. Pottier, E.; Ferro-Famil, L.; Allain, S.; Cloude, S.; Hajnsek, I.; Papathanassiou, K.; Moreira, A.; Williams, M.; Minchella, A.; Lavalle, M.; *et al.* Overview of the PolSARpro V4.0 software. The open-source toolbox for polarimetric and interferometric polarimetric SAR data processing. *IEEE Proc. Int. Geosci. Remote Sens. Symp.* **2009**, *2009*, 936–939.
11. Geospatial Information Authority of Japan. Available online: http://nlftp.mlit.go.jp/ksj-e/index.html (accessed on 18 March 2015).
12. Chen, M.F.; Fung, A.K. A numerical study of the regions of validity of the Kirchhoff and small-perturbation rough surface scattering model. *Radio Sci.* **1988**, *23*, 163–170. [CrossRef]

13. Ulaby, F.T.; Kouyate, F.; Fung, A.K.; Sieber, A.J. A backscatter model for a randomly perturbed periodic surface. *IEEE Trans. Geosci. Remote Sens.* **1982**, *GRS-20*, 518–528. [CrossRef]

14. Story, M.; Congalton, R.G. Accuracy assessment: A user's perspective. *Photogramm. Eng. Remote Sens.* **1986**, *52*, 397–399.

![remote sensing logo] *remote sensing* **MDPI**

Article

Analysis of Landslide Evolution Affecting Olive Groves Using UAV and Photogrammetric Techniques

Tomás Fernández [1,2,*], José Luis Pérez [1,2], Javier Cardenal [1,2], José Miguel Gómez [1,2], Carlos Colomo [1] and Jorge Delgado [1]

1 Department of Cartographic, Geodetic and Photogrammetric Engineering, University of Jaén, Campus de las Lagunillas s/n, 23071 Jaén, Spain; jlperez@ujaen.es (J.L.P.); jcardena@ujaen.es (J.C.); jmgl0003@red.ujaen.es (J.M.G.); cmcj0002@red.ujaen.es (C.C.); jdelgado@ujaen.es (J.D.)
2 Centre for Advanced Studies in Earth Sciences (CEACTierra), University of Jaén, Campus de las Lagunillas s/n, 23071 Jaén, Spain
* Correspondence: tfernan@ujaen.es; Tel.: +34-53-212-843

Academic Editors: Zhenhong Li, Roberto Tomas, Norman Kerle and Prasad S. Thenkabail
Received: 30 June 2016; Accepted: 29 September 2016; Published: 13 October 2016

Abstract: This paper deals with the application of Unmanned Aerial Vehicles (UAV) techniques and high resolution photogrammetry to study the evolution of a landslide affecting olive groves. The last decade has seen an extensive use of UAV, a technology in clear progression in many environmental applications like landslide research. The methodology starts with the execution of UAV flights to acquire very high resolution images, which are oriented and georeferenced by means of aerial triangulation, bundle block adjustment and Structure from Motion (SfM) techniques, using ground control points (GCPs) as well as points transferred between flights. After Digital Surface Models (DSMs) and orthophotographs were obtained, both differential models and displacements at DSM check points between campaigns were calculated. Vertical and horizontal displacements in the range of a few decimeters to several meters were respectively measured. Finally, as the landslide occurred in an olive grove which presents a regular pattern, a semi-automatic approach to identifying and determining horizontal displacements between olive tree centroids was also developed. In conclusion, the study shows that landslide monitoring can be carried out with the required accuracy—in the order of 0.10 to 0.15 m—by means of the combination of non-invasive techniques such as UAV, photogrammetry and geographic information system (GIS).

Keywords: Unmanned Aerial Vehicle (UAV); photogrammetric techniques; Structure from Motion (SfM); landslide evolution; olive grove

1. Introduction

Remote sensing and geographic information system (GIS) techniques are basic landslide analysis tools [1] which give us the possibility of studying areas of different size and scale with adequate resolution and accuracy, as well as the ability to develop 3D and multi-temporal approaches. These techniques range from space remote sensing, in optical spectrum or Differential Synthetic Aperture Radar Interferometry (DInSAR) [2,3], to airborne photogrammetry and/or Light Detection and Ranging (LiDAR), which are frequently integrated. The selection of the appropriate technique depends on the scale and spatial resolution, the typology and evolution of the area and the availability of data.

In the case of medium to high resolution studies, with landslides of a diachronic evolution in which processes of reactivation separated in time take place [4,5], aerial photogrammetric techniques are very suitable and therefore their use is increasingly widespread [5–17], sometimes combined with LiDAR [5,10,12,16] or global navigation satellite system (GNSS) techniques [8]. In these studies, the image block orientation is based on conventional aerial triangulation techniques and bundle adjustment, using a reduced number of ground control points (GCP) measured with global navigation satellite system (GNSS) techniques [18]. After image orientation, Digital Terrain Models (DTMs) or Digital Surface Models (DSMs, terrain plus the objects on it, both natural and artificial) are calculated using automatic matching techniques. Based on models from successive epochs, some quantitative approaches such as differential DSM/DTMs between campaigns, topographic profiles and volumetric calculations are addressed in practically all these studies; furthermore, in some of them, 3D displacement vectors are calculated [6–8,11,15] and also observations for qualitative characterization of movements are made [5,6,8,10,13,14,16].

On very high resolution studies, the last decade has seen an extensive use of Unmanned Aerial Vehicles (UAV), a technology in clear progression in many environmental applications. The use of UAVs or Remotely Piloted Aircraft Systems (RPAS), has recently been expanded to civil applications [19,20] including precision agriculture [21,22], civil protection and fires [23], and more recently to engineering, environmental sciences and surveying [24]. The present extensive use of UAVs has been stimulated by falling prices, the increasing miniaturization and improved performance of these systems, the recent developments with respect to GNSS and inertial systems, advances in autopilot guiding, etc. The current state of applicability is also due to the use of new algorithms in computer vision. Indeed, a new generation of low cost instruments and user-friendly photogrammetric software based on dense matching and Structure from Motion (SfM) approaches [25,26] has also contributed to this dramatic increase of UAV applications in environmental and Earth sciences.

In landslide research, different types of UAVs and methodologies have been used. Heavy equipment, usually fixed-wing drones [27–31], but also helicopters, has been employed in high resolution studies. However, in most cases, light equipment such as multi-copters [32–46] has been employed in very high resolution studies.

UAVs have been used in preliminary studies supporting other techniques such as remote sensing from satellites [47–49], photogrammetry from historical flights [27,29,31,39–41] and airborne LiDAR [31]. They have also been applied to inventory data collection studies by means of photo interpretation [49,50], change analysis [27] and object-oriented analysis [30], as well as the study of the effects of catastrophic events [51]. In some cases, the aim of the study is the elaboration of susceptibility and stability maps [33,48,49] and the assessment of the exposure of buildings or infrastructure to landslide risk [33], but in other cases, the UAV techniques are integrated as subsystems for observation, material transport and rescue services in larger systems of quick response to emergency events [29,49,52,53].

Many approaches are based on the development of DSM/DTMs [30–45] and orthophotographs [27–45]. In most cases, the starting point are flights oriented by aerial triangulation, based on ground control points (GCP) measured with GNSS [30–32,34–41] or registered from previous photogrammetric [29] or LiDAR models [31,39]. Structure from Motion (SfM) methods have also been tested [26,34–41,43]. While direct orientation from in-flight parameters (GNSS/INS) is routinely used in conventional aerial photogrammetry, such direct orientation techniques are not usually possible with present micro-UAVs when absolute orientation accuracy is a critical factor, although some experiments for direct orientation without GCP have also been undertaken [26,29,46].

DTMs or, alternatively, DSMs are frequently used to obtain differential models for estimating the vertical displacements of the terrain surface [30,31,34–41,45] and for volumetric calculations and profiles [31]. Although DSMs introduce errors and uncertainties in terrain surface calculations, most authors use these models due to the difficulties in obtaining true DTMs from photogrammetric point clouds [34–41,45], as in areas covered by dense vegetation. In other cases, the analysis of these models of very high resolution allows the automatic detection of scarps and other features of landslides [43].

Orthophotographs—besides being used for landslide inventories, as in the aforementioned studies—allow the calculation of the horizontal displacements between significant points with great accuracy, given their high resolution [30,34–36,39–41,45]. In some studies, methods and algorithms for the automatic detection of movements of the terrain surface from the orthoimages have been proposed, as well as techniques of image co-registration [30,37,38,42]. Finally, some authors have used orthoimages for textural analysis of the fissures formed in the landslides [36] and to determine vegetation indices such as GVI (Green Vegetation Index) for purposes of land classification or landslide inventory [27,30].

Moreover, when the measurement and monitoring of terrain deformation is the aim, analyses of the positional accuracy of the results have to be conducted. These studies can be carried out by means of error analysis of the GCPs (error propagation, distribution of points, field GNSS measurements, etc.) as well as the analysis of the resultant products such as DSM/DTMs or orthophotographs [26,28,31,32,34,41,42,45]. Generally, satisfactory results can be achieved, similar to those calculated with conventional aerial photogrammetry [29,44].

This paper deals with the use of UAV techniques for very high accuracy and resolution field data collection. Present UAVs allow fast, low cost and effective data acquisition. The study case is a multi-temporal analysis of an earth flow affecting an olive grove. Autopilot guiding missions allowed the acquisition of blocks of high overlapping stereo-models, which led to the detailed observation and mapping of terrain features. Furthermore, analysis from DSMs and orthophotographs could be carried out by comparing surfaces (differential DSMs) or measuring displacements between points. Given the difficulties of automatic identification and matching of points between multi-temporal images due to changes in vegetation, sun illumination and the landslide movement itself, the points were selected manually, but an approach was also tested for the semi-automatic calculation of displacements of the olive trees that cover most of the study area.

2. Study Area and Landslides

The area is located in an olive grove on a hillslope with a moderate gradient of 15% (5° to 20°) above the A-44 highway in La Guardia de Jaén, Southern Spain (Figure 1a,b). In this area there are many landslide indications [54], most of them corresponding to mud or earth flows [55,56], such as the study case of this paper, although slide-type movements have also been identified (Figure 1c). Some of these have been studied by means of photogrammetric and LiDAR techniques [57], but also with UAVs [39–41]. The study movement occurred as a result of the heavy rains of winter 2012–2013, and had an approximate extension of 500 m length, 50–150 m width and a height difference of 80 m. The landslide seriously affected the roads in the area (Figures 1c and 2).

The study area is located on the geological Guadalquivir Units [58] (Figure 1d), a set of materials with a complex structure and diverse lithology, in which the following formations predominate: Triassic evaporites and shales, Cretaceous-Paleogene marls and clays (from the Subbetic Units), and Miocene loamy-clay sediments belonging to the Guadalquivir basin. In this area, the Guadalquivir Units are overthrusted by Intermediate Betic Units represented by thick stratigraphic successions of folded Middle Jurassic limestones [59,60], which form a prominent relief. This thrust is affected by normal faults with a NNW-SSE direction. Both types of tectonic structures present geomorphological and seismic evidence of recent activity [61,62], which may be related to the regional uplift and subsequent slope instability processes. The contact is dotted with small alluvial fans, thick piedmont deposits and travertines. Travertines are associated with springs at the foot of the Jurassic limestones. Most of the travertine areas are involved in old movements, indicating a direct relationship between them and the springs. In the study case, travertine crusts were observed over the clays and loams, showing a continuous water discharge [39–41].

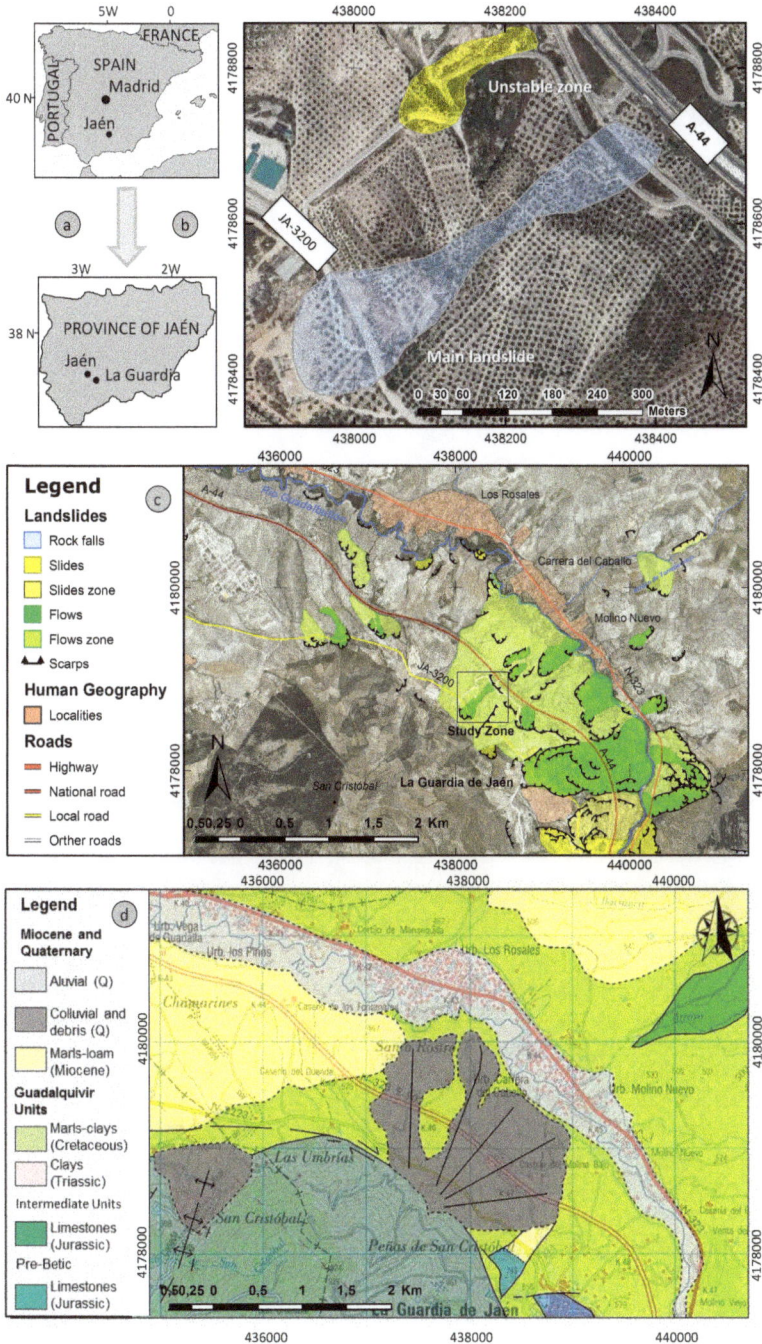

Figure 1. Geographical location and geological setting: (**a**) Geographical location; (**b**) Landslide; (**c**) Landslide evidence in the surrounding area (taken from our own study) [54]; (**d**) Geological setting (taken from the Geological Map of Spain at scale 1:50,000) [59]. Coordinates are in ETRS89-UTM-30.

Figure 2. Landslide and affected roads: (**a**) Landslide general view; (**b**) Landslide head and the road JA-3200 interrupted; (**c**) Road JA-3200 after repair work; (**d**) Temporary road affected by an incipient unstable zone at the north of main landslide; (**e**) Landslide foot and affected access road to highway.

3. Materials and Methods

The methodology was based on digital photogrammetry techniques, used in previous studies by the authors [5,9,16,17], although with some variants due to the use of a UAV [39–41] and the semi-automatic method developed to recognize common points (olive trees) between flights in order to measure displacements. The methodology can be summarized in the following steps that will be explained in Figure 3 and the next sections:

1. Data capture: UAV equipment, mission planning and execution of the flights.
2. Georeferencing and flight orientation.
3. DSM, DTM and orthophotograph generation.
4. Comparison between observation campaigns: Measurement of displacements.

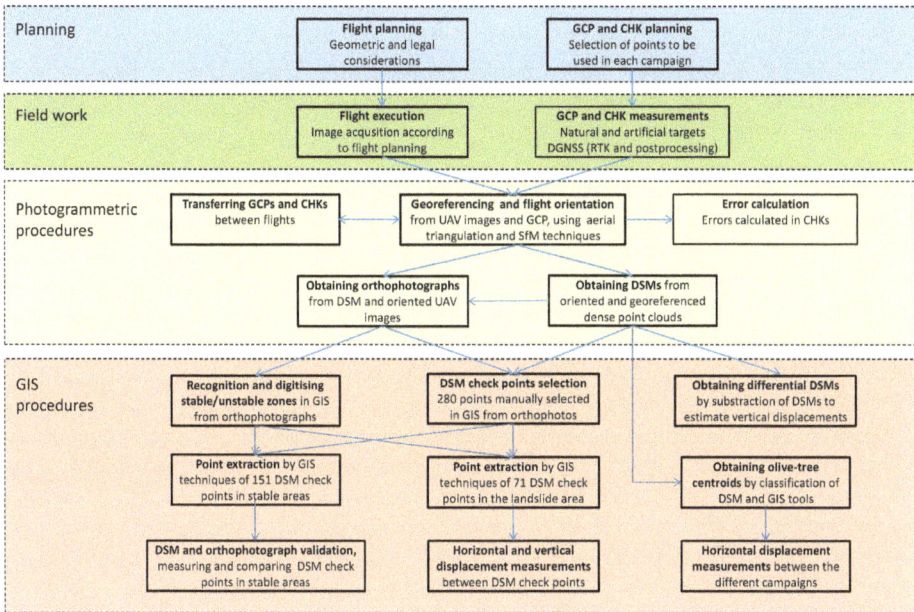

Figure 3. Methodology flow chart.

3.1. Data Capture: Unmanned Aerial Vehicle (UAV) Equipment, Mission Planning and Execution of the Flights

For this study, data from five UAV flights of very high resolution and accuracy undertaken between November 2012 and February 2016 were available. The flights were executed with two UAVs, shown in Figure 4. The flight characteristics are shown in Table 1 and their location and extension in Figure 5.

Figure 4. UAV equipment used in this study: (**a**) AscTec Falcon 8; (**b**) ATyges FV-8 Drone.

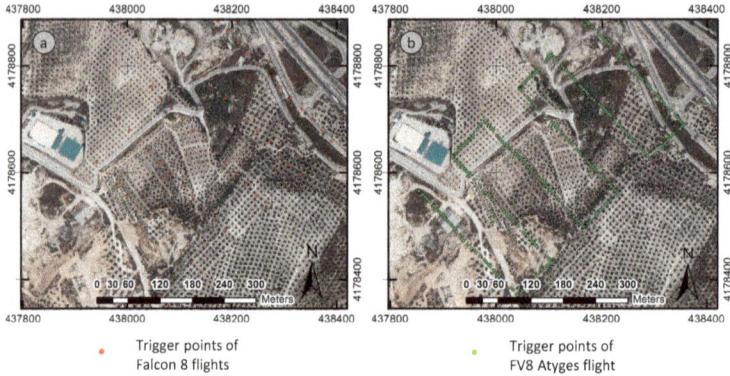

Trigger points of
Falcon 8 flights

Trigger points of
FV8 Atyges flight

Figure 5. Mission planning and location of flights: (**a**) AscTec Falcon 8 flights; (**b**) ATyges FV-8 Drone flight. Coordinates are in ETRS89, UTM-30.

Table 1. Properties of photogrammetric flights.

Date	UAV	Camera	Strips	Photog.	Overlap	Sidelap	GSD (m)
19 November 2012	Falcon 8 Asctec	Sony Nex 5N	6	72	70%	40%	0.029
16 April 2013	Falcon 8 Asctec	Sony Nex 5N	6	72	70%	40%	0.030
16 July 2014	Falcon 8 Asctec	Sony Nex 5N	6	72	70%	40%	0.034
31 July 2015	FV-8 Atyges	Canon G12	6	364	90%	40%	0.037
17 February 2016	Falcon 8 Asctec	Sony Nex 5N	6	72	70%	40%	0.044

The first three flights (2012, 2013 and 2014) were performed with an eight-rotor AscTec Falcon 8 [63] (Figure 4) coupled with GNSS/INS and equipped with a MILC (mirrorless interchangeable lens camera) Sony NEX 5N (APS-C format, 16 Mpx, pixel size 4.9 μm). The lens used was a Sony E 16 mm f/2.8 fixed focal length (24 mm in equivalent 35 mm format). The maximum takeoff weight (MTOW) was 2.3 kg and the speed was in the range of 4.5 to 15 m/s depending on the flight mode (manual, height or GPS mode). The flight times were as long as 20 min. For the fifth flight (2016), the UAV was upgraded to the new Asctec Trinity system with improved performance regarding the redundant inertial measurement unit, flight control and flight dynamic [63].

The fourth campaign (2015) was carried out with another eight-rotor multicopter, ATyges FV-8 Drone [64] (Figure 4) due to the unavailability of the other UAV as it was in repair and maintenance at the time of the field survey. It is formed by a structure in carbon fiber of eight rotors, control electronics by Mikrokopter and a gyro-stabilized bench sensor holder ROLL-TILT. The ATyges FV-8 Drone is capable of autonomous flight for up to 30 min and the payload is up to 2 kg. It is equipped with a compact Canon Powershot G12 camera (CCD sensor 1/1.7, 10 Mpx, pixel size 2 μm). The lens used was a 6.1–30.5 mm f/2.8 zoom lens (28–140 mm in equivalent 35 mm format), although only the wide-angle position was set.

The flying heights were between 100 and 120 m above the terrain, which guaranteed a ground sample distance (GSD) lower than 5 cm. In any case, the flying height needed to be kept at those values since 400 feet (approximately 120 m) is the maximum height limit allowed by the Spanish regulations on the use of Remotely Piloted Aircraft Systems. Due to the moderately steep gradient of the hillslope, the strip flights were planned at different flying heights above the terrain in order to maintain a constant GSD for all the images.

The flight missions were planned with the AscTec Navigator software for the Falcon 8 and the MikroKopter-Tool free software for the ATyges FV-8 drone. The whole study area (500 m × 300 m) was covered with six strips. In the case of the Falcon 8, UAV 72 images were necessary (70% overlap and 40% sidelap), but with the ATyges system, 364 images were taken (90% overlap and 40%

sidelap). The differences were given by the image acquisition mode of the ATyges system, which used a continuous shooting mode with the Canon G12 camera. Figure 5 and Table 1 show these flight plans.

3.2. Georeferencing and Flight Orientation

For orientation of the UAV image blocks, a GCP (ground control points) network was planned following the conventional distribution used for aerial triangulation [18]. Blocks were oriented in the ETRS89/UTM 30N coordinate reference system (CRS). The field-surveyed GCPs were distributed all around the block periphery and additional GCP chains were measured across the flight lines [18].

Due to the problems involved with setting up a permanent network in an unstable area, the GCPs were defined as artificial circular targets and some well-defined natural points (Figure 6). These GCPs were surveyed with differential GNSS methods (DGNSS) in the range of centimeter accuracy with LeicaSystem 1200 and Leica Viva systems. A set of these points was not used in the orientation process but they were reserved as check points (CHKs) to validate the accuracy of this process. The number of GCP and check points are included in Table 2.

Nevertheless, as had been tested in previous studies [5,16,17,39–41], it was possible to transfer points between different campaigns (second-order GCPs), taking one of the surveys as common reference, thereby reducing the field work in a cost-saving technique, if the points are stable, unambiguously identifiable and well distributed. This procedure involves measuring points in the reference flight and calculating their coordinates in the corresponding CRS by means of spatial intersection (PhotoScan employs a collinearity model, as other conventional photogrammetric software), and transferring them as GCPs to the other flights. The use of these points helps to improve the georeferencing of both sets of images in a common reference system, facilitating the comparison between the models and orthoimages generated from them. In this case, a sample of these points has been also used as check points. The error rates of transferred points (both GCPs and check points) are usually higher than the conventional GCPs measured by means of GNSS techniques in the field (Table 2), although the accuracies obtained are enough for the requirements of this work, as will be discussed in the next sections.

This procedure allowed the orientation of the flight of 2013 transferring points from the flight of 2012, in addition to the flight of 2015 transferring points from the flight of 2014. Therefore, no field GNSS surveys were carried out for the flights of 2013 and 2015, in which transferred GCPs and check points were selected only in stable areas. However, it was not possible to transfer enough points to the flights of 2014 and 2016 from the previous flights as GCPs because too many of them changed or disappeared—particularly in the landslide area that dramatically modified the terrain or were difficult to identify. Consequently, GCPs were measured again in the field by means of GNSS for the flights of 2014 and 2016. For all the flights, those with GNSS-based GCPs as well as those with transferred GCPs, the points network followed a similar configuration that ensured the quality of block adjustment [18,26], although it did not always materialize exactly, and there were changes in the location and number of the GCPs in the different flights (Table 2).

The images were processed and aligned by means of dense matching and Structure from Motion (SfM) techniques, which implied the automatic measurement of some 10,000–15,000 common tie points. After the measurement of the GCPs, the final photogrammetric orientations were computed using a global bundle block adjustment [18,25,26]. All processes were implemented with Agisoft PhotoScan software.

The final accuracies of all adjustments are shown in Table 2. RMS (root mean square) pixel is the reprojection error and has values usually lower than the pixel size that informs of the general quality of adjustment. The errors in the check points (CHKs), expressed also as RMS, refer to the residuals calculated in these points after the bundle adjustment. In the case of the flights of 2012, 2014 and 2016, these errors only refer to those field-surveyed (GNSS-based check points). But in the flights of 2013 and 2015, since both the GCPs and the check points were second-order points transferred from a previous flight, the errors in the transferred check points are influenced by the propagation errors (the 2012 flight on the 2013 flight and the 2014 flight on the 2015 flight). These propagation errors have been added in Table 2. The expressions for calculating these propagation errors are:

$$PE_{XY}(2013/2015) = (RMS_{XY}(2012/2014)^2 + RMS_{XY}(2013/2015)^2)^{0.5} \tag{1}$$

$$PE_Z(2013/2015) = (RMS_Z(2012/2014)^2 + RMS_Z(2013/2015)^2)^{0.5} \tag{2}$$

Figure 6. (a) GCP targets and their measurement by DGNSS; (b) Detail of the targets.

Table 2. Orientation errors.

Date	GCP/CHK Number	Tie Point Number	RMS Pixel	CHK RMS (m)		Propag. Error (m)	
				RMS$_{XY}$	RMS$_Z$	PE$_{XY}$	PE$_Z$
19 November 2012	11/5	17,795	0.75	0.017	0.015	-	-
16 April 2013	12/5	14,322	0.39	0.036	0.020	0.040	0.025
16 July 2014	8/4	14,105	0.86	0.030	0.032	-	-
31 July 2015	11/5	17,591	0.50	0.039	0.029	0.049	0.043
17 February 2016	13/5	9045	1.07	0.034	0.025	-	-

Errors at the check points do not exceed 0.04 m in XY and 0.03 in Z in practically any case, as they are usually higher in flights with transferred points. Thus the propagation errors have values lower than 0.05 in XY and Z. Another quality control from derived products (DSMs and orthophotographs) was performed calculating the relative displacements in 151 stable points (DSM check points) distributed throughout the overall zone. The procedures are explained in the next section (DSM/DTM and orthophotograph generation) and the results are presented in the corresponding section.

3.3. Digital Surface Model (DSM), Digital Terrain Model (DTM) and Orthophotograph Generation

The Digital Surface Models (DSMs) of all flights were generated from a densification of the initial sparse point cloud (using the dense point cloud tool of PhotoScan) after the block orientation. These dense point clouds involved some 15,000,000 points as average for all flights. Then orthoimages for each campaign were also generated. Finally, both products were exported as raster files (TIFF) to be incorporated into GIS analysis.

The resolution of DSMs was 0.10 m, given that the resolution of the orthoimages was 0.05 m and the GSD even lower (0.029–0.044 m). In conventional photogrammetry the resolution of models are several times that the GSD of the images and the resolution of orthoimages generated from them. But in this case, given the new global dense matching techniques used by the current SfM photogrammetric software and taking into account the dimensions of the area and the characteristics of the flights, the pixel size of DSMs are twice the orthoimage.

Unlike previous studies [39], in this paper the landslide evolution was monitored with DSMs instead of DTMs, because the study area presented a high density of vegetation in some sectors with grass, scrubs and bushes that had grown intensely in these last years. Under this circumstance the DTMs obtained from DSMs, using the tools of point clouds classification and filtering of PhotoScan, had only partially removed the olive trees. Stereo-model edition using photogrammetric workstations and the corresponding software was not performed because it did not ensure good results—it could not eliminate the grass and dense scrub—and would have been time consuming.

The orthophotographs, shown in Figure 7, were obtained with a resolution of 5 cm (0.05 m), higher than GSD (0.029–0.044 m). Therefore, taking into account that the XY errors and uncertainties calculated from the flight orientation process at check points—and at the DSM check points as will be presented and discussed—are of the same order of magnitude than the resolution, the measurements made on the orthoimages can be considered as reliable.

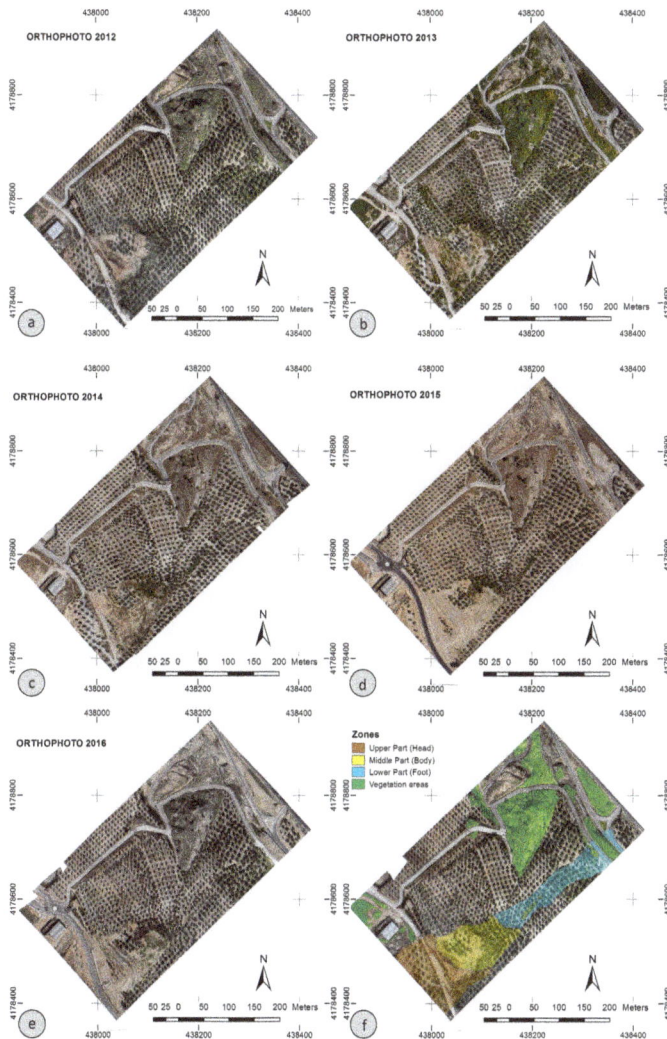

Figure 7. Orthophotographs corresponding to the photogrammetric UAV flights: (**a**) 2012 flight; (**b**) 2013 flight; (**c**) 2014 flight; (**d**) 2015 flight; (**e**) 2016 flight; (**f**) Zonation of the studied landslide and the surrounding area. Coordinates are in ETRS89, UTM-30.

After all DSMs and orthophotographs were available, a network of 280 well-defined points visible in all orthoimages was established (Figure 8) as DSM check points. First of all, those points located in stable areas (151) were selected as check points in order to analyze the relative adjustment between all DSMs and orthophotographs and also to validate the errors at the check points (GNSS-based and transferred) of the orientation process given in Table 2. The results of this quality control are shown in Table 3. Secondly, the points located in unstable areas (93), and specifically in the main landslide (71) were used to measure horizontal and vertical displacements. All these DSM check points (stable and unstable) were selected manually and placed carefully on the bare ground, which was easy because they were located in olive groves—in which the vegetation is usually removed by farming tracks and roads. Points located in areas covered by vegetation (36) have finally been excluded from the analysis.

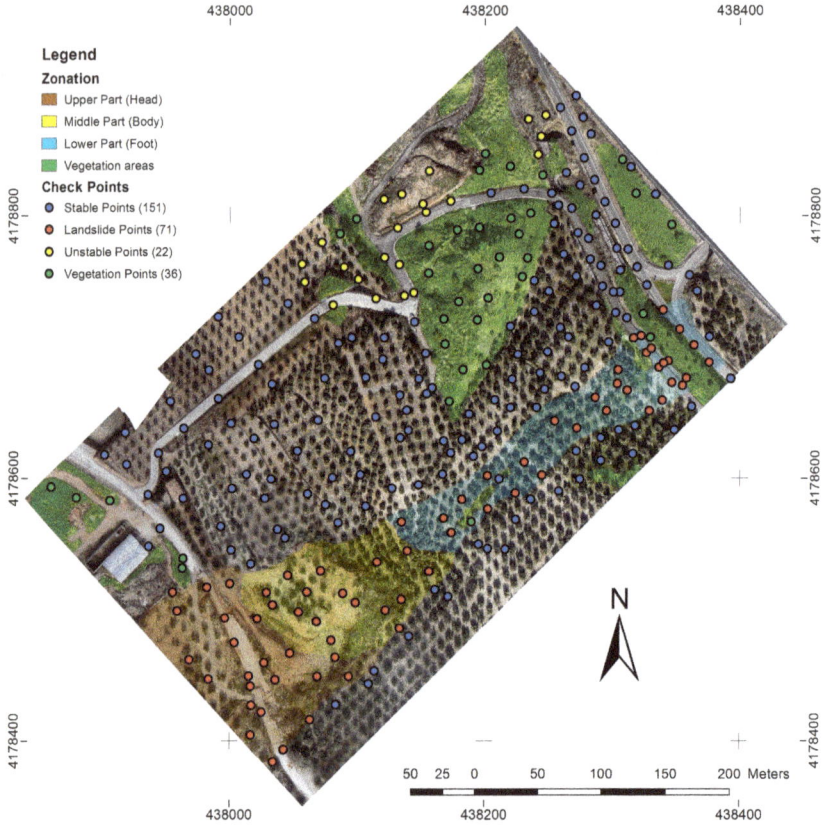

Figure 8. Distribution of DSM check points in the study area. Coordinates are in ETRS89, UTM-30.

Table 3. Relative errors between flights. Displacements calculated between stable DSM check points.

XY Error	Reference Flight											
	19 November 2012			16 April 2013			16 July 2014			31 July 2015		
Flight to Compare	M [1]	SD [2]	R [3]	M	SD	R	M	SD	R	M	SD	R
16 Aril 2013	0.010	0.083	0.071									
16 July 2014	0.014	0.095	0.078	0.008	0.084	0.071						
31 July 2015	0.006	0.097	0.067	0.008	0.098	0.071	0.015	0.096	0.089			
17 February 2016	0.007	0.104	0.069	0.017	0.108	0.092	0.019	0.112	0.095	0.012	0.095	0.075

Table 3. *Cont.*

Z Error	Reference flight											
Flight to Compare	19 November 2012			16 April 2013			16 July 2014			31 July 2015		
	M	SD	R	M	SD	R	M	SD	R	M	SD	R
16 Aril 2013	0.038	0.085	0.079									
16 July 2014	−0.010	0.094	0.069	−0.048	0.096	0.088						
31 July 2015	−0.029	0.139	0.104	−0.066	0.158	0.128	−0.019	0.125	0.092			
17 February 2016	0.013	0.094	0.086	−0.033	0.177	0.102	−0.023	0.090	0.079	0.046	0.140	0.113

[1] M: Mean error; [2] SD: Standard Deviation; [3] R: RMS error. All the error data are in meters (m).

3.4. Comparison between Observation Campaigns: Measurement of Displacements

The comparison between the different epochs has been carried out in this paper based on the products described below:

- From the DSMs the vertical displacements were obtained by subtracting the pixel values between two sequential models (differential models).
- From the 3D coordinates of the DSM check points extracted manually from the orthophotographs and DSMs, the vertical and horizontal displacements between points were calculated.
- From the plane coordinates of the centroid of olive trees, the horizontal displacements between these features were calculated.

The comparisons between DSMs (differential models) were performed with ArcGIS and the open source software QGIS. These comparisons mainly allow the visual identification of landslide features (head and depletion area, main and secondary scarp, body, toe and accumulation area) and the estimation of vertical displacements in certain areas not covered by vegetation. But given the difficulty in obtaining a true DTM, mean values of displacements and the volumes involved were not calculated. More precise calculations of vertical and horizontal displacements were made from the coordinate measurements at DSM check points in the landslide area. These points were selected manually and located carefully on the bare ground.

Furthermore, although some techniques of automatic recognition of features based on image analysis were tested, they did not offer satisfactory results because of the dramatic changes in the orthophotographs corresponding to the different epochs. These changes were caused by different lighting conditions, deep shadows, the landslide itself, repair works and the changing vegetation cover. In some limited areas, such as roads, some results were obtained but they were considered inconclusive for this paper. In this regard, this study deals with another approach based on the automatic recognition of olive trees that cover most of the area. These elements are sufficiently distinguishable from the background to provide reliable results, but at the same time have the disadvantage of suffering changes such as growing, pruning or removal (especially in the landslide area). Being a methodological approach, the analysis was only performed between the first (2012) and the last epoch (2016), in which the displacements had a larger magnitude.

The approach uses the dense point clouds generated with PhotoScan, which are classified within the LAStools software suite. The classification process is carried out semi-automatically. The first step was the automatic classification of the dense points cloud with the Ultra Quality option in the LASground tool, using the default parameters for forest or hills areas. Then a manual edition is performed to refine the classified data. These classified point clouds are transformed into normalized clouds, which means that the points classified as terrain appear in a horizontal plane at a constant level (0) and only the points classified as non-terrain have an assigned height. Then the normalized clouds are transformed to raster images, so image processing filters can be applied in order to extract information.

The normalized raster images are segmented using a threshold value of 0.3 m. This corrects classification errors and so all elements with Z higher than 0.3 m are assigned with value 1 and the other elements take value 0. Then the centroids of the olive trees are extracted semi-automatically using GIS tools: the extracted elements are vectorized so polygons are obtained; then those polygons too small to be considered as olive trees are filtered (a threshold of 2 m^2 is used); finally, the centroids of the different polygons are calculated.

This operation is performed for the point clouds corresponding to the initial (2012) and final (2016) campaigns. In order to reference the olive trees of different years and so calculate the displacements between them, it is necessary that the olive trees present the same identification in the two campaigns analyzed. This is accomplished by building buffers around the centroids, merging those buffer layers for the same olive tree in different campaigns and renaming them, thus allowing the centroids of both campaigns to be named with the same identifier.

The last phase of the process, the calculation of displacement vectors, is performed using a JAVA application developed by the authors, in which both the displacement module and direction are calculated for each of the points considered in the analysis. In total, 1100 olive trees were identified, those that remained throughout the whole period, although some of them underwent shape changes in the last campaigns.

4. Results

The results are analyzed in different sections, which essentially coincide with the different methodologies followed for comparison between campaigns: analysis of differential models, analysis of displacements at selected points and analysis of the horizontal displacements of the centroids of olive trees. At the beginning, two brief sections are presented about the results of displacements measured at stable DSM check points), and the landslide recognition and zonation.

4.1. Displacements between Models at Stable DSM Check Points

The results of displacements between models obtained at the stable DSM check points are shown in Table 3. It can be observed that the mean errors are below ±0.02 m in XY (in the order of the GSD values) and below ±0.06 m in Z. Moreover, the RMS errors—employed in most studies [26]—are below 0.10 m in XY and 0.13 m in Z. Finally, the standard deviation (SD) is usually less than 0.10 m in XY and Z, although in this last case some values reach 0.15 m.

4.2. Landslide Recognition and Zonation

A visual inspection and photo-interpretation of the models in the photogrammetric software, and the orthoimages (Figure 8) and differential models (Figure 9) in the GIS allowed the clear recognition and delineation of a landslide in the south sector. This inspection has been helped by field observations. This landslide covered the hillslope from the upper part, affecting and ruining the A-3200 road—which connects the town of Jaén with the village of La Guardia—to the lowest part in the vicinity of the A-44 highway, affecting the access way from the highway to the village of La Guardia and another road constructed to avoid traffic interruption of the A-3200 road (Figure 2).

In the landslide mass, a head area in the upper part—very chaotic and with several scarps—and a foot area in the lower part could be clearly identified [65]. A transitional or body area in the middle part was also distinguished. This landslide zonation is shown in Figure 7e.

Moreover, towards the northern part of the area, a second unstable zone is identified, corresponding to an incipient landslide in which the deformation is lower and the roads only suffer cracks and bumps (Figure 2d). The remaining areas are initially stable but these zones where there are changes in the vegetation state are also highlighted.

4.3. Differential Models

Differential models calculated from the DSMs are shown in Figure 9. In the first two differential models (2012–2013 and 2013–2014), ignoring the effect of the zones of grass and scrub vegetation located in the central-northern sector, areas where the height of the ground surface descends and areas where the height of the ground surface ascends were observed. The first areas predominated in the upper part of the slope (zone of depletion or head) associated with the main scarp and also with secondary scarps; the latter dominated the middle and lower part of the slope, associated with the landslide foot (zone of accumulation), but also appearing at the head, alternating with depletion zones in the area where the secondary scarps were formed. Therefore, in these first periods, descents up to 0.8 m and also ascents up to 0.3 m, in alternating bands, were observed in the upper part or head zone; descents up to 0.4 m and ascents up to 1.3 m were observed in the middle part, predominating the ascents toward the lower part; finally, descents up to 0.3 m and ascents up to 0.7 m were observed in the foot area, reaching the maximum values of both cases near the access roads.

Since the 2014–2015 period, the landslide dynamic has changed dramatically at the upper and even the middle part of the hillslope, where large vertical displacements of the terrain surface were observed, both in terms of depletion and uplift zones. In the lower part of the landslide, the situation is completely different, and only smooth terrain ascents were observed. Thus, in the period of 2014–2015, we were able to observe descents of up to 1.2 m and ascents up to 1.5 m in the upper part, above and below the new road A-3200; descents up to 1.0 m and ascents up to 1.2 m in the middle part, with the maximum absolute values in this case towards the upper part; and descents up to 0.15 m and ascents up to 0.2 m in the lower part, with the maximum values also in the zone near the access roads. Finally, in the last period studied (2015–2016), scarcely any changes occurred on the surface of the ground in the landslide area and its surroundings.

4.4. Analysis of Displacements at Unstable DSM Check Points

This analysis is shown graphically in a map of displacement vectors in the Figure 10 and it is summarized in Table 4 for the different parts of the main landslide area. The mobilized and stable areas can be clearly distinguished on the map of displacement vectors.

In this case, the largest displacements towards the northeast are located in the main landslide area, while the remaining zone presents much smaller displacements without a clear direction. Also, in the northern sector, an incipient unstable zone is observed, although the displacements are not so large.

Table 4 shows the mean values (absolute and rates) of the displacement vector modules in the vertical and horizontal component, as well as the mean direction and the mean length of the resultant vector (MLRV). MLRV is a measure of circular dispersion, the directions being more uniform when it is close to 1 and more dispersed when it is far from 1 [66]. The horizontal displacements observed in Table 4 are an order of magnitude higher than those in the vertical displacements. From the absolute values of displacements, the rates of displacements are calculated, dividing the absolute values by the period of time between campaigns (expressed in years).

The vertical displacements measured at landslide DSM check points were usually negative, so the points tend to descend as the movement progresses. Thus the descent rates in the first period reached 1.428 m/year in the upper part of the hillslope, 1.920 m/year in the middle part and 0.129 m/year at the lower and less steep part. These descent rates were of 0.338 m/year, 0.732 m/year and 0.160 m/year at the upper, middle and lower parts respectively, in the second period. In the third period (2014–2015), most points in the upper and middle part disappeared, while in the lower part, the vertical displacements of the points were not significant. The same happened in the fourth period (2015–2016) in this case for all the sectors of the landslide.

The analysis of horizontal displacements shows that in the first period considered (2012–2013), large displacement rates higher than 4 m/year and 13 m/year m occurred, respectively, in the upper and middle areas, while in the lower part, the rate was about 1.4 m/year. In the second period (2013–2014), horizontal displacement rates decreased to 1.146 m/year in the upper part, 3.304 m/year in the middle part and 0.778 m/year in the lower part. In the third period (2014–2015), the points in the upper and middle part disappeared again, whereas in the lower part, horizontal displacements with rates of about 0.268 m/year remained. Finally, in the period 2015–2016, the horizontal displacements were practically null or not significant in all parts of the hillslope.

The mean direction of displacement vectors is northeast (NE). This direction remained practically constant (between 048° and 057°) throughout the first (2012–2013) and second period (2013–2014).

The mean length of the resultant vector presented values close to 1 in the first and second periods, although in the upper part, values of around 0.8 were observed. These parameters decreased to below 0.5 in the third (2014–2015) and fourth periods (2015–2016).

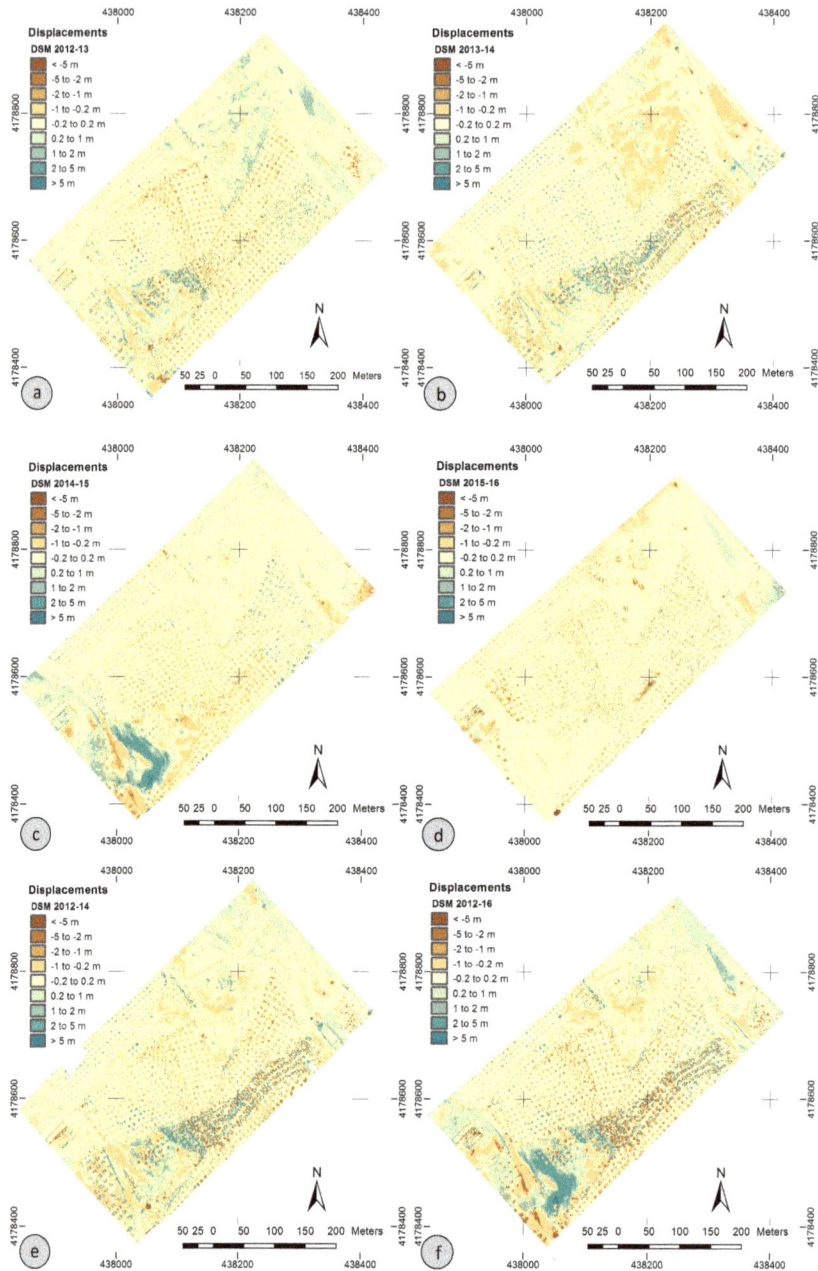

Figure 9. Differential models calculated from DSM: (**a**) 2012–2013; (**b**) 2013–2014; (**c**) 2014–2015; (**d**) 2015–2016; (**e**) 2012–2014; (**f**) 2012–2016. Coordinates are in ETRS89, UTM-30.

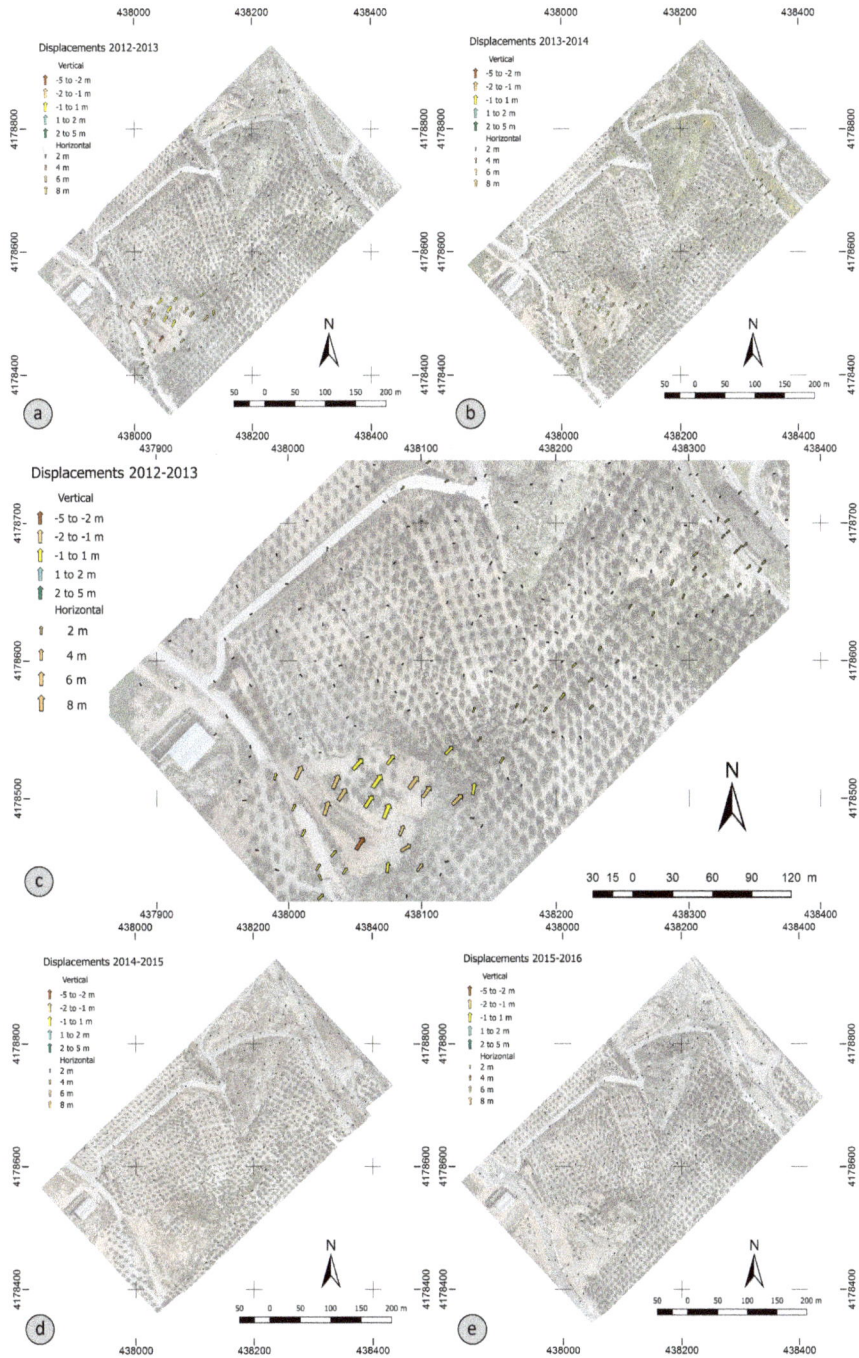

Figure 10. Displacements at DSM check points: (**a**) 2012–2013; (**b**) 2013–2014; (**c**) Detail of 2012–2013 (**a**); (**d**) 2014–2015; (**e**) 2015–2016. Coordinates are in ETRS89, UTM-30.

Table 4. Vertical and horizontal displacements calculated at DSM check points in the landslide area.

	Vertical Displacements					
Periods	Upper Part		Middle Part		Lower Part	
	Absolute [1]	Rate [2]	Absolute	Rate	Absolute	Rate
2012–2013	−0.595	−1.428	−0.794	−1.906	−0.054	−0.129
2013–2014	−0.424	−0.338	−0.919	−0.732	−0.201	−0.160
2014–2015	−0.027	-0.025	−0.112	−0.106	−0.103	−0.097
2015–2016	−0.003	−0.005	0.052	0.093	0.049	0.089
	Horizontal displacements					
Periods	Upper Part		Middle Part		Lower Part	
	Absolute	Rate	Absolute	Rate	Absolute	Rate
2012–2013	1.779	4.273	5.457	13.104	0.576	1.383
2013–2014	1.438	1.146	4.146	3.304	0.976	0.778
2014–2015	0.270	0.255	0.292	0.276	0.284	0.268
2015–2016	0.060	0.108	0.109	0.196	0.085	0.152
	Directions					
Periods	Upper Part		Middle Part		Lower Part	
	Mean	MLRV [3]	Mean	MLRV	Mean	MLRV
2012–2013	48.477	0.856	53.112	0.955	47.813	0.963
2013–2014	57.468	0.739	54.358	0.980	48.892	0.969
2014–2015	3.366	0.253	50.511	0.361	58.493	0.520
2015–2016	−48.089	0.328	17.281	0.354	25.473	0.370

[1] Data of displacements are in meters (m); [2] Data of displacement rates are in m/year; [3] MLRV: Mean length of the resultant vector.

4.5. Analysis of the Horizontal Displacements of the Olive Tree Centroids

This analysis, based on a semi-automatic approach, also allows a mapping of the deformation, although restricted only to the horizontal component of displacement. Figure 11 shows a map of deformation based on this approach where the mobilized area is clearly differentiated from the surrounding stable area. The mean displacement modules and directions of the period 2012–2016 are shown in Table 5. The larger horizontal displacements were observed in the middle part (2.832 m), followed by the lower part (2.452 m) and the upper part (1.336 m), and the shorter values were in the stable area (0.719 m). The mean direction was towards NE in the different parts of the landslide and towards the SSW in the stable area. Finally, the MLRV was at its maximum in the lower part (0.907 m), the middle part (0.815 m) and the upper part (0.716 m) and minimum in the stable area (0.176 m).

Table 5. Horizontal displacements calculated from olive centroids in the stable and landslide area.

	Stable Area		Upper Part		Middle Part		Lower Part	
Horizontal	Absol. [1]	Rate [2]	Absol.	Rate	Absol.	Rate	Absol.	Rate
displacements	0.719	0.196	1.366	0.373	2.832	0.772	2.452	0.669
Directions	Mean	MLRV [3]	Mean	MLRV	Mean	MLRV	Mean	MLRV
	201.038	0.176	48.687	0.716	59.151	0.815	53.180	0.907

[1] Data of displacements are in meters (m). [2] Data of displacement rates are in m/year. [3] MLRV: Mean length of the resultant vector.

However, since olive trees are irregular features that can change their physical appearance because of pruning, growth, etc., different types of filters were tested in order to identify displacements due only to landsliding. Firstly, several local statistics of the displacement vectors (mean module, mean direction and mean length of resultant vector) were calculated in a buffer of 15 m around each centroid. It can be noticed that in mobilized areas the vector modules have clearly higher values than in stable areas. In fact, in stable areas, the displacements due to changes in the shape of the olive treetops are less than

1 m, while in landslide, the displacement values are usually up to 2 m and higher. Moreover, in the mobilized areas, the vectors have a more uniform direction (NE) and the mean length of the resultant vector has higher values (close to 1) than in stable areas.

The module and direction of the displacement vectors were therefore selected as filters and, subsequently, those vectors with modules lower than 1 m were eliminated. Moreover, those vectors with a direction value of ±60° with respect to the mean direction of the landslide displacement (rounded to 050°), which means directions between −10° (350°) and 110°, were also eliminated. The result of the application of these filters is also shown in Figure 11.

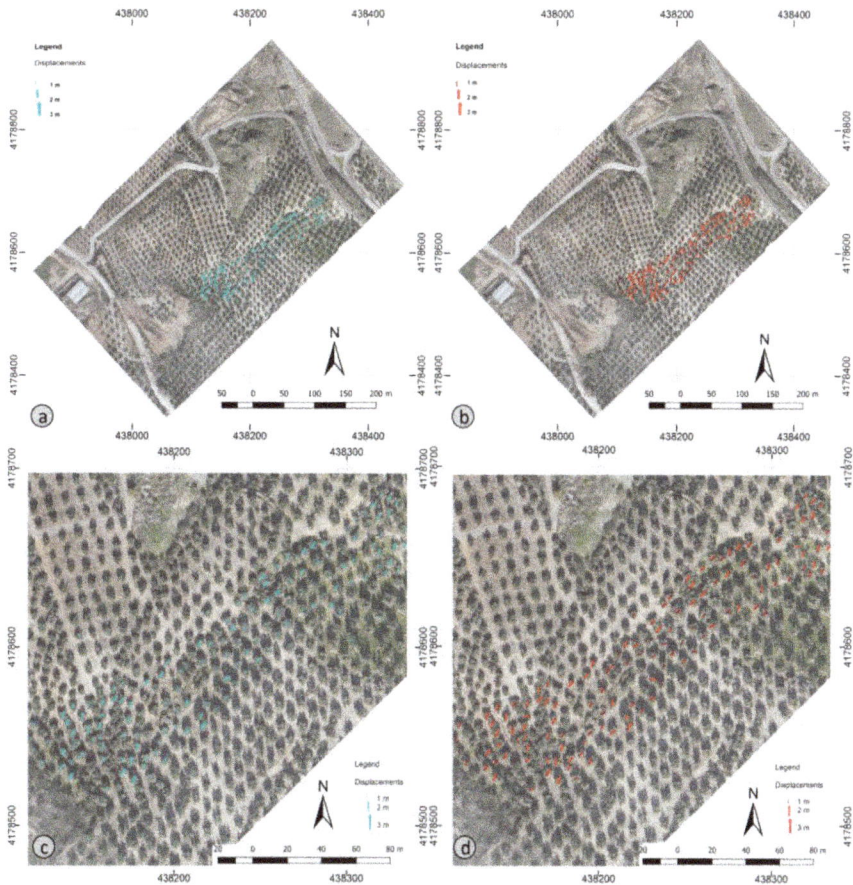

Figure 11. Displacement between olive trees: (**a**) All the olive trees extracted from DSMs; (**b**) Olive trees filtered by module and direction; (**c**) Detail of (**a**); (**d**) Detail of (**b**). Coordinates are in ETRS89, UTM-30.

5. Discussion

The discussion deals with different aspects of the previous sections. Firstly, some considerations about the limits of accuracy of the study will be presented, based on the results of the orientation and georeferencing process and the analysis of check points. Secondly, the results of the analysis of differential models, analysis of displacements at selected DSM check points and analysis of the horizontal displacements of the olive tree centroids will be discussed. Then a brief description of qualitative aspects extracted from these analyses will be given. The section will finish with some conclusions regarding the relationships between the trigger mechanism of the landslide (rainfalls) and its kinematics.

5.1. Accuracies, Errors and Uncertainties

The RMS errors referring to the GNSS-based and transferred check points after the bundle adjustment do not exceed 0.04 m in XY and 0.03 m in Z in practically any case (Table 2), neither in the flights oriented by GNSS-based GCPs (2012, 2014 and 2016) nor the flights oriented by second-order points (2013 and 2015) transferred from a previous flight (2012 and 2014, respectively). In the last cases, the propagation errors are also taken into account showing values lower than 0.05 m. These errors are of the same order of magnitude than the image resolution, as it has been described in previous comparable studies [26,32,34,38,42,44,45] with similar properties (equipment, flight altitude and resolution).

From the displacements measured between the 151 DSM check points placed on the bare ground in stable zones, the mean errors, RMS errors and standard deviation (SD) were calculated (Table 3). RMS errors also show values of the same order of magnitude of those calculated in this work at GNSS-based and transferred check points, and those found in the aforementioned studies with similar properties. In these studies, the RMS calculated in GNSS-based check points are in the range of 0.05–0.10 m [32,38,45] or even higher [34]. In a lower resolution study [31], a value of 0.50 m is found in the comparison of UAV and LiDAR DSMs, where DSM check points are used in the same way as in this study. Other studies, also of lower resolution, in which the errors in ortophotographs and DSMs are around 2–3 times the resolution of images, agree with these observations [28,29].

Furthermore, it can be observed that the mean errors at the DSM check points are below ±0.02 m in XY (in the order of the GSD value) and below ±0.06 m in Z, which confirms the good general adjustment or centering of the models, without systematic horizontal or vertical displacements; however, there can be local misalignments as the higher values of RMS and SD reveal. In this way, the standard deviation, which is actually a measure of the uncertainty of the data, is usually lower than ±0.10 m in XY and Z, although in this last case, some values reach ±0.15 m. It can be established that the general adjustment models between flights are very fine (better than 0.06 m in XY and Z), while the uncertainty, and so the accuracy, is 0.10 cm in XY and 0.15 cm in Z. Thus higher displacement values with respect to these thresholds allow us to deduce changes in the ground surface, while lower values are inconclusive.

5.2. Differential Models

Differential models (from DTMS or DSMs) are calculated in many studies in order to study the landslide evolution [30,31,34–36,39–41,45]. In this study, DTMs are obtained by classifying and filtering DSMs, but only olive trees could be partially eliminated, while the grass and scrubs remained. In these conditions, DTMs and the differential DTMs do not improve significantly in respect to DSM and differential DSMs. Therefore, the use of differential DTMs does not allow more precise evaluations of displacements and volumes. Thus differential DSMs are used in this work to make visual observations of the landslide features and evolution and to calculate approximate and local vertical displacements by means the GIS tools in some areas not covered by vegetation.

Differential DSMs between 2012 and 2014 allowed the clear delineation of the main landslide, with notable changes in the terrain surface, regarding the stable zone in which these changes were not significant. Moreover, the depletion area in the upper part (head) of the landslide, where descents of the terrain surface predominated, could be distinguished from the accumulation area in the middle (body) or lower part (foot), where ascents of the terrain surface predominated. In reality, the ascent of terrain surface in these zones was not due to an uplift of the terrain surface but to an advance of the landslide mass in the accumulation area (body and toe). This type of evolution has been observed in other studies [6–8,10–15,31,34–41,45] with different types of landslides and it is related to displacements of the terrain surface in both the vertical and in the horizontal direction.

The absolute values of vertical displacements were similar in the upper and middle part where the maximum deformation concentrated. In the head, the alternating bands of terrain descents and ascents were due to the formation of secondary scarps and the mass advance. This area also presented higher slope-angles, alternating between very steep scarps and flatter zones, although always inclined downslope without rotational evidences. In the middle part or landslide body, the maximum ascents of terrain surface could be observed, evidencing an accumulation zone where the landslide foot began. This zone also presented higher slope-angles, as corresponds to the upper part of the foot. In this point, there is a narrowing of the landslide shape due to the original relief—in part modeled by older landslides—with zones of higher slope-angles at the flanks and a lower inclined area between them throughout which the landslide was able to advance. Nevertheless, this morphology could have stopped the landslide partially, giving rise to the formation of this steep foot at the middle of the hillslope, although the movement continued downslope. So, in the lower part, the foot had lower slope-angles and could open moderately after the narrowing. The observed ascents of the terrain surface were lower, as corresponds to a smooth foot where there is a greater horizontal than vertical development. In fact, in this zone, the vertical displacements were of the order of Z uncertainty, but in some sectors—such as the affected roads—an advance of the mass and a certain uplift of these features, due to the push on a rigid structure, could be clearly observed.

The landslide kinematics changed along the full time period studied. The largest vertical displacement rates thus occurred during the first period (2012–2013), slowing in the second period (2013–2014), though retaining the same pattern. In the third period (2014–2015), the evolution of the upper part of the hillslope underwent great changes as the result of the repair works for stabilization and construction of the new road JA-3200, while in the middle and lower part, a residual movement still remained, although the values in the order of uncertainty are not conclusive. In the last period (2015–2016), the landslide finally stopped and significant changes in the ground surface could not be observed.

Very interesting are the effects of the displacements of olive trees observed in the differential DSMs because of the landslide movement. Thus, in certain areas, some strips with surface uplifts and descents are observed. The vertical displacements range between 2 and 3 m (approximately the height of the olive trees) and they do not correspond to vertical movements but rather horizontal displacements of the tree lines.

5.3. Analysis of Displacement at Unstable DSM Check Points

The displacement vectors both in the vertical and the horizontal components allowed us to distinguish between mobilized and stable zones, as has been described in previous studies [6,8,15,34,35,39–41]. In general terms, the largest vector modules and the maintenance of uniform directions are indicative of those zones where movements occur.

In the study area, two unstable zones were identified, one corresponding to a more developed landslide in the southern part, with larger displacements, and the other one in the northern part, in an incipient state with shorter displacements.

Focusing on the main landslide, it could be observed that the horizontal displacements were an order of magnitude higher than the vertical displacements, which is consistent with the kinematics of a flow type movement where the horizontal component is greater than the vertical [55,56,67,68]. Moreover, the usually negative values of vertical displacements show that the points tend to descend as the landslide progress, unlike what happened in differential models where the observed uplifts are due to the mass advance and accumulation.

As in the analysis of differential models, two different stages could be observed before and after the stabilization works. The maximum descent rates were reached in the first period (2012–2013), being larger in the upper (head) and middle part (body) than in the lower part (foot), where the horizontal component of the movement predominated, according to a typical flow-type movement with formation of scarps in the head. According to the average data, the vertical displacement rates of

DSM points slowed in the second period (2013–2014), except in the foot, which indicates transmission of the movement downslope, after the movement in the head and body was stopped by the narrowing and also by the rainfall decrease. In both periods, the absolute values of vertical displacements were significantly higher than those values found in the stable zones (Table 3), where mean values did not exceed 0.06 m and the RMS and SD were lower than 0.10–0.15 m. In both periods, the movement can be catalogued as slow, according to the suggested classification of WP/WLI [69] and the available data (one campaign per year), although it is possible that it reached moderate velocities in some rainy event, like in some movements analyzed in surrounding zones [39–41].

In the third period (2014–2015), the disappearance of points in the upper and middle part was due to the repair works (the landscape changed completely). Meanwhile, in the lower part, the vertical displacements of the points were not significant, as the absolute values were below the accuracy limit (0.15 m) estimated from the standard deviation of stable points. The same occurred in the fourth period (2015–2016) for the whole landslide, as a consequence of stabilization works.

Similar considerations could be established from the analysis of horizontal displacements. The largest displacement rates of several meters per year were reached in the first period (2012–2013) in the upper (head) and middle (body) part, where the larger deformations occur, and shorter (about 1.5 m/year) in the lower part (foot). The velocity was from slow to even moderate (in the upper parts). Again, the more active zones were the head, the body and even the starting point of the foot, where the slope-angles were higher, while in the lower part of the foot—after the narrowing—the slope angles were lower and the movement more residual. The rates slowed in the second period (2013–2014) to values close to 1 m/year (slow velocity). In any case, in both periods, the mean absolute horizontal displacements were much larger than those calculated in stable areas (0.02 m), the orthoimages resolution (0.05 m) and the accuracy limit (0.10 m). In the third period (2014–2015), ignoring the points that disappeared in the upper and middle part, residual displacement rates in the lower part could be observed. As in the previous analyses, the horizontal movements were practically null in all parts of the hillslope in the fourth period.

An average direction of the horizontal displacement towards downslope (NE), and high values of the mean length of resultant vector—low dispersion of directions—were found in the landslide area in the first periods (2012–2014). This low dispersion of the directions of displacement vector suggests a fairly coherent movement, despite being a flow in which the movements tend to have a chaotic behavior [55,56,67]. It confirms the earth-flow typology in which the movement is more coherent than other flow types, including mud flows or debris flows. Nevertheless, the higher dispersion observed at the head suggests a more chaotic pattern in this area of greater deformation.

On the contrary, directions different to the general trend of the hillslope and lower values of the mean length resultant vector—high dispersion of the directions—appeared in stable areas. In these areas, the displacement should really be null and without a set direction. So the short modules and the high dispersion in the directions found in these stable areas seems due to the limit of precision in the positioning of measurement points and not to a real displacement of the terrain. The same can be said about the landslide area when the movement stopped (since 2014).

5.4. Analysis of the Horizontal Displacements of the Olive Tree Centroids

This analysis, based on a semi-automatic approach that allows a mapping of the horizontal deformation and identification of potential unstable areas, is original of this study. However, the methodology has its limitations because the elements selected to evaluate the displacements—the olive trees—are irregular features that can change their physical appearance because of pruning, growth, etc. Some approaches based on techniques of expertise classification of point clouds to detect objects such as trees and their parts (trunks, branches, etc.) [70] or taking into account properties of DSMs (heights, slopes . . .) and the orthoimages (radiometric bands) could refine the methodology. In this work, however, we have followed a simple approach based on applying filters in order to distinguish true displacements due only to landsliding from other possible changes.

The mean displacement modules and directions calculated for the whole period (Table 5) show differences between stable areas and the different parts of the landslide. The maximum horizontal displacements are observed in the middle part followed by the lower and the upper part, coinciding roughly with the analysis of the DSM check points. However, the displacements between olive centroids are not as large as between check points, especially in the upper and middle parts, because the olives in these areas of maximum deformation have disappeared (in fact, in the upper part, there are only six olives and the measurements are not significant). In stable areas, the displacements are shorter than in the landslide areas, but higher than those observed in the DSM check points, due probably to the irregular shape of olive trees and the low accuracy in the centroid position. It agrees with the mean values of directions of displacement and MLRV; the directions in the landslide area are quite uniform (high values of MLRV) and, towards NE (downslope), they coincide with the values observed in the DSM check points. However, the stable areas are very irregular (low values of MLRV) and even upslope. In these cases, the irregularity is larger than in the DSM check points, due to less accuracy in the positioning of olive tree centroids.

Thus, in order to distinguish the unstable areas from the stable ones, different filters were tested based on local statistics of the displacement vectors (mean module, mean direction and mean length of resultant vector) calculated in a buffer of 15 m around each olive centroid. The mean module and direction of the displacement filters operate efficiently and so those vectors with a module lower than 1 m and with a direction value of ±60° with respect to the mean direction of the landslide displacement were eliminated. The remaining vectors highlight the main landslide area in the south part and also the incipient unstable area in the north.

With respect to the other statistical parameter, the mean length of the resultant vector also shows the mobilized areas clearly, but this coefficient presents some problems since its distribution is highly irregular near the boundary between stable and unstable areas. Moreover, in areas affected by flow type landslides with a chaotic behavior, the use as filter of the mean length of the resultant vector could lead to error.

It can be thus concluded that the application of these filters results in the delimitation of the movement (Figure 11), thereby validating this semi-automatic recognition method that can be applied to the detection of movements in areas of olive groves or other groves that follow a regular pattern.

5.5. Qualitative Characterization of the Landslide

A visual inspection in the field as well as the previous analyses show a slope movement of earth-flow type [55,56,67]. This type of slope movement is defined as "rapid or slower, intermittent flow-like movement of plastic, clayey soil, facilitated by a combination of sliding along multiple discrete shear surfaces, and internal shear strains (with) long periods of relative dormancy alternating with more rapid surges" [56]. The velocity (catalogued in this case as slow to moderate), the plastic marls and clays of the Guadalquivir Units [54,58–62] outcropping in the study area and the slope-angles (ranging from 5° to 20°) are characteristic of this landslide typology. Additionally, the water sources from the carbonate materials of the surrounding reliefs—and also by the colluvial deposits—during the rainy periods contributed to the landslide origin and evolution.

The landslide presented a crown area [65] in the upper part located in the surrounding area of the road JA-3200. This road, which connects the village of La Guardia and the city of Jaén, was affected by the landslide and had to be repaired (Figure 2). The shape of this crown was rounded and the head inside had several successive moderate scarps up and down the road, ending in a back-scarp at the road cut. Among the scarps some flat areas appeared as relics of the original relief. These areas were not inclined counter-slope, but moderately downslope, so there was no evidence of rotational mechanisms. As the landslide progressed, the tension cracks grew upslope and eventually affected the constructions of the top of the hillslope (an olive oil mill and roads). All these features were formed mainly in the colluvial or piedmont deposits coming from the surrounding reliefs that cover a high proportion of the area. These materials are more coherent and facilitate the formation of scarps and tension cracks.

The landslide mass had a rather disorganized sector in the head or upper part of the landslide (Figures 7–11), where the largest deformation occurred between 2012 and 2014. This upper area—now dismantled after the repair works—was in general a depleted zone and contained abundant transverse cracks. It also had a well-defined lateral boundary and ended in an accumulation zone that extended throughout the body and the foot of the landslide. This accumulation zone formed a shoulder in the terrain surface characteristic of a foot zone, but the movement did not finish at this point and continued downslope after a narrowing in this middle area (Figure 7), that can be still observed in 2016, although with a much smaller deformation. This narrowing in the middle part of the landslide as a consequence of the original terrain surface could stop or slow the movement and produce the shoulder. After the narrowing, the mass ran along a gentle slope to the tip of the landslide located near highway A-44, seriously affecting the access road to La Guardia from the A-44 highway (Figure 2). In the lower part, the lateral limits were more moderate and only appeared as a crack.

The estimated mass thickness is quite small, probably no more than 10 m. Taking into account the length, about 500 m, and the depth-length ratio, about 1%–2%, the landslide belongs to a flow type movement [68].

As well as the main landslide, some other unstable areas can be observed in the northern part of the study area (Figures 7–11). In most cases, only the development of cracks and steps occurred. These affected the temporary road constructed in order to avoid the traffic disruption of the JA-3200 road (Figure 2d). At the bottom of the slope a differentiated movement in an incipient state with a well-formed scarp appeared.

5.6. Relationship between Rainfall and Landslide Activity

The previous analyses agree that between November 2012 (first UAV flight) and July 2014 (third UAV flight) a slope movement of flow type remained active on a hillslope near the village of La Guardia, which affected the JA-3200 road connecting this village with the city of Jaen. From July 2014 to February 2016, a stabilization of the movement was observed and the landslide only maintained some residual movements in localized areas.

Rainfall has been considered one of the main triggering factors of landslides [71,72] all over the world but also in the Mediterranean countries [73], as has been demonstrated in regions close to the study area [74]. Figure 12 shows the monthly rainfall measured for this period and also the monthly average. The monthly average presents an absolute maximum in autumn and winter months (October to January) and also a relative maximum in spring (April), while the minimum values occur in summer. Moreover, the monthly rainfall shows an irregular distribution typical of the Mediterranean climate of southern Spain where cycles alternating between dry and wet years occur [75,76]. At the beginning of the period, between 2012 and 2013, the rainfall values were significantly higher than the average, whereas at the end of the period, especially between 2015 and 2016, the rainfall decreased below the average.

More precisely, between the flights of November 2012 and April 2013, the landslide was very active and this activity coincided closely with two periods of heavy rains where values close to 200% of the average rainfall in the area were reached. The first period occurred between September and November 2012 and the second one between February and April 2013, as described in previous studies of neighboring areas [39–41]. The landslide activity was maintained—although reduced—for the next period between April 2013 and July 2014, perhaps discontinuously as described in the aforementioned studies. Therefore, after April 2013, a drier period began, only interrupted in August by an extraordinary peak near 300% of the average value, and between December 2013 and February 2014 in which 120% of the average value was reached.

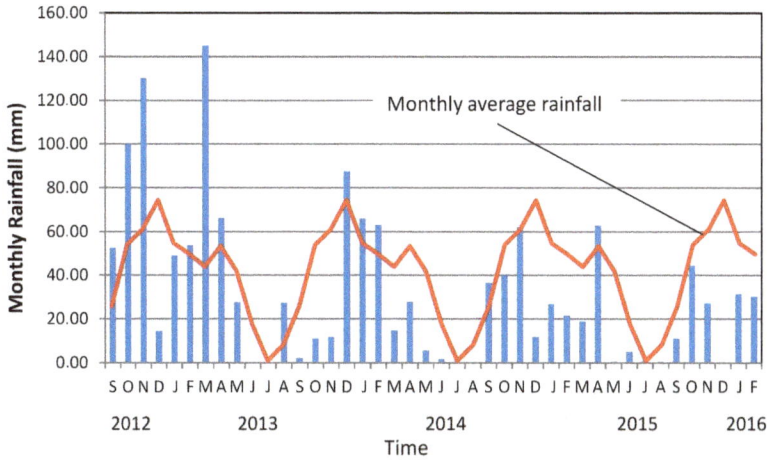

Figure 12. Rainfalls in the studied period in a meteorological station near the zone.

After these periods of more intense activity, the movement had little activity, coinciding with the works of stabilization and restoration of the hillslope and the road and, at the same time, with a fairly dry period with values below 50% of the average rainfall except for a few months where the rainfall occasionally exceeded 100%. This intermittent character is typical of the earth flows in Mediterranean or arid climates [56]. In this sense, the landslide that can be catalogued as slow to moderate following velocity classifications is also characterized as type VII of the diachroneity classification (landslides active in a period between one and ten years) [4].

In conclusion, a certain relationship between activity and rainfall is observed, although the execution of works of stabilization of the slope interrupted the analysis and prevented more definitive conclusions from being reached.

6. Conclusions

The use of Unmanned Aerial Vehicles (UAV) is a useful tool for fast, very high resolution and precise surveys in areas of about 0.01 to 100 km^2. This paper proposes a methodology for the analysis of slope instability processes. In this sense, light equipment, such as that used in this study, is well suited for updates and landslide monitoring, and thus for disaster management. It allows working with intermediate scales between terrestrial techniques (global navigation satellite system (GNSS), classic surveying, terrestrial laser scanner, terrestrial photogrammetry, etc.) and aerial or space surveys (conventional aerial photogrammetry and light detection and ranging (LiDAR), very high resolution satellite remote sensing, Differential Synthetic Aperture Radar Interferometry (DInSAR)). At this scale, detailed morphological features can still be observed, while allowing relatively large areas to be covered. Light UAVs are very agile and easy to operate compared to other heavier equipment. Also, field surveys can be performed within a few hours of work, including the measurement of ground control points (GCPs).

Since these light UAVs can also be equipped with different sensors (infrared cameras, LiDAR, Radar, etc.), their potential will increase. The use of conventional photogrammetric and Structure from Motion (SfM) techniques [26] allows us to orient UAV flights and to obtain Digital Surface Models (DSMs)/Digital Terrain Models (DTMs) and orthophotographs from which vertical and horizontal displacements and volumetric changes can be calculated. As it has been presented in the introduction section, some studies have made advances in co-registration methodologies of multi-temporal image datasets by means of points or surfaces, the development of algorithms for detecting surface motions, and the implementation of methods of direct orientation from in-flight parameters. However, the elimination or reduction of GCP (a time-consuming task) is not yet possible in all circumstances unless more accurate positioning and inertial systems can be used on board light UAVs, although these new systems will soon be available.

In this study, both DSMs and orthophotographs at different epochs have been obtained from aerial photographs of the UAV surveys, using a methodology based on conventional photogrammetry and SfM techniques. The results obtained have allowed the characterization of the slope movement flow rate and some morphological features (crown, scarps, head, lateral limits, tension cracks, foot, etc.) on a hillslope of hectometric dimensions. In this way, the movement has been classified as an earth flow, based on the observations made on the field, the DSMs and the orthophotographs. The technique also allows a monitoring analysis by calculating differential DSMs in order to measure vertical displacements and identify depletion and accumulation areas inside the landslide. At the same time, the measurement of displacements between points selected in a manual way from the orthophotographs and models (DSM check points) allows us to analyze landslide kinematics and activity with an accuracy of about 10 cm in XY and 15 cm in Z. This accuracy has been established by means of an analysis of displacements measured in the GIS between the DSMs at 151 check points placed on the bare ground in stable zones. The uncertainties correspond to the standard deviation of these measured displacements at stable points.

In this sense, the observed displacements in the main landslide area reached values from decimeters to meters in the vertical component and several meters in the horizontal component, significantly larger than the accuracy limit. As the periods between UAV campaigns are more or less one year, the displacement rates ranged from decimeters to meters by year in the vertical direction and from 1 m/year to 10 m/year in the horizontal direction, so the landslide can be catalogued as slow (to moderate). Nevertheless, the displacement rates varied along the whole studied period and depended on the part of the landslide. Therefore, the maximum rate was reached at the beginning (2012–2013), but the landslide was gradually slowing until it stopped completely by 2015. Moreover, the areas with larger deformation were the upper part (head) and the middle part (body and the starting of the foot) of the landslide, while the area less deformed was the lower part (the smooth foot downslope).

The displacements have been correlated to the rain events that occurred in the region between the different epochs. These first analyses show a higher activity of the landslide—larger vertical and horizontal displacements—in the rainy periods at the beginning of the study, then a residual activity and finally a stop of the landslide in the following dry period, coinciding with the stabilization works of the landslide slope.

Automatic techniques of feature recognition, proposed in some studies, have been tested in this study, but they have not worked because of the changes in solar illumination, deep shadows, seasonal and inter-annual vegetation, the landslide itself and the repair works that have occurred in this zone. Nevertheless, an approach based on olive tree recognition by means of DSM classification and filtering has been applied in a groundbreaking way. The horizontal displacement measured between olive tree centroids coincide roughly with those observed between DSM check points but the disappearance of many trees in the areas of maximum deformation (the upper and middle zones) and the lower accuracy in the positioning of the centroids make the difference between stable and landslide areas less clear than in the analysis of DSM check points. In this way, the application of filters using displacement modules and directions allows a better differentiation between stable and unstable areas and thus the recognition of landslide areas. This methodology can be employed in the study of landslide activity in olive groves and other similar areas.

Future work will deal with advances in the methodology, regarding the reduction of ground control points and applying methods of direct orientation and other techniques of automatic detection of surface motions based on expert classification techniques from DSMs and images even in complex areas. Other sensors can be incorporated both in the spectral domain (i.e., near and thermal infrared sensors) and in the geometric domain (LiDAR and RADAR). Even terrestrial probes (rain, humidity, movement) in wireless networks (WSN) can be used. Other areas of interest will include the mapping and analysis of morphological features, such as cracks, scarps, steps and landslide limits.

Acknowledgments: This work has been financed by the Photogrammetric and Topometric Systems (TEP-213) Research Group, the research project ISTEGEO (RNM-06862, Andalusian Research Plan) from the Regional Andalusian Government, and the Centre for Advanced Studies on Earth Sciences (CEACTierra) of the University of Jaén.

Author Contributions: T.F., J.L.P. and J.C. conceived and organized the research activity. J.L.P., J.C., J.M.G. and C.C. made the UAV surveys. J.L.P., J.C. and J.M.G. processed the photogrammetric data. C.C. made the olive-trees extraction and processing in GIS. T.F. made the GIS analysis and interpreted the resulting data. J.D. supervised the research activity. All authors contributed to the writing of this manuscript.

Conflicts of Interest: The authors declare no conflict of interest.

References

1. Chacón, J.; El Hamdouni, R.; Irigaray, C.; Fernández, T. Engineering geology maps: Landslides and GIS. *Bull. Eng. Geol. Environ.* **2006**, *65*, 341–411. [CrossRef]
2. Metternicht, G.; Hurni, L.; Gogu, R. Remote sensing of landslides: An analysis of the potential contribution to geo-spatial systems for hazard assessment in mountainous environments. *Remote Sens. Environ.* **2005**, *98*, 284–303. [CrossRef]
3. Tofani, V.; Hong, Y.; Singhroy, V. Introduction: Remote Sensing Techniques for Landslide Mapping and Monitoring. In *Landslide Science for a Safer Geoenvironment*; Sassa, K., Canuti, P., Yin, Y., Eds.; Springer International Publishing: Cham, Switzerland, 2014; Volume 2, pp. 301–303.
4. Chacón, J.; Irigaray, C.; El Hamdouni, R.; Jiménez, J.D. Diachroneity of landslides. In *Geologically Active*; Williams, A.L., Pinches, G.M., Chin, C.Y., McMorran, T.J., Massey, C.I., Eds.; Taylor & Francis Group: London, UK, 2010; pp. 999–1006.
5. Fernández, T.; Pérez, J.L.; Delgado, J.; Cardenal, J.; Irigaray, C.; Chacón, J. Evolution of a diachronic landslide by comparison between different DEMs obtained from digital photogrammetry techniques in las Alpujarras. In Proceedings of the Geoinformation for Disaster Management (GI4DM), Antalya, Turkey, 3–8 May 2011.
6. Walstra, J.; Chandler, J.H.; Dixon, N.; Dijkstra, T.A. Time for change—Quantifying land-slide evolution using historical aerial photographs and modern photogrammetric methods. *Int. Arch. Photogramm. Remote Sens. Spatial Inf. Sci.* **2004**, *XXXV Part B4*, 475–480.
7. Casson, B.; Delacourt, C.; Allemand, P. Contribution of multitemporal remote sensing images to characterize landslide slip surface—Application to the "La Clapière" landslide (France). *Nat. Hazards* **2005**, *5*, 425–437. [CrossRef]
8. Brückl, E.; Brunner, F.K.; Kraus, K. Kinematics of a deep-seated landslide derived from photogrammetric, GPS and geophysical data. *Eng. Geol.* **2006**, *88*, 149–159. [CrossRef]
9. Cardenal, J.; Delgado, J.; Mata, E.; González, A.; Olague, I. Use of historical flight for landslide monitoring. In Proceedings of the Spatial Accuracy 2006, Lisbonne, Portugal, 5–7 July 2006; pp. 129–138.
10. Dewitte, O.; Jasselette, J.C.; Cornet, Y.; Van Den Eeckhaut, M.; Collignon, A.; Poesen, J.; Demoulin, A. Tracking landslide displacement by multi-temporal DTMs: A combined aerial stereophotogrammetric and LiDAR approach in Belgium. *Eng. Geol.* **2008**, *99*, 11–22. [CrossRef]
11. Baldi, P.; Cenni, N.; Fabris, M.; Zanutta, A. Kinematics of a landslide derived from archival photogrammetry and GPS data. *Geomorphology* **2008**, *102*, 435–444. [CrossRef]
12. Corsini, A.; Borgatti, L.; Cervi, F.; Dahne, A.; Ronchetti, F.; Sterzai, P. Estimating mass-wasting processes in active earth slides-Earth flows with time-series of High-Resolution DEMs from photogrammetry and airborne LiDAR. *Nat. Hazards Earth Syst. Sci.* **2009**, *9*, 433–439. [CrossRef]
13. Kasperski, J.; Delacourt, C.; Allemand, P.; Potherat, P. Evolution of the Sedrun landslide (Graubünden, Switzerland) with ortho-rectified air images. *Bull. Eng. Geol. Environ.* **2010**, *69*, 421–430. [CrossRef]
14. Prokešová, R.; Kardoš, M.; Medved'ová, A. Landslide dynamics from high-resolution aerial photographs: A case study from W Carpathians, Slovakia. *Geomorphology* **2010**, *115*, 90–101. [CrossRef]
15. Fabris, M.; Menin, A.; Achilli, V. Landslide displacement estimation by archival digital photogrammetry. *Ital. J. Remote Sens.* **2011**, *43*, 2, 23–30. [CrossRef]

16. Fernández, T.; Jiménez, J.; Delgado, J.; Cardenal, F.J.; Pérez, J.L.; El Hamdouni, R.; Irigaray, C.; Chacón, J. Methodology for Landslide Susceptibility and Hazard Mapping Using GIS and SDI. In *Intelligent Systems for Crisis Management, Lecture Notes in Geoinformation and Cartography*; Zlatanova, S., Peters, R., Dilo, A., Scholten, H., Eds.; Springer: Berlin/Heidelberg, Germany, 2012; pp. 185–198.

17. González-Díez, A.; Fernández-Maroto, G.; Doughty, M.W.; Díaz de Terán, J.R.; Bruschi, V.; Cardenal, J.; Pérez, J.L.; Mata, E.; Delgado, J. Development of a methodological approach for the accurate measurement of slope changes due to landslides, using digital photogrammetry. *Landslides* **2014**, *11*, 615–628. [CrossRef]

18. Kraus, K. *Photogrammetry: Geometry from Images and Laser Scans*; Walter de Gruyter: Berlin, Germany, 2007.

19. Cox, T.H.; Nagy, C.J.; Skoog, M.A.; Somers, I.A. *Civil UAV Capability Assessment*; Draft version, NASA Report; NASA Dryden Flight Research Center: Edwards, CA, USA, 2004; p. 103.

20. Nonami, K. Prospect and recent research and development for civil use autonomous unmanned aircrafts as UAV and MAV. *J. Syst. Des. Dyn.* **2007**, *1*, 120–128. [CrossRef]

21. Jiménez-Berni, J.A.; Zarco-Tejada, P.J.; Suárez, J.; Fereres, E. Thermal and narrowband multispectral remote sensing for vegetation monitoring from Unmanned Aerial Vehicle. *IEEE Trans. Geosci. Remote Sens.* **2009**, *47*, 722–738. [CrossRef]

22. Grenzdörffer, G.J.; Engel, A.; Teichert, B. The photogrammetric potential of low-cost UAVs in forestry and agriculture. *Int. Arch. Photogramm. Remote Sens. Spatial Inf. Sci.* **2008**, *XXXVII Part B1*, 1207–1214.

23. Maza, I.; Caballero, F.; Capitán, J.; Martínez, J.R.; Ollero, A. Experimental results in Multi-UAV coordination for disaster management and civil security applications. *J. Intell. Robot. Syst.* **2010**, *61*, 563–583. [CrossRef]

24. Longo, A.A.; Pace, P.; Marano, S. A system for monitoring volcanoes activities using high altitude platform stations. In Proceedings of the 55th International Astronautical Congress, Vancouver, BC, Canada, 4–8 October 2004; pp. 1203–1210.

25. Hartley, R.; Zisserman, A. *Multiple View Geometry in Computer Vision*; Cambridge University Press: Cambridge, UK, 2004; p. 655.

26. Eltner, A.; Kaiser, A.; Castillo, C.; Rock, G.; Neugirg, F.; Abellán, A. Image-based surface reconstruction in geomorphometry—Merits, limits and developments. *Earth Surf. Dyn.* **2016**, *4*, 359–389. [CrossRef]

27. Rau, J.Y.; Jhan, J.P.; Lo, C.F.; Lin, Y.S. Landslide mapping using imagery acquired by a fixed-wing UAV. *Int. Arch. Photogramm. Remote Sens. Spatial Inf. Sci.* **2011**, *XXXVIII-1-C22*, 195–200. [CrossRef]

28. Tahar, K.N.; Ahmad, A.; Akib, W.A.A.W.M.; Mohd, W.M.N.W. Unmanned Aerial Vehicle Photogrammetric Results Using Different Real Time Kinematic Global Positioning System Approaches. In *Developments in Multidimensional Spatial Data Models*; Lecture Notes in Geoinformation and Cartography; Raman, A.A., Bogulawski, P., Gold, C., Said, M.N., Eds.; Springer: Berlin/Heidelberg, Germany, 2013; pp. 123–134.

29. Liu, C.; Chen, P.; Matsuo, T.; Chen, C. Rapidly responding to landslides and debris flow events using a low cost Unmanned Aerial Vehicle. *J. Appl. Remote Sens.* **2015**, *9*, 096016. [CrossRef]

30. Shi, B.; Liu, C. UAV for Landslide Mapping and Deformation Analysis. *Proc. SPIE* **2015**, *9808*, 98080P.

31. Hsieh, Y.C.; Chan, Y.; Hu, J. Digital elevation model differencing and error estimation from multiple sources: A case study from the Meiyuan Shan landslide in Taiwan. *Remote Sens.* **2016**, *8*, 199. [CrossRef]

32. Carvajal, F.; Agüera, F.; Pérez, M. Surveying a landslide in a road embankment using Unmanned Aerial Vehicle photogrammetry. *Int. Arch. ISPRS* **2011**, *38*, 201–206.

33. Liu, C.; Li, W.; Lei, W.; Liu, L.; Hu, H. Architecture planning and geo-disasters assessment mapping of landslide by using airborne LiDAR data and UAV images. *Proc. SPIE* **2011**, *8286*, 82861Q.

34. Niethammer, U.; James, M.R.; Rothmund, S.; Travelletti, J.; Joswig, M. UAV-based remote sensing of the Super-Sauze landslide: Evaluation and results. *Eng. Geol.* **2012**, *128*, 2–11. [CrossRef]

35. Niethammer, U.; Rothmund, S.; Schwaderer, U.; Zeman, J.; Joswig, M. Open source image-processing tools for low-cost UAV-based landslide investigations. *Int. Arch. Photogramm. Remote Sens. Spat. Inf. Sci.* **2011**, *XXXVIII-1-C22*, 161–166. [CrossRef]

36. Stumpf, A.; Malet, J.P.; Kerle, N.; Niethammer, U.; Rothmund, S. Image-based mapping of surface fissures for the investigation of landslide dynamics. *Geomorphology* **2013**, *186*, 12–27. [CrossRef]

37. Turner, D.; Lucieer, A. Using a micro Unmanned Aerial Vehicle (UAV) for ultra-high resolution mapping and monitoring of landslide dynamics. In Proceedings of the IEEE International Geoscience and Remote Sensing Symposium, Melbourne, Australia, 21–26 July 2013.

38. Turner, D.; Lucieer, A.; de Jong, S.M. Time series analysis of landslide dynamics using an Unmanned Aerial Vehicle (UAV). *Remote Sens.* **2015**, *7*, 1736–1757. [CrossRef]

39. Fernández, T.; Pérez, J.L.; Arenas, A.; Gómez, J.M.; Sánchez, M.; Cardenal, F.J.; Delgado, J.; Pérez, A. Empleo de una plataforma aérea ligera no tripulada (UAV) y técnicas fotogramétricas para el estudio de una zona inestable en la Guardia de Jaén. In Proceedings of the VIII Simp. Nacional sobre Taludes y Laderas Inestables, Palma de Mallorca, Spain, 11–14 June 2013; pp. 869–880.

40. Fernández, T.; Pérez, J.L.; Arenas, A.; Gómez, J.M.; Cardenal, F.J.; Delgado, J. Monitorización de laderas inestables mediante técnicas fotogramétricas a partir de vuelos UAV e históricos. In Proceedings of the XVI Congreso de Tecnologías de la Información Geográfica, Alicante, Spain, 25–27 June 2014.

41. Fernández, T.; Pérez, J.L.; Cardenal, F.J.; López, A.; Gómez, J.M.; Colomo, C.; Sánchez, M.; Delgado, J. Use of a light UAV and photogrammetric techniques to study the evolution of a landslide. *Int. Arch. Photogramm. Remote Sens. Spatial Inf. Sci.* **2015**, *XL-3-W3*, 241–248. [CrossRef]

42. Peterman, V. Landslide activity monitoring with the help of Unmanned Aerial Vehicle. *Int. Arch. Photogramm. Remote Sens. Spatial Inf. Sci.* **2015**, *XL-1-W4*, 215–218. [CrossRef]

43. Al-Rawabdeh, A.; He, F.; Moussa, A.; El-Sheimy, N.; Habib, A. Using an Unmanned Aerial Vehicle-based digital imaging system to derive a 3D point cloud for landslide scarp recognition. *Remote Sens.* **2016**, *8*, 95–126. [CrossRef]

44. Vrublova, D.; Kapica, R.; Jirankova, E.; Strus, A. Documentation of landslides and inaccessible parts of a mine using an unmanned UAV system and methods of digital terrestrial photogrammetry. *GeoSci. Eng.* **2015**, *LXI*, 8–19.

45. Lindner, G.; Schraml, K.; Mansberger, R.; Hübl, J. UAV monitoring and documentation of a large landslide. *Appl. Geomat.* **2016**, *8*, 1–11. [CrossRef]

46. Daakir, M.; Pierrot-Deseilligny, M.; Bosser, P.; Pichard, F.; Thom, C. UAV onboard photogrammetry and GPS positioning for earthworks. *Int. Arch. Photogramm. Remote Sens. Spatial Inf. Sci.* **2015**, *XL-3/W3*, 293–298. [CrossRef]

47. Gong, J.; Wang, D.; Li, Y.; Zhang, L.; Yue, Y.; Zhou, J.; Song, Y. Earthquake-induced geological hazards detection under hierarchical stripping classification framework in the Beichuan area. *Landslides* **2010**, *7*, 181–189. [CrossRef]

48. Yang, Z.; Lan, H.; Liu, H.; Li, L.; Wu, Y.; Meng, Y.; Xu, L. Post-earthquake rainfall-triggered slope stability analysis in the Lushan area. *J. Mt. Sci.* **2015**, *12*, 232–242. [CrossRef]

49. Yang, Z.; Lan, H.; Gao, X.; Li, L.; Meng, Y.; Wu, Y. Urgent landslide susceptibility assessment in the 2013 Lushan earthquake-impacted area, Sichuan Province, China. *Nat. Hazards* **2015**, *75*, 2467–2487. [CrossRef]

50. Lin, J.; Tao, H.P.; Huang, Z. Practical Application of Unmanned Aerial Vehicles for Mountain Hazards Survey. In Proceedings of the 18th International Conference on Geoinformatics, Beijing, China, 18–20 June 2010; pp. 1–5.

51. Yeh, M.L.; Hsiao, Y.C.; Chen, Y.H.; Chung, J.C. A study on Unmanned Aerial Vehicle applied to acquire terrain information of landslide. In Proceedings of the 32 Asian Conf. Remote Sensing, Taipei, Taiwan, 3–7 October 2011; Volume 3, pp. 2210–2215.

52. Huang, Y.; Yi, S.; Lia, Z.; Shao, S.; Qin, X. Design of highway landslide warning and emergency response systems based on UAV. *Proc. SPIE* **2011**, *8203*, 820317.

53. Nedjati, A.; Vizvari, B.; Izbirak, G. Post-earthquake response by small UAV helicopters. *Nat. Hazards* **2016**, *80*, 1669–1688. [CrossRef]

54. Fernández, T.; Sánchez, M.; García, F.; Pérez, F. Cartografía de movimientos de ladera en el frente montañoso de la Cordillera Bética en el sector de Jaén. In Proceedings of the VIII Congreso Geológico de España, Oviedo, Spain, 17–19 July 2012.

55. Varnes, D.J. Slope movement, types and processes. In *Landslides: Analysis and Control*; Transportation Research Board Special Report; Schuster, R.L., Krizek, R.J., Eds.; National Academy of Sciences: Washington, DC, USA, 1978; Volume 176, pp. 12–33.

56. Hungr, O.; Leroueil, S.; Picarelli, L. The Varnes classification of landslide types, an update. *Landslides* **2014**, *11*, 167–194. [CrossRef]

57. Fernández, T.; Pérez, J.L.; Cardenal, J.; Colomo, C.M.; Moya, F.; Sánchez-Gómez, M.; Tovar, J.; Carpena, R. Estimación de la actividad y la peligrosidad a los movimientos de ladera en la cuenca del río Guadallbullón (Jaén) mediante técnicas fotogramétricas y LiDAR. In Proceedings of the Congreso Geológico de España, Huelva, Spain, 12–14 September 2016.

58. García-Rosell, L. Estudio Geológico de la Transversal Úbeda-Huelma y Sectores Adyacentes (Cordilleras Béticas, Provincia de Jaén). Ph.D. Thesis, Universidad de Granada, Granada, Spain, 1973.

59. Roldán, F.J.; Lupiani, E.; Jerez, L. *Mapa Geológico de España, Escala 1:50.000, Mapa y Memoria Explicativa*; Instituto Geológico Nacional: Madrid, Spain, 1988.

60. Navarro, V.; Ruiz-Ortiz, P.A.; Molina, J.M. Birth and demise of a Middle Jurassic isolated shallow-marine carbonate platform on a tilted fault block: Example from the Southern Iberian continental palaeomargin. *Sediment. Geol.* **2012**, *269*, 37–57. [CrossRef]

61. Peláez, J.A.; Sánchez Gómez, M.; López Casado, C. La serie sísmica de Mancha Real de 1993. *Bol. Inst. Estud. G.* **2005**, *191*, 169–183.

62. Sánchez-Gómez, M.; Peláez, J.A.; García-Tortosa, F.J.; Torcal, F.; Soler, P.; Ureña, M.A. Aproximación geológica, geofísica y geomorfológica a la actividad tectónica en el valle del alto Guadalquivir. In Proceedings of the 6th Asamblea Hispano Portuguesa de Geodesia y Geofísica, Tomar, Portugal, 5–8 September 2008.

63. AscTec Falcon 8 + AscTec Trinity. Ascending Technologies: Munich, Germany, 2016. Available online: http://www.asctec.de/en/asctec-trinity/ (accessed on 10 September 2016).

64. Atyges FV8-Drone. ATyges Drones UAV: Málaga, Spain, 2016. Available online: http://www.atyges.es/drones (accessed on 10 September 2016).

65. IAEG Commission on Landslides. Suggested nomenclature for landslides. *Bull. IAEG* **1990**, *41*, 13–16.

66. Mardia, K.V.; Jupp, P. *Directional Statistics*, 2nd ed.; John Wiley and Sons Ltd.: Chichester, UK, 2000; p. 428.

67. Hutchinson, J.N. General Report: Morphological and Geotechnical Parameters of Landslides in Relation to Geology and Hydrogeology. In Proceedings of the 5th Intern. Symposium on Landslides, Lausanne, Switzerland, 10–15 July 1988.

68. Crozier, M.J. Techniques for the morphometric analysis of landslips. *Z. Geomorphol* **1973**, *17*, 78–101.

69. International Union of Geological Sciences Working Group on Landslides. A suggested method for describing the rate of movement of a landslide. *Bull. Eng. Geol. Environ.* **1995**, *52*, 75–78.

70. Conner, J.C.; Olsen, M.J. Automated quantification of distributed landslide movement using circular tree trunks extracted from terrestrial laser scan data. *Comput. Geosci.* **2014**, *67*, 31–39. [CrossRef]

71. Crozier, M.J. *Landslides: Causes, Consequences and Environment*; Routledge: London, UK; New York, NY, USA, 1995.

72. Finlay, P.J.; Fell, R.; Maguire, P.K. The relationship between the probability of landslide occurrence and rainfall. *Can. Geotech. J.* **1997**, *34*, 811–824. [CrossRef]

73. Guzzeti, F. Landslide hazard assessment and risk evaluation: Limits and prospectives. In Proceedings of the 4th EGS Plinius Conference, Mallorca, Spain, 2–4 October 2002.

74. Irigaray, C.; Lamas, F.; El Hamdouni, R.; Fernández, T.; Chacón, J. The importance of the precipitation and the susceptibility of the slopes for the triggering of landslides along the roads. *Nat. Hazards* **2000**, *21*, 65–81. [CrossRef]

75. Trigo, R.M.; Pozo, D.; Timothy, C.; Osborn, J.; Castro, Y.; Gámiz, S.; Esteban, M.J. North Atlantic Oscillation influence on precipitation, river flow and water resources in the Iberian Peninsula. *Int. J. Climatol.* **2004**, *24*, 925–994. [CrossRef]

76. Pozo, D.; Esteban, M.J.; Rodrigo, F.S.; Castro, Y. An analysis of the variability of the North Atlantic Oscillation in the time and the frequency domains. *Int. J. Climatol.* **2000**, *20*, 1675–1692. [CrossRef]

remote sensing

MDPI

Article

Landslide Mapping in Vegetated Areas Using Change Detection Based on Optical and Polarimetric SAR Data

Simon Plank *, André Twele and Sandro Martinis

German Aerospace Center (DLR), German Remote Sensing Data Center (DFD), Muenchener Str. 20, 82234 Oberpfaffenhofen, Germany; Andre.Twele@dlr.de (A.T.); Sandro.Martinis@dlr.de (S.M.)
* Correspondence: simon.plank@dlr.de; Tel.: +49-8153-28-3460

Academic Editors: Roberto Tomas, Zhenhong Li, Zhong Lu and Prasad S. Thenkabail
Received: 29 January 2016; Accepted: 31 March 2016; Published: 6 April 2016

Abstract: Mapping of landslides, quickly providing information about the extent of the affected area and type and grade of damage, is crucial to enable fast crisis response, *i.e.*, to support rescue and humanitarian operations. Most synthetic aperture radar (SAR) data-based landslide detection approaches reported in the literature use change detection techniques, requiring very high resolution (VHR) SAR imagery acquired shortly before the landslide event, which is commonly not available. Modern VHR SAR missions, e.g., Radarsat-2, TerraSAR-X, or COSMO-SkyMed, do not systematically cover the entire world, due to limitations in onboard disk space and downlink transmission rates. Here, we present a fast and transferable procedure for mapping of landslides, based on change detection between pre-event optical imagery and the polarimetric entropy derived from post-event VHR polarimetric SAR data. Pre-event information is derived from high resolution optical imagery of Landsat-8 or Sentinel-2, which are freely available and systematically acquired over the entire Earth's landmass. The landslide mapping is refined by slope information from a digital elevation model generated from bi-static TanDEM-X imagery. The methodology was successfully applied to two landslide events of different characteristics: A rotational slide near Charleston, West Virginia, USA and a mining waste earthflow near Bolshaya Talda, Russia.

Keywords: landslide; change detection; SAR polarimetry; PolSAR; object-based image analysis; OBIA; TanDEM-X

---

1. Introduction

Large landslides are a global phenomenon, causing damage and casualties [1]. Landslides arouse emergency situations when urban areas or man-made constructions, such as buildings, bridges, railroads, and roads, are affected. Rapid mapping of landslides is crucial to detect the extent of the affected area, including grade and type of damage. Rapid mapping is a key element of fast crisis response, e.g., to support rescue, humanitarian, and reconstruction operations in the crisis area [2]. Therefore, Earth Observation (EO) based on satellite remote sensing plays a key role due to its fast response, wide field of view, and relatively low cost [3]. (Semi)-automatic landslide mapping based on satellite EO data provides an important information source to support field surveys. Furthermore, during rapid to very rapid events, *i.e.*, deformation rates in the order of several meters per hour [4], access to the landslide area may be too difficult, making field surveys too dangerous [5].

The most frequently used EO data for rapid mapping of landslides is very high-resolution (VHR) optical satellite imagery [6–8]. A common way for landslide detection is the mapping of rapid changes of the vegetation layer derived from vegetation indices calculated for pre- and post-event optical EO imagery, e.g., [9–15].

However, as heavy rain events are one of the most frequent triggers for landslides, *i.e.*, there is a high probability of cloud coverage right after the event, optical EO data is not always useful for rapid mapping applications [16,17]. The advantages of SAR compared to optical EO sensors are (I) the day and night availability of this active sensor, and (II) its almost complete weather independency due to its longer wavelength. In most cases, SAR EO data of a given crisis area is available earlier than cloud-free optical imagery. Therefore, faster disaster response is enabled by SAR-based rapid mapping procedures. However, for both optical and SAR sensors, the revisit time of the satellite has to be taken into account [18].

In addition to satellite-based methods for landslide mapping, there are also other approaches using for instance airborne laser altimetry (LiDAR), e.g., [19], or advanced field mapping techniques such as the combination of a laser rangefinder binocular combined with a GPS receiver [20], are described in the literature. Despite of their promising results, and contrary to satellite data-based landslide mapping methodologies, those approaches are not suited for a world-wide use due to their limited availability.

According to Czuchlewski *et al.* [21], one post-event single-polarized SAR image is insufficient for distinguishing and mapping landslides. The investigation of the temporal development of the interferometric coherence by analyzing a time-series of SAR imagery, including pre- and post-event imagery in the data stack, enables landslide detection, e.g., [22,23]. While multi-temporal SAR interferometry enables long-term monitoring of extremely slow and very slow movements, e.g., [24–26], speckle tracking approaches are able to measure higher deformation rates (up to tens of meters) [27].

Polarimetric SAR (PolSAR) data of at least two polarizations (dual-pol) provides more information on the ground, which enables a better land cover classification and landslide mapping. Quad-pol data, containing the full polarimetric backscattering (*i.e.*, all four combinations of horizontal (H) and vertical (V) polarized waves) allows the most accurate land cover mapping [28] using SAR data.

Based on airborne L- and P-band quad-pol imagery, Rodriguez *et al.* [29] analyzed changes in the pedestal height and the Radar Vegetation Index (RVI) over time to detect landslides in Taiwan, which were triggered by the 1999 Chi-Chi earthquake.

Cui *et al.* [30] investigated landslides in earthen levees by means of a multi-classifier decision framework for textural features (grey level co-occurrence matrix) derived from multi-polarized SAR imagery.

Plank *et al.* [31] compared object-based landslide detection methods based on PolSAR (dual-pol TerraSAR-X) and VHR optical imagery for a case study in Taiwan. The PolSAR procedure is based on a textural analysis with focus on the Normalized Difference Standard Deviation (NDSD) of the calibrated intensities of both polarimetric channels.

Decomposition procedures based on quad-pol SAR imagery, such as the Freeman-Durden decomposition [32] and the further enhancement of it, the Yamaguchi decomposition [33,34], allow the derivation of surface (e.g., bare surfaces), volume (e.g., vegetation) and double-bounce (e.g., man-made objects and at tree trunks) scattering components. Watanabe *et al.* [35], Yamaguchi [36], Shibayama and Yamaguchi [16,37] report landslide detection procedures by detecting changes of the polarimetric scattering components. Landslides in vegetated areas cause a decrease of volume scattering, *i.e.*, a loss of vegetation, and an increase of surface scattering (bare surfaces). Shibayama *et al.* [38] found that the local incidence angle has high influence on landslide detection based on polarimetric scattering analysis.

In addition to the aforementioned Freeman-Durden decomposition, Yonezawa *et al.* [39] investigated also the change of the entropy/anisotropy/alpha (H/A/α) decomposition [40] in pre- and post-event ALOS/PALSAR imagery. *H* showed lower values for landslide areas than for forested areas. However, farmlands showed similar low values of *H* as landslides, making the differentiation of these classes very difficult.

Except for [30,31], all aforementioned landslide detection procedures are based on change detection approaches of pre- and post-event VHR SAR imagery, requiring identical imaging geometries of both acquisitions. However, VHR archive SAR imagery acquired shortly before a landslide event, especially at the same image acquisition geometry as the next possible post-event imagery, is in most cases not available. Modern VHR SAR missions, such as TerraSAR-X, COSMO-SkyMed, or RADARSAT-2, do not systematically cover the entire world. Each acquisition has to be programmed manually. Furthermore, due to limited disk space on board the satellites and especially due to limited downlink transmission rates, these sensors are not able to provide worldwide coverage within a short time period—*i.e.*, commonly, no archive image recorded shortly before the event is available. Here, we present a fast and transferable landslide detection methodology based only on post-event VHR PolSAR imagery supported by freely available and systematically-acquired pre-event high-resolution (HR) optical data. The post-event VHR PolSAR acquisition can be programmed before the next overpass of the satellite after the landslide event, independent of any geometrical restrictions by a pre-event SAR imagery. The proposed landslide mapping procedure is a semi-automatic change detection approach based on pre-event HR optical imagery of Landsat-8 or Sentinel-2 and post-event VHR PolSAR data (e.g., TerraSAR-X) acquired shortly after the event.

The methodology was successfully applied to two case studies of different characteristics: first, a rotational slide, which occurred on 12 March 2015 at the Yeager Airport near Charleston, West Virginia, USA was investigated. Second, the methodology was tested at a mining waste landslide event, which occurred on 1 April 2015 near Bolshaya Talda, Kemerovo Oblast, Russia (*cf.* Section 2).

Section 3 describes the developed landslide mapping methodology. Section 4 describes and discusses the results of both test sites. Finally, a conclusion and outlook is given in Section 5.

2. Study Sites and Data

Two landslide events are studied. The first one occurred at the Yeager Airport landslide, Charleston, West Virginia, USA. The second one is a mining waste landslide near Bolshaya Talda, Kemerovo Oblast, Russia.

On 12 March 2015, a large-scale landslide occurred at an artificial slope at the Yeager Airport. The airport, completed in 1947, was constructed atop seven semi-connected hilltops. In 2005, due to new Federal Aviation Administration (FAA) safety regulations, the construction of an Engineered Material Arrestor System (EMAS) was necessary. Therefore, the Yeager Airport had to be extended, leading to the construction of a large artificial slope, being the tallest geosynthetic reinforced slope in North America (horizontal/vertical ratio of 1:1). The construction of the slope was finished in 2007 [41]. The construction of this large artificial slope is described in detail in [42]. The functionality of the EMAS was successfully put to the test on 19 January 2010: a US Airways flight bound for Charlotte aborted takeoff. The CRJ 200 aircraft could not stop before the end of the runway. Fortunately, the jet was stopped approximately 45 m from the edge of the slope by the EMAS. All 34 passengers and crew survived the incident with only minor injuries reported.

First movements at the slope were noticed in June 2013. In the following time, the deformation increased. On 12 March 2015, the slope failed. A secondary failure of the slope occurred on 13 April 2015 (Figure 1). The landslide can be described as rotational debris slide [4]. Further details on this landslide event, as well as a video of the first slope failure, and a drone flight video recorded after the first slope failure are available at the AGU landslide blog of Dave Petley [43]. These videos are also attached to this article to guarantee long-term accessibility.

The available SAR data are two post-event TerraSAR-X HighResolution SpotLight (HS) dual-pol (HH/VV) imagery, acquired on 25 March 2015 (after the first slope failure) and on 16 April 2015 (after the second slope failure). In addition, one pre-event Landsat-8 imagery acquired on 15 January 2015 was used (*cf.* Table 1).

Figure 1. (**a**) Pre-failure photograph of the Yeager Airport [42]; (**b**) top of the landslide, showing the remains of the EMAS (white) and the geosynthetic reinforcements (Marcus Constantino, Charleston Gazette 12 March 2015 [44]); (**c**) the remains of the Keystone Apostolic Church, which was damaged by the landslide (Tyler Bell, Charleston Gazette 13 March 2015 [45]); (**d**) after the 2nd slope failure (F. Brian Ferguson, Charleston Gazette 13 April 2015 [46]); and (**e**) rolling up of asphalt at the toe of the landslide (Rusty Marks, Charleston Gazette 13 April 2015 [46]).

The second landslide, to which the developed landslide mapping procedure is applied on, is a large mining waste landslide, which occurred near a road between Novokuznetsk and Bolshaya Talda, Kemerovo Oblast, Russia. This landslide occurred on 1 April 2015 at 1 p.m. local time [47]. It can be described as a very rapid earthflow [4]. Unfortunately, no further details about this landslide event are available. However, an interesting video of the landslide is available and provided with this article

(Source: [47]). Two post-event TerraSAR-X HS dual-pol (HH/VV) SAR imagery acquired on 26 April and 14 August 2015, as well as one pre-event optical Landsat-8 image acquired on 14 September 2014, were available for the analysis (*cf.* Table 1).

Table 1. Timeline of the landslide events and used satellite imagery.

Study Site	Date of Event	Acquisition Date	Satellite [1]	Relative Orbit [2]	Polarization
Yeager Airport	-	2 April 2014	TanDEM-X	226/Asc.	HH
	-	15 January 2015	Landsat-8	18/33	-
	1st failure: 12 March 2015	25 March 2015	TerraSAR-X	44/Asc.	HH/VV
	2nd failure 13 April 2015	16 April 2015	TerraSAR-X	44/Asc.	HH/VV
Bolshaya Talda	-	14 September 2014	Landsat-8	146/22	-
	1 April 2015	26 April 2015	TerraSAR-X	14/Desc.	HH/VV
		14 August 2015	TerraSAR-X	14/Desc.	HH/VV

[1] TanDEM-X data was acquired in bi-static mode used for DEM generation; TerraSAR-X imagery was acquired in HighResolution SpotLight (HS) mode: 1.2 m (range) × 2.2 m (azimuth) spatial resolution; Landsat-8 provides imagery in 15 m (pan) and 30 m (multi-spectral) spatial resolution; [2] relative orbit: For Landsat-8: WSR-Path/WSR-Row; for TerraSAR-X: Relative orbit/Path direction with ascending (Asc.) and descending (Desc.).

3. Method

The basic principle of the methodology proposed in this article is to detect landslides via change detection of freely available, systematically-acquired HR optical pre-event and VHR PolSAR post-event imagery. Figure 2 shows the flowchart of the procedure. Assuming land cover changes due to a landslide event, *i.e.*, destruction and removal of the vegetation cover the first step (I) of the object-oriented procedure is the pre-selection of formerly-vegetated areas based on the Normalized Difference Vegetation Index (NDVI) of the multispectral pre-event imagery (e.g., Landsat-8 or Sentinel-2). (II) Next, after polarimetric speckle filtering using the edge-preserving refined Lee filter, (III) the H/α decomposition is applied to the post-event polarimetric SAR image to detect, within the pre-selected areas, regions characterized by low entropy (H) values, *i.e.*, evidence of bare soil or rock (landslide material). (IV) Then, assuming a minimum slope value $\delta \geqslant 20°$ (*cf.* Section 3.4) as a necessary requirement for a landslide event, the landslide detection map is refined accordingly. (V) Finally, to reduce the number of false classifications, all detected landslides smaller than a minimum mapping unit (MMU) are excluded (*cf.* Section 3.4).

3.1. Pre-Event Imagery: Selection of Vegetated Areas

Cloud-free optical (multispectral, MS) pre-event imagery is used to derive vegetated areas prior to the landslide event. Landsat-8 or the recently launched Sentinel-2 sensor are the preferred sources as their imagery is freely available and systematically acquired with a high repetition rate of five days (Sentinel-2 constellation) to 16 days (Landsat-8). In the ideal case, the optical imagery is acquired shortly before the landslide event. However, due to too high cloud coverage no useful imagery might be available, and optical imagery acquired one year before the event could be used. To minimize seasonal effects on the change detection procedure described below, it is important to use optical data acquired in the same season as the PolSAR imagery.

By using data of MS sensors working in the visible and near infrared (NIR) region of the electromagnetic spectrum, one is able to calculate the NDVI, being a proxy for the site's vegetation density and greenness [48–50]. The NDVI uses the difference of the vegetation signature between the RED (0.6–0.7 μm) and NIR (0.7–1.1 μm) channel (Equation (1)) [51]:

$$NDVI = \frac{NIR - RED}{NIR + RED} \tag{1}$$

The rationing concept makes the NDVI independent of the illumination, atmospheric effects, topography, *etc.* Consequently, NDVI images acquired at different dates can be compared. The NDVI ranges from −1 to +1. As water has commonly no reflection in infrared, its NDVI is −1. The NDVI value of bare areas (rock, sand, and snow) is less than +0.1. The NDVI increases with vegetation density (NDVI range +0.1 to +0.7) [50]. Vegetated areas in the MS pre-event imagery are selected by setting NDVI > +0.1 as threshold.

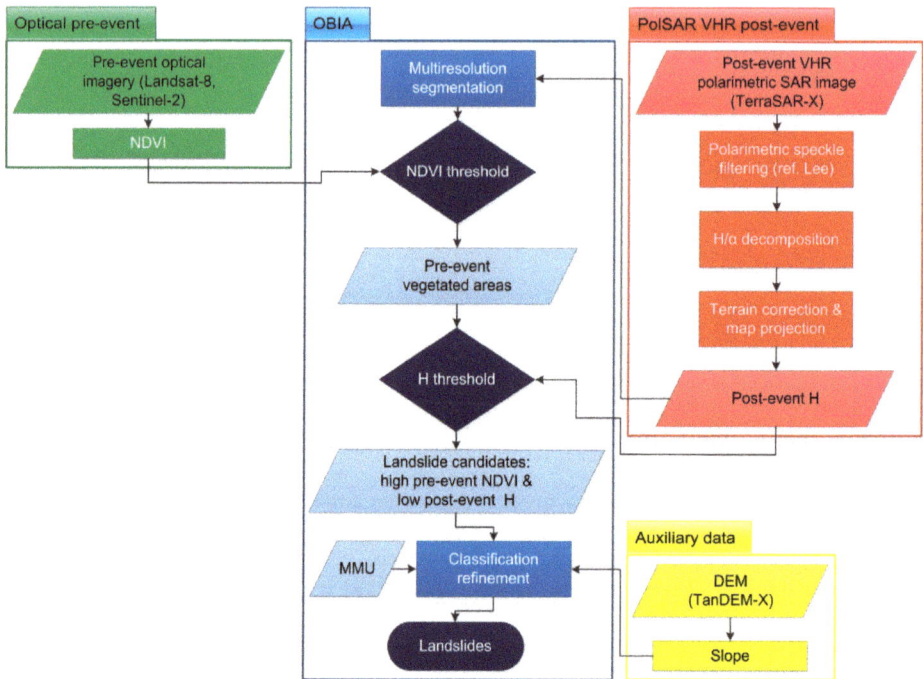

Figure 2. Workflow of the landslide detection procedure based on post-event polarimetric SAR and pre-event optical imagery.

3.2. Post-Event Imagery: Selection of Bare Areas

The selection of bare areas (*i.e.*, possible landslide areas) by means of the post-event VHR PolSAR imagery requires several pre-processing steps, which are described in the following.

3.2.1. Polarimetric Speckle Filtering

The speckle effect, which is caused by the interference of the coherent reflected SAR waves of many individual scatterers within a resolution cell, complicates visual interpretation and classification of SAR images. The natural environment, characterized by distributed targets, is mainly affected by the speckle effect. To reduce this effect, polarimetric speckle filtering using the refined Lee filter is applied [52,53]. This filter aims to preserve the structure of the image, *i.e.*, the edges, while filtering homogenous areas. The correlation between the different polarizations is conserved. The refined Lee filter searches for edges in eight directions: in the vertical, horizontal, and two diagonal directions. A kernel window of 7×7 pixels is used. Then, the covariance matrix is filtered.

3.2.2. Polarimetric Decomposition

Objects with different geometric and structural properties show different SAR backscatter. Based on physical assumptions, polarimetric decomposition procedures aim to separate these different backscatter types [54,55]. As dual-pol SAR imagery was available for this study, the H/α decomposition proposed by Cloude and Pottier [40] was applied.

This decomposition is based on the eigenvalues λ and eigenvectors of the covariance matrix C, which is shown in Equation (2) for the current dual-pol case (HH/VV):

$$\langle C_2 \rangle = \left\langle \begin{bmatrix} |S_{HH}|^2 & S_{HH}S_{VV}^* \\ S_{VV}S_{HH}^* & |S_{VV}|^2 \end{bmatrix} \right\rangle \tag{2}$$

with the Sinclair-Matrix S_{xy} representing the two combinations of transmitted (index y) and received polarization (index x). The superscript * denotes the complex conjugate.

α describes the type of backscattering. α values close to zero indicate domination of surface scattering (single bounce scattering). α values around 45° show domination of volume scattering, caused by multiple scattering inside a volume, such as the crown of a tree or dense vegetation [56]. High α values (up to 90°) represent domination of double-bounce scattering (e.g., in urban area). α_m represents the mean of α_1 and α_2. The former describes the backscattering type of the dominant scatterer and the latter the backscattering type of the second dominant one (Equation (3)):

$$\alpha_m = \frac{1}{\lambda_1 + \lambda_2} \begin{bmatrix} \lambda_1 & \lambda_2 \end{bmatrix} \begin{bmatrix} \alpha_1 \\ \alpha_2 \end{bmatrix} \tag{3}$$

The heterogeneity of the scattering is represented by the entropy H, which ranges from 0 to 1 (Equation (4)). $H = 0$ indicates a dominant scatterer such as a corner reflector. $H \ll 1$ indicates natural areas free of vegetation, *i.e.*, bare soil/rocks and landslide material. High H values with H close to 1 represent a random mixture of scattering mechanisms, e.g., multiple scattering inside the crown of a tree. Therefore, high H values are an indicator of vegetated areas such as forests.

$$H = \frac{-1}{\lambda_1 + \lambda_2} \begin{bmatrix} \lambda_1 & \lambda_2 \end{bmatrix} \log_2 \left(\frac{1}{\lambda_1 + \lambda_2} \begin{bmatrix} \lambda_1 \\ \lambda_2 \end{bmatrix} \right) \tag{4}$$

Rapid landslides [4] remove the vegetation cover. Therefore, the entropy H can be used to detect possible landslides, which are characterized by low H values.

The use of α for landslide detection is more critical, especially in the dual-pol case, which is investigated in this study. As shown for the Yeager Airport landslide, the landslide is indistinguishable in the α image (Figure 3). Landslide material is very heterogeneous. Depending on the geological and environmental setting, landslide material may consist of rocks and debris of different size, as well as trunks and branches of fallen trees. Consequently, within a landslide body, α could show high variable values: α close to zero for rocks/debris and α close to 90° for tree trunks. Therefore, in the following only H is used to differ between landslides and areas not affected by landslides.

The change detection described in Section 3.3 requires all imagery to be in the same coordinate system. Therefore, the H image is orthorectified and map projected, *i.e.*, transformed from the typical SAR geometry (range/azimuth) into a projected coordinate system.

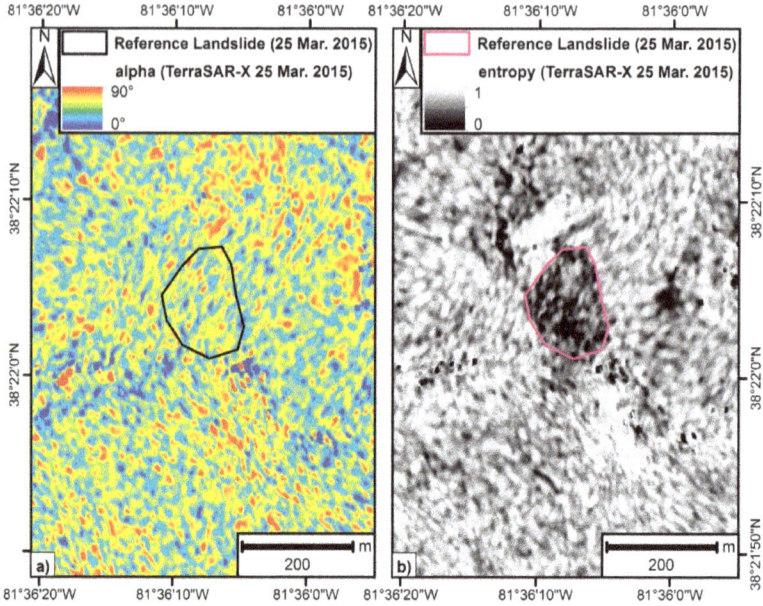

Figure 3. Yeager Airport: (**a**) α angle and (**b**) polarimetric entropy *H* computed from the post-event dual-pol TerraSAR-X (acquired on 25 March 2015). As described above, the landslide is well recognizable in the *H* image but not in the α image. The black/pink polygon marks the reference landslide derived by visual interpretation and manual digitization of the TerraSAR-X imagery.

3.3. Change Detection: Mapping of the Landslides

The landslides are detected by change detection of the aforementioned pre-event NDVI (*cf.* Section 3.1) and the post-event *H* derived from the VHR PolSAR imagery. The basic concept of the proposed methodology is to detect areas free of vegetation based on low *H* values at time t_{post}, which were previously covered by vegetation, *i.e.*, NDVI > +0.1 at time t_{pre}. This change detection is executed in an object-based image analysis (OBIA) environment using the Cognition Network Language (CNL). First, the *H* image is segmented using the multiresolution approach based on the Fractal Net Evolution Approach (FNEA) [57,58]. The developed procedure described above uses a scale parameter of 10. The scale parameter is an abstract value to determine the maximum possible change of heterogeneity with no direct correlation to the object size measured in pixels [59]. A compactness (ranging from 0 to 1) value of 0.5 is chosen. The features of interest are natural ones. Therefore, the shape parameter, ranging from 0 to 1 was set to 0.1.

All thresholds mentioned in the following are mean values for the generated objects, with $\overline{X}$ representing the mean value of all pixels within a certain object for variable X. There are three possible cases (Equation (5)). Based on empirical tests, $\overline{H} \leqslant 0.8$ turned out to be best suited to detect landslide areas.

$$\begin{cases} \overline{NDVI}_{t_{pre}} \leqslant +0.1 & \text{areas free of vegetation at time } t_{pre} \\ \overline{NDVI}_{t_{pre}} > +0.1 \wedge \overline{H}_{t_{post}} \leqslant 0.8 & \text{landslide candidate} \\ \overline{NDVI}_{t_{pre}} > 0.1 \wedge \overline{H}_{t_{post}} > 0.8 & \text{no landslide candidate} \end{cases} \quad (5)$$

Areas of $\overline{NDVI} \leqslant +0.1$ are free of vegetation at time t_{pre} and are not considered in the following. Based on a threshold of $\overline{NDVI} > +0.1$ vegetated areas at the time t_{pre} before the landslide event are selected (*cf.* Section 3.1). Then, the polarimetric entropy $\overline{H}$ (*cf.* Section 3.2.2) is investigated for the areas, which were vegetated at time t_{pre}. Low H values indicate areas free of vegetation, such as landslide material. Consequently, objects are selected as landslide candidates if they are covered by vegetation at t_{pre} but are free of vegetation at t_{post}, *i.e.*, if $\overline{NDVI}_{t_{pre}} > +0.1 \wedge \overline{H}_{t_{post}} \leqslant 0.8$ is true (Equation (5)).

3.4. Refinement of Classification by Topographic Information

Assuming a minimum slope δ value as a necessary requirement for a landslide event, the landslide detection map, described in Section 3.3, is refined as follows: only landslide candidates located at $\overline{\delta} \geqslant 20°$ are selected as final landslides (Equation (6)).

$$\begin{cases} \overline{NDVI}_{t_{pre}} > +0.1 \wedge \overline{H}_{t_{post}} \leqslant 0.8 \wedge \overline{\delta} \geqslant 20° & \text{final landslide} \\ \overline{NDVI}_{t_{pre}} > +0.1 \wedge \overline{H}_{t_{post}} \leqslant 0.8 \wedge \overline{\delta} < 20° & \text{no landslide} \end{cases} \tag{6}$$

To detect the foot of a landslide region-growing into neighboring areas of $\overline{H} \leqslant 0.8$ and $\overline{\delta} > 12°$ is executed. Next, all neighboring landslide objects are merged to a common object.

As the Yeager Airport landslide took place on an artificial slope, which was constructed in the year 2007 (*cf.* Section 2), the Shuttle Radar Topography Mission (SRTM) DEM [60] from the year 2000 is too old and could not be used for the slope analysis at this site. Therefore, we used a bi-static TanDEM-X [61] dataset acquired on 2 April 2014 to generate via SAR interferometric (InSAR) analysis an up-to-date DEM of 12 m spatial resolution. Therefore, a more accurate measurement of the pre-failure slope was obtained.

Analysis of optical imagery (Landsat, as well as GoogleEarth) showed that the second study site, *i.e.*, the mining waste landslide in Russia, is a very dynamic area with lots of changes that occurred after the last TanDEM-X acquisition on 23 August 2012 over this area. Therefore, neither the SRTM DEM nor a TanDEM-X DEM could be used for slope analysis in this area.

Finally, to decrease the number of false classifications, all detected landslides smaller than a minimum mapping unit (MMU) of 30 m × 30 m are excluded.

4. Results and Discussion

4.1. The Yeager Airport Landslide, Charleston, West Virginia, USA

First, the methodology is applied to Test Case 1: the Yeager Airport landslide, located near Charleston, West Virginia, USA. Figures 4–6 show the results of the developed landslide detection procedure based on pre-event optical Landsat-8 imagery and post-event polarimetric VHR TerraSAR-X imagery acquired on 25 March 2015 (after the first failure of the slope) and 16 April 2015 (after the second slope failure). Figures 5 and 6 demonstrate that the landslide is very well detected by the classification. In the SAR image acquired on 25 March 2015, one can also see a small over classification east of the landslide. At the SAR image acquired after the second slope failure the landslide detection procedure was able to detect the main part of the landslide. However, the scarp area of the landslide is not detected. Here, the second slope failure extended the scarp area of the landslide. At this very steep part, geometric distortions such as layover and foreshortening occurred, changing the SAR backscattering values, which influence the landslide detection procedure.

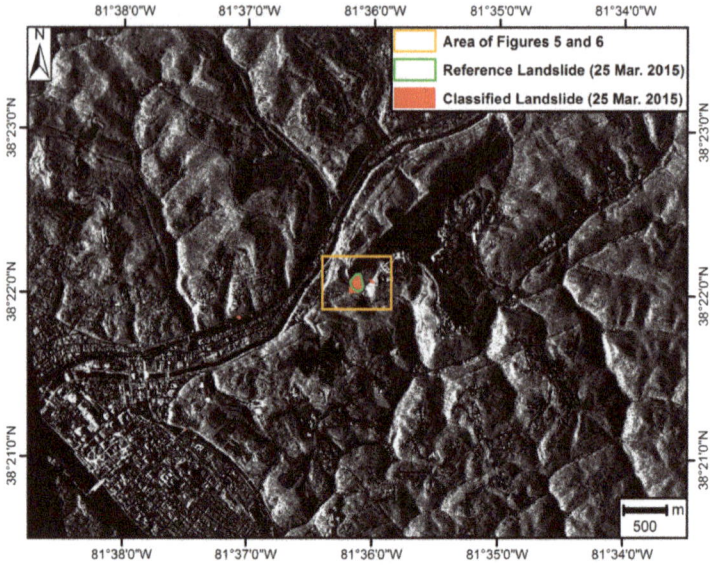

Figure 4. Overview image of the Yeager Airport landslide. TerraSAR-X data acquired on 25 March 2015 serves as background image. The orange box shows the area of the Figures 5 and 6. TerraSAR-X © 2015 German Aerospace Center (DLR), 2015 Airbus Defence and Space/Infoterra GmbH.

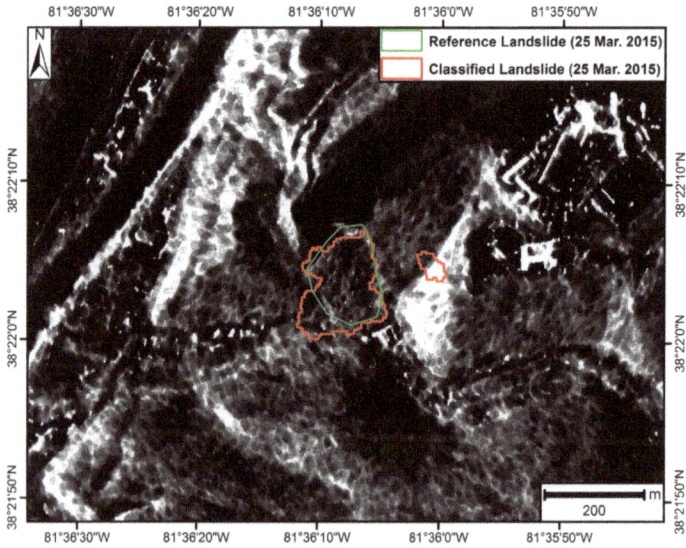

Figure 5. Results (red) of the landslide detection procedure at the Yeager Airport based on post-event dual-pol TerraSAR-X (acquired on 25 March 2015; background image) and pre-event optical Landsat-8 (15 January 2015) imagery. The green polygon marks the reference landslide derived by visual interpretation and manual digitization of the TerraSAR-X imagery. TerraSAR-X © 2015 German Aerospace Center (DLR), 2015 Airbus Defence and Space/Infoterra GmbH.

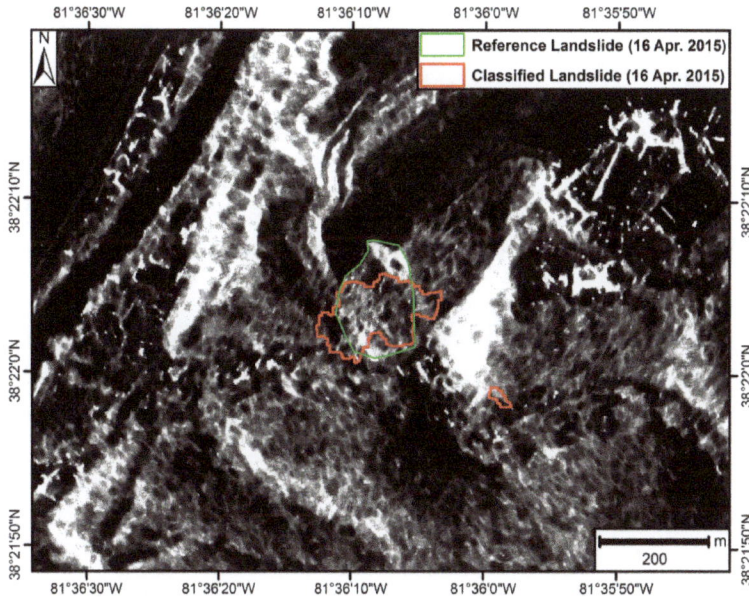

Figure 6. Result (red) of the landslide detection procedure at the Yeager Airport based on post-event dual-pol TerraSAR-X (acquired on 16 April 2015; background image) and pre-event optical Landsat-8 (15 January 2015) imagery. The green polygon marks the reference landslide derived by visual interpretation and manual digitization of the TerraSAR-X imagery. TerraSAR-X © 2015 German Aerospace Center (DLR), 2015 Airbus Defence and Space/Infoterra GmbH.

Table 2 shows the accuracy values of the methodology applied to the Yeager Airport landslide at the two stages of the landslide event. We applied an area-based accuracy assessment. The classification results are compared to a polygon of the landslide derived by visual interpretation of the SAR image and manual digitization. The landslide boundary is clearly visible in the SAR image due to its different roughness compared to the surroundings. Furthermore, photographs taken in the field or by airplanes (Figure 1), as well as the drone video (Supplementary Materials S2), were used to refine the reference polygon of the landslide.

Table 2. Classification accuracies for the Yeager Airport landslide (12 March and 13 April 2015). Overall accuracy (OA), user's (UA), and producer's accuracy (PA).

Date	OA [%]	PA Landslide [%]	UA Landslide [%]	PA Other [%]	UA Other [%]	KHAT
25 March 2015	99.9	87.0	67.4	99.9	100.0	+0.759
16 April 2015	99.9	64.3	66.9	99.9	99.9	+0.655

It is important to note that not only the relatively small areas shown in Figures 3–9 were considered for the validation procedure, but the entire TerraSAR-X HS scene (5 km azimuth × 10 km ground range). The values of the confusion matrix are described as follows: the overall accuracy (OA), ranging from 0%–100%, describes the ratio of correctly classified area units (e.g., pixels) to the total number of pixels of the satellite scene. The producer's accuracy (PA), ranging from 0%–100%, gives the percentage of reference data detected by the classification, while the user's accuracy (also ranging from 0%–100%) describes the percentage of the classification matching with the reference data. Finally, the KHAT statistics gives information about the strength of the correlation between the classification result and the reference data. The KHAT coefficient ranges from 0, representing a completely random match between classification result and reference data, to +1, representing no random match between classification result and the reference [62].

The OA, as well as the values UA and PA, of the class other are very high (99.9%). This is due to a two-class problem, *i.e.*, landslide and other (areas not affected by landslides), with the percentage area of the latter being much higher than the percentage area coverage of landslide. Consequently, the interesting parameters of the accuracy assessment are the UA and PA values of the landslides class.

The good matching between the landslide classification result and the reference data, shown in Figure 5 for the SAR image acquired after the first slope failure, is reflected by the high UA and PA values of the landslide class of *ca.* 64% and 87%, respectively (*cf.* Table 2). Contrary to this, the classification results of the image acquired after the second slope failure clearly show lower values for the landslide class PA while the landslide class UA is stable. The KHAT coefficient of the two SAR acquisitions shows a similar behavior.

4.2. Mining Waste Landslide near Bolshaya Talda, Kemerovo Oblast, Russia

Second, the methodology is applied to Test Case 2: the mining waste landslide near Bolshaya Talda, Kemerovo Oblast, Russia. Figure 7 shows a false color composite pre-event Landsat-8 imagery acquired on 14 September 2014 of the mining waste site.

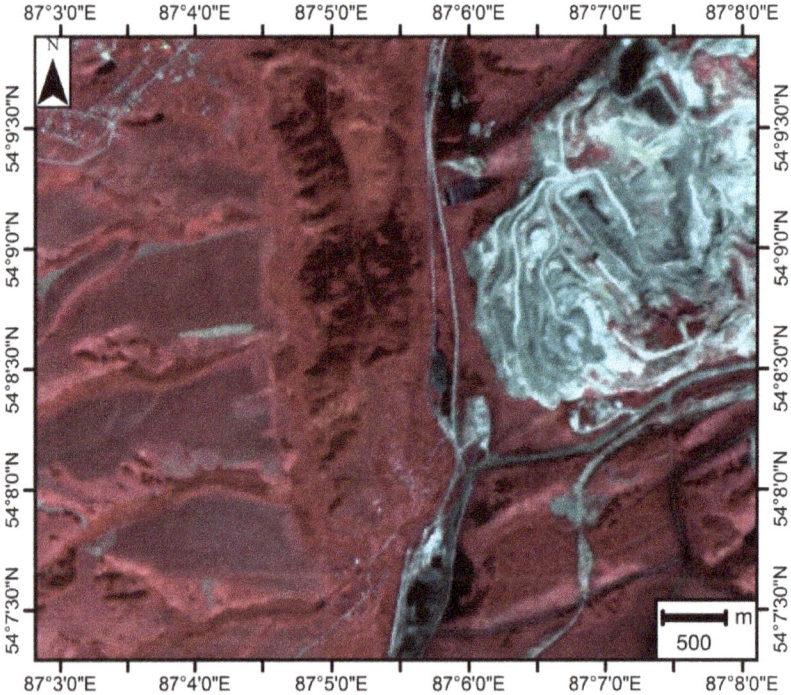

Figure 7. Landsat-8 pre-event imagery (14 September 2014) of the Russian mining waste site (NIR/Red/Green). The mining area (cyan) is free of vegetation. Landsat-8 © USGS 2014.

The landslide was covered twice by dual-pol TerraSAR-X HS imagery: On 26 April 2015 and 14 August 2015. Figures 8 and 9 show the corresponding classification results of the developed landslide detection procedure as well as the reference data derived by visual interpretation, *i.e.*, visual change detection between the pre-event optical HR and post-event SAR image data, and manual digitization.

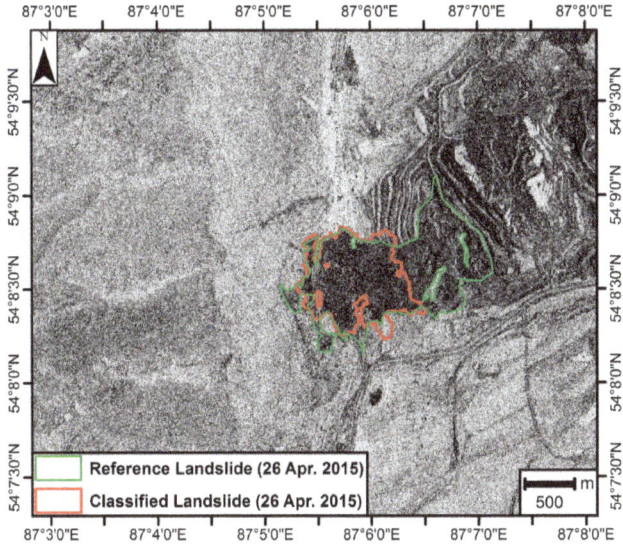

Figure 8. Result (red) of the landslide detection procedure at the Russian mining waste site based on post-event dual-pol TerraSAR-X (acquired on 26 April 2015; background image) and pre-event optical Landsat-8 (14 September 2014) imagery. Green: Reference derived by visual interpretation and manual digitization of the TerraSAR-X imagery. TerraSAR-X © 2015 German Aerospace Center (DLR), 2015 Airbus Defence and Space/Infoterra GmbH.

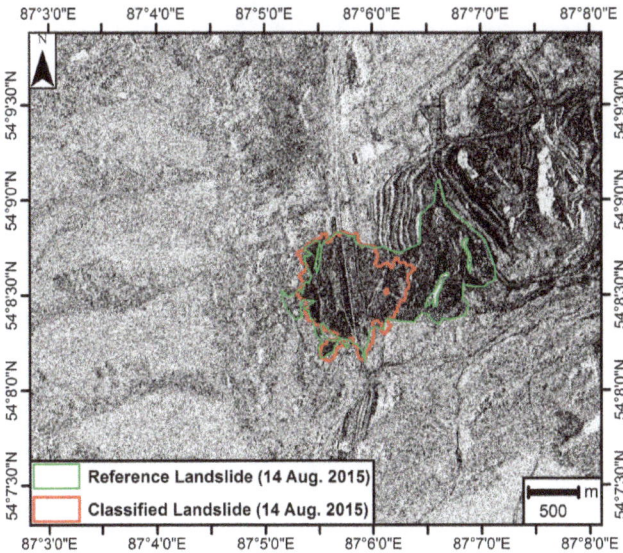

Figure 9. Result (red) of the landslide detection procedure at the Russian mining waste site based on post-event dual-pol TerraSAR-X (acquired on 14 August 2015; background image) and pre-event optical Landsat-8 (14 September 2014) imagery. Green: reference derived by visual interpretation and manual digitization of the TerraSAR-X imagery. TerraSAR-X © 2015 German Aerospace Center (DLR), 2015 Airbus Defence and Space/Infoterra GmbH.

The figures clearly show that the developed methodology is only able to detect parts of the landslide. UA values of the landslide class are very high, with *ca.* 90%, *i.e.*, most of the classification results are correct. However, the low PA values (landslide class) show that only *ca.* 50% of the real landslide area is detected (Table 3). The reason for this is that the methodology assumes that the landslide area was covered by vegetation before the landslide event. However, this is only true for the area at the foot of the landslide, where the landslide ran over an area outside of the mining waste area (formerly vegetated area). The landslide moved from east to west. The mining waste area itself is free of vegetation and, therefore, shows low NDVI values in the pre-event optical imagery (Figure 7). Therefore, the developed methodology is not able to detect the part of the landslide which occurred inside the mining waste area. Furthermore, the entropy H, alone, would only detect the entire mining waste area and not only the landslide area, as the entire mining waste area is characterized by low H values, *i.e.*, the area is free of vegetation (bare soil).

Table 3. Classification accuracies for the Russian mining waste landslide (01 April 2015). Overall accuracy (OA), user's (UA), and producer's accuracy (PA).

Date	OA [%]	PA Landslide [%]	UA Landslide [%]	PA Other [%]	UA Other [%]	KHAT
26 April 2015	96.8	48.2	89.6	99.7	97.0	+0.612
14 August 2015	96.8	49.7	90.0	99.7	97.0	+0.625

When considering only the area outside of the original mining site as affected landslide area, the PA of the landslide class increases to 83%–90%, while the UA of the landslide class slightly decreases (Table 4).

The developed landslide detection methodology based on post-event polarimetric VHR SAR imagery showed promising accuracy values. However, the limitation of this methodology is that only landslides at slopes previously covered by vegetation can be detected.

Table 4. Classification accuracies for the Russian mining waste landslide (1 April 2015). Only landslide material outside the original mining site is treated as a landslide. Overall accuracy (OA), user's (UA), and producer's accuracy (PA).

Date	OA [%]	PA Landslide [%]	UA Landslide [%]	PA Other [%]	UA Other [%]	KHAT
26 April 2015	98.8	83.1	76.4	99.3	99.5	+0.790
14 August 2015	99.1	90.0	81.6	99.4	99.7	+0.850

4.3. General Discussion

Other landslide detection methodologies reported in the literature are based on change detection approaches using pre- and post-event PolSAR imagery [16,29,35,37,39,63]. As the side-looking geometry of SAR systems causes geometric distortions (e.g., shadowing, foreshortening, *etc.*) identical acquisition geometries for pre- and post-disaster imagery are required for change detection applications. However, as mentioned in Section 1, VHR archive SAR imagery acquired shortly before a disaster event, in particularly at the same image geometry as the next possible post-event acquisition, is in most cases not available. The advantage of the methodology presented in this article is that in addition to freely available and systematically acquired pre-event HR optical imagery only post-event VHR PolSAR is required, which can be programmed before the next overpass of the satellite, independent of any geometrical restrictions by a pre-event SAR imagery. Nevertheless, to guarantee useful image acquisition geometries, *i.e.*, to decrease the influence of layover and shadowing effects, the terrain and the slope's orientation in space should be considered [64]. Also, the procedure proposed in the current article is based on change detection. Contrary to the aforementioned studies, we use an optical

HR imagery, which is systematically acquired and freely available for the entire Earth's landmass (*cf.* Section 3.1).

Compared to other published procedures, which apply a pixel-based classification of the PolSAR imagery, e.g., [16,29,35,37], the proposed OBIA methodology enables an incorporation of optical pre-event imagery and also the additional use of DEM data, which strongly increases the classification accuracy.

According to Shimada *et al.* [17] and Dabbiru *et al.* [65], L-band is better suited for landslide detection than X-band, as due to its longer wavelength, L-band is characterized by a better penetration through the forest canopy. On the other hand, the shorter wavelength of X-band enables the acquisition of higher spatial resolution imagery compared to L-band. Consequently, smaller and thinner landslides are only detectable by X-band. In summary, one has to find a compromise between the degree of penetration through vegetation (*i.e.*, detection of landslide mass under tree canopy) and spatial resolution (*i.e.*, detection of smaller landslides, not covered by vegetation)—and data availability. L-band imagery is currently only provided by ALOS-2/PALSAR-2. However, its imagery is acquired at a pre-planned acquisition schedule. Contrary to this, X-band satellite missions, such as TerraSAR-X, acquire imagery after an on demand tasking, enabling a very flexible acquisition and fast reaction in case of a disaster, e.g., a landslide event. Therefore, this article presented a methodology based on VHR X-band PolSAR imagery.

The most important information for the landslide identification is the polarimetric entropy *H* derived from the post-event SAR imagery. As shown in Table 1, the proposed methodology is based on TerraSAR-X HS imagery of 1.2 m (range) × 2.2 m (azimuth) spatial resolution. Based on our experience we can report that landslides with a minimum size of 30 m × 30 m are detectable using only VHR PolSAR imagery. The pre-failure information on vegetation cover can be derived from 10 m to 30 m spatial resolution optical imagery (e.g., Sentinel-2 or Landsat-8). The aforementioned MMU of 30 m × 30 m was also chosen to consider the spatial resolution of Landsat-8. Since very small landslides as mentioned above cover only 1 to 9 pixels in the optical imagery, it is very difficult to identify such landslides in the optical imagery itself. However, this does not influence the results of the methodology, as the optical data is only used to derive NDVI information for selecting pre-failure vegetated slopes. The segmentation of the OBIA approach is based on the higher spatial resolution PolSAR imagery (*cf.* Section 3.3).

The proposed landslide detection methodology is based on change detection between pre-event optical and post-event PolSAR imagery. The change we focus on is the removal of the vegetation cover by the landslide. However, as slow-moving slides and slide-earth flows often still preserve entire portions of undisturbed vegetation cover, the proposed methodology is limited to the detection of rapid and faster movements [4], *i.e.*, to the detection of landslides where the vegetation cover was removed.

Czuchlewski *et al.* [21] used VHR L-band airborne SAR imagery of full polarization to map landslides after the 1999 Chi-Chi earthquake in Taiwan. Despite of their promising results, a worldwide applicability of airborne PolSAR sensors is not feasible. Only satellite-based remote sensing enables fast response to disasters with global applicability.

Quad-pol imagery provides more information on the backscattering characteristics than dual-pol imagery. In general, the former achieves higher classification accuracies [28]. The methodology proposed in the current article is based on TerraSAR-X imagery. This SAR sensor provides only dual-pol imagery in operational mode (quad-pol data are only available in an experimental mode). Nevertheless, Sections 4.1 and 4.2 showed that high accuracy values can be achieved even with dual-pol imagery.

5. Conclusions

This article presented a fast and transferable methodology for landslide detection. The procedure combines post-event Very High Resolution (VHR) Polarimetric Synthetic Aperture Radar (PolSAR) imagery of TerraSAR-X with pre-event multispectral imagery of Landsat-8 or Sentinel-2.

First, vegetated slopes are selected in the pre-event imagery using the Normalized Difference Vegetation Index (NDVI) and a Digital Elevation Model (DEM). Second, based on the post-event VHR PolSAR imagery, areas of low polarimetric entropy H, derived by the entropy/alpha (H/α) polarimetric decomposition, are extracted. Low H values represent areas free of vegetation, e.g., areas covered by landslide material, whereas high H values are characteristic for densely vegetated areas, such as forests. Landslides are detected by change detection between the pre-event optical and the post-event PolSAR imagery. More precisely, possible landslides are areas, which are characterized by a high NDVI value in the pre-event imagery and a low H value in the post-event acquisition.

The developed landslide detection procedure is characterized by the following advantages:

1. The utilization of SAR imagery allows fast response in a crisis situation due to the day/night availability and almost complete weather independency of the SAR system. As heavy rain events are an important trigger for landslides, optical sensors, relying on a cloud-free sky to be able to provide a useful imagery are, in many cases, not suited.

2. The presented methodology requires only freely-available and systematically-acquired pre-event optical high resolution imagery and post-event VHR PolSAR imagery. Other landslide mapping procedures, which are based on change detection using SAR imagery, require pre- and post-event VHR SAR imagery. However, the VHR archive SAR imagery acquired shortly before a landslide event are, in most cases, not available. This is especially true when a certain imaging geometry is required determined by the next possible SAR acquisition over the crisis area. Modern VHR SAR missions, such as COSMO-SkyMed, TerraSAR-X, or RADARSAT-2 do not systematically cover the entire Earth's landmass.

3. The methodology proposed in this article is also based on change detection. However, high-resolution optical imagery of Landsat-8 or Sentinel-2 is used as pre-event information. As these imagery are freely available and systematically acquired on the entire Earth's landmass at high repetition rates (*cf.* Section 3.1), it is guaranteed that useful, *i.e.*, cloud-free, pre-event imagery is available for the entire Earth's landmass. In the ideal case, the optical imagery is acquired shortly before the landslide event. However, in cases where cloud coverage is too high, cloud-free optical imagery acquired at the same season one year before could be used.

The methodology was successfully applied to two landslide case studies of different characteristics: a rotational slide near Charleston, West Virginia, USA, which occurred on 12 March 2015, and a mining waste earthflow near Bolshaya Talda, Russia, which occurred on 1 April 2015.

In the future, the developed methodology will be applied and tested on further upcoming landslide events, also including applicability tests during rapid mapping activities of DLR's Center for Satellite Based Crisis Information (ZKI). In addition to the Landsat-8 data utilized in this study, imagery of the recently launched Sentinel-2 will also be employed as pre-event information.

Supplementary Materials: The following are available online at www.mdpi.com/2072-4292/8/4/307/s1, Video S1: First slope failure of the Yeager Airport landslide (from AGU landslide blog of Dave Petley [43]), Video S2: Drone flight over the Yeager Airport landslide (AGU landslide blog of Dave Petley [43]), Video S3: Moving landslide mass at Bolshaya Talda mining waste landslide (It is originally from [47]).

Acknowledgments: The authors would like to thank Dave Petley, who provided with his AGU landslide blog valuable background information about the studied landslides (http://blogs.agu.org/landslideblog/). The authors thank the three anonymous reviewers for their very constructive remarks. This work has been funded by ESA in the framework of the ASAPTERRA project (Contract No.: 4000112375/14/I-NB). TerraSAR-X and TanDEM-X data was provided by DLR through the MTH2790 and HYDR6913 projects.

Author Contributions: Simon Plank, the principle author, wrote the paper and developed and tested the methodology. André Twele and Sandro Martinis supported the application of the technique and provided suggestions for its improvement. All authors read, revised and approved the final manuscript.

Conflicts of Interest: The authors declare no conflict of interest.

Abbreviations

The following abbreviations are used in this manuscript:

ALOS	Advanced Land Observing Satellite
CNL	Cognition Network Language
DEM	Digital Elevation Model
DLR	German Aerospace Center
EMAS	Engineered Material Arrestor System
EO	Earth Observation
FAA	Federal Aviation Administration
FNEA	Fractal Net Evolution Approach
HR	High Resolution
HS	HighResolution SpotLight
InSAR	Synthetic Aperature Radar Interferometry
LiDAR	Light Detection And Ranging
MS	Multispectral
NDVI	Normalized Difference Vegetation Index
NIR	Near-Infra-Red
OA	Overall Accuracy
OBIA	Object-Based Image Analysis
PA	Producer's Accuracy
PALSAR	Phased Array type L-band Synthetic Aperture Radar
PolSAR	Polarimetric Synthetic Aperture Radar
RVI	Radar Vegetation Index
SAR	Synthetic Aperture Radar
SRTM	Shuttle Radar Topography Mission
UA	User's Accuracy
USGS	United States Geological Survey
VHR	Very High Resolution

References

1. Petley, D. Global patterns of loss of life from landslides. *Geology* **2012**, *40*, 927–930. [CrossRef]
2. Voigt, S.; Kemper, T.; Riedlinger, T.; Kiefl, R.; Scholte, K.; Mehl, H. Satellite image analysis for disaster and crisis-management support. *IEEE Trans. Geosci. Remote Sens.* **2007**, *45*, 1520–1528. [CrossRef]
3. Singleton, A.; Li, Z.; Hoey, T.; Muller, J.-P. Evaluating sub-pixel offset techniques as an alternative to D-InSAR for monitoring episodic landslide movements in vegetated terrain. *Remote Sens. Environ.* **2014**, *147*, 133–144. [CrossRef]
4. Cruden, D.M.; Varnes, D.J. Landslides Types and Processes. In *Landslides: Investigation and Mitigation. Special Report 247, Transportation Research Board, National Research Council*; Turner, A.K., Schuster, R.L., Eds.; National Academy Press: Washington, DC, USA, 1996; pp. 36–75.
5. Manconi, A.; Casu, F.; Ardizzone, F.; Bonano, M.; Cardinali, M.; de Luca, C.; Gueguen, E.; Marchesini, I.; Parise, M.; Vennari, C.; *et al.* Brief Communication: Rapid mapping of landslide events: The 3 December 2013 Montescaglioso landslide, Italy. *Nat. Hazards Earth Syst. Sci.* **2014**, *14*, 1835–1841. [CrossRef]
6. Tralli, D.M.; Blom, R.G.; Zlotnicki, V.; Donnellan, A.; Evans, D.L. Satellite remote sensing of earthquake, volcano, flood, landslide and coastal inundation hazards. *ISPRS J. Photogramm. Remote Sens.* **2005**, *59*, 185–198. [CrossRef]
7. Scaioni, M.; Longoni, L.; Melillo, V.; Papini, M. Remote Sensing for Landslide Investigations: An Overview of Recent Achievements and Perspectives. *Remote Sens.* **2014**, *6*, 9600–9652. [CrossRef]

8.   Joyce, K.E.; Samsonov, S.V.; Levick, S.R.; Engelbrecht, J.; Belliss, S. Mapping and monitoring geological hazards using optical, LiDAR, and synthetic aperture RADAR image data. *Nat. Hazards* **2014**, *73*, 137–163. [CrossRef]

9.   Martha, T.R.; Kerle, N.; Jetten, V.; van Westen, C.J.; Kumar, K.V. Characterising spectral, spatial and morphometric properties of landslides for semi-automatic detection using object-oriented methods. *Geomorphology* **2010**, *116*, 24–36. [CrossRef]

10.   Mondini, A.C.; Chang, K.-T.; Yin, H.-Y. Combining multiple change detection indices for mapping landslides triggered by typhoons. *Geomorphology* **2011**, *134*, 440–451. [CrossRef]

11.   Behling, R.; Roessner, S.; Segl, K.; Kleinschmit, B.; Kaufmann, H. Robust Automated Image Co-Registration of Optical Multi-Sensor Time Series Data: Database Generation for Multi-Temporal Landslide Detection. *Remote Sens.* **2014**, *6*, 2572–2600. [CrossRef]

12.   Behling, R.; Roessner, S.; Kaufmann, H.; Kleinschmit, B. Automated Spatiotemporal Landslide Mapping over Large Areas Using RapidEye Time Series Data. *Remote Sens.* **2014**, *6*, 8026–8055. [CrossRef]

13.   Hölbling, D.; Füreder, P.; Antolini, F.; Cigna, F.; Casagli, N.; Lang, S. A Semi-Automated Object-Based Approach for Landslide Detection Validated by Persistent Scatterer Interferometry Measures and Landslide Inventories. *Remote Sens.* **2012**, *4*, 1310–1336. [CrossRef]

14.   Othman, A.A.; Gloaguen, R. Automatic Extraction and Size Distribution of Landslides in Kurdistan Region, NE Iraq. *Remote Sens.* **2013**, *5*, 2389–2410. [CrossRef]

15.   Mondini, A.C.; Guzzetti, F.; Reichenbach, P.; Rossi, M.; Cardinali, M.; Ardizzone, F. Semi-automatic recognition and mapping of rainfall induced shallow landslides using optical satellite images. *Remote Sens. Environ.* **2011**, *7*, 1743–1775. [CrossRef]

16.   Shibayama, T.; Yamaguchi, Y. An application of polarimetric radar analysis on geophysical phenomena. In Proceedings of the IEEE IGARSS, Melbourne, Australia, 21–26 July 2013; pp. 3191–3194.

17.   Shimada, M.; Watanabe, M.; Motooka, T.; Ohki, M.; Wada, Y. PALSAR-2 and Pi-SAR-L2—Multi frequency Polarimetric Sensitivity on Disaster. In Proceedings of the EUSAR Berlin, Germany, 3–5 June 2014; pp. 93–94.

18.   Plank, S. Rapid Damage Assessment by Means of Multi-Temporal SAR—A Comprehensive Review and Outlook to Sentinel-1. *Remote Sens.* **2014**, *6*, 4870–4906. [CrossRef]

19.   McKean, J.; Roering, J. Objective landslide detection and surface morphology mapping using high-resolution airborne laser altimetry. *Geomorphology* **2004**, *57*, 331–351. [CrossRef]

20.   Santangelo, M.; Cardinali, M.; Rossi, M.; Mondini, A.C.; Guzzetti, F. Remote landslide mapping using a laser rangefinder binocular and GPS. *Nat. Hazards Earth Syst. Sci.* **2010**, *10*, 2539–2546. [CrossRef]

21.   Czuchlewski, K.R.; Weissel, J.K.; Kim, Y. Polarimetric synthetic aperture radar study of the Tsaoling landslide generated by the 1999 Chi-Chi earthquake, Taiwan. *J. Geophys. Res.* **2003**, *108*, 6006. [CrossRef]

22.   Christophe, E.; Chai, A.S.; Yin, T.; Kwoh, L.K. 2009 Earthquakes in Sumatra: The Use of L-band Interferometry in a SAR-Hostile Environment. In Proceedings of the IEEE IGARSS, Honolulu, HI, USA, 25–30 July 2010; pp. 1202–1205.

23.   Kawamura, M.; Tsujino, K.; Tsujiko, Y.; Tanjung, J. Detection Method of Slope Failures Due to the 2009 Sumatra Earthquake by Using TerraSAR-X Images. In Proceedings of the IEEE IGARSS, Vancouver, BC, Canada, 24–29 July 2011; pp. 4292–4295.

24.   Ferretti, A.; Prati, C.; Rocca, F. Nonlinear Subsidence Rate Estimation Using Permanent Scatterers in Differential SAR Interferometry. *IEEE Trans. Geosci. Remote Sens.* **2000**, *38*, 2202–2012. [CrossRef]

25.   Ferretti, A.; Fumagalli, A.; Novali, F.; Prati, C.; Rocca, F.; Rucci, A. A New Algorithm for Processing Interferometric Data-Stacks: SqueeSAR. *IEEE Trans. Geosci. Remote Sens.* **2011**, *49*, 3460–3470. [CrossRef]

26.   Berardino, P.; Fornaro, G.; Lanari, R.; Sansosti, E. A new algorithm for surface deformation monitoring based on small baseline differential SAR interferograms. *IEEE Trans. Geosci. Remote Sens.* **2002**, *40*, 2375–2383. [CrossRef]

27.   Raspini, F.; Ciampalini, A.; Del Conte, S.; Lombardi, L.; Nocentini, M.; Gigli, G.; Ferretti, A.; Casagli, N. Exploitation of Amplitude and Phase of Satellite SAR Images for Landslide Mapping: The Case of Montescaglioso (South Italy). *Remote Sens.* **2015**, *7*, 14576–14596. [CrossRef]

28.   Ainsworth, T.L.; Kelly, J.P.; Lee, J.S. Classification comparisons between dual-pol, compact polarimetric and quad-pol SAR imagery. *ISPRS J. Photogramm. Remote Sens.* **2009**, *64*, 464–471. [CrossRef]

29.   Rodriguez, K.M.; Weissel, J.K.; Kim, Y. Classification of Landslide Surfaces Using Fully Polarimetric SAR: Examples from Taiwan. In Proceedings of the IEEE IGARSS, Toronto, Canada, 24–28 July 2002; pp. 2918–2920.

30. Cui, M.; Prasad, S.; Mahrooghy, M.; Aastoos, J.V.; Lee, M.A.; Bruce, L.M. Decision Fusion of Textural Features Derived From Polarimetric Data for Levee Assessment. *IEEE J-STARS* **2012**, *5*, 970–976. [CrossRef]

31. Plank, S.; Hölbling, D.; Eisank, C.; Friedl, B.; Martinis, S.; Twele, A. Comparing object-based landslide detection methods based on polarimetric SAR and optical satellite imagery—A case study in Taiwan. In Proceedings of the 7th International Workshop on Science and Applications of SAR Polarimetry and Polarimetric Interferometry, POLinSAR 2015, Frascati, Italy, 26–30 January 2015; p. 5.

32. Freeman, A.; Durden, S.L. A three-component scattering model for polarimetric SAR data. *IEEE Trans. Geosci. Remote Sens.* **1998**, *36*, 963–973. [CrossRef]

33. Yamaguchi, Y.; Moriyama, T.; Ishido, M.; Yamada, H. Four-component scattering model for polarimetric SAR image decomposition. *IEEE Trans. Geosci. Remote Sens.* **2005**, *43*, 1699–1706. [CrossRef]

34. Yamaguchi, Y.; Sato, A.; Boener, W.-M.; Sato, R.; Yamada, H. Four-Component Scattering Power Decomposition with Rotation of Coherency Matrix. *IEEE Trans. Geosci. Remote Sens.* **2011**, *49*, 2251–2258. [CrossRef]

35. Watanabe, M.; Yonezawa, C.; Iisaka, J.; Sato, M. ALOS/PALSAR full polarimetric observations of the Iwate-Miyagi Nairiku earthquake of 2008. *Int. J. Remote Sens.* **2012**, *33*, 1234–1245. [CrossRef]

36. Yamaguchi, Y. Disaster Monitoring by Fully Polarimetric SAR Data Acquired With ALOS-PALSAR. *IEEE Proc.* **2012**, *100*, 2851–2860. [CrossRef]

37. Shibayama, T.; Yamaguchi, Y. A landslide detection based on the change of scattering power components between multi-temporal PolSAR data. In Proceedings of the IEEE IGARSS, Quebec, Canada, 13–18 July 2014; pp. 2734–2737.

38. Shibayama, T.; Yamaguchi, Y.; Yamada, H. Polarimetric Scattering Properties of Landslides in Forested Areas and the Dependence on the Local Incidence Angle. *Remote Sens.* **2015**, *7*, 15424–15442. [CrossRef]

39. Yonezawa, C.; Watanabe, M.; Saito, G. Polarimetric decomposition analysis of ALOS PALSAR observation data before and after a landslide event. *Remote Sens.* **2012**, *4*, 2314–2328. [CrossRef]

40. Cloude, S.R.; Pottier, E. An entropy based classification scheme for land applications of polarimetric SAR. *IEEE Trans. Geosci. Remote Sens.* **1997**, *35*, 68–78. [CrossRef]

41. Charleston Gazette. 22 March 2015. Available online: http://www.wvgazettemail.com/article/20150322/GZ01/150329837 (accessed on 23 March 2015).

42. STGEC 2010. Available online: https://stgec.org/presentations/STGEC_2010/2010%20STGEC%20-%20Yeager%20Airport%20-%20Tallest%20Reinforced%20Slope%20in%20N%20America.pdf (accessed on 21 December 2015).

43. Petley, D. AGU Landslide Blog. Yeager Airport Landslide. 2015. Available online: http://blogs.agu.org/landslideblog/2015/04/14/yeager-airport-landslide-next/ (accessed on 15 April 2015).

44. Charleston Gazette. 12 March 2015. Available online: http://www.wvgazettemail.com/article/20150312/DM05/150319672 (accessed on 13 March 2015).

45. Charleston Gazette. 13 March 2015. Available online: http://www.wvgazettemail.com/article/20150313/DM01/150319516/2007062715 (accessed on 14 March 2015).

46. Charleston Gazette. 13 April 2015. Available online: http://www.wvgazettemail.com/article/20150413/GZ01/150419806 (accessed on 14 April 2015).

47. Petley, D. AGU Landslide Blog. The Bolshaya Talda Earthflow in Russia Was a Mine Waste Failure. 2015. Available online: http://blogs.agu.org/landslideblog/2015/04/20/bolshaya-talda-1/ (accessed on 21 April 2015).

48. Lillesand, T.M.; Kiefer, R.W. *Remote Sensing and Image Interpretation*, 4th ed.; John Wiley & Sons: New York, NY, USA, 2000.

49. Gupta, R.P. *Remote Sensing Geology*; Springer: Heidelberg, Germany, 2003.

50. Albertz, J.; Wiggenhagen, M. *Guide for Photogrammetry and Remote Sensing*, 5th ed.; Wichmann: Paderborn, Germany, 2009.

51. Rouse, J.W.; Haas, R.H.; Schell, J.A.; Deering, D.W. Monitoring vegetation systems in the Great Plains with ERTS. In *Third Earth Resources Technology Satellite–1 Syposium, Volume I: Technical Presentations, NASA SP-351*; Freden, S.C., Mercanti, E.P., Becker, M., Eds.; NASA: Washington, DC, USA, 1974; pp. 309–317.

52. Lee, J.S. Refined filtering of image noise using local statistics. *Comput. Graph. Image Process.* **1981**, *15*, 380–389. [CrossRef]

53. Lee, J.S.; Jurkevich, I.; Dewaele, P.; Wambacq, P.; Oosterlinck, A. Speckle filtering of synthetic aperture radar images: A review. *Remote Sens. Rev.* **1994**, *8*, 313–340. [CrossRef]
54. Cable, J.W.; Kovacs, J.M.; Shang, J.; Jiao, X. Multi-Temporal polarimetric RADARSAT-2 for land cover monitoring in northeastern Ontario, Canada. *Remote Sens.* **2014**, *6*, 2372–2392. [CrossRef]
55. Qi, Z.; Yeh, A.G.-O.; Li, X.; Lin, Z. A novel algorithm for land use and land cover classification using RADARSAT-2 polarimetric SAR data. *Remote Sens. Environ.* **2012**, *118*, 21–39. [CrossRef]
56. Jagdhuber, T.; Stockamp, J.; Hajnsek, I.; Ludwig, R. Identification of soil freezing and thawing states using SAR polarimetry at C-Band. *Remote Sens.* **2014**, *6*, 2008–2023. [CrossRef]
57. Baatz, M.; Schäpe, A. Object-oriented and multi-scale image analysis in semantic networks. In Proceeding of the 1999 International Symposium on Operationalization of Remote Sensing, Enschede, The Netherlands, 16–20 August 1999.
58. Baatz, M.; Schäpe, A. Multiresolution Segmentation: An Optimization Approach for High Quality Multi-Scale Image Segmentation. Available online: http://www.ecognition.com/sites/default/files/technology.pdf (accessed on 12 January 2016).
59. Willhauck, G. Comparison of object oriented classification techniques and standard image analysis for the use of change detection between SPOT multispectral satellite images and aerial photos. *Int. Arch. Photogramm. Remote Sens.* **2000**, *33*, 214–221.
60. Rabus, B.; Eineder, M.; Roth, A.; Bamler, R. The shuttle radar topography mission—A new class of digital elevation models acquired by spaceborne radar. *ISPRS J. Photogramm. Remote Sens.* **2003**, *57*, 241–261. [CrossRef]
61. Krieger, G.; Zink, M.; Bachmann, M.; Bräutigam, B.; Schulze, D.; Martone, M.; Rizzoli, P.; Steinbrecher, U.; Antony, J.W.; DeZan, F.; *et al.* TanDEM-X: A radar interferometer with two formation-flying satellites. *Acta Astronautica* **2013**, *89*, 83–98. [CrossRef]
62. Cohen, J. A coefficient of agreement for nominal scales. *Educ. Psychol. Meas.* **1960**, *20*, 37–46. [CrossRef]
63. Furuta, R.; Sawada, K. Case Study of Landslide Recognition using Dual/Quad Polarization data of ALOS/PALSAR. In Proceedings of the Asia-Pacific Conference on Synthetic Aperture Radar (APSAR), Tsukuba, Japan, 23–27 September 2013; pp. 481–484.
64. Plank, S.; Singer, J.; Minet, C.; Thuro, K. Pre-survey suitability evaluation of the differential synthetic aperture radar interferometry method for landslide monitoring. *Int. J. Remote Sens.* **2012**, *33*, 6623–6637. [CrossRef]
65. Dabbiru, L.; Aanstoos, J.V.; Hasan, K.; Younan, N.H.; Li, W. Landslide Detection on Earthen Levees with X-band and L-band Radar Data. In Proceedings of the Applied Imagery Pattern Recognition Workshop Sensing for Control and Augmentation 2013 IEEE (AIPR), Washington, DC, USA, 23–25 October 2013; p. 5.

remote sensing

|MDPI|

Article

Digital Elevation Model Differencing and Error Estimation from Multiple Sources: A Case Study from the Meiyuan Shan Landslide in Taiwan

Yu-Chung Hsieh [1,2], Yu-Chang Chan [3,*] and Jyr-Ching Hu [2]

[1] Central Geological Survey, MOEA, Taipei 235, Taiwan; hsiehyc@moeacgs.gov.tw
[2] Department of Geosciences, National Taiwan University, Taipei 106, Taiwan; jchu@ntu.edu.tw
[3] Institute of Earth Sciences, Academia Sinica, Taipei 115, Taiwan
*   Correspondence: yuchang@earth.sinica.edu.tw; Tel.: +886-227839910 (ext. 411)

Academic Editors: Zhenhong Li, Roberto Tomas, Zhong Lu and Prasad S. Thenkabail
Received: 14 January 2016; Accepted: 24 February 2016; Published: 29 February 2016

Abstract: In this study, six different periods of digital terrain model (DTM) data obtained from various flight vehicles by using the techniques of aerial photogrammetry, airborne LiDAR (ALS), and unmanned aerial vehicles (UAV) were adopted to discuss the errors and applications of these techniques. Error estimation provides critical information for DTM data users. This study conducted error estimation from the perspective of general users for mountain/forest areas with poor traffic accessibility using limited information, including error reports obtained from the data generation process and comparison errors of terrain elevations. Our results suggested that the precision of the DTM data generated in this work using different aircrafts and generation techniques is suitable for landslide analysis. Especially in mountainous and densely vegetated areas, data generated by ALS can be used as a benchmark to solve the problem of insufficient control points. Based on DEM differencing of multiple periods, this study suggests that sediment delivery rate decreased each year and was affected by heavy rainfall during each period for the Meiyuan Shan landslide area. Multi-period aerial photogrammetry and ALS can be effectively applied after the landslide disaster for monitoring the terrain changes of the downstream river channel and their potential impacts.

Keywords: Airborne LiDAR (ALS); unmanned aerial vehicles (UAV); photogrammetry; digital elevation model (DEM) differencing; swath profile

1. Introduction

Multi-source, multi-period DTM data are not only a critical tool for research on the mechanism of landslides, but also considered as greatly useful information regarding active faults, earthquake disasters [1–4], and flooding/river bank erosion [5]. When it comes to the comparison of this type of terrain information, error estimation of data obtained in various periods using various techniques becomes crucial. Almost every technology, including theodolites, GPS, photogrammetry, InSAR, airborne LiDAR (ALS), and ground LiDAR, has its application ranges and restrictions in terms of spatial and time scales when employed to obtain 3-D terrain data [6,7]. By comparing the information with various precision and resolution methods that are collected during different periods, these multi-method, multi-period data of digital terrain models (DTM) have been used in landslide mechanism-related research [8], observations of surficial activity [9–11], landslide volume calculations [12–14], and volume variation estimations [15], as well as disaster scale assessment and simulation [16].

Most of the previous comparative studies on multi-period DTM data have adopted ground control points, e.g., the discussion of landslide and river terrain variation by comparing the DTM data from ground measurements and those from aerial photogrammetry and ALS measurements [10,17–19], analysis of river terrain variation and slope earthflows by contrasting between two-period [20,21] or three-period [22] ALS data and ground measurement data, evaluation of earthflow terrain variation using airborne and ground LiDAR data, and error assessment of these two techniques [23], using InSAR-derived DTM to discuss terrain changes before and after the landslide event [7], and error assessment based on the two LiDAR datasets of Taiwan [24]. All of these studies mentioned above were conducted based on accessible ground measurement points. However, difficulties associated with ground measurements might occur if the target areas were located on high and steep mountains or densely vegetated areas, which could potentially affect the assessment of errors. Moreover, Tseng *et al.* calculated errors based on the undisturbed areas [25] and regarded the standard deviation of distribution of elevation difference as the average difference and error range of the two-period digital elevation model (DEM), respectively. Such an assessment method tended to be affected by the selected undisturbed areas. Hence, data obtained using different DTM generating techniques under various terrains or landscapes could lead to varied errors, which would eventually impact the rationality of the subsequent analysis or assessment.

Due to the highly unpredictable feature of landslide incidents, few practically measured landslide volume values exist. Most research work conducted in the past evaluated slide volume by measuring the landslide area on the obtained image and then estimating the average depth [26–28]. However, the certainty of landslide depth is often difficult to evaluate due to the lack of terrain information before the disaster, which would affect the accuracy of landslide volume estimation. As a result, there also exist a number of studies where statistical/empirical equations, *i.e.*, relations between area and volume derived by compiling a large quantity of landslide cases, are used to approximate the potential landslide volume based on a certain landslide area. Nevertheless, this type of method is also restricted by various factors such as area features, landslide type, slope morphology, and rock properties [29]. With the development of DTM data, it has become feasible to adopt various techniques such as photogrammetry, ALS, and ground LiDAR, along with multi-period terrain information to calculate landslide volume [10,12–16,19]. This work used the multi-period aerial photogrammetry, UAV, and ALS techniques to explore the application of data generated from different techniques in DEM differencing analysis after landslide hazards, as well as to calculate the earthflow volume produced by massive landslides. Specifically, through error assessment of multi-source, multi-period DTM data obtained from a densely-vegetated mountain area, this research attempted to examine the applicability of various techniques in massive landslides under such terrain conditions. The landslide volume and the subsequent volume variation were also calculated using DTM data from multiple periods. This research enables researchers who apply multi-period DEMs to know more about the characters, activities, and potential effects of landslides and, thereby, offering effective guidance on the assessment of landslide hazards.

2. Materials and Methods

2.1. Study Area

In 2008, Taiwan region was struck by the massive rainfall brought by Typhoon Sinlaku. The mountainous area in Central Taiwan received over 1000 mm of rainfall, with a maximum accumulated precipitation of more than 1600 mm [29], which caused severe geological hazards in this area. The restaurant buildings in the Lushan hot spring area in Nantou County collapsed and the Houfeng Bridge in Taichung City was damaged by floods [30]. The Meiyuan Shan landslides (Figure 1) occurred near Ren'ai Township, in Nantou County in Central Taiwan, near Huisun Experimental Forest Station, within the river basin of Meitangan River of the Beigang River system. The significantly heavy rain brought by Typhoon Sinlaku on 14–15 September 2008 caused the rain gauge at the Qingliu station to reach 761 mm. As a result, a large-scale landslide with an area of about 0.9 km^2 occurred on the southeastern slope of Meiyuan Shan (1785 m), which led to a large landslide. Subsequently, the large body of rocks and sediments propagated downstream with water, depositing at the confluence of the Beigang and Meitangan Rivers (Figure 2). The base of the Tou 80 County Road and the slope next to the camping area within the Huisun Experimental Forest Station were seriously hollowed out, and the collapsed sediments congested the Meitangan River, forming a landslide dam. The main geomorphological features of this area (marked by the blue boundary in Figure 1) are described here:

- Ridges stretch from Baxianshan in the northeast to Meiyuan shan in the southwest, with the dominant southeast slope aspect;
- The flow of the Meitangan River system is naturally bounded by the ridge boundaries, *i.e.*, flowing from the northeast toward the southwest and then converging into the major river in this area, the Beigang River;
- In addition to the Huisun Experimental Forest Station, villages such as Qingliu, Chungyuan, and Meiyuan are all located along the bank of the Beigang River.
- While these villages can communicate with the outside area through the county road, the study area cannot be reached through any roads other than walking in the countercurrent direction.

Near the study area, there is a rain gauge station close to Qingliu Village set by the Central Weather Bureau and a flowmeter set by the Water Resource Agency, Ministry of Economic Affairs can also be found downstream (Figure 1). The data from the rain gauge station and the flow monitoring station during 2005 to 2013 are shown in Figure 3.

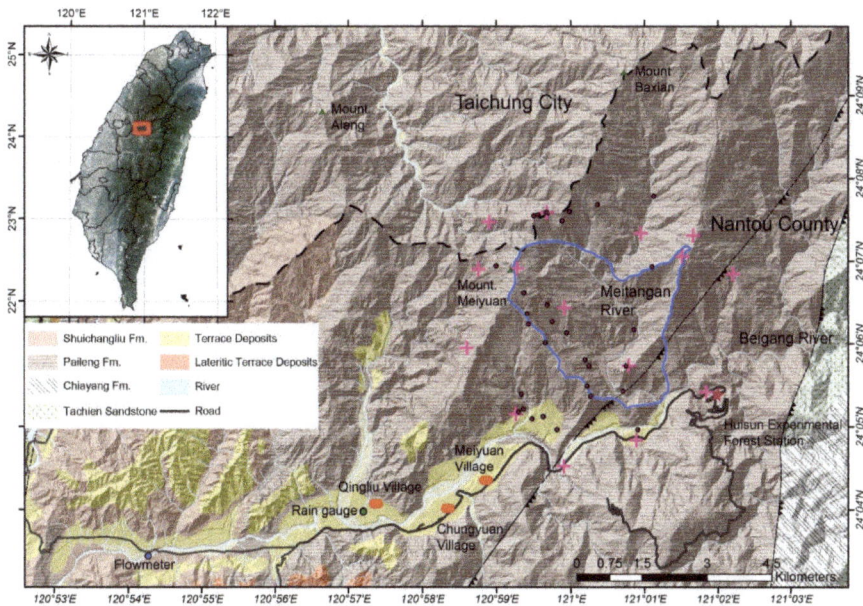

Figure 1. Geological map and geographical position of the study area. The blue line shows the range of orthoimage and digital terrain model (DTM) data generated in this research. There is a rain gauge (green point) and a flowmeter (blue point) in the adjacent area. Multiple residential villages are located downstream of the study area. The study area is shown in red square within the upper-left index map. Three data sets derived from aerial photogrammetry in this research used the same 15 ground control points shown as pink cross marks. The UAV dataset used the 30 ground control points shown as dark brown dots.

Figure 2. Panorama of the Meiyuan Shan landslide. The left image, a view from the south of Meiyuan shan, illustrates that the Meiyuan shan terrain along the northeast–southwest direction is a dip slope. It also shows the alluvial fan generated by rock slides and debris flows. The Meitangan River feeds into the Beigang River, the main river in this area, which is shown at the right side of this image. The right image, a view from the east of the Meiyuan Shan landslide, displays the relatively smooth surface of the zone of depletion revealed after the landslide. A large body of rock and sediments can still be clearly observed at the foot of the slope.

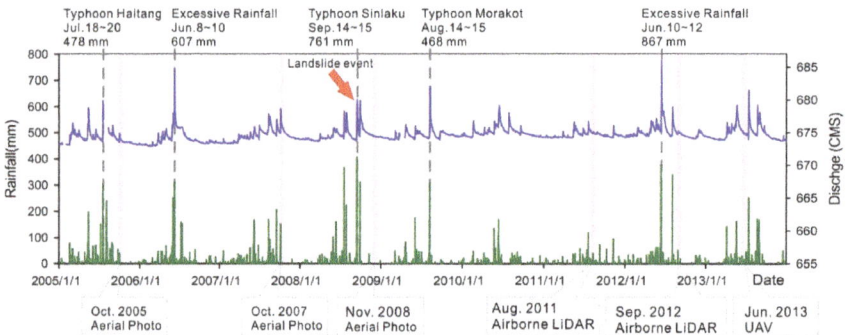

Figure 3. The green bars show the rainfall data recorded by the Qingliu rainfall monitoring station from 2005 to 2013. The blue lines are the flowrate data measured in Beigang River, and the dash lines representing typhoon and torrential rain incidents during this period of time. The purple lines indicate the times when aerial photography and ALS measurements were conducted in this study.

2.2. Geological Setting

This area is part of the Hsueshan Range zone in terms of geological setting. The exposure strata of its adjacent areas contain the Tachien Sandstone, Chiayang Formation, Paileng Formation, Shuichangliu Formation, terrace deposits, and lateritic terrace deposits (Figure 1). The study area was mainly in the range of the Paileng Formation, with exposure rocks of white or grey, fine or coarse quartzitic sandstone, and interbedding of grey dense sandstone and dark grey argillite or slate. With high rock strength and weather resistance, this type of sandstone is a major feature of the main northeast–southwest ridge in this area, forming the above-mentioned dip slope terrain. The main geological structures in this area are the Meiyuan Fault and the Guandaosan Fault. The Meiyuan Fault stretches from northeast to southwest, coincidently passing through the confluence of the Yangan River and the Beigang River within the study area. It is an eastwardly tilted, high-angle thrust fault, located west of the study area. The Guandaosan Fault lays southeast of the Meiyuan Fault, running almost south to north. It is also an eastwardly tilted, high-angle thrust fault, with the Paileng Formation to its west and the Tachien Sandstone and slate of Chiayang Formation to its east.

2.3. Aerial Photogrammetry

Aerial photogrammetry is one of the commonly adopted techniques for obtaining 3-D terrain information, which can be used to generate DTM data from aerial images through digital photogrammetry. It is often employed in studies of landslide geomorphology. In most cases, complete DTM data prior to the landslide event are not necessarily available. In a situation like this, however, historical aerial images can be used to generate DTM through digital aerial photogrammetry. The Aerial Survey Office, Forest Bureau (ASO), has conducted aerial photography for Taiwan forests on a regular basis, the images obtained from which could be used as orthoimage and DTM data before landslide events for analysis. Each photo had to be checked for factors, such as cloud content and the existence of fog to ensure the image quality for subsequent processing. Photos would not be usable with blurry images, fog, excessive shadows, or clouds (shadows). By searching the historical images taken before the 2008 Meiyuan Shan landslide incident, three periods of datasets are suitable for post-processing, *i.e.*, those taken on 22 October 2005, 23 October 2007, 11 November 2007, 10 October 2008, and 30 November 2008 were selected and obtained from Geoforce Company and the ASO historical aerial photogrammetry database [31]. The set from 2005 contained 203 images taken by ULTRACAM. The set from 2007 was composed of four images taken by RMK TOP and seven taken by an Intergraph digital mapping camera (DMC). The set from 2008 had seven images taken by DMC and an air route covered by a Leica airborne digital sensor 40 (ADS40). Ground sample distances (GSD) were 0.17 m, 0.19 m, and 0.18 m, respectively. All three datasets used the same 15 ground control points and were processed by SimActive software and aerial photogrammetry work stations. The processing of the aerial photogrammetry data is briefly described here:

- After appropriate aerial images were chosen and aerotriangulation mapping was done, models based on the accurate parameters of exterior orientation of every image and superimposition of high-precision GPS coordinates were made.
- When conducting aerotriangulation adjustment, the 15 ground control points were also taken into consideration to improve the accuracy of the parameters of exterior orientation. The control point errors of three data sets after aerotriangulation adjustment are shown in Table 1. The distribution of the 15 ground control points are shown in Figure 1.
- Then stereo image pairs, DTM, as well as orthoimages, could eventually be generated. The resolutions of orthoimages and DTM were 25 cm and 2 m, respectively.

Table 1. The control point errors of three datasets after aerotriangulation adjustment.

	Maximum Changes (m) at Control Points			RMS of Changes (m) at Control Points		
	X	Y	Z	X	Y	Z
2005	0.4045	0.3814	−0.3926	0.2087	0.2043	0.1563
2007	0.0636	0.1395	−1.1654	0.0315	0.0519	0.3373
2008	−0.0355	0.0437	−0.0197	0.0230	0.0260	0.0108

Unmanned aerial vehicles (UAV) have been applied more and more widely in national defense, military affairs, and civil applications in recent years. It has also become one of the main DEM techniques due to its advantages of low operation costs, high data processing speed, low flying height, and convenient flying preparation. UAVs are an important technique in post-disaster response when it comes to a confined environment, particular terrain, or the need for consistent monitoring after disasters. For the reasons mentioned above, this study also adopted UAV techniques to survey the Meiyuan Shan landslide area in June 2013, using a fixed wing drone with a flying height of 2500–2750 m, a Canon 500D camera with a lens with focal length of 50 mm, and planned ground sample distance (GSD) of 8–19 cm. During the flight operation, 868 images were taken, which were then processed using the same method applied for aerial photogrammetry data with the Pix4D software. Thirty ground control points shown in Figure 1 were used for UAV aerial photogrammetry. The RMS of changes at control points were 0.055 m (x), 0.018 m (y), and 0.362 m (z). Resolutions of the generated orthoimage and DTM were 25 cm and 2 m, respectively.

2.4. Airborne LiDAR (ALS)

For regional wide-range topographic surveying and mapping, ALS is currently considered as the technique able to obtain terrain data with the best resolution and precision. The data used in this study were measured by the Geological Survey in August 2011 and September 2012, respectively. The former was a surveying and mapping program for DTM data over the whole island of Taiwan conducted by the Geological Survey. The program was aimed to obtain DTM data with a resolution of 1 m; that is, a point cloud with a density of 1.5 points/m^2 was expected at a monitoring height of 800 m. Flight parameters are shown in Table 2. The latter was the repeated LiDAR DTM survey work for the Meiyuan Shan landslide, data which were assumed to be the same as the former. Flight parameters of this program are also shown in Table 2. These two ALS datasets were processed using the same method, as given here:

- Data obtained from the flight operation were calculated to get information about each air route for flight strip adjustment.
- Original point clouds were gained after adjustment, followed by point cloud classification.
- The point clouds of these two operations were divided into four categories: ground points, non-ground points, water body points, and outlier point clouds.
- Then the digital surface model (DSM) and digital elevation model (DEM) were generated. The DSM represents the Earth's surface, including all objects on it. The DEM represents bare ground surface without any objects, like vegetation and buildings. The term DTM comprises both DSM and DEM in this study. The two LiDAR survey programs took aerial photos, which were used to generate orthoimages with a resolution of 25 cm. This study adopted data from both 2011 and 2012. LiDAR data of 2011 were set as the benchmark for precision assessment due to wide coverage and completeness.

For comparison purposes, the TWD 97 ground coordinate system [32] was employed for all of the data sets, and the Taiwan Vertical Datum 2001 (TWVD 2001) [33] was selected as the vertical coordinate system.

Table 2. Flight parameters and scanning system settings.

	2011	2012
Date	2011/8	2012/9
Sensor system	Leica ALS60	Riegl/LMS-Q680i
Flying height (m)	2700	3000
Flying speed (kts) [1]	100	100
Pulse frequency (KHz)	53.4	150
Field of view (FOV) (°)	40	40
Scanning swath width (m)	1175	1223
Strips overlapping (%)	52	50
Laser pulse density (pt/m^2)	1.8	2. 2
Resolution (m)	1.0	1.0

[1] 1 kts = 1.852 km/h.

3. Results

3.1. Orthoimage

This study employed orthoimages generated using various aerial photogrammetry cameras from six different periods, and the results are shown in Figure 4. These aerial photogrammetry orthoimages show changes before and after the Meiyuan Shan landslide. Figure 4a,b, obtained prior to the disaster, show that only occasional shallow landslides occurred on some upstream slopes. Figure 4c–f were obtained after the disaster. It can be clearly seen that, in 2008, a landslide dam was formed due to a large body of landslide rocks, and a blue waterbody existed in the river channel at the toe of the landslide. Parts of the waterbody are still visible in Figure 4d, and the area of the landslide dam has reduced to less than half, though still existing in this basin. By comparing Figures 3 and 4 it can be inferred that the landslide dam might have existed for almost four years before being damaged by the torrential rain in June 2012.

Figure 4. Orthoimages generated using various aerial photogrammetry cameras from six periods. (**a–c**) historical aerial photogrammetry images; (**d,e**) taken under the ALS survey task; and (**f**) a photograph by an unmanned aerial vehicle (UAV).The number is ground sample distance (GSD). The large-scale landslide area in (**c–f**) is the Meiyuan Shan landslide area, in which specific landslide characteristics can be observed from top to middle and then to the foot of the slope. The blue waterbodies distributed in (**c,d**) are where the landslide dams are located.

3.2. Digital Terrain Model

This study used six periods of DTM data, which comprise both DEM and DSM data, obtained from aerial photogrammetry, ALS, and UAV techniques. All three DTM techniques can be used to generate DSM and DEM data. Since the actual ground points under plants can be scanned by the infrared laser, ALS is able to obtain the DEM data of actual ground elevation without being affected by ground vegetation [34,35]. However, aerial photogrammetry is only able to gain the actual ground elevations of exposed areas because of the limited ability of only mapping ground elevation displayed in the images. In plant-covered areas, potential real elevations can be determined from DSM data and elevations of the exposed areas, which might produce relatively large errors. In this study, the discussed landslide area and affected river channel were both considered as exposed areas without plant coverage. Hence, DSM data were used to analyze terrain change before and after the landslide disaster. Results of this comparison are shown in Figure 5.

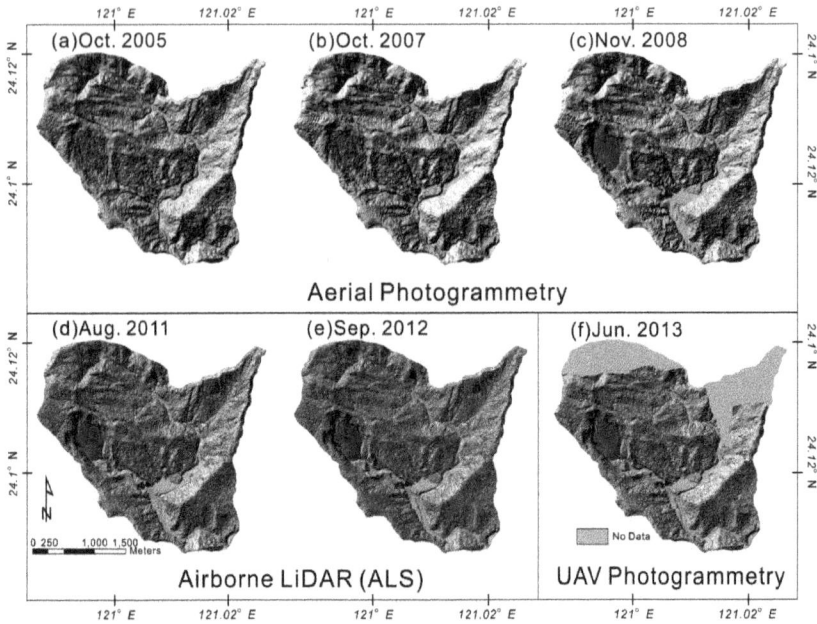

Figure 5. Digital Surface Model (DSM) images generated from six periods. (**a–c**) are the traditional aerial photogrammetry images; (**d,e**) were taken by the ALS technique; and (**f**) obtained through UAV aerial photogrammetry. The uniform grey areas shown in the images represent lack of information due to clouds.

3.3. Error Estimation

Error assessment is critical information for DTM data users. Most users do not necessarily participate in the production of DTM data. When it comes to comparative analysis of DTM data, especially for those areas with terrain variation, differences of data obtained from different levels of precision would affect the results of any comparison. Most of the previous comparative studies as aforementioned in the introduction on multi-period DTM data have adopted ground control points to evaluate errors of individual DTM data sets. From the perspective of general users, this study conducted error assessments for mountain and forest areas with poor traffic accessibility using limited information, including error reports obtained from the data generation process and comparison errors of terrain elevations.

Generally speaking, users can get aerotriangulation errors and interpolation errors from the data generation process. Aerotriangulation errors are the errors in X, Y, and Z directions after the aerotriangulation adjustment, which can be obtained using survey adjustment software and calculating the difference between aerotriangulation matching points and applied ground control points, *i.e.*, the aerotriangulation σ shown in Table 3. Interpolation errors refer to the errors originated during interpolation, which is used to generate gridded DTM data after stereo pairs are matched. This type of error can be obtained by calculating the difference between the estimated elevation of gridded data and the actual ground elevation, as interpolation σ shown in Table 3. These two errors then can be used to calculate the total error based on the simple error propagation theory, as shown in Equation (1) [36]. Among these periods, the gridded DTM data from 2011 and 2012 were generated based on point clouds using the ALS technique. Therefore, their total errors were achieved directly from the difference between the actual terrain measured elevation and the calculated elevation:

$$\sigma_{total} = \sqrt{\sigma_1{}^2 + \sigma_2{}^2} \tag{1}$$

Ground control points or image characteristic points are usually assigned to those identifiable, flat, hardly-vegetated spots or characteristic points of buildings for error estimation. They can be used in terrain GPS measurements to represent the real ground elevation. However, most landslide disasters occur in areas with high mountains and forests that are rugged and hardly accessible. Hence, selection of the image characteristic points would be greatly restricted, and the representativeness of the error assessment would be affected due to the insufficiency of measurement points. The LiDAR data used in this study were from 2011 and 2012. In order to make error assessments for each period complete and representative, the 2011 DEM data were adopted as the terrain benchmark due to their wide coverage. The elevation values at selected characteristic points of each period were mapped with the elevation values at the same points of the 2011 DEM. Then their differences were the errors [37], as shown in Table 3. Errors in this paper were all given in the form of root-mean-square error (RMSE), which is a commonly used error expression [38,39]. The terrain change could be found by comparing data from different periods. Therefore, the errors between the two successive periods were also calculated using the described error relation. The results are shown in Table 4.

Table 3. Accuracy evaluation of datasets obtained from different periods.

	2005	2007	2008	2011	2012	2013
Aerotriangulation σ (m)	0.156	0.337	0.108			
Interpolation σ (m)	0.242	0.218	0.195			
σ_{total} (m)	0.288	0.402	0.223	0.150	0.170	0.247
$\sigma_{2011\ DEM}$ (m)	0.750	0.510	0.430		0.180	0.580

Table 4. Accuracy evaluation of two terrain datasets obtained from two successive periods.

	2005–2007	2007–2008	2008–2011	2011–2012	2012–2013
σ_{total} (m)	0.494	0.459	0.269	0.227	0.300
$\sigma_{2011\ DEM}$ (m)	0.907	0.667	0.455	0.234	0.607

3.4. DEM Differencing

The method of using multi-period DTM data to investigate terrain evolution has recently been applied in various studies regarding river bed sediment transport, landslide, and earthflow [16,25,40–42]. One of the most straightforward ways is to subtract an earlier terrain elevation from a later one. As six periods of DTM data were employed in this work, five values of terrain elevation change could be successively calculated in chronological order. Data from 2005–2007, as shown in Figure 6a, demonstrated the terrain change of this basin before the landslide disaster, whereas the other four periods, as shown in Figure 6b–e, demonstrated the terrain changes observed in each year after the disaster. Warm colors with negative values stand for a decrease in terrain elevation in these images, *i.e.*, the occurrence of surface erosion; and cold colors with positive values represent an increase in terrain elevation, *i.e.*, the occurrence of surface deposition. Through the color changes, the terrain evolution and sediment transportation of the river channel prior and subsequent to the landslide incident can be observed directly.

Figure 6a mainly shows sediment deposition in the river channel from the upstream catchment before the landslide. Figure 6b demonstrates the terrain change before and after the incident, in which the zone of depletion of the Meiyuan Shan landslide (shown in yellow and red) is the main source of sediment and the blue area is the zone of sliding sediment deposition. Obviously in the blue area, a large body of sediments deposited from the middle of the slope all the way to the foot, creating a river blocking at the foot of the slope, which is typical for landslide terrain. Figure 6c–e illustrate sediment erosion and deposition in different areas within the landslide affected range after the incident. There was small-scale erosion in the zone of depletion after the incident, but the majority of the area did not change. Erosion was the main feature in the zone of sediment deposition, which was the main source of the sediments downstream after the disaster. Regarding the river channel range evolution after the incident, deposition, and partial erosion has happened since the disaster until the year of 2011 and elevation only changed by several meters; whereas, river channel erosion played a critical role from 2011 to 2013, and elevation changes due to river bank erosion were be more than 10 m in some parts. As for the unchanged areas, it is difficult to homogenize the DTMs from different time periods because in these areas vegetation was dominant. The DTMs from these areas tended to show differencing noises that were either resulted from tree growth, wind effects, or forest farming. For this reason, we have intentionally avoided the unchanged areas and masked them out before our landslide change analysis. As for the landslide affected areas, they are generally bare grounds and the DTMs from these areas show much better quality in comparison with those from the unchanged areas. Thus, we mainly focused on the landslide and river channel areas where vegetation was limited.

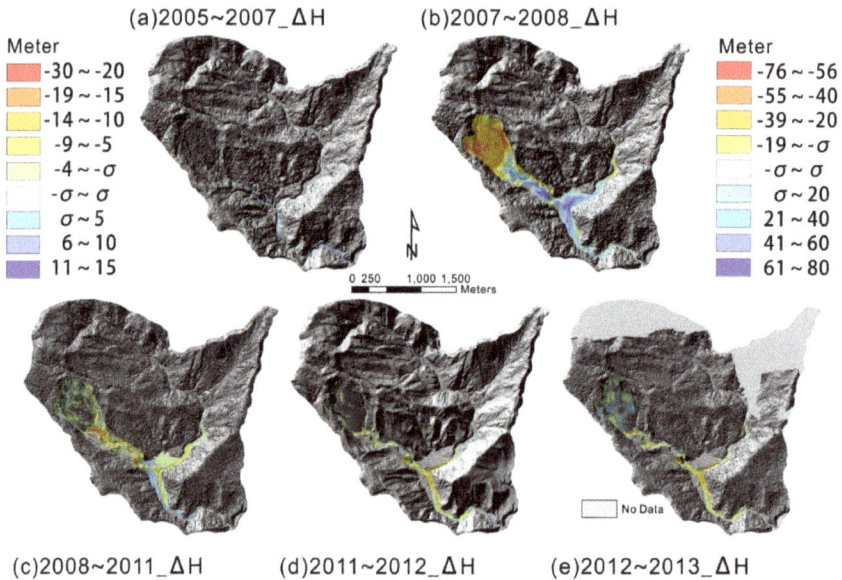

Figure 6. Images of terrain change over five periods calculated from multi-period digital terrain model (DTM) data in the Meiyuan Shan area. The color scale illustrates values of terrain change, where warm colors stand for an elevation decrease due to erosion, and cold colors represent an elevation increase led by the effects of deposition. (**a**) Terrain change (October 2005~October 2007) before the September 2008 landslide event; (**b**) Terrain change (October 2007~November 2008) right after the September 2008 landslide event. Please notice the markedly difference in elevation changes due to the fresh landslide event; (**c**) Terrain change (November 2008~August 2011); (**d**) Terrain change (August 2011~September 2012); (**e**) Terrain change (September 2012~June 2013). The scale for (**a**,**c**–**e**) is shown on the left; whereas, the one for (**b**) is shown on the right. σ is the total error of the two periods of DTM data shown in Table 4.

3.5. Volume Estimation

Volume calculations were implemented in the Spatial Statistics module of ArcGIS, and a brief description is given here:

- Calculate the difference between the elevation values at a characteristic point from a later DTM dataset and an earlier dataset, which could be either positive or negative.
- If resolution discrepancy of DTM data exists between the two periods, the higher resolution scale should be converted to the lower one.
- Then the deposition volume (positive result) and erosion volume (negative result) can be calculated based on the same resolution level.
- The summation of these two values is the net rock volume change over this period.
- Furthermore, with the calculated deposition area and erosion area and the previously obtained error value, the volume estimation error can also be gained, as shown in Table 5 [15].

Volume change showed a positive value in 2005–2007, but it remained negative in the other four periods. This suggested that earth and rocks brought from upstream of this basin deposited in the area before the Meiyuan Shan landslide; on the other hand, the earthflow produced from the landslide became the main source of sediments in Beigang River after the disaster, as indicated by the negative sign. The volume loss reached its peak in the disaster year and has decreased every year since then. During 2007–2008, the erosion and deposition volumes were 16.7 and 12.6 million m^3, respectively. Neglecting rock volume expansion and supplemental debris from upstream, at least 16.7 million m^3 of sediment were moved during the Meiyuan Shan landslide.

Table 5. Volumes of erosion, deposition, and volume change within the Meiyuan Shan landslide area estimated using digital terrain model (DTM) data obtained from multiple periods.

Event		Area (m^2)	Volume (m^3)	Volume Change (m^3)
2005–2007	Erosion	27,768	35,800 ± 25,185	272,700 ± 139,939
	Deposition	126,520	308,500 ± 114,754	
2007–2008	Erosion	642,851	16,773,400 ± 428,782	−4,132,500 ± 788,122
	Deposition	539,641	12,640,900 ± 359,940	
2008–2011	Erosion	767,337	2,708,100 ± 349,738	−2,392,000 ± 432,115
	Deposition	182,367	316,100 ± 82,977	
2011–2012	Erosion	492,829	1,797,400 ± 115,322	−1,644,400 ± 218,042
	Deposition	438,973	153,000 ± 102,720	
2012–2013	Erosion	605,620	817,700 ± 407,284	−705,500 ± 534,870
	Deposition	150,562	112,200 ± 127,586	

3.6. Profiles

It is a common analytical method to discuss terrain evolution using terrain profiles. Most traditional methods generate terrain profiles by extracting elevation values passing through a straight line. However, pure line-extracting generated profiles often fail to truly represent the real terrain [43]. When generating profiles using a single straight line, the selection position is not likely to be objective, and it also might be subject to stochastic terrain or deviation. Thus, it is possible to generate misleading or incorrect information. In order to avoid or mitigate these stochastic deviations and their miscomprehension, more profile lines can be added. As a result, a profiling method that stack elevation values on parallel, equally-spaced tangents has been developed, *i.e.*, the swath profile method. Applications of the swath profile method can often be found in the analysis of large-scale regional terrain issues, such as tectonic structures or landslides [44–46].

To discuss terrain and river channel changes before and after the large-scale landslide incident, both conventional straight line profiling and swath profiling techniques were used in this study to generate profiles. For the conventional straight line profiling method, the starting and ending points were first selected, and elevation values between the points were calculated. A terrain profile could be eventually generated based on the distribution of elevations. The spacing between elevation values was set at the DEM resolution in this study. In addition to terrain profiles calculated using the straight line method (Figure 7), based on artificial selection, the profiling calculation was also conducted using polyline profiles over the critical range (Figure 7). For a swath profile, a rectangular area is usually selected as the target. Then DEM data within this rectangle are rotated so that the side to be profiled is parallel with the south to north direction. Then, the maxima, minima, average, and standard deviation of each elevation row can be calculated and drawn in profiles. The minima scenario was employed in this work to generate profiles (Figure 8). Additionally, irregular areas were also selected to generate swath profiles. Similarly, DEM data within this irregular area were rotated so that the side to be profiled was parallel with the south to north direction before generating profiles (Figure 8) using the above-mentioned calculation approach.

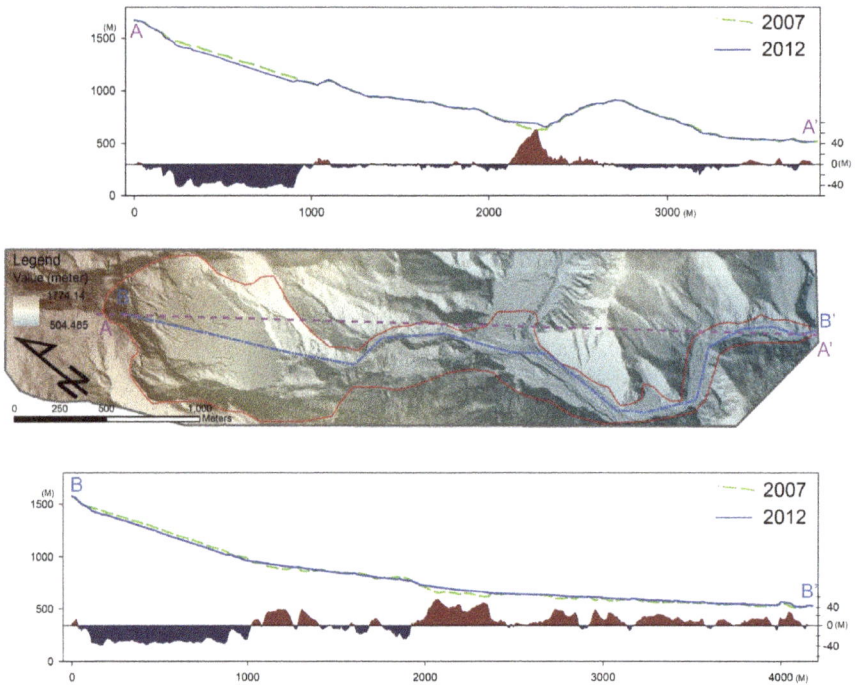

Figure 7. Straight line and polyline profiles based on terrain data obtained in 2007 (**before landslide**) and 2012 (**after landslide**). The purple A–A′ dashed section shown in the middle represents the straight line profile, whose results are shown in the top; the blue B–B′ section in the middle represents the polyline profile, whose results are shown in the bottom. The bottoms of the profile subfigures illustrate the elevation changes during 2007–2012, where dark blue denotes elevation decrease, while dark red denotes elevation increase.

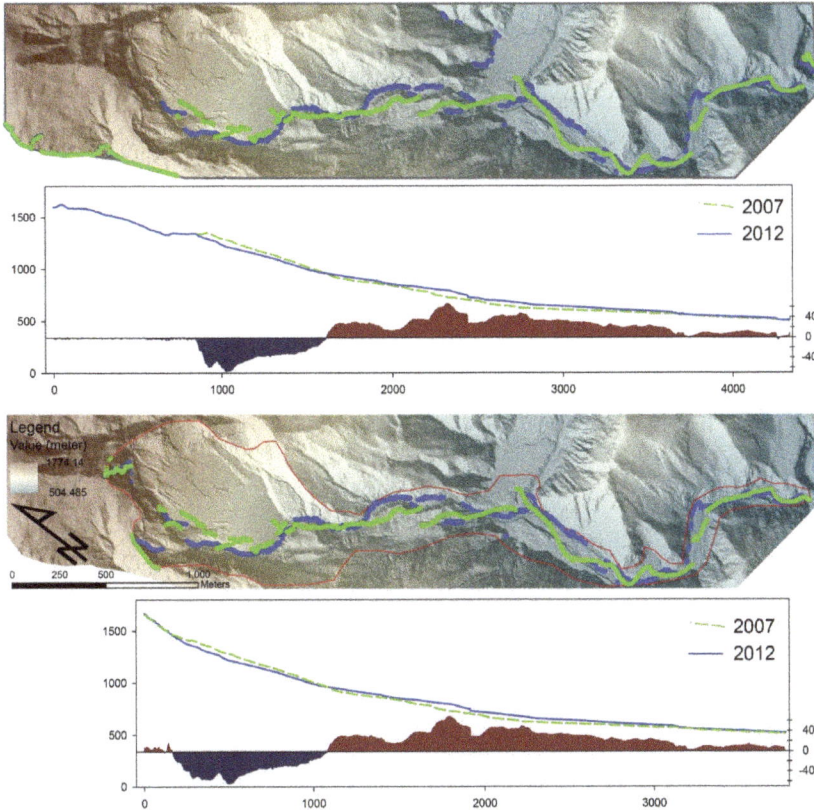

Figure 8. Swath profile calculation based on the terrain data obtained from 2007 (**before landslide**) and 2012 (**after landslide**). The first and the third subfigures show the swath profiles calculated using different target areas, *i.e.*, the entire rectangular range and the zone of depletion and the river channel, respectively. The results are, respectively, shown in the second and the fourth subfigures, where the green dashed line represents the elevation of 2007 while the blue line represents the elevation of 2012. The positions of sampling points are shown with the same colors in their corresponding maps. The bottoms of the profile subfigures illustrate the elevation changes during 2007–2012, where dark blue denotes elevation decrease, while dark red denotes elevation increase.

4. Discussion

4.1. Use of Data and Error Estimates

The six periods of DTM and orthoimage data used in this study were obtained from various aerial vehicles using multiple techniques such aerial photogrammetry and ALS. The generated data all had different errors that originated from their data generation methods and flight operations. Therefore, in a DTM application, the data acquisition procedure needs to be carefully determined once the data quality requirement is known. This is especially true for those flight parameters regarding the target area that would affect data resolution and errors, such as flying height, flying speed, flight direction, and overlapping strips. For terrain data produced using aerial photogrammetry, the elevation error is, generally speaking, about 0.2~1.3 m [37,38]. In a normal processing approach with sufficient ground control points, errors are mostly determined by the value of GSD [38]. Elevation error of ALS, on the other hand, is about 0.15~0.3 m [24,39,47]. Under regular processing, errors tend to be impacted by slope angle, plant type, and vegetation density; though, generally, errors of bare flat areas would be superior to those of bare slope areas and errors of hardly vegetated areas would be superior to those of densely vegetated areas. For the six periods of DTM data generated in this work, with the difference between the elevation of set control points and calculated elevation, the statistical data errors were expressed in the form of RMSE. For the four periods of data generated using aerial photogrammetry, the errors of σ_{total} ranged from 0.22 m to 0.40 m (Table 3). The comparison with the image GSD values indicated that σ_{total} error was 0.98~2.1 times that of GSD, which was in good agreement with similar previous studies [37,38]. This suggested that the precision of generated DTM data in this work based on historical aerial images and UAV images using aerial photogrammetry was suitable for landslide analysis.

The method of estimating errors via ground control points could provide reference for applications. However, due to the facts that most landslide areas are often located in areas with high mountains and forests featuring rugged terrain and poor traffic accessibility, with only a limited number of image characteristic points, the representativeness of error assessment might be adversely influenced due to the lack of sufficient ground measurement points, leading to potential underestimation. There are reported studies in which aerotriangulation or aerial photogrammetry operations were conducted by using high-precision LiDAR DEM data as control points or a reference terrain surface [48–50]. With the six periods of DTM data generated using various techniques in this study, in order to avoid the impacts of insufficient ground control points on error estimation, the DEM data from 2011 were used as the reference elevation. Then the elevation difference between other data sets and the reference at the same image characteristic point could be calculated as $\sigma_{2011\ DEM}$. The error range was 0.43~0.75 m (Table 4), which was higher than σ_{total}. This might be attributed to the fact that in such a densely vegetated area with rugged terrain, the originally measured errors from the ground control points tend to underestimate the elevation. Another possibility was that errors might get magnified when using the 2011 DEM data as reference due to the errors of the reference itself. However, it should be kept in mind that those identifiable spots in bare and flat areas can be selected as characteristic points and only minor errors will be produced when applying ALS in such areas. Moreover, it would be helpful to have a common terrain reference for comparisons between various data sets from different periods. Therefore, it is legitimate to choose at least one period of ALS DEM data as the benchmark for error assessment of multi-period DTM data. Similar concepts could also be applied to forest or mountainous areas with only a few or no control points. For example, after an earthquake or hurricane, when rapid and efficient DTM and orthoimage generation is required or subsequent post-disaster comparisons are needed, one period of ALS data can be used as a terrain reference to select ground control points as needed.

The basic idea of the multi-period DTM error estimation method employed in this paper was to calculate errors using known elevation check points and then to estimate errors between two successive periods using the error propagation equation. The advantage of this approach was that fairly accurate error values could be gained by comparing the known elevation inspection point data with the measured results. However, considering the fact that those identifiable and measurable points are mostly selected in flat, bare, and easily accessible areas, errors might be underestimated due to neglecting the practical slope terrain and vegetation. Moreover, in densely vegetated mountain areas, practical measurable control points are difficult to assign due to the impacts of plants and poor accessibility. An alternative DTM error estimation method is to calculate errors based on the undisturbed terrain areas [25]. The core of this method is to find the elevation differences of the undisturbed area and regard the average value and standard deviation of distribution of elevation difference as the average difference and error range of the two-period DEM, respectively. However, this assessment method tends to be affected by the selected undisturbed areas as the variant areas in mountains/forests might be included in calculations. Hence, one period of ALS data is utilized as the terrain reference for selecting ground control points. In addition to the advantages of the high resolution and precision of the DTM data obtained using ALS, this method can also solve the problem of insufficiency of elevation check points in areas with high mountains and forests due to measurement difficulties. The work by Podobnikar [51] also proposed the use of a known and high-quality DTM dataset as an important reference for combining different DTM datasets into a better DTM dataset. It applied more rigorous ideas of regionalization and parameter weighting to help improve the quality of the combined DTM data set. In our study, we similarly utilized one period of ALS data as a reference for error estimation but emphasized its convenience and cost-effectiveness for estimating landslide volume under users' demands. As shown in Table 3, the estimated errors of various DTM data sets based on the reference of 2011 DEM do reflect the potential error range of the DEM data.

UAVs have been applied more and more widely in various applications. It has also became one of the common tools for landslide disaster research due to its advantages of low operating cost, high data processing speed, low flying height, convenient flying preparation, and the ease of generating orthoimage and DTM data. Error range evaluation associated with the generated image and DTM data also attracts attention in the research community. The results obtained from this work (Table 3) showed that the error range of DTM data obtained from UAV, which was not equipped with conventional aerial photogrammetry camera, was not significantly different from those of the other three DTM datasets obtained from aircraft equipped with conventional aerial photogrammetry cameras. It should be considered as the optimal choice for investigations over small areas due to its advantages of portability, low cost, and high data processing speed.

Due to the highly unpredictable nature of landslide incidents, few measured landslide volume values exist. It is common practice to evaluate slide volume by measuring the landslide area on the obtained image and then estimating the average depth [26–28]. However, the certainty of landslide depth is often difficult to evaluate due to the lack of terrain information before the disaster, which would affect the accuracy of landslide volume estimation. As a result, there also exist a number of studies where statistical/empirical equations, *i.e.*, relations between area and volume derived by compiling a large quantity of landslide cases, were used to approximate the potential landslide volume based on a certain landslide area. Nevertheless, this type of method is also restricted by various factors such as area features, landslide type, slope morphology, and rock properties [28].

With the utilization of multi-period DTM data in this work, the appearance of the entire landslide area, both before and after, the disaster could be clearly observed thanks to the terrain data obtained from the historical aerial photogrammetry images. The Meiyuan Shan landslide incident with a sliding area of about 0.55 km^2 was analyzed as an example in this paper. In 2008, the disaster produced 16 million m^3 of rock volume, 12 million m^3 of which was deposited in the landslide dam, and 4 million m^3 of which entered the mainstream of Beigang River. The volume changes could be clearly calculated. The use of high-resolution DTM data obtained from multi-period aerial photogrammetry, UAV aerial measurement, and ALS has been proven to be an efficient method to calculate landslide volumes. Specifically, small-area investigations can be implemented using UAV by taking advantage of its portability and speed, whereas large area investigations can employ aerial photogrammetry or ALS to obtain data. Additionally, ALS is also applicable for regional investigations since it possesses the best data precision.

4.2. Profile Comparison

Profiling is a commonly used method in terrain research, among which straight line profiling is the most straightforward, convenient, and rapid approach. As the section AA' shows in Figure 7, the elevation values located on this straight line were selected to generate profiles. Although easy and fast, this approach would also pass non-target areas. Incorporation of unnecessary terrain into the profile would lead to abnormal or irregular terrain information, which might be misleading. Polyline profiling could extract necessary elevation points to generate profiles using a polyline, as the BB' section shows in Figure 8. Compared with the AA' straight line profiling, this approach requires manual selection of polyline sections. Although selection of non-target area data can be mitigated, misleading abnormal, or irregular profiles are still not completely avoidable.

The method of swath profiling has been mostly employed in large-scale terrain applications, such as tectonic structures. However, it was adopted in this work to analyze a small scale slope. Rectangle and the irregular ranges, such as the Meiyuan Shan landslide and the river channel, were used for swath profiling with the minimum amount of horizontal rows. Shown in the top of Figure 8 is a profile based on rectangles, while the irregular shapes of the Meiyuan Shan landslide and the river channel are shown in the bottom of the figure. The major differences between these two were the surroundings of the original landslide on the left and the vicinity of the slope foot, with the rest of the selected elevation values remaining the same. As the profile was based on the minimum values of each row, it could reflect the real landslide situation through the DTM elevation change before and after the disaster, including obvious descent of the original area and clear ascent of the slope middle and the river channel. The trend of elevation change also demonstrated the variations of erosion of the collapse origin and deposition of the rock accumulation area with the profile position. There is a clear elevation change at around 1900 m in the bottom subfigure of Figure 8, which should be the location between the landslide deposition toe and the river channel deposition. This change cannot be observed from the two profiles in Figure 7. Multiple parallel straight line profiling had been previously utilized in landslide discussions [16,52]. Instead, the swath profile method is able to express the landslide terrain using the maxima, minima, or average of each row, which is helpful in landslide interpretation and analysis.

Usually after a landslide event, the sliding sediments would not entirely propagate downstream at once. The remaining body in most cases would be moved downstream due to a subsequent heavy rainfall and, thus, the remaining rocks and the moving earthflow are threats to the downstream area. The ratio between the sediment volume flowing out of the landslide dam and the total sediment volume within the landslide dam can be defined as the sediment delivery rate [21,53,54]. This can represent degrees of erosion and sediment outflow in the water accumulation area and, thereby, provide references for hydraulic engineering or landslide disaster prevention planning. Since we were able to calculate five terrain variations from six DTM data sets, along with the fact that the data we have in this work covered an entire water accumulation area, sediment delivery rate calculations could be conducted. Thus, the sediment volume and the delivery rate of earthflow propagation from this area to the downstream main river channel (Beigang River) could be obtained.

Using the volume around the incident of the Meiyuan Shan landslide as a calculation benchmark, the delivery rates up to 2008, 2011, 2012, and 2013 were 24.6%, 14.3%, 9.8%, and 4.2%, respectively, declining year by year. In fact, the heavy rainfall that occurred during these years also played a critical role. By looking into the total sediment volume generated after the landslide, there were more than six years during which 47.1% of the total volume remained within the landslide dam. This indicated that the sliding sediments would not entirely propagate downstream in a short period of time but, instead, the remaining bulk of these sediments would be moved downstream by subsequent heavy rainfall over time. Therefore, after a landslide, these sediments should be carefully monitored as potential post disaster hazards. Aerial photogrammetry and ALS can be continuously applied after the disaster subsequent to a massive landslide to gather terrain information around the landslide area in order to monitor the terrain changes of the downstream river channel and their potential impacts.

5. Conclusions

For the six periods of DTM data generated in this work, the error of ALS was about 0.15 m–0.18 m, the smallest among these DTM generation techniques. For the four periods of data generated using aerial photogrammetry (including UAV images), the errors of σ_{total} ranged from 0.22 m to 0.40 m. The comparison with the image ground sample distance (GSD) values indicated that σ_{total} error was 0.98–2.1 times that of GSD. This suggested that the precision of the DTM data generated in this work using different aircrafts and generation techniques is suitable for landslide analysis. Especially in mountainous and densely vegetated areas, data generated by ALS can be used as a benchmark to solve the problem of insufficient control points. The error range of DTM data obtained from UAVs was not

Remote Sens. **2016**, *8*, 199

significantly different from those of the other three DTM datasets obtained using the traditional aerial photogrammetry technique. This suggested that UAV aerial photogrammetry should be considered as the optimal choice for investigations over small areas due to its advantages of portability, low costs, and speedy data processing.

DTM data were compared in this study to calculate the sediment volume and terrain change before and after the landslide incident. The comparison between the traditional straight line profile and swath profile suggested that the swath profile method is able to express landslide terrain using the maxima, minima, or average of each row, which is helpful in landslide interpretation and analysis. Sediment delivery rate can be determined based on sediment volume change calculated by contrasting DTM data. It was shown that sediment delivery rate decreased each year and was affected by heavy rainfall during each period. Since the Meiyuan Shan landslide in 2007 to 2013, there were more than six years during which approximately half of the total sediment volume remained within the landslide dam. This indicated that the sliding sediments would not entirely propagate downstream in a short period of time. Therefore, after a landslide, these sediments should be carefully monitored as potential post disaster hazards. Aerial photogrammetry and ALS can be applied after the disaster as often as necessary to gather terrain information around the landslide area in order to monitor the terrain changes of the downstream river channel and their potential impacts subsequent to a massive landslide.

Acknowledgments: This study was supported by the Project of Investigation and Analysis for Geologically Sensitive Areas under the Program of National Land Preservation from the Central Geological Survey, MOEA, Taiwan; Grant No. MOST103-2116-M-001-002, MOST104-2116-M-001-022, MOST103-2116-M-002-024, and MOST104-2116-M-002-013 from the Ministry of Science and Technology, Taiwan; and Institute of Earth Sciences, Academia Sinica. We would like to thank Kuo-Jen Chang of the National Taipei University of Technology for improving the UAV data processing of this paper and GeoForce Technologies Co. for aerial-photo processing. We very much appreciate the three reviewers whose constructive and insightful comments have greatly improved the content of the manuscript. Yi-Zhung Chen and Ruo-Ying Wu from the National Cheng-Kung University assisted with the data analysis. We also received generous support from our colleagues at the LiDAR Group of the Environmental Engineering Division in the Central Geological Survey, MOEA. We sincerely appreciate all of the aforementioned help, which ensured the smooth completion of this study.

Author Contributions: Yu-Chung Hsieh and Yu-Chang Chan developed the scientific concept of the study. Yu-Chung Hsieh carried out data acquisition, analysis and preliminary writing. Yu-Chang Chan coordinated the project together with Jyr-Ching Hu, and contributed to the final writing and revisions of the paper.

Conflicts of Interest: The authors declare no conflicts of interest.

References

1. Oskin, M.E.; Arrowsmith, J.R.; Corona, A.H.; Elliott, A.J.; Fletcher, J.M.; Fielding, E.J.; Gold, P.O.; Garcia, J.J.G.; Hudnut, K.W.; Liu-Zeng, J.; *et al.* Near-Field Deformation from the El Mayor–Cucapah Earthquake Revealed by Differential LIDAR. *Science* **2012**, *335*, 702–705. [CrossRef] [PubMed]
2. Nissen, E.; Krishnan, A.K.; Arrowsmith, J.R.; Saripalli, S. Three-dimensional surface displacements and rotations from differencing pre- and post-earthquake LiDAR point clouds. *Geophys. Res. Lett.* **2012**, *39*. [CrossRef]
3. Glennie, C.L.; Hinojosa-Corona, A.; Nissen, E.; Kusari, A.; Oskin, M.E.; Arrowsmith, J.R.; Borsa, A. Optimization of legacy LiDAR data sets for measuring near-field earthquake displacements. *Geophys. Res. Lett.* **2014**, *41*. [CrossRef]
4. Duffy, B.; Quigley, M.; Barrell, D.J.A.; Van Dissen, R.; Stahl, T.; Leprince, S.; McInnes, C.; Bilderback, E. Fault kinematics and surface deformation across a releasing bend during the 2010 MW 7.1 Darfield, New Zealand, earthquake revealed by differential LiDAR and cadastral surveying. *Geol. Soc. Am. Bull.* **2013**, *125*, 420–431. [CrossRef]
5. Grove, J.R.; Croke, J.; Thompson, C. Quantifying different riverbank erosion processes during an extreme flood event. *Earth Surf. Process. Landf.* **2013**, *38*, 1393–1406. [CrossRef]
6. Heritage, G.; Hetherington, D. Towards a protocol for laser scanning in fluvial geomorphology. *Earth Surf. Process. Landf.* **2007**, *32*, 66–74. [CrossRef]

7. Huang, Y.; Yu, M.; Xu, Q.; Sawada, K.; Moriguchi, S.; Yashima, A.; Liu, C.; Xue, L. InSAR-derived digital elevation models for terrain change analysis of earthquake-triggered flow-like landslides based on ALOS/PALSAR imagery. *Environ. Earth Sci.* **2014**, *73*, 7661–7668. [CrossRef]

8. Tarolli, P. High-resolution topography for understanding Earth surface processes: Opportunities and challenges. *Geomorphology* **2014**, *216*, 295–312. [CrossRef]

9. Ghuffar, S.; Székely, B.; Roncat, A.; Pfeifer, N. Landslide displacement monitoring using 3D range flow on airborne and terrestrial LiDAR data. *Remote Sens.* **2013**, *5*, 2720–2745. [CrossRef]

10. Dewitte, O.; Jasselette, J.C.; Cornet, Y.; Van Den Eeckhaut, M.; Collignon, A.; Poesen, J.; Demoulin, A. Tracking landslide displacements by multi-temporal DTMs: A combined aerial stereophotogrammetric and LiDAR approach in western Belgium. *Eng. Geol.* **2008**, *99*, 11–22. [CrossRef]

11. Peyret, M.; Djamour, Y.; Rizza, M.; Ritz, J.F.; Hurtrez, J.E.; Goudarzi, M.A.; Nankali, H.; Chéry, J.; Le Dortz, K.; Uri, F. Monitoring of the large slow Kahrod landslide in Alborz mountain range (Iran) by GPS and SAR interferometry. *Eng. Geol.* **2008**, *100*, 131–141. [CrossRef]

12. Chen, R.-F.; Chang, K.-J.; Angelier, J.; Chan, Y.-C.; Deffontaines, B.; Lee, C.-T.; Lin, M.-L. Topographical changes revealed by high-resolution airborne LiDAR data: The 1999 Tsaoling landslide induced by the Chi–Chi earthquake. *Eng. Geol.* **2006**, *88*, 160–172. [CrossRef]

13. Chen, Z.; Zhang, B.; Han, Y.; Zuo, Z.; Zhang, X. Modeling accumulated volume of landslides using remote sensing and DTM data. *Remote Sens.* **2014**, *6*, 1514–1537. [CrossRef]

14. Chan, Y.-C.; Chang, K.-J.; Chen, R.-F.; Liu, J.-K. Topographic changes revealed by airborne LiDAR surveys in regions affected by the 2009 Typhoon Morakot, southern Taiwan. *West. Pac. Earth Sci.* **2012**, *12*, 67–82.

15. Corsini, A.; Borgatti, L.; Cervi, F.; Dahne, A.; Ronchetti, F.; Sterzai, P. Estimating mass-wasting processes in active earth slides—Earth flows with time-series of High-Resolution DEMs from photogrammetry and airborne LiDAR. *Nat. Hazards Earth Syst. Sci.* **2009**, *9*, 433–439. [CrossRef]

16. Iverson, R.M.; George, D.L.; Allstadt, K.; Reid, M.E.; Collins, B.D.; Vallance, J.W.; Schilling, S.P.; Godt, J.W.; Cannon, C.M.; Magirl, C.S.; *et al.* Landslide mobility and hazards: Implications of the 2014 Oso disaster. *Earth Planet. Sci. Lett.* **2015**, *412*, 197–208. [CrossRef]

17. Milan, D.J.; Heritage, G.L.; Hetherington, D. Application of a 3D laser scanner in the assessment of erosion and deposition volumes and channel change in a proglacial river. *Earth Surf. Process. Landf.* **2007**, *32*, 1657–1674. [CrossRef]

18. Brasington, J.; Langham, J.; Rumsby, B. Methodological sensitivity of morphometric estimates of coarse fluvial sediment transport. *Geomorphology* **2003**, *53*, 299–316. [CrossRef]

19. Ventura, G.; Vilardo, G.; Terranova, C.; Sessa, E.B. Tracking and evolution of complex active landslides by multi-temporal airborne LiDAR data: The Montaguto landslide (Southern Italy). *Remote Sens. Environ.* **2011**, *115*, 3237–3248. [CrossRef]

20. Lallias-Tacon, S.; Liébault, F.; Piégay, H. Step by step error assessment in braided river sediment budget using airborne LiDAR data. *Geomorphology* **2014**, *214*, 307–323. [CrossRef]

21. DeLong, S.B.; Prentice, C.S.; Hilley, G.E.; Ebert, Y. Multitemporal ALSM change detection, sediment delivery, and process mapping at an active earthflow. *Earth Surf. Process. Landf.* **2012**, *37*, 262–272. [CrossRef]

22. Daehne, A.; Corsini, A. Kinematics of active earthflows revealed by digital image correlation and DEM subtraction techniques applied to multi-temporal LiDAR data. *Earth Surf. Process. Landf.* **2013**, *38*, 640–654. [CrossRef]

23. Bremer, M.; Sass, O. Combining airborne and terrestrial laser scanning for quantifying erosion and deposition by a debris flow event. *Geomorphology* **2012**, *138*, 49–60. [CrossRef]

24. Peng, M.H.; Shih, T.Y. Error assessment in two LiDAR-derived TIN datasets. *Photogramm. Eng. Remote Sens.* **2006**, *72*, 933–947. [CrossRef]

25. Tseng, C.-M.; Lin, C.-W.; Stark, C.P.; Liu, J.-K.; Fei, L.-Y.; Hsieh, Y.-C. Application of a multi-temporal, LiDAR-derived, digital terrain model in a landslide-volume estimation. *Earth Surf. Process. Landf.* **2013**, *38*, 1587–1601. [CrossRef]

26. Guzzetti, F.; Ardizzone, F.; Cardinali, M.; Rossi, M.; Valigi, D. Landslide volumes and landslide mobilization rates in Umbria, central Italy. *Earth Planet. Sci. Lett.* **2009**, *279*, 222–229. [CrossRef]

27. Guzzetti, F.; Ardizzone, F.; Cardinali, M.; Galli, M.; Reichenbach, P.; Rossi, M. Distribution of landslides in the Upper Tiber River basin, central Italy. *Geomorphology* **2008**, *96*, 105–122. [CrossRef]

28. Larsen, I.J.; Montgomery, D.R.; Korup, O. Landslide erosion controlled by hillslope material. *Nat. Geosci.* **2010**, *3*, 247–251. [CrossRef]

29. Central Weather Bureau (CWB); R.O.C. Typhoon Database. Available online: http://rdc28.cwb.gov.tw/ (accessed on 15 December 2015).

30. Water Resources Agency; Ministry of Economic Affairs; R.O.C. Water Resources Agency. Available online: http://eng.wra.gov.tw/ (accessed on 15 December 2015).

31. The Aerial Survey Office; Forestry Bureau; R.O.C. The ASO Historical Aerial Photogrammetry Database. Available online: http://www.afasi.gov.tw/ (accessed on 15 December 2015).

32. Ministry of th Interior; R.O.C. Establishment of The National Coordinate System. Available online: http://gps.moi.gov.tw/SSCenter/Introduce_E/IntroducePage_E.aspx?Page=GPS_E8 (accessed on 15 December 2015).

33. Ministry of th Interior; R.O.C. Taiwan Vertical Datum. Available online: http://gps.moi.gov.tw/SSCenter/Introduce_E/IntroducePage_E.aspx?Page=Height_E4 (accessed on 15 December 2015).

34. Baltsavias, E.P. A comparison between photogrammetry and laser scanning. *ISPRS J. Photogramm. Remote Sens.* **1999**, *54*, 83–94. [CrossRef]

35. Wehr, A.; Lohr, U. Airborne laser scanning—An introduction and overview. *ISPRS J. Photogramm. Remote Sens.* **1999**, *54*, 68–82. [CrossRef]

36. Taylor, J.R. *An Introduction to Error Analysis: The Study of Uncertainties in Physical Measurements*; Univ. Sci. Books: Mill Valley, CA, USA, 1982; p. 327.

37. Prokešová, R.; Kardoš, M.; Medved'ová, A. Landslide dynamics from high-resolution aerial photographs: A case study from the Western Carpathians, Slovakia. *Geomorphology* **2010**, *115*, 90–101. [CrossRef]

38. Müller, J.; Gärtner-Roer, I.; Thee, P.; Ginzler, C. Accuracy assessment of airborne photogrammetrically derived high-resolution digital elevation models in a high mountain environment. *ISPRS J. Photogramm. Remote Sens.* **2014**, *98*, 58–69. [CrossRef]

39. Hodgson, M.E.; Bresnahan, P. Accuracy of airborne LiDAR-derived elevation: Empirical assessment and error budget. *Photogramm. Eng. Remote Sens.* **2004**, *70*, 331–339. [CrossRef]

40. Orem, C.A.; Pelletier, J.D. Quantifying the time scale of elevated geomorphic response following wildfires using multi-temporal LiDAR data: An example from the Las Conchas fire, Jemez Mountains, New Mexico. *Geomorphology* **2015**, *232*, 224–238. [CrossRef]

41. Benjamin, M.J.; Jason, M.S.; Ann, E.G.; Guido, G.; Vladimir, E.R.; Thomas, A.D.; Nicole, E.M.K.; Bruce, M.R. Quantifying landscape change in an arctic coastal lowland using repeat airborne LiDAR. *Environ. Res. Lett.* **2013**, *8*, 925–932.

42. Rumsby, B.; Brasington, J.; Langham, J.; McLelland, S.; Middleton, R.; Rollinson, G. Monitoring and modelling particle and reach-scale morphological change in gravel-bed rivers: Applications and challenges. *Geomorphology* **2008**, *93*, 40–54. [CrossRef]

43. Telbisz, T.; Kovács, G.; Székely, B.; Szabó, J. Topographic swath profile analysis: A generalization and sensitivity evaluation of a digital terrain analysis tool. *Z. Geomorphol.* **2013**, *57*, 485–513. [CrossRef]

44. Burbank, D.W.; Blythe, A.E.; Putkonen, J.; Pratt-Sitaula, B.; Gabet, E.; Oskin, M.; Barros, A.; Ojha, T.P. Decoupling of erosion and precipitation in the Himalayas. *Nature* **2003**, *426*, 652–655. [CrossRef] [PubMed]

45. Clarke, B.A.; Burbank, D.W. Bedrock fracturing, threshold hillslopes, and limits to the magnitude of bedrock landslides. *Earth Planet. Sci. Lett.* **2010**, *297*, 577–586. [CrossRef]

46. Robl, J.; Hergarten, S.; Stüwe, K. Morphological analysis of the drainage system in the Eastern Alps. *Tectonophysics* **2008**, *460*, 263–277. [CrossRef]

47. Reutebuch, S.E.; McGaughey, R.J.; Andersen, H.-E.; Carson, W.W. Accuracy of a high-resolution LiDAR terrain model under a conifer forest canopy. *Can. J. Remote Sens.* **2003**, *29*, 527–535. [CrossRef]

48. Liu, X.; Zhang, Z.; Peterson, J.; Chandra, S. LiDAR-derived high quality ground control information and DEM for image orthorectification. *GeoInformatica* **2007**, *11*, 37–53. [CrossRef]

49. Gneeniss, A.S.; Mills, J.P.; Miller, P.E. In-flight photogrammetric camera calibration and validation via complementary LiDAR. *ISPRS J. Photogramm. Remote Sens.* **2015**, *100*, 3–13. [CrossRef]

50. Gneeniss, A.S.; Mills, J.P.; Miller, P.E. Reference LiDAR surfaces for enhanced aerial triangulation and camera calibration. *Int. Arch. Photogramm. Remote Sens. Spat. Inf. Sci.* **2013**, *1*, 111–116. [CrossRef]

51. Podobnikar, T. Production of integrated digital terrain model from multiple datasets of different quality. *Int. J. Geogr. Inf. Sci.* **2005**, *19*, 69–89. [CrossRef]

52. Kasperski, J.; Delacourt, C.; Allemand, P.; Potherat, P.; Jaud, M.; Varrel, E. Application of a terrestrial laser scanner (TLS) to the study of the Séchilienne Landslide (Isère, France). *Remote Sens.* **2010**, *2*, 2785–2820. [CrossRef]

53. Tseng, C.-M.; Lin, C.-W.; Chang, K.-J. The Sediment Budgets Evaluation in a Basin Using LiDAR DTMs. In *Engineering Geology for Society and Territory*; Springer: Berlin, Germany; Heidelberg, Germany, 2015; Volume 3, pp. 37–41.

54. Mackey, B.H.; Roering, J.J. Sediment yield, spatial characteristics, and the long-term evolution of active earthflows determined from airborne LiDAR and historical aerial photographs, Eel River, California. *Geol. Soc. Am. Bull.* **2011**, *123*, 1560–1576. [CrossRef]

remote sensing

MDPI

Article

Using an Unmanned Aerial Vehicle-Based Digital Imaging System to Derive a 3D Point Cloud for Landslide Scarp Recognition

Abdulla Al-Rawabdeh [1,*], Fangning He [2,†], Adel Moussa [1,3,†], Naser El-Sheimy [1,†] and Ayman Habib [2,†]

1   Department of Geomatics Engineering, University of Calgary, Calgary, AB T2N 1N4, Canada;
    amelsaye@ucalgary.ca (A.M.); elsheimy@ucalgary.ca (N.E.S.)
2   Lyles School of Civil Engineering, Purdue University, West Lafayette, IN 47907, USA;
    he270@purdue.edu (F.H.); ahabib@purdue.edu (A.H.)
3   Department of Electrical Engineering, Port-Said University, Port Said 42523, Egypt
*   Correspondence: amalrawa@ucalgary.ca; Tel.: +1-587-897-8051; Fax: +1-403-284-1980
†   These authors contributed equally to this work.

Academic Editors: Zhenhong Li, Gonzalo Pajares Martinsanz and Prasad S. Thenkabail
Received: 21 October 2015; Accepted: 18 January 2016; Published: 27 January 2016

Abstract: Landslides often cause economic losses, property damage, and loss of lives. Monitoring landslides using high spatial and temporal resolution imagery and the ability to quickly identify landslide regions are the basis for emergency disaster management. This study presents a comprehensive system that uses unmanned aerial vehicles (UAVs) and Semi-Global dense Matching (SGM) techniques to identify and extract landslide scarp data. The selected study area is located along a major highway in a mountainous region in Jordan, and contains creeping landslides induced by heavy rainfall. Field observations across the slope body and a deformation analysis along the highway and existing gabions indicate that the slope is active and that scarp features across the slope will continue to open and develop new tension crack features, leading to the downward movement of rocks. The identification of landslide scarps in this study was performed via a dense 3D point cloud of topographic information generated from high-resolution images captured using a low-cost UAV and a target-based camera calibration procedure for a low-cost large-field-of-view camera. An automated approach was used to accurately detect and extract the landslide head scarps based on geomorphological factors: the ratio of normalized Eigenvalues (*i.e.*, $\lambda 1/\lambda 2 \geqslant \lambda 3$) derived using principal component analysis, topographic surface roughness index values, and local-neighborhood slope measurements from the 3D image-based point cloud. Validation of the results was performed using root mean square error analysis and a confusion (error) matrix between manually digitized landslide scarps and the automated approaches. The experimental results using the fully automated 3D point-based analysis algorithms show that these approaches can effectively distinguish landslide scarps. The proposed algorithms can accurately identify and extract landslide scarps with centimeter-scale accuracy. In addition, the combination of UAV-based imagery, 3D scene reconstruction, and landslide scarp recognition/extraction algorithms can provide flexible and effective tool for monitoring landslide scarps and is acceptable for landslide mapping purposes.

Keywords: landslides scarps; geomorphology; slope; surface roughness; Semi-Global dense matching (SGM); unmanned aerial vehicles (UAVs)

1. Introduction

High urbanization rates and unplanned settlements expose the residents of urban areas to natural hazards-related risks. Earthquakes, floods, and landslides are among the major threats to urban areas [1,2]. During the 1990s, nearly nine percent of the world's natural disasters were landslides, which are difficult to predict [3–5]. Landslides are a natural geological phenomenon that is widely recognized as an important process in the transport of sediment. They occur on a wide variety of spatial and temporal scales in many mountainous areas [6–12], and identifying the smallest detectable area is important because small landslides are very likely to broaden under heavy rainfall conditions. The accurate detection and quick identification of small landslides are crucial for adopting appropriate mitigation measures and efficient decision-making strategies [13–15]. In recent years, an increasing number of studies have been conducted worldwide regarding geohazard susceptibility mapping and risk assessment. Traditional landslide mapping methods are typically based on interpretations of aerial photography and field survey inspections, which are used to better understand the characteristics of landslides [11,16]. The photographic, or image interpretation, approach can be adapted and implemented manually, automatically, or semi-automatically [17,18]. Depending on the scale and spatial resolution of the image, the details of the extracted geomorphologic features can be significantly affected. Singhroy [19] asserted that an aerial image scale of 1:25,000 is considered the largest scale at which one can properly interpret slope instability phenomena from aerial photographs. Mantovani *et al.* [20] suggested that the best image scale is 1:15,000 because the disrupted topography of a landslide scarp can be clearly identified at this scale. To achieve a landslide map with high accuracy and reliability, manual interpretation requires a well-trained geologist to delineate landslides in a stereoscopic environment. This process is time- and labor-intensive [17,18]. Alternatively, ground surveys with GPS and a total station are also time-consuming and have sparse spatial coverage, which results in the omission of fine-scale terrain structures in the resulting digital surface models (DSMs) [21]. Aside from being inaccessible and time-consuming, ground surveys and fieldwork are also subjective and prone to human and instrumental errors. Traditional strategies are unlikely to provide a satisfactory solution, especially for small-area investigations where mass movement rates are slower [16,22]. For these reasons, researchers have adopted remote sensing techniques in their research methodology [8,9,20,23–26].

A number of studies have explored the feasibility of an automated approach using satellite or airborne imagery and object-based image analysis [27–30]. The performance of this approach is limited due to the method's inability to distinguish the Earth's surface changes. In addition, improper processing strategies impact the adjacent pixels, resulting in low efficiency, and require machine learning theories, such as image classification and segmentation. This process requires large and remarkably different learning samples or a priori knowledge. It is also difficult to detect the causes of changes in information, which can lead to incorrect identification. Li *et al.* [31] used multi-resolution segmentation and object features from a digital terrain model and high-resolution satellite images. They encountered problematic results due to the presence of landslide and non-landslide pixels present along the landslide borders, which can directly affect classification accuracy.

Others have attempted to use a combination of satellite imagery and digital elevation models (DEMs). Barlow *et al.* [32] combined Landsat enhanced thematic mapper plus (ETM+) imagery and the use of DEM-derived geomorphometric data to detect and classify fresh translational landslide scarps within an area of the Cascade Mountains in British Columbia, Canada. They attempted to overcome the problem of unreliable detection of most types of landslides by segmenting the images and using the geomorphometric data. This method had a 75% overall accuracy in the detection of landslides that were over 1 ha^2. The abovementioned methods are less effective than Light Detection and Ranging (LiDAR) at detecting large landslides that have experienced significant historical activity [33]. The use of LiDAR technology for landslide identification has become increasingly popular in digital terrain modelling [12,34–40]. LiDAR is a valuable tool in geology [41], geomorphology [12], and hazard reduction efforts. Only a few studies have attempted to develop computer-assisted methods for extracting landslides from single or multiple pulse LiDAR data via pixel-based analysis [4,12,42]. Schulz [33,43] used a LiDAR DEM to identify the topography of the Puget Sound region of Seattle, USA. They found LiDAR data to be useful in identifying complex large-scale landslides and for locating potential landslides in the area. Schulz [33] used a LiDAR DEM with a 2 m resolution to produce slope and aspect-shading maps with azimuth angles of 45°, 135°, and 315° to identify landslides. Miner *et al.* [44] used a LiDAR DEM to interpret old landslides and identify potential landslides in the Victoria Otway Ranges near the Johanna area of Australia and successfully extracted vegetation from real ground elevation and shaded maps with different aspect directions to verify the landslide detection. Airborne LiDAR data and imagery are safe, accurate, and able to achieve a valuable top-view. However, the extremely high cost associated with the use of aircraft and its time-consuming nature makes this strategy an impractical solution, especially for investigations of small areas. Leshchinsky *et al.* [45] used LiDAR DEMs and head scarps to perform a semi-automated landslide inventory comparing three different study areas. They noted an increase in computational cost associated with post-processing data, which became especially prominent when one or more of the following parameters increased: study area size, input parameters, and resolution of the datasets.

To overcome the previous drawbacks of airborne LiDAR, the development of other active remote sensing techniques, such as terrestrial laser scanning (TLS), are making it possible to collect dense 3D point cloud for sites of interest. TLS has attracted interest for use in landslide studies, including (1) landslide detection and characterization [46,47]; (2) hazard assessment and susceptibility mapping [41]; and (3) modelling and monitoring [40,48–52]. A comprehensive review of laser scanning technology and its applications in landslide investigations can be found in the work of Jaboyedoff *et al.* [53].

This technology is not without limitations. These limitations include orientation biases that occur when the TLS line of sight is nearly parallel to the orientation of the discontinuity, occlusions that occur when parts of a rock face cannot be sampled because it is obscured by protruding surfaces, and truncation that occurs when the exposure of the discontinuity is less than the available resolution of the TLS point cloud. The TLS results can have occluded areas that leave gaps in the data due to shadows, truncated data, and orientation biases from sensor positioning. TLS does not eliminate the need for visiting the landslide site because these devices have a limited range, and a larger ranges (*i.e.*, the greater distances between the scanner and the object) lead to lower accuracies.

The precision of the measurements, the amount of time, and money required to conduct these measurements are important considerations when creating a landslide inventory [54]. It is not necessary to reach sub-centimeter precision when monitoring shallow landslides because the effort is directed towards evaluating the entire landslide body [55]. The need for high-accuracy measurements cannot be satisfied by traditional mapping methods due to the limitations in financial and technical resources. High-resolution topographic data are necessary for the morphological analysis of landslides [4]. Identifying the smallest detectable area possible is important because a small landslide has a high probability of broadening as a result of heavy rainfall. Recent advances in low-cost digital imaging, navigation systems, and software development have made it possible to accurately reconstruct 3D surfaces without using costly mapping-grade data acquisition systems. Furthermore, advancements in mobile mapping systems, such as unmanned aerial vehicles (UAVs), have made accurate 3D surface reconstruction more feasible whenever and wherever it is required. When small target areas need to be examined, a UAV is a better and more cost-effective choice than other platforms because UAVs are highly portable and are dynamic acquisition platforms. UAVs combine the advantages of vertical aerial photogrammetry and the high resolution of ground-based images. UAVs have the ability to fly at low altitudes and slow speeds and to reach areas that are not accessible from the ground or with manned aircrafts [56]. Moreover, these systems can be stored and deployed at a minimal expense [56–58]. These characteristics make UAVs an optimal platform for affordable rapid responses in mapping applications; consequently there is great interest in utilizing them in natural hazard applications.

Recently, 3D object reconstruction using UAV systems have become a popular area of research. This technology is currently used in a wide variety of applications that span across many fields such as, soil erosion [59,60], landslides or rocky surfaces inventory [56,61], landslide detection [62], landslide dynamics and monitoring [57,63–67], and natural disaster monitoring and evaluation [68,69].

A comprehensive review of remote sensing applications based on UAVs equipped with specific sensor-based technologies has been accomplished to varying degrees across a range of applications. A detailed summary, specifically focused on the progress made in environmental science applications, is provided in Pajares [56] and Colomina *et al.* [58].

The purpose of this research is to present a comprehensive system that uses a low-cost UAV, Semi-Global dense Matching (SGM) techniques, and automated approaches to detect and extract the geomorphological features of landslide scarps. This system reduces the many limitations that decrease the accuracy, completeness, and reliability of the previously described methods, thereby increasing the effectiveness of the landslide mapping.

2. Study Area

2.1. Location of the Study Area

This study was conducted in a 266 m × 185 m area situated along the main highway from Amman to Irbid in the Salhoub/Al-Juaidieh area of north-central Jordan. The study area is in a mountainous region that receives an average annual precipitation of 200 mm and contains creep landslides induced by heavy rainfall, [70]. The map in Figure 1 illustrates the location and extent of the study area, which is 21 km northwest of Amman and located at the following coordinates: 32.11430°N–35.85689°E. The terrain elevation varies from 650 m to 720 m above sea level and has a slope inclination of between 10° and 70°. The study area is dominated by two significant landslide features, which are known as the Al-Juaidieh slides. Abderahman [71] determined that the Al-Juaidieh landslide was a reactivation of an old landslide. Various field observations across the slope body and deformation along the highway and existing gabions indicate that the slope is still active and scarp features across the slope will continue to open and develop new tension crack features; leading to further drift of rocks downward. Substantial first failure activity in the area dates back to August 1992 during the construction of the Irbid-Amman highway. This landslide affected approximately 250 m of the highway, destroyed two houses, caused the evacuation of 10 others, and delayed the opening of the highway for several months [71].

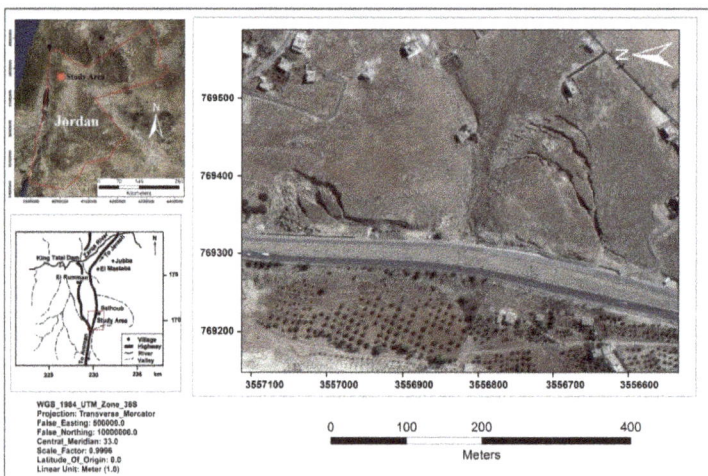

Figure 1. Location of study area and distribution of the landslides.

2.2. Geologic Setting of the Study Area

The land-cover types include landslide areas, roads, bare soil, and a few manmade buildings. The vegetation cover on the landslide is predominantly grass with occasional isolated bushes. The geology of the study area consists mainly of the Ajlun Group, which is subdivided into the five formations [72] shown in Figure 2. From bottom to top, these formations are the Na'ur, Fuheis, Hummar, Shu'ayb, and Wadi Es-Sir. The Na'ur limestone formation forms the majority of the outcrops in the study area, which starts with a few meters of a very hard limestone overlain by a medium-hard to somewhat medium-marly limestone, soft marl, very hard dolomitic limestone, soft marl, and intermixed gravel and boulders of limestone and marl at the surface [71]. The Hummar formation forms outcrops of a few meters of marly limestone followed by successive layers of medium-thick bedded grey to whitish crystalline limestone, dolomitic limestone, and dolomite that changes near the top to marly limestone [73]. The thickness of this formation is approximately 65 m within the Wadi Shu'ayb, Suweileh, and Hummar areas [74]. The Fuheis formation consists of soft marl and marly limestone interbedded with thin to medium-bedded shally to nodular limestone. As can be seen in Figure 3, the geological cross section taken along (A-A' (in the study includes a marlstone layer the bottom and a dolomite layer, limestone layer, and caliche layer at the top [75]. Saket [76] concluded that most of Jordan's landslides take place within the Lower Cretaceous Kurnub Sandstone, and the Upper Cretaceous Fuheis and Na'ur formations. These formations are characterized by the presence of marls and clays that act as weak interlayers, rendering the slope vulnerable to instability and sliding.

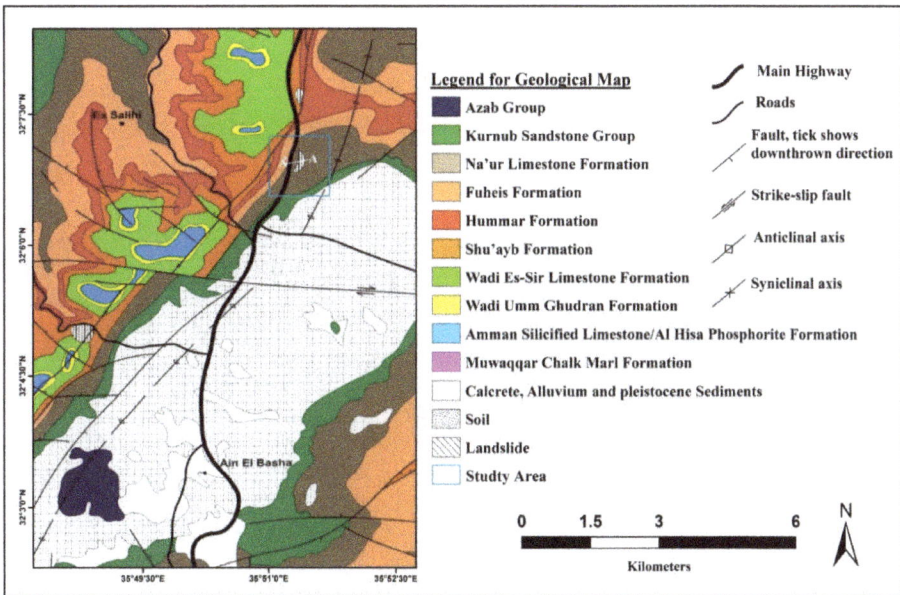

Figure 2. Geological map of the study area (modified after Sawariah and Barjous [73]).

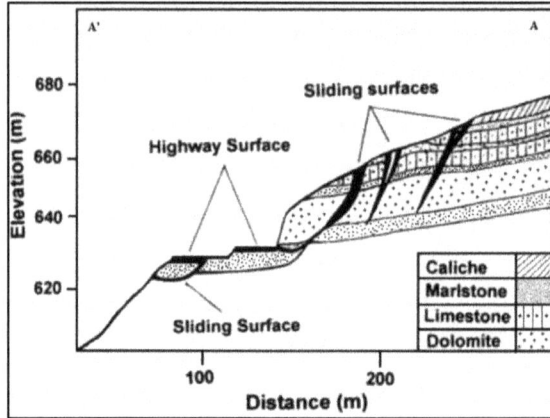

Figure 3. Schematic geologic cross section of the landslide ([71,77]).

From a structural point of view, the outcropping rocks in the study area show well-developed joints and cracks. Their widths of these cracks range from a few centimeters to multiple meters, indicating that discontinuity sliding could take place. Furthermore, all the landslides within the area are above the highway ridge and the mass movement along these landslides is rotational in a westerly to northwesterly direction.

3. Methodology

Figure 4 illustrates the proposed framework of the implemented methodology in this research work. This research aims to analyze and examine the automated approaches for landslide scarp detection and extraction. This is achieved first through mission planning and data collection using a UAV equipped with an off-the-shelf large-field-of-view (LFOV) camera. The camera calibration and stability analysis performed for this step can be expressed through the analysis approach proposed by Habib *et al.* [78,79]. After calibration 3D surface reconstruction was completed using the SGM approach to generate dense image-based point cloud [80,81]. The next step in achieving the research goal is to develop automated approaches for detecting and extracting the geomorphological features of landslide scarps using a ratio of Eigenvalues based on principal component analysis (PCA), examining topographic surface roughness index, and measuring the variability in slope in the local neighborhood of dense 3D image-based point cloud. As proposed by Varnes [82] (see Figure 5), the detection of landslides is based on the identification of the following geomorphological features: (1) a crown with tension cracks; (2) a main scarp, which tends to be the easiest feature to recognize because it is semi-circular (long and narrow) with a steep slope and convex planform and has a main direction perpendicular to the flow direction; (3) a minor scarp; (4) related slide blocks; and (5) a toe bulge or accumulation area, which is higher than the surrounding area due to the accumulation of debris.

Root mean square error (RMSE) analysis and a confusion (error) matrix is performed in order to validate and accuracy examination of the assessment in order to quantify errors as well as to assess the quality difference between the two datasets obtained for the manually digitized landslide scarps (the reference data) and the automated extraction scrap segments using the proposed approaches.

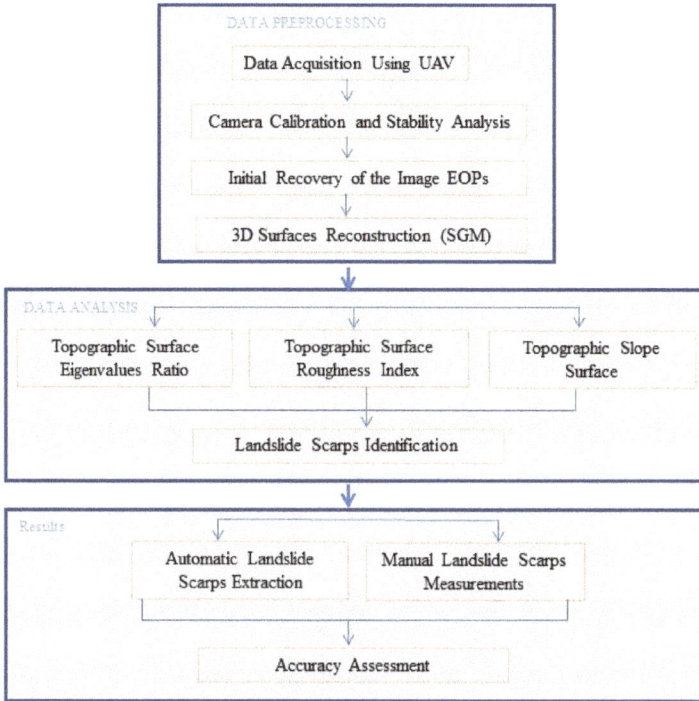

Figure 4. Methodology flow chart.

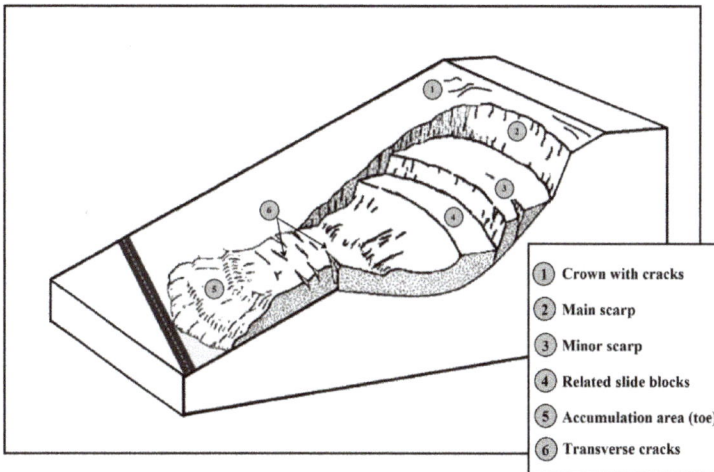

Figure 5. Geomorphologic characteristics of large-scale landslide (modified after Varnes [82]).

3.1. Mission Planning and Data Acquisition

To apply a cost-effectively collect data and detect landslide detection in hazardous, unstable areas a low-cost remote sensing approach can be applied using a UAV and LFOV digital camera. In this study, a GoPro camera was chosen because the UAV's gimbaled camera mount is specifically compatible with GoPro cameras. The Hero 3+ Black Edition was the highest quality camera available at the time of purchase. This camera had several innate advantages for this study, such as a fixed focal length and a robust design. GoPro cameras are built for use in rugged environments, which makes them an ideal choice for a UAV platform. The accompanying UAV is the DJI Phantom 2 (see Figure 6). The DJI Phantom 2 is designed for hobbyist use and has user-friendly controls. It also offers several other functions, such as an autopilot, no-fly zones, and auto-return home. This UAV offers approximately 25 min of flight duration on a single battery charge, can carry less than 1 kg of payload, and fly up to 1 km from the controller. This design is sufficient to cover a relatively small to medium area. According to Van Den Eeckhaut *et al.* [83] small landslide area are less than 0.02 km^2 in size and a medium to large landslide area size is greater than 0.02 km^2. However, it lacks several features that are necessary to obtain a photogrammetric product, such as time synchronization between the camera and the navigation unit, control of image exposure, and the overlap and sidelap of imagery. Furthermore, the camera has a short focal length that causes considerable barrel lens distortion.

Figure 6. The DJI Phantom 2 unmanned aerial vehicles (UAV) drone equipped with the GoPro camera used in this study.

The mission planning (flight and data acquisition carried out in September 2014) for the study area involved determining a detailed flight plan, including the flight direction, number of flight lines, flying height, and knowledge of the interior orientation parameters (IOPs) of the mounted digital camera. This configuration depends on the desired overlap, sidelap, and ground sampling distance (GSD) or footprint. The DJI Phantom 2 is equipped with an autopilot package that allows for pre-programmed flight paths, enabling the user to choose horizontal waypoints and the desired height of flight above the ground. The desired overlap (in flying direction) and sidelap (between adjacent flight lines) were set considering the array width to be parallel to the flying direction. The mission planning was designed to achieve the optimum configuration for the area of interest, as shown in Figure 7. Figure 7a illustrates the mission flight plan for image capture that covers the area of interest using north-south flight paths with 80% overlap and 60% sidelap. Figure 7b shows the east-west flight paths with 50% overlap and 30% sidelap which were used to obtain minimum data redundancy and to fill in data gaps caused by shadowed areas, occlusions, and blurry images. This step guarantees success in the automated point matching process and application of the SGM approach. Table 1 summarizes the designed north-south flight configuration for the area affected by the creeping landslide in the study area.

Figure 7. Graphical interface of the image capturing mission flight plan designed to cover the area of interest: (**a**) north-south flight path and (**b**) east-west flight path.

Table 1. North-south flight path configuration for area of interest.

Flight Configuration	
Average flying height (AGL)/speed	50 m/5 m/s
Autopilot	Available
Camera Specs	GoPro Hero3 + Black
Image format (pixels)	3000 × 2250
Pixel size	1.55 μm
Focal length (nominal)	3 mm
Time lapse	~2 s
Image Block Specs	
GSD (nominal)	~2.0 cm
Overlap/sidelap %	80/60
Image footprint	83 m × 62 m
Distance between images	14 m
Distance between lines	39.8 m
Number of strips	6
Number of Images	370
Total Area Covered	
Study area	48,981 m² (12.73 min)

To overcome the difficulties of time synchronization, limited image exposure control, and barrel lens distortion, the following steps are proposed. (a) An accurate analytical camera calibration of a low-cost LFOV camera followed by stability analysis using the ROT (Rotation) procedure was performed to evaluate the similarities and differences between the derived camera calibration parameters [78]; (b) The camera was set to automatic exposure using the camera's built-in timer. The selected time lapse is a function of the image footprint and flying speed, as shown in Equation (3). The GSD can be estimated based on the assumption that the terrain is relatively flat (see Equation (2)); (c) The exterior orientation parameters (EOPs) of the images can be indirectly recovered as part of the image-based point cloud generation process, which will be discussed later.

$$Pixel\ Size = \frac{Sensor\ width\ (mm)}{Image\ width\ (px)} \tag{1}$$

$$GSD = \frac{Pixel\ Size\ .\ Elevation\ above\ ground\ (H)}{Focal\ length\ (f)} \tag{2}$$

$$Time\ lapse = \frac{Sensor\ width \cdot \frac{H}{f} \cdot (1 - overlap)}{Flying\ speed} \tag{3}$$

where time lapse is the time between two successive exposures, sensor width is the number of pixels, f is the camera focal length, H is the flying height, overlap is a user-defined value that represents the overlap percentage between two successive images, and flying speed is a user-defined value.

3.2. Automated Surface Reconstruction

To generate a dense 3D image-based point cloud both Structure from Motion (SfM) and SGM techniques are utilized. The proposed SfM approach is first performed for the automated recovery of image EOPs. Then, SGM [80,81] is adapted for the dense image matching using the derived image orientation from the SfM approach from the scanned scene using the captured images. An overview of the proposed procedure is shown in Figure 8.

Figure 8. Proposed dense 3D reconstruction procedure.

This procedure was initiated by determining the camera's EOPs at each imaging station via the SfM approach developed by He and Habib [84]. This procedure is based on a two-step linear solution for the initial recovery of the image EOPs First, point feature correspondences are identified, followed by calculation of the relative orientation parameters relating stereo-images from derived conjugate points. A local reference coordinate frame is established in the second stage. Then, the EOPs of the remaining images are sequentially recovered through an incremental augmentation process. In this approach, tie points are identified using the scale-invariant feature transform (SIFT) feature detection algorithm in individual images as described by Lowe [85]. The acquired images are then normalized based on epipolar geometry while considering the camera IOPs and estimated EOPs, to minimize the search space for corresponding features in overlapping imagery. The algorithm used to determine the epipolar geometry and to define the normalized image plane is described in detail by Cho *et al.* [86]. This algorithm effectively removes the y-parallax in each stereo-pair of images. The images are projected onto a normalized image plane in which the rows of pixels in one image lie on the same row as in the normalized stereo-pair image.

An SGM procedure is then utilized to find the pixel-wise correspondence between the stereo-paired images. The algorithm used here starts by minimizing a cost function along eight different directions (two horizontally, two vertically, and four diagonally) from a pixel to determine an initial disparity value. The cost function is the sum of all the costs of the disparity along a given path. The path (direction) that has the least cost for a given pixel is used to determine the final disparity value, which is used to relate the pixels in one image to the conjugate pixels in the stereo-pair image.

The next step is a tracking procedure that identifies the matched pixels in all images. In this step, individual pixels are tracked across multiple images using the disparity values from the previous step. Starting with a stereo-pair of images, which are adjacent, and their relative orientation, which is known (determined with the EOPs), a corresponding disparity image can be computed. This image can be used to determine matching pixels in other images. Once a pixel is tracked through the disparity images until a valid disparity value cannot be found, the tracking algorithm for that pixel stops and the coordinates are transformed back into the ground coordinate system for use in the next step.

Finally, a dense 3D point cloud is generated based on the spatial intersection of light rays from all matched pixels. Using the EOPs, IOPs, and pixel coordinates of the corresponding pixels, the collinearity equations can be used to compute the object coordinates of each point. This procedure is conducted as a least squares estimation process, which is updated until the coordinates of the point do not change between iterations. There is a possibility that a pixel may be incorrectly tracked among images (known as a blunder) or that the intersection contains very poor geometry (low precision), and these situations should be avoided. To avoid low precision, the ground coordinates that are not sufficiently precise are eliminated. Blunders can be eliminated by projecting the ground coordinates back onto an image plane and computing the residuals, which should be within one pixel of the original coordinates. In addition, the ground coordinates are not computed if fewer than three pixels contribute to the intersection because this will not create a reliable solution. The resulting point cloud uses an arbitrary ground coordinate system with an arbitrary scale. The next step is to transform the point cloud from the arbitrary coordinate system into a meaningful one using surveyed ground control points.

3.3. Automated Landslide Scarp Features Detection and Extraction

The rapid development of high-resolution DEM data has provided more detailed topographic information. Higher resolution topographic data result in the more accurate detection of landslide scarps, thereby assisting in the monitoring and assessment of landslides as well as the development of mitigation plans. In addition, an in-depth understanding of topographic variations within hazardous landslide terrains is vital for companies developing construction plans for new or existing projects. Identifying landslide-specific spatial features from single surface models is important because not all the changes can be detected through a temporal analysis of landslide-susceptible areas. Previous landslides are the key predictors of the distribution of future landslides [87]. The following methodological subsections focus on detecting and extracting landslide scarps from a single surface using an image-based point cloud that is generated using the SGM approach. High-resolution topographic data was used to extract data on landslide scarps by examining local topographic variability through an analysis of the Eigenvalue ratio, slope, and surface roughness.

3.3.1. Eigenvalue Ratios

Eigenvalue ratios represent the degree of three-dimensional roughness or the crease edge of land surfaces [88–90]. Point clouds that result from the SGM approach are comprised of massive amounts of 3D coordinates. The well-known KD-tree data structure [91] was used in this study to handle the point cloud. The KD-tree is a tool that organizes the point cloud and allows for different query processes in the 3D space during the Eigenvalue estimating procedure. The KD-tree is used to search for neighbors within a specified search radius from the query point. A simple method for establishing this local neighborhood is to select the closest points to the query point according to a fixed Euclidian distance [92]. In this study, a 0.5 m radius was selected because of the size of the relevant geomorphic features. Selecting a smaller radius would have resulted in an increase in recognition of non-scarp landscape features, such as shrub vegetation, tree stumps or boulders. The Eigenvalue ratio methodology begins by utilizing PCA to determine the geometric properties of the local neighborhood of image-based points. To check whether a certain point (query point) belongs to a rough surface or a crease edge, the following steps are taken. First, a local neighborhood (P_n) is defined to enclose the (n) neighbors nearest to the query point. Then, a covariance matrix (Cov) is formed based on the dispersion of the points (P_n) from their centroid $(\overline{P}_c)$, as given by Equation (4). An Eigenvalue analysis is then performed to decompose the covariance matrix into two matrices (Equation (5)). The first matrix (*W*) is comprised of thee eigenvectors $(\vec{e}_1, \vec{e}_2, \vec{e}_3)$, and the other matrix ($\Lambda$) provides their corresponding Eigenvalues ($\lambda_1, \lambda_2, \lambda_3$).

$$Cov_{3\times3} = \frac{1}{n}\sum_{i=1}^{n}\left(\begin{bmatrix} P_{ix} \\ P_{iy} \\ P_{iz} \end{bmatrix} - \begin{bmatrix} \overline{P}_{cx} \\ \overline{P}_{cy} \\ \overline{P}_{cz} \end{bmatrix}\right)\left(\begin{bmatrix} P_{ix} \\ P_{iy} \\ P_{iz} \end{bmatrix} - \begin{bmatrix} \overline{P}_{cx} \\ \overline{P}_{cy} \\ \overline{P}_{cz} \end{bmatrix}\right)^{T} \tag{4}$$

where

$$\overline{P}_c = \frac{1}{n}\sum_{i=1}^{n}\begin{bmatrix} P_{ix} \\ P_{iy} \\ P_{iz} \end{bmatrix}$$

$$Cov_{3\times3} = W\,\Lambda\,W^T = \begin{bmatrix} \vec{e}_1 & \vec{e}_2 & \vec{e}_3 \end{bmatrix}\begin{bmatrix} \lambda_1 & 0 & 0 \\ 0 & \lambda_2 & 0 \\ 0 & 0 & \lambda_3 \end{bmatrix}\begin{bmatrix} \vec{e}_1^{\,T} \\ \vec{e}_2^{\,T} \\ \vec{e}_3^{\,T} \end{bmatrix} \tag{5}$$

The Eigenvectors/Eigenvalues are quite advantageous in determining the geometric nature of the established neighborhood. The Eigenvectors represent the orientation of the neighborhood in 3D space, whereas the Eigenvalues define the extent of the neighborhood along the directions of their corresponding eigenvectors [93]. The relative sizes of the Eigenvalues and the Eigenvectors' directions can indicate the type of primitive feature. For a rough surface/crease edge point, two of the estimated Eigenvalues will be much smaller than to the third Eigenvalue, for which the conventional equations (Equations (6a) and (6b)) were used in this study. The three normalized Eigenvalues denoted by "λ" were sorted from largest to smallest values as λ_3, λ_2, and λ_1,

$$\lambda_1 \approx \lambda_2 \tag{6a}$$

$$\frac{\lambda_1}{\lambda_2} \geqslant \lambda_3 \tag{6b}$$

3.3.2. Slope

The slope angle is the most important parameter in a slope stability analysis [94] because the slope angle is directly related to landslide probability. Other parameters contributing to landslide probability include geology, soil type, and hydrology [95], but this study specifically utilizes sudden change detection via examination of changes in slope angle. This method can also be useful in the identification and extraction of cliff information for other site-specific needs. The variability in slope is used to detect and extract landslide scarps [96–98]. The slope is often initially employed to identify landslides [99], based on the assumption that the slope changes abruptly between two successive scarps and that scarps become more distinct from their surroundings as they evolve. The slope angle for each point in an image-based point cloud can be estimated using Equation (7) because the Eigenvector for each point is calculated based on the PCA analysis in the previous step.

$$|\theta| = tan^{-1}\frac{\sqrt{Nx^2 + Ny^2}}{Nz} * \frac{180}{\pi} \tag{7}$$

where θ is the slope angle, and Nx, Ny, and Nz are the normal vector components of the plane.

3.3.3. Surface Roughness Index

Surface roughness can be defined as the irregularity of a topographic surface [96]. The surface roughness index is based on the calculation of deviations between the elevation model's surfaces fitted in the local range of a moving 0.5 m search radius window. The surface of most landslides is rougher, at the local scale of a few meters, than adjacent stable slopes, which are relatively smoother.

Tarolli [100] states that as long as erosion associated with precipitation and time is not severe, old landslides can be identified because there will still be a distinct difference in surface roughness between the landslide area and the local terrain. On the other hand, low average rainfall and semiarid climatic conditions dominating the study area result in poor susceptibility to weathering effects, thus maintaining associated surface roughness features.

This characteristic can be exploited to automatically detect and map landslides captured in high-resolution 3D point clouds. As illustrated in Figure 9, the surface roughness of landslide terrain (bottom image) features higher topographic variability than stable terrain (top image). McKean and Roering [12] and Glenn *et al.* [4] examined surface roughness and confirmed that landslide surfaces are rougher than the neighboring stable terrain due to the landslide mechanics, surface deformation, and subsidence of material. The algorithm developed in this study utilizes surface roughness information to detect and extract landslide scarps based on the measurement of the variability in local topographic surfaces.

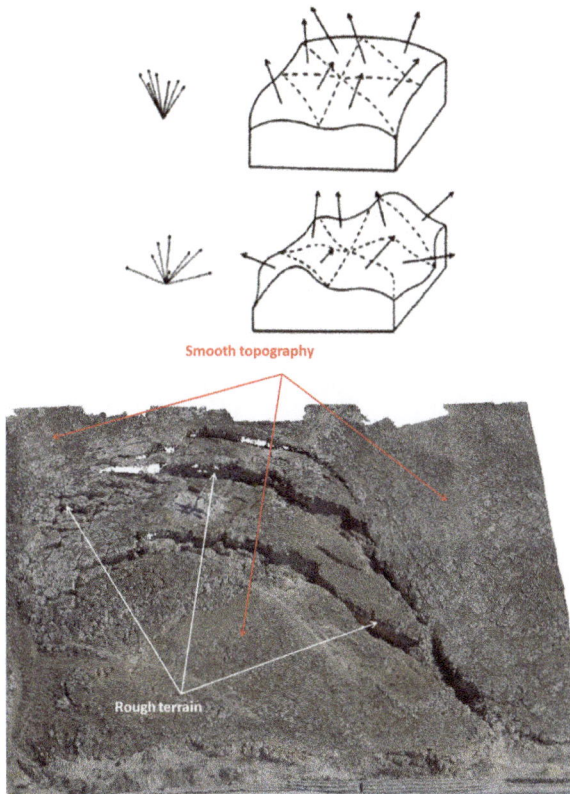

Figure 9. Surface normal representation of topographic surface roughness in a digital elevation model (DEM). The difference between smooth terrain (top) and rough terrain (bottom), illustrates greater variability in the latter [12,101].

The surface roughness index is the standard deviation of the object height (*h*) within 0.5 m local sampling window was used to calculate the surface roughness in this study in Equation (8).

$$\gamma = \sqrt{\frac{\sum (h - h\prime)^2}{n - 1}} \tag{8}$$

where γ is the surface roughness index, $h\prime$ is the mean height of all points within the local window, and n is the number of points within the local search window.

4. Results and Discussion

In this study, three sets of results were produced. The first set of results consists of the camera calibration parameters that were obtained by calibrating the LFOV digital camera. The second set of results consists of the image-based point cloud and orthophotos produced by implementing the image processing using the SGM algorithm for the study area. Finally, the third set of results consists of the landslide scarps detected and extracted using (a) the ratio of the Eigenvalues ($\lambda 1/\lambda 2 \geqslant \lambda 3$) based on PCA analysis; (b) the topographic surface roughness index; and (c) the approach based on measuring the variability in slope in the local neighborhood of the 3D image-based point cloud.

4.1. Dataset Description

The study area (Salhoub/Al-Juaidieh, Jordan), yielded a set of 370 nearly vertical images with overlap of 80% and sidelap of 60% along the north-south flight path direction. These images were captured with a calibrated GoPro Hero 3+ Black Edition camera mounted to a DJI Phantom 2 UAV over a total flight time of 12.73 min. To obtain the lowest data redundancy while filling in data gaps caused by shadowed areas, occlusions, and blurred images, a set of 160 images was collected along east-west flight paths with an overlap of 50% and a sidelap of 30%. The utilized GoPro camera was equipped with several imaging modes; however, in this study, only the medium field-of-view mode was utilized. Each image covered an area of approximately 83 m × 62 m on the ground, resulting in imagery with a resolution of approximately 2 cm (GSD). Figure 10 shows an example of a strip of aerial photographs in the study area.

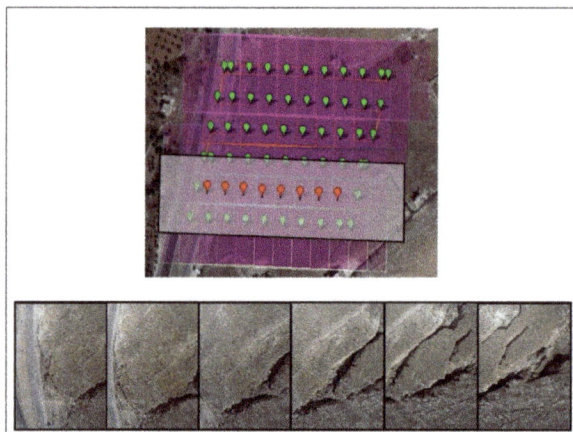

Figure 10. Strip of aerial photographs of the study area captured from the areas indicated in red along of the east-west flight path shown in the upper image.

4.2. Camera Calibration and Stability Analysis

The camera was calibrated and checked for stability over a two-month period from January 2014 to March 2014. It should be noted that, over the two-month the camera is mounted on a UAV that is flown in different environments, difficult terrain, and have been subjected to rough landings, significant vibrations from the UAV, and even a few crashes. For each image dataset, a total of 12 convergent images of a 2D calibration test field with 116 targets were captured with the same image network configuration. The ultimate goal was to decide whether the two sets of IOPs were equivalent. If the sets of IOPs are similar, then a stability analysis of the camera is ensured and the IOP parameters have not changed over time. Thus, the camera can be deemed stable, which is essential for achieving an accurate 3D model later on. The experimental results were examined using the United States Geological Survey (USGS) Simultaneous Multi-frame Analytical Calibration (SMAC) model with R_0 values of 0 and 3 mm with K_1, K_2, and K_3 as the distortion parameters. The derived square roots of posteriori variance (σ_0) values are all smaller than one pixel in size (1.55 μm) according to Habib *et al.* [102], and the image coordinate measurement had an accuracy range of one-half pixel, which indicates that no blunders were present, that the utilized SMAC model was appropriate and that the derived IOPs were acceptable.

The stability of the camera was analyzed using the ROT method [78]. In the case of the GoPro Hero 3+ Black Edition camera used in this study, the pixel size was 1.55 μm. Therefore, if the square root of the variance component was less than 1.55 μm, which is approximately around one pixel, then the IOP sets were considered similar and stable according to the ROT method [102–104] and was the correct choice for an accurate estimation of the 3D coordinates.

4.3. Automated Point Cloud Generation

To generate a point cloud using the SGM methodology, a total of 530 images were captured. The image orientation (EOP) is a fundamental prerequisite parameter in any image-based reconstruction. Orienting the images was accomplished by applying the SfM approach developed by He and Habib [83]. The estimated image position and orientation and the reconstructed sparse point cloud from the UAV image dataset are shown in Figure 11.

Once a set of images was oriented, the surface was digitally reconstructed by implementing the SGM algorithm starting from the known exterior orientation and camera calibration parameters. A dense point cloud with more than 13.65 million points was constructed for the study area using all of the captured images with an average point spacing of 1.5 cm and an average point density of approximately 4431.52 points/m^2. The dense point cloud obtained from SGM is illustrated in Figure 12.

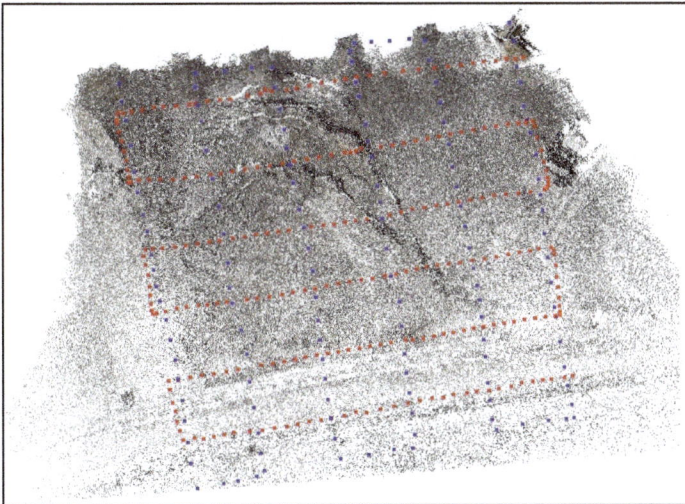

Figure 11. Perspective view of the constructed sparse point cloud. The red (North to South direction) and blue (East to West direction) dots over the sparse points show the camera positions and orientations during image acquisition by the UAV.

Figure 12. The dense 3D point cloud generated from the UAV image dataset collected in September 2014. The blue inset square represents the point density within 1 m^2.

4.4. Detection and Extraction of Landslide Scarp Features

The landslide geomorphological analysis in this study was based on a 1.5 cm point spacing in a dense 3D point cloud generated for the study area. The dense image-based point cloud proved to be a very useful tool for rapidly creating profiles along the slope direction. Using these profiles the vertical walls of the landslide scarps were measured and observed to have a range of 3 m to 5 m. The cross section along the profile A-A′ is shown in Figure 13.

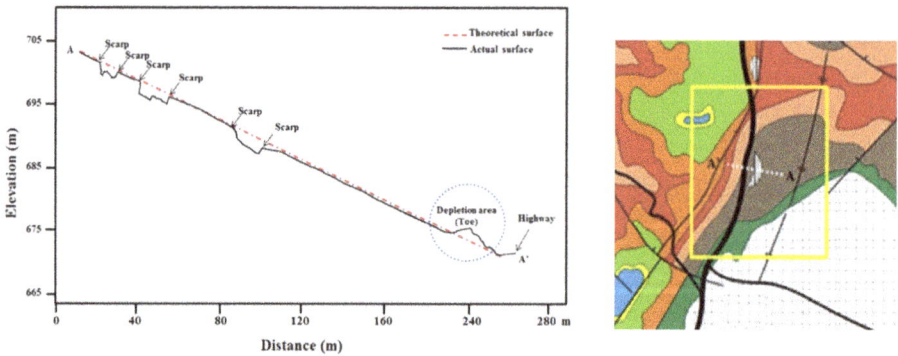

Figure 13. Cross-section (left image) along the profile A-A′ (right image is a subset of the geological map) of the landslide in the central part of the study area.

4.4.1. Topographic Eigenvalue Ratios

Figure 14 illustrates the variation in the topographic parameters using the normalized Eigenvalue ratio of $\lambda1/\lambda2$ computed in a 0.5 m moving sampling window on a dense 3D image-based point cloud. The map in Figure 14 clearly shows that the boundary of the landslide scarps (yellow-red color in the map) were recognized in detail, which means the normalized Eigenvalue ratio values inside and surrounding the landslide scarps were much less clustered than those in the nearby terrain outside the slide.

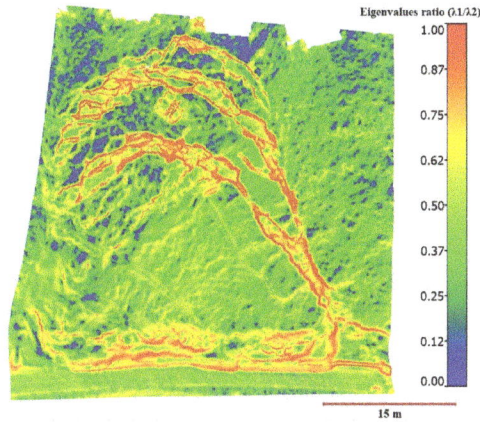

Figure 14. Eigenvalue ratio ($\lambda1/\lambda2$) computed with a 0.5 m moving sampling window in the dense 3D image-based point cloud. Higher Eigenvalue ratios indicate landslide scarp features because the values of these areas are less clustered than the smooth terrain surface.

The extracted features were filtered, and only the areas with $\lambda1/\lambda2$ Eigenvalue ratios greater than $\lambda3$ were selected as a candidate scarp points. The landslide scarps that were automatically extracted using the proposed threshold values ($\lambda1/\lambda2 \geqslant \lambda3$) are shown in Figure 15. Some open areas associated with the spacing/opening between two successive landslide scarps were observed to have been filled with recently fractured rocks and sediments. Therefore, after the initial landslide scarp extraction, these areas were also classified as scarps because they presented similar characteristics (*i.e.*, top and bottom scarps). Further refinement to filter out noisy extracted points (*i.e.*, points enclosed between higher candidate scar points circled in red in Figure 16) based on the height profile (cross section) along the slide direction was required to obtain more accurate results. In Figure 16, the height profile along the slide direction is in blue, the preliminary scarp points are in green, and the accepted scarp points are circled in red.

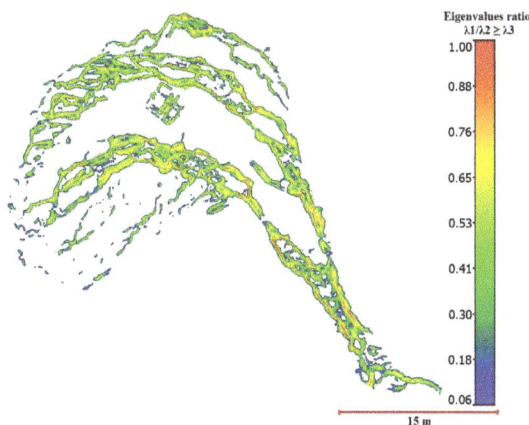

Figure 15. Extraction of landslide scarp features for the of Salhoub/Al-Juaidieh landslide based on variation in the local topography's Eigenvalue ratio.

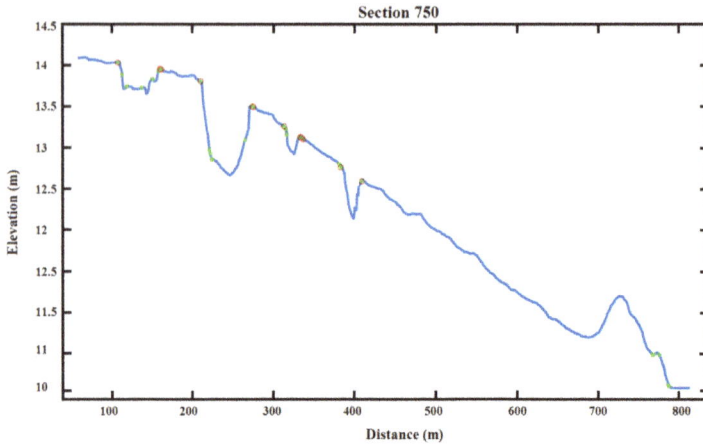

Figure 16. Height profile (cross section) along the slide direction in blue, preliminary scarp points in green, and accepted scarp points circled in red.

The extracted features based on the proposed threshold were overlaid on the orthophoto image (Figure 17). Visual inspection indicates that the features related to landslide scarps (blue segments) were accurately extracted and accurately overlaid on the corresponding landslide scarps.

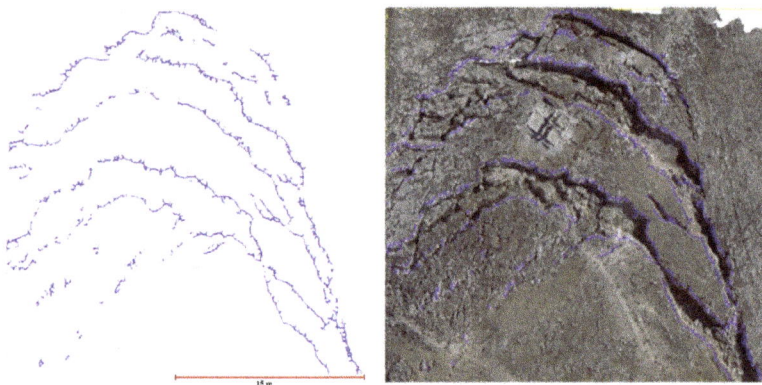

Figure 17. Final landslide scarp detection results (blue) based on Eigenvalues ratios of the topographic surface (**left image**) overlaid on the generated orthophoto using UAV images of the study area (**right image**).

4.4.2. Topographic Slope Surface

The 3D topographic image-based point cloud data were used to generate maps of the local topographic slopes. The slope angle for each point of the image-based point cloud was derived using a 0.5 m moving sampling window and was determined by calculating the slope of the normal vector for each point. As shown in Figure 18, the resulting slopes of the study area were distributed between 0° and 90°. The slope calculation results show that the active landslide scarps had slopes up to 50° associated with the abrupt changes in the Earth's surface compared with their local neighborhood and, to a lesser degree, they formed a ridge of the depletion area (toe area) that is comprised of marlstone sediments.

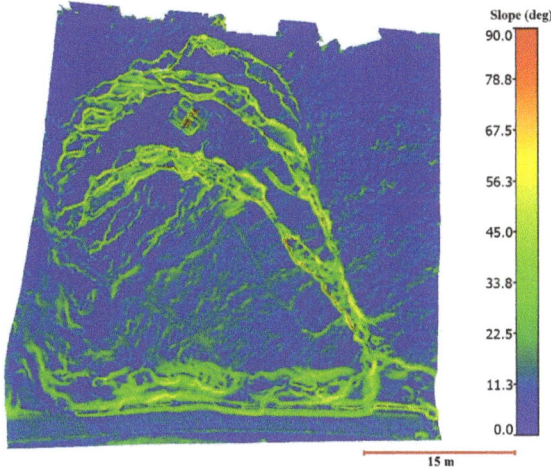

Figure 18. Slope map of the creeping Salhoub/Al-Juaidieh landslide calculated using a 0.5 m moving sampling window on a dense 3D dense image-based point cloud.

For landslide scarp detection using the slope method, the point cloud classified the ground objects into either landslide or non-landslide classes based on the assumption that the slope changes abruptly between two successive scarps and that the scarps become more distinguishable from their surroundings as they evolve. In the analysis of the slope distribution in this study, the classification of landslide scarps usually occurred above a mean slope value of 22°, which corresponds to the internal frictional angle at which most slope instabilities in the area occur [28]. The statistics for the slope distribution, derived from the point cloud, are shown in Figure 19.

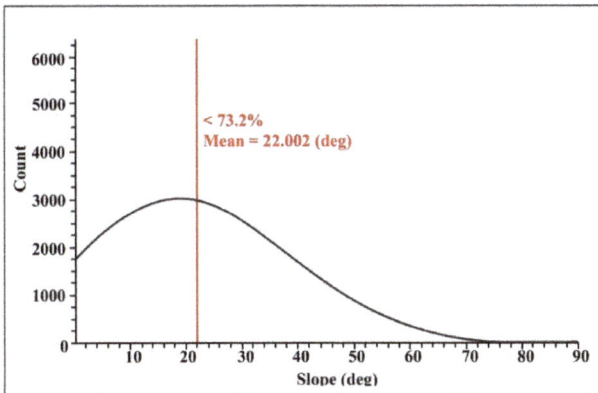

Figure 19. Statistics of the slope distribution.

An estimate of the threshold was based on the fact that the histogram of the slope value was set to a mean slope value of 22°. The obtained threshold provided a filtering process to extract the most likely landslide scarps in the study area. The extracted potential landslide scarps from the derived slope variation based on the defined threshold are shown in Figure 20.

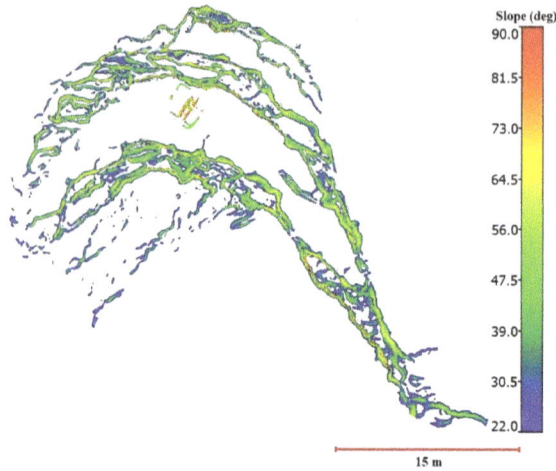

Figure 20. Distribution of the identified Salhoub/Al-Juaidieh landslide scarp features based on local slope variations.

The extraction smooth landslide head scarps from the previous results were based on the assumption that the detected candidate points should only decrease in height. In this study, the height profile (cross section) in the slide direction was analyzed to ensure that the smoothed head scarps fit the established criteria (*i.e.*, Figure 16). As seen in Figure 21, the final landslide scarp detection results were overlaid on the orthophoto image. Visual inspection revealed that the detection results were accurately overlaid onto the corresponding landslide scarps.

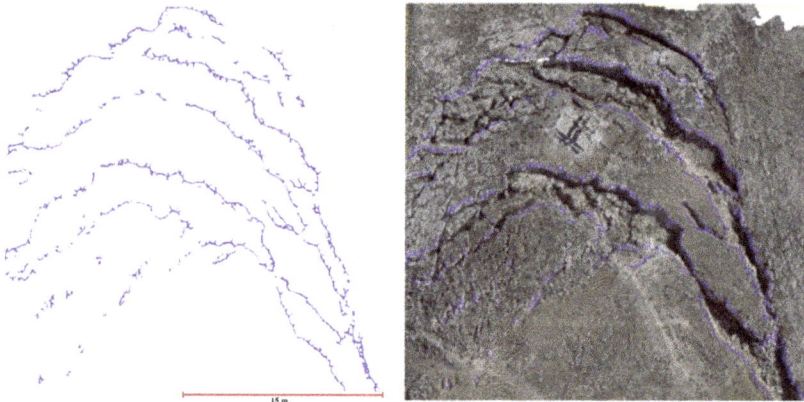

Figure 21. Final landslide scarp detection results (blue) overlaid on the orthophoto generated used UAV images of the study area.

4.4.3. Topographic Surface Roughness Index

The surface of a landslide is usually rougher on a local scale than that neighboring stable terrain, which means that the local vector orientations of the rougher surfaces more highly variable than those of the smooth topographic surfaces, which have similar orientations. The roughness of a landslide surface is calculated using point clouds without interpolation. To provide a more reliable roughness indicator, the standard deviation of the height point (h) within a 0.5 m local moving sampling window is used to calculate surface roughness of each point in the dense 3D point cloud, using Equation (8) in Section 3.3.3. As illustrated in Figure 22, the surface roughness is higher in the landslide terrain in this study than in the stable terrain. The surface roughness varies from approximately 5 cm to 14 cm, with greater roughness near active landslide scarps and along the higher steeper slopes due to landslide mechanics, surface deformation, and subsidence of material. The smoother topographic surface roughness, which covered the flat surfaces and stable areas, was within a maximum range of 5 cm.

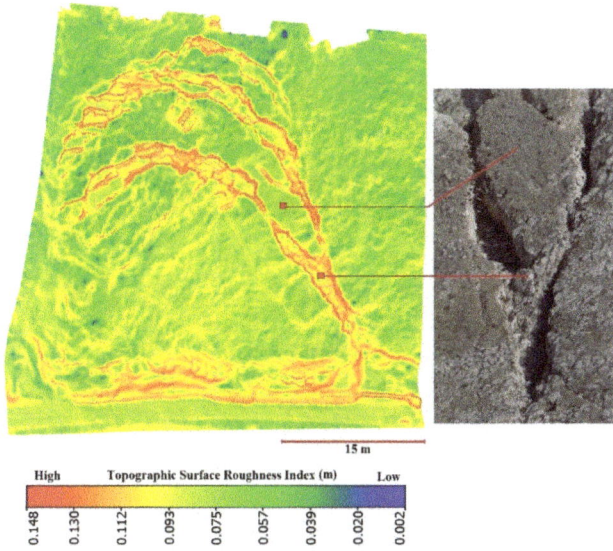

Figure 22. Topographic surface roughness index map of the creeping Salhoub/Al-Juaidieh landslide calculated using a 0.5 m moving sampling window in the dense 3D point cloud.

The topographic surface roughness map (Figure 22) indicates that the dataset exhibits a good correlation with the landslide scarp boundaries using the topographic surface roughness index approach. To extract the candidate landslide scarps, the histogram of the topographic surface roughness index distribution (Figure 23) was used to estimate the thresholds.

A defined threshold, which was equal to 2σ (0.046 m), was used as a filtering process. The filtering process selected areas with a surface roughness value greater than 0.046 m. Figure 24 shows the extracted features based on the 2σ threshold value.

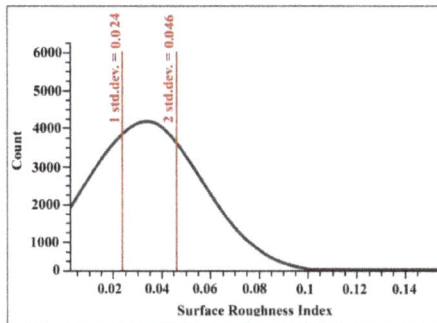

Figure 23. Statistics of the topographic surface roughness index distribution.

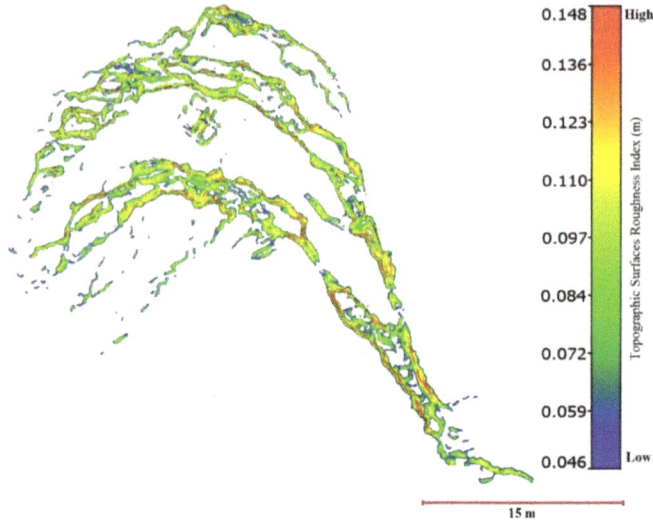

Figure 24. Extraction of landslide scarp features of the Salhoub/Al-Juaidieh landslide based on local topographic surface roughness values.

The final results of the landslide scarp extraction process using the surface roughness index approach are shown in Figure 25a. The extracted landslide scarps are overlaid on the orthophoto image for verification in Figure 25b, and is the result shows that the extraction results are accurately overlaid on the corresponding landslide scarps.

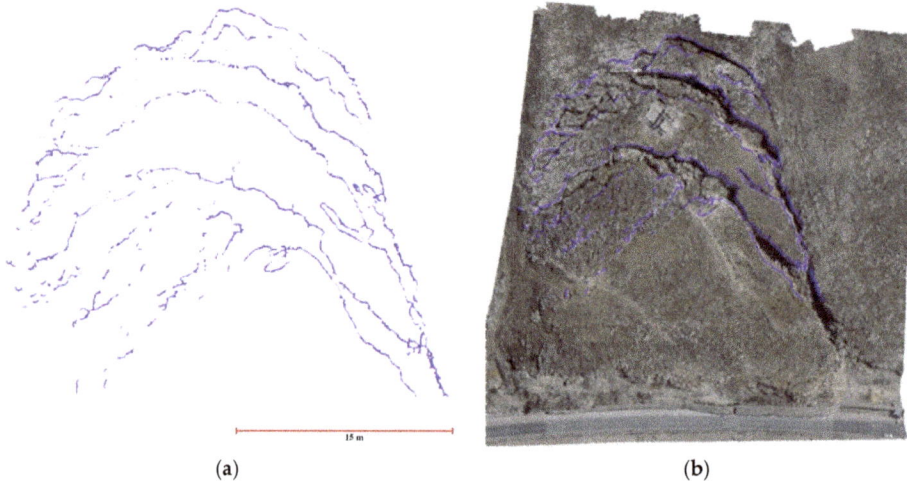

(a)   (b)

Figure 25. (a) Final landslide scarp detection results (blue) based on the topographic surface roughness index and (b) the scarps overlaid on the orthophoto using UAV images of the study area.

4.5. Accuracy Assessment

The accuracy assessment of the identified results consisted of a comparison between the automatically extracted scarp segments obtained from the different topographic surface analyses and manually digitized scarp segments. The manual measurements were accomplished using GIS software, whereby the active landslide scarps were manually digitized (screen digitizing) by visual recognition of the landslide scarp edges in the orthophoto image of the study area generated using the dense 3D image-based point cloud, and this manual image was treated as ground truth data (reference data) (Figure 26).

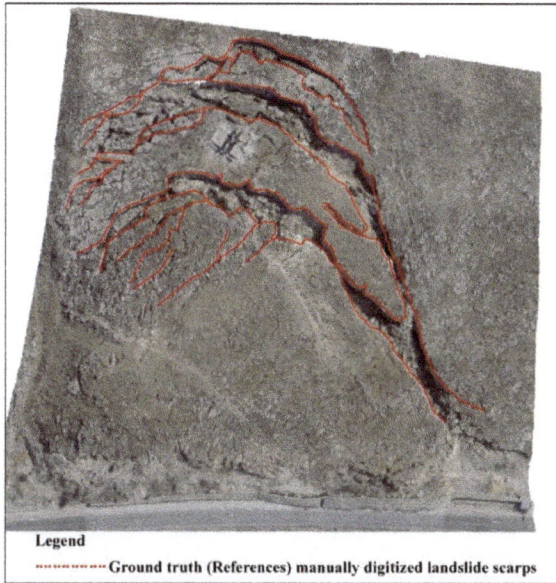

Figure 26. Manually digitized ground truth (reference) data for the landslide scarps overlaid on the orthophoto image of the study area.

A per-pixel-based approach is used to validate and examine the accuracy of the extracted landslide scarps. The results were compared to the reference data by calculating the RMSE between the automatically extracted landslide scarp segments and the manually digitized landslide reference segments (ground truth). If the average RMSE was above a threshold, then the entire landslide scarp segment was regarded as an unmatched segment. The maximum RMSE tolerance was set to 6 pixels, which is equal to 30 cm (number of pixels multiplied by the pixel size equal to 5 cm) due to the horizontal accuracy in the orthophoto when visually and manually digitizing. This maximum tolerance buffer of 6 pixels was chosen to account for the manually digitized reference accuracy and the minimum spacing between neighboring scarps (~70 cm) observed in this dataset. Using lower tolerance values might miss true matches because of manual reference inaccuracies, and using a higher tolerance value might cause detected scarps to be double matched to neighboring manually referenced scarps.

The RMSE analysis was achieved using the proposed approaches shown in Figure 27a–c. No significant differences in accuracy were found between the different automatic extraction methods in this paper and the manually digitized landslide scarps. The RMSE values, which represent an accuracy assessment, of the Eigenvalue ratio, topographic surface slope, and topographic surface roughness index methods are 11.98 cm, 9.05 cm, and 10.45 cm, respectively. Thus, the positions of the extracted scarp segments may have shifted by approximately two pixels on average from their positions in the reference data. In general, the automatically extracted results were comparable to the results obtained through manual digitization of the orthophoto image of the study area for landslide scarp identification. Therefore, the scarps in this study were accurately extracted.

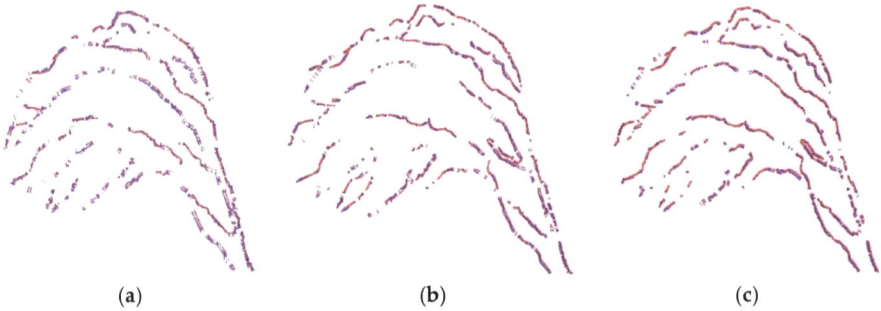

(a)	(b)	(c)

Figure 27. The average distance (RMSE) between the ground truth (red segments) and the extracted landslide scarp data (blue segments); Scarp segments extracted using (a) Eigenvalues ratios; (b) topographic surface slopes; and (c) topographic surface roughness index values.

Further validation and examination of the accuracy of the assessment was accomplished by performing a confusion (error) matrix to quantify the errors as well as to assess the quality difference between the two datasets obtained by visual interpretation (the reference data) and the map data (extracted scarp segments) using a per-pixel approach. Each variable of the two datasets was organized in a matrix format populated by the number of agreements and disagreements according to the following four classes: (1) true positive (TP), or matched, which indicates that an extracted landslide scarp correspond to a reference landslide scarp; (2) true negative, which indicates that the extracted non-landslide pixels correspond to reference non-landslide pixels; (3) false positive (FP), or unmatched, which indicates an extracted landslide scarp correspond to reference non-landslide pixels; and (4) false negative, which indicates that an extracted non-landslide pixel corresponds to a reference landslide scarp Matching was performed based on the constant in the pre-defined buffer method. The maximum tolerance of the buffer around the reference landslide scarp data was set at 6 pixels. The parts of the extracted data within the buffer were considered to match the manual reference. The accuracies of the extracted landslide scarp segments based on the different approaches and the reference datasets are shown in Tables 2–4. The error matrix was obtained for the per-pixel assessments of quality based on a comparison of the extracted scarp segments as presented in Figures 17a, 21a, and 25a. The reference map is shown in Figure 26.

The overall accuracy of the confusion matrix was calculated by dividing the total number of agreements (*i.e.*, the sum of the diagonal cells of the matrix) by the total number of samples. The user's accuracies, producer's accuracies, and Cohen's kappa coefficient statistics were calculated from the confusion matrix to obtain a more in-depth perspective of the uncertainty analysis. The user's accuracy (row values) was based on the agreement of a particular class to the summation of all classes in each row. Similarly, the producer's accuracy (column values) was computed considering the agreement of a particular class to the summation of that column. In many cases, according to Zhan *et al.* [105] and Tuermer *et al.* [106], the user's accuracy represents a measure of correctness (Equation (9)), and the producer's accuracy represents as a measure of completeness (Equation (10)).

$$\text{Correctness} = \left(\frac{TP}{TP + FP}\right) \times 100 \tag{9}$$

$$\text{Completness} = \left(\frac{TP}{TP + FN}\right) \times 100 \tag{10}$$

Cohen's kappa coefficient was calculated from the confusion matrix and was estimated for the performance evaluation of landslide extraction. This coefficient is a measure of the agreement between the extracted and reference data. In other words, the kappa statistics are a measure of true agreement, which is represented by the following relationship [107]:

$$\text{Kappa coefficient (k)} = \frac{N\sum_{i=1}^{r}X_{ii} - \sum_{i=1}^{r}(X_{i+} * X_{+i})}{N^2 - \sum_{i=1}^{r}(X_{i+} * X_{+i})} \tag{11}$$

where (k) is the Kappa value, r is the number of rows in the confusion (error) matrix, X_{ii} is the number of observations in row i and column i on the major diagonal of the matrix, X_{i+} is the total observations in row i, X_{+i} is the total observations in the column, and N is the total number of observations included in the matrix.

Table 2. Confusion (error) matrix for assessing the quality of extracted landslide scarps based on the topographic Eigenvalue ratios approach.

		Ground Truth (Reference)				
		Positive	NegativeTotal		User's Accuracy/Correctness (%)	Error of Commission (%)
Extracted Features	Positive	3917	846	4763	82.24	17.76
	Negative	1306	26698	28004	95.37	4.66
	Total	5223	27544	32767		
	Producer's Accuracy/Completeness (%)	75.0	96.93	Overall accuracy 93.43%; kappa 74.58%.		
	Error of Omission (%)	25.00	3.07			

Table 3. Confusion (error) matrix for assessing the quality of extracted landslide scarps based on the topographic surface slope approach.

		Ground Truth (Reference)				
		Positive	Negative	Total	User's Accuracy/Correctness (%)	Error of Commission (%)
Extracted Features	Positive	3325	781	4106	80.98	19.02
	Negative	1468	27193	28661	94.88	5.12
	Total	4793	27974	32767		
	Producer's Accuracy/Completeness (%)	69.37	97.20	Overall accuracy 93.14%; kappa 70.78%.		
	Error of Omission (%)	30.63	2.79			

Table 4. Confusion (error) matrix for assessing the quality of extracted landslide scarps based on the topographic surface roughness approach.

		Ground Truth (Reference)			User's Accuracy/Correctness (%)	Error of Commission (%)
		Positive	Negative	Total		
Extracted Features	Positive	3168	450	3618	87.56	12.44
	Negative	1669	27480	29149	94.27	5.73
	Total	4837	27930	32767		
	Producer's Accuracy/Completeness (%)	65.50	98.39		Overall accuracy 93.53%; kappa 71.31%.	
	Error of Omission (%)	34.50	1.61			

The overall qualities of the landslide scarp segments extracted from the Salhoub/Al-Juaidieh landslide test site using different approaches were calculated to be 93.43% for the Eigenvalue ratio approach, 93.14% for the slope approach, and 93.53% for the surface roughness index approach. The producer's accuracies (completeness) and user's accuracies (correctness) were found to be 75.0% and 82.24%, respectively, for the Eigenvalue ratio approach; 69.37% and 80.98%, respectively, for the slope approach; and 65.50% and 87.56%, respectively, for the surface roughness index approach. The main causes of error, in terms of the extracted landslide scarp segments, were due to a few manmade buildings and isolated bushes within the study area. These segments were not extracted because they were directly connected to adjacent bushes and do not show the desired characteristics of landslide scarps. Additional incomplete extracted scarp segments were associated with errors in the visual and manual digitizing due to horizontal inaccuracy present in the orthophoto.

Kappa statistics arrange in from value of 1 or (100%), indicating a strong agreement, to 0 (or 0%), indicating that any agreement is entirely due to chance (*i.e.*, incorrectly extracted). Kappa values greater than 75% indicate very good to strong agreement [108–110]. The values of the kappa coefficient associated with the confusion matrices in Tables 2–4 were 74.58%, 70.78%, and 71.31%, respectively, thereby denoting good agreement between the extracted results and the reference results.

5. Conclusions and Recommendations for Future Work

This paper presents a practical approach for the detection of landslide scarps using point clouds that have been derived from captured imagery by low-cost unmanned aerial vehicles (UAVs). The main advantages of the proposed methodology is allowing for the derivation of accurate information for landslide characterization while alleviating the inherent risk in surveying hazardous landslide-prone areas and reducing the incurred cost. Due to the inherent excessive lens distortions in the utilized imaging system, GoPro Hero 3+ Black Edition onboard a DJI Phantom 2, a camera calibration and stability analysis procedure is essential. Derived interior orientation parameters (IOPs) for the GoPro during the time period from January to March, 2014 revealed that the camera maintains the stability of its internal characteristics and is suitable for landslide mapping. The paper also introduces procedures for automated recovery of the exterior orientation parameters (EOPs) of the images that have been captured over the investigated landslide location as well as generation of dense point clouds representing the surface within the mapped area. More specifically, a point cloud comprised of more than 13.65 million points—whose average inter-point spacing and local point density values are 1.5 cm and 4431.5 points/m^2, respectively—have been derived from a set of 530 overlapping images.

The paper presented three automated approaches for the detection and extraction of scarps using the morphometric characteristics of the derived point cloud. The first approach uses ratios among the principal component analysis PCA-based Eigenvalues at local neighborhoods to extract salient geo-morphometric features that represent scarps. The second approach is based on slope variability within local neighborhoods, which are defined by a 0.5 m moving window, while assuming that abrupt slope changes will take place between two successive scarps. This slope change will become more distinct as the scarps evolve. Finally, the last approach is based on evaluating the surface roughness index for the derived point cloud at local neighborhoods represented by a moving window with 0.5 m search radius.

The experimental results show that the proposed approaches accurately identify and extract landslide scarps with a recognition accuracy of approximately 72%. The results also indicate that the scarp detection is 70% complete and 84% correct with an overall quality of 93.4%. These measures have been evaluated by calculating the Root Mean Square Error (RMSE) between the automatically extracted landslide scarp segments and manually-digitized ones. No significant differences in accuracy are observed among the different approaches. The RMSE analysis also revealed that the accuracy of the topographic surface slope, topographic surface roughness index, and Eigenvalue-based approaches are 9.05 cm, 10.45 cm, and 11.98 cm, respectively. Thus, the proposed approaches can accurately identify and extract landslide scarps at the decimeter-level accuracy. Such ability to detect landslide scarps will lead to better understanding of the landslide mechanisms for a given area. This in turn will lead to better identification of the most likely failure site within a landslide prone area and estimation of the volume of potential sliding rock mass. In summary, the developed approaches are fast, economical, labor-saving, and safe tools for detecting and recognizing landslide scarps. They can be also used to monitor and assess the rate of horizontal displacement between the extracted landslide scarps at different times.

The experimental results are based on datasets covering bare earth with little to no vegetation, where the proposed approaches will lead to accurate scarp detection and extraction. Future work will be extended to deal with areas covered by bushes, boulders, and tree stumps. More specifically, filtering techniques will be developed to remove such features before the application of the proposed approaches. In addition, rather than having three independent approaches, future research will integrate the three proposed approaches to explore the possibility of improving the scarp detection and extraction accuracy, especially in areas with higher levels of vegetation cover than the current study site. Finally, the proposed methodologies will be incorporated for monitoring and change detection evaluation while using temporal UAV-based datasets over a given site.

Acknowledgments: The authors are grateful to Yarmouk University in Jordan for financing the first author's scholarship at the Department of Geomatics Engineering in University of Calgary, Canada. We also express our gratitude to the Natural Sciences and Engineering Research Council of Canada (NSERC) and the University of Calgary for the financial support given through human resources. The authors would like to express their appreciation for the anonymous reviewers and editors, whose comments have helped to improve the overall quality of this paper.

Author Contributions: All the authors contributed extensively to the work presented in this paper. In particular, Abdulla Al-Rawabdeh developed the comprehensive system that uses unmanned aerial vehicles (UAVs), collected all the UAV datasets; developed the algorithms, performed the experiments and experimental analysis, and wrote most of the manuscript. Fanging He implemented the Semi-Global dense Matching algorithm for the image-based point cloud generation. Adel Moussa developed portions of the algorithms and experimental analysis. Habib and El-Sheimy provided editorial contributions and constructive comments for the improvement and revision of the manuscript. All authors read and approved the final manuscript.

Conflicts of Interest: The authors declare no conflict of interest.

References

1. Nadim, F.; Kjekstad, O.; Peduzzi, P.; Herold, C.; Jaedicke, C. Global landslide and avalanche hotspots. *Landslides* **2006**, *3*, 159–173. [CrossRef]
2. Pesci, A.; Teza, G.; Casula, G.; Loddo, F.; de Martino, P.; Dolce, M.; Obrizzo, F.; Pingue, F. Multitemporal laser scanner-based observation of the Mt. Vesuvius crater: Characterization of overall geometry and recognition of landslide events. *ISPRS J. Photogramm. Remote Sens.* **2011**, *66*, 327–336. [CrossRef]

3. Gokceoglu, C.; Sonmez, H.; Nefeslioglu, H.A.; Duman, T.Y.; Can, T. The 17 March 2005 Kuzulu landslide (Sivas, Turkey) and landslide-susceptibility map of its near vicinity. *Eng. Geol.* **2005**, *81*, 65–83. [CrossRef]

4. Glenn, N.F.; Streutker, D.R.; Chadwick, D.J.; Thackray, G.D.; Dorsch, S.J. Analysis of LiDAR-derived topographic information for characterizing and differentiating landslide morphology and activity. *Geomorphology* **2006**, *73*, 131–148. [CrossRef]

5. Westoby, M.J.; Brasington, J.; Glasser, N.F.; Hambrey, M.J.; Reynolds, J.M. "Structure-from-motion" photogrammetry: A low-cost, effective tool for geoscience applications. *Geomorphology* **2012**, *179*, 300–314. [CrossRef]

6. Schuster, R.L.; Fleming, R.W. Economic losses and fatalities due to landslides. *Bull. Assoc. Eng. Geol.* **1986**, *23*, 11–28. [CrossRef]

7. Dikau, R.; Brunsden, D.; Schrott, L.; Ibsen, M.L. *Landslide Recognition: Identification, Movements and Causes*; John Wiley & Sons Ltd.: Chichester, UK, 1996.

8. Hovius, N.; Stark, C.P.; Allen, P.A. Sediment flux from a mountain belt derived by landslide mapping. *Geology* **1997**, *25*, 231–234. [CrossRef]

9. Hovius, N.; Stark, C.P.; Hao-Tsu, C.; Jiun-Chuan, L. Supply and removal of sediment in a landslide-dominated mountain belt: Central range, Taiwan. *J. Geol.* **2000**, *108*, 73–89. [CrossRef] [PubMed]

10. Jakob, M. The impacts of logging on landslide activity at Clayoquot Sound, British Columbia. *Catena* **2000**, *38*, 279–300. [CrossRef]

11. Guzzetti, F.; Cardinali, M.; Reichenbach, P.; Carrara, A. Comparing landslide maps: A case study in the upper Tiber River basin, Central Italy. *Environ. Manag.* **2000**, *25*, 247–263. [CrossRef]

12. McKean, J.; Roering, J. Objective landslide detection and surface morphology mapping using high-resolution airborne laser altimetry. *Geomorphology* **2004**, *57*, 331–351. [CrossRef]

13. Zhu, J.J.; Ding, X.L.; Chen, Y.Q. Dynamic landsliding model with integration of monitoring information and mechanic information. *Acta Geod. Cartogr. Sin.* **2003**, *32*, 261–266.

14. Mahler, C.; Varanda, E.; de Oliveira, L. Analytical model of landslide risk using GIS. *Open J. Geol.* **2005**, *2*, 182–188. [CrossRef]

15. Zhang, J. 3S-aided landslide hazard monitoring and modeling. *Eng. Surv. Mapp.* **2015**, *14*, 1–5.

16. Booth, A.M.; Roering, J.J.; Perron, J.T. Automated landslide mapping using spectral analysis and high-resolution topographic data: Puget Sound lowlands, Washington, and Portland Hills, Oregon. *Geomorphology* **2009**, *109*, 132–147. [CrossRef]

17. He, Y.P.; Xie, H.; Cui, P.; Wei, F.Q.; Zhong, D.L.; Gardner, J.S. GIS-based hazard mapping and zonation of debris flows in Xiaojiang Basin, southwestern China. *Environ. Geol.* **2003**, *45*, 286–293. [CrossRef]

18. Donati, L.; Turrini, M.C. An objective method to rank the importance of the factors predisposing to landslides with the GIS methodology: Application to an area of the Apennines (Valnerina; Perugia, Italy). *Eng. Geol.* **2002**, *63*, 277–289. [CrossRef]

19. Singhroy, V. *Landslide Hazards: CEOS, the Use of Earth Observing Satellites for Hazard Support: Assessments and Scenarios*; Final Report of the CEOS. NOAA: Silver Spring, MD, USA, 2002; p. 98.

20. Mantovani, F.; Soeters, R.; van Westen, C.J. Remote sensing techniques for landslide studies and hazard zonation in Europe. *Geomorphology* **1996**, *15*, 213–225. [CrossRef]

21. Martha, T.R.; Kerle, N.; Jetten, V.; van Westen, C.J.; Kumar, K. Landslide volumetric analysis using Cartosat-1-derived DEMs. *IEEE Geosci. Remote Sens. Lett.* **2010**, *7*, 582–586. [CrossRef]

22. Galli, M.; Ardizzone, F.; Cardinali, M.; Guzzetti, F.; Reichenbach, P. Comparing landslide inventory maps. *Geomorphology* **2008**, *94*, 268–289. [CrossRef]

23. Lin, P.; Lin, J.; Hung, H.C.; Yang, M. Assessing debris-flow hazard in a watershed in Taiwan. *Eng. Geol.* **2002**, *66*, 295–313. [CrossRef]

24. Martin, Y.; Rood, K.; Schwab, J.W.; Church, M. Sediment transfer by shallow landsliding in the Queen Charlotte Islands, British Columbia. *Can. J. Earth Sci.* **2002**, *39*, 189–205. [CrossRef]

25. Westen, C.J.; Getahun, F. Analyzing the evolution of the Tessina landslide using aerial photographs and digital elevation models. *Geomorphology* **2003**, *54*, 77–89. [CrossRef]

26. Hervás, J.; Barredo, J.I.; Rosin, P.L.; Pasuto, A.; Mantovi, F.; Silvano, S. Monitoring landslides from optical remotely sensed imagery: The case history of Tessina landslide, Italy. *Geomorphology* **2003**, *54*, 63–75. [CrossRef]

27. Sauchyn, D.; Trench, N. Landsat applied to landslide mapping. *Photogramm. Eng. Remote Sens.* **1978**, *44*, 735–741.

28. Connors, K.F.; Gardner, T.W.; Petersen, G.W. Classification of geomorphic features and landscape stability in northwestern New Mexico using simulated SPOT imagery. *Remote Sens. Environ.* **1987**, *22*, 187–207. [CrossRef]

29. Epp, H.; Beaven, L. Mapping slope failure tracks with digital thematic mapper data. In Proceedings of the 1988 International Geoscience and Remote Sensing Symposium on Remote Sensing: Moving Towards the 21st Century, Edinburgh, UK, 12–16 September1988.

30. McKean, J.; Buechel, S.; Gaydos, L. Remote sensing and landslide hazard assessment. *Photogramm. Eng. Remote Sens.* **1991**, *57*, 1185–1193.

31. Li, X.; Cheng, X.; Chen, W.; Chen, G.; Liu, S. Identification of forested landslides using LiDar data, object-based image analysis, and machine learning algorithms. *Remote Sens.* **2015**, *7*, 9705–9726. [CrossRef]

32. Barlow, J.; Martin, Y.; Franklin, S.E. Detecting translational landslide scars using segmentation of Landsat ETM+ and DEM data in the northern Cascade Mountains, British Columbia. *Can. J. Remote Sens.* **2003**, *29*, 510–517. [CrossRef]

33. Schulz, W.H. *Landslides Mapped Using LIDAR imagery, Seattle, Washington*; Open File Report 04-1396. USA Department of the Interior: Seattle, WA, USA.

34. Dou, J.; Chang, K.; Chen, S.; Yunus, A.; Liu, J.; Xia, H.; Zhu, Z. Automatic case-based reasoning approach for landslide detection: Integration of object-oriented image analysis and a genetic algorithm. *Remote Sens.* **2015**, *7*, 4318–4342. [CrossRef]

35. Kraus, K.; Pfeifer, N. Determination of terrain models in wooded areas with airborne laser scanner data. *ISPRS J. Photogramm. Remote Sens.* **1998**, *53*, 193–203. [CrossRef]

36. Vosselman, G. Slope-based filtering of laser altimetry data. *Int. Arch. Photogramm. Remote Sens. Spat. Inf. Sci.* **2000**, *33*, 935–942.

37. Gold, R.D.; Wegmann, K.W.; Palmer, S.P.; Carson, R.J.; Spencer, P.K. A Comparative Study of Aerial Photographs and LiDAR Imagery for Landslide Detection in the Puget Lowland, Washington. In Proceedings of the 99th Annual Meeting, Puerto Vallarta, Jalisco, Mexico, 1–3 April 2003.

38. Rowlands, K.A.; Jones, L.D.; Whitworth, M. Landslide laser scanning: A new look at an old problem. *Q. J. Eng. Geol. Hydrogeol.* **2003**, *36*, 155–157. [CrossRef]

39. Sithole, G.; Vosselman, G. Experimental comparison of filter algorithms for bare-earth extraction from airborne laser scanning point clouds. *ISPRS J. Photogramm. Remote Sens.* **2004**, *59*, 85–101. [CrossRef]

40. Hsiao, K.; Lu, J.; Yu, M.; Tseng, G.Y. Change detection of landslide terrains using ground-based Lidar data. In Proceedings of the 20th ISPRS Congress, Istanbul, Turkey, 12–23 July 2004.

41. Metternicht, G.; Hurni, L.; Gogu, R. Remote sensing of landslides: An analysis of the potential contribution to geo-spatial systems for hazard assessment in mountainous environments. *Remote Sens. Environ.* **2005**, *98*, 284–303. [CrossRef]

42. Bull, J.M.; Miller, H.; Gravley, D.M.; Costello, D.; Hikuroa, D.C.H.; Dix, J.K. Assessing debris flows using LIDAR differencing: 18 May 2005 Matata event, New Zealand. *Geomorphology* **2010**, *124*, 75–84. [CrossRef]

43. Schulz, W.H. *Landslide Susceptibility Estimated from Mapping Using Light Detection and Ranging (LIDAR) Imagery and Historical Landslide Records*; Open File Report 2005-1405. USA Department of the Interior: Seattle, WA, USA.

44. Miner, A.S.; Flentje, P.; Mazengarb, C.; Windle, D.J. Landslide recognition using LiDAR derived digital elevation models-lessons learnt from selected Australian example. In Proceedings of the 11th IAEG Congress of the International Association of Engineering Geology and the Environment, Auckland, New Zealand, 5–10 September 2010; pp. 5–9.

45. Leshchinsky, B.A.; Olsen, M.J.; Tanyu, B.F. Contour connection method for automated identification and classification of landslide deposits. *Comput. Geosci.* **2015**, *74*, 27–38. [CrossRef]

46. Gibson, A.; Forster, A.F.; Poulton, C.; Rowlands, K.; Jones, L.; Hobbs, P.; Whitworth, M. An integrated method for terrestrial laser-scanning subtle landslide features and their geomorphological setting. In Proceedings of the Remote Sensing and Photogrammetry Society 2003: Scales and Dynamics in Observing the Environment, Nottingham, England, UK, 10–12 September 2003.

47. Rosser, N.J.; Petley, D.N.; Lim, M.; Dunning, S.A.; Allison, R.J. Terrestrial laser scanning for monitoring the process of hard rock coastal cliff erosion. *Q. J. Eng. Geol. Hydrogeol.* **2005**, *38*, 363–375. [CrossRef]

48. Bauer, A.; Paar, G.; Kaltenböck, A. Mass movement monitoring using terrestrial laser scanner for rock fall management. In Proceedings of the 1st International Symposium on Geoinformation for Disaster Management, Delft, The Netherlands, 21–23 March 2005.

49. Lim, M.; Petley, D.N.; Rosser, N.J.; Allison, R.J.; Long, A.J.; Pybus, D. Combined digital photogrammetry and time-of-flight laser scanning for monitoring cliff evolution. *Photogramm. Rec.* **2005**, *20*, 109–129. [CrossRef]

50. Dunning, S.A.; Massey, C.I.; Rosser, N.J. Structural and geomorphological features of landslides in the Bhutan Himalaya derived from terrestrial laser scanning. *Geomorphology* **2009**, *103*, 17–29. [CrossRef]

51. Jaboyedoff, M.; Couture, R.; Locat, P. Structural analysis of Turtle Mountain (Alberta) using digital elevation model: Toward a progressive failure. *Geomorphology* **2009**, *103*, 5–16. [CrossRef]

52. Sturzenegger, M.; Stead, D. Quantifying discontinuity orientation and persistence on high mountain rock slopes and large landslides using terrestrial remote sensing techniques. *Nat. Hazards Earth Syst. Sci.* **2009**, *9*, 267–287. [CrossRef]

53. Jaboyedoff, M.; Oppikofer, T.; Abellán, A.; Derron, M.; Loye, A.; Metzger, R.; Pedrazzini, A. Use of LiDAR in landslide investigations: A review. *Nat. Hazards* **2012**, *61*, 5–28. [CrossRef]

54. Ahmad, A. Digital photogrammetry: An experience of processing aerial photograph of UTM acquired using digital camera. In Proceedings of the AsiaGIS, Johor, Malaysia, 3–5 March 2006.

55. Marek, L.; Miřijovský, J.; Tuček, P. Monitoring of the shallow landslide using UAV photogrammetry and geodetic measurements. In *Engineering Geology for Society and Territory*; Springer Verlag International Publishing: Gewerbestrasse, Switzerland, 2015; Volume 2, pp. 113–116.

56. Pajares, G. Overview and current status of remote sensing applications based on unmanned aerial vehicles (UAVS). *Photogramm. Eng. Remote Sens.* **2015**, *81*, 281–330. [CrossRef]

57. Niethammer, U.; James, M.R.; Rothmund, S.; Travelletti, J.; Joswig, M. UAV-based remote sensing of the super-Sauze landslide: Evaluation and results. *Eng. Geol.* **2012**, *128*, 2–11. [CrossRef]

58. Colomina, I.; Molina, P. Unmanned aerial systems for photogrammetry and remote sensing: A review. *ISPRS J. Photogramm. Remote Sens.* **2014**, *92*, 79–97. [CrossRef]

59. Morillas, L.; García, M.; Nieto, H.; Villagarcia, L.; Sandholt, I.; Gonzalez-Dugo, M.P.; Zarco-Tejada, P.J.; Domingo, F. Using radiometric surface temperature for surface energy flux estimation in Mediterranean drylands from a two-source perspective. *Remote Sens.Environ.* **2013**, *136*, 234–246. [CrossRef]

60. Frankenberger, J.R.; Huang, C.; Nouwakpo, K. Low-altitude digital photogrammetry technique to assess ephemeral gully erosion. In Proceedings of the IEEE International Geoscience and Remote Sensing Symposium (IGARSS 2008), Boston, MA, USA, 7–11 July 2008; pp. 117–120.

61. Agüera, F.; Carvajal, F.; Pérez, M. Measuring sunflower nitrogen status from an unmanned aerial vehicle-based system and an on the ground device. *ISPRS Int. Arch. Photogramm. Remote Sens. Spat. Inf. Sci.* **2011**, *22*, 33–37. [CrossRef]

62. Rau, J.; Jhan, J.; Lob, C.; Linb, Y. Landslide mapping using imagery acquired by a fixed-wing UAV. *ISPRS Int. Arch. Photogramm. Remote Sens. Spat. Inf. Sci.* **2011**, *22*, 195–200. [CrossRef]

63. Turner, D.; Lucieer, A.; de Jong, S.M. Time series analysis of landslide dynamics using an unmanned aerial vehicle (UAV). *Remote Sens.* **2015**, *7*, 1736–1757. [CrossRef]

64. Stumpf, A.; Lampert, T.A.; Malet, J.; Kerle, N. Multi-scale line detection for landslide fissure mapping. In Proceedings of the IEEE International Geoscience and Remote Sensing Symposium (IGARSS), Munich, Germany, 22–27 July 2011; pp. 5450–5453.

65. Stumpf, A.; Malet, J.P.; Kerle, N.; Niethammer, U.; Rothmund, S. Image-based mapping of surface fissures for the investigation of landslide dynamics. *Geomorphology* **2013**, *186*, 12–27. [CrossRef]

66. Lin, Y.; Hyyppä, J.; Jaakkola, A. Mini-UAV-borne LIDAR for fine-scale mapping. *IEEE Geosci. Remote Sens. Lett.* **2011**, *8*, 426–430. [CrossRef]

67. Niethammer, U.; Rothmund, S.; Schwaderer, U.; Zeman, L.; Joswig, M. Open source image-processing tools for lowcost UAV-based landslide investigations. *ISPRS Int. Arch. Photogramm. Remote Sens. Spat. Inf. Sci.* **2011**, *22*, 57–62.

68. Towler, J.; Krawiec, B.; Kochersberger, K. Radiation mapping in post-disaster environments using an autonomous helicopter. *Remote Sens.* **2012**, *4*, 1995–2015. [CrossRef]

69. Qian, Y.; Shengbo, C.; Peng, L.; Tengfei, C.; Ming, M.; Yanli, L.; Chao, Z.; Liang, Z. Application of low-altitude remote sensing image by unmanned airship in geological hazards investigation. In Proceedings of the Image and Signal Processing 5th International Congress, Chongqing, China, 16–18 October 2012; pp. 1015–1018.

70. WAJ—Water Authority of Jordan, WAJ Internal Files for Groundwater. Basins in Jordan. 2006. Available online: http://www. mwi.gov.jo/sites/en-us (accessed on 15 September 2015).

71. Abderahman, N.S. Landslide at km 56.4 along the Irbid-Amman Highway, Northern Jordan. *Environ. Geosci.* **1998**, *5*, 103–114. [CrossRef]

72. Masri, M. *Report on the Geology of the Amman-Zarqa Area*; Central Water Authority: Amman, Jordan, 1963.

73. Sawariah, A.; Barjous, M. *Geological Map of Suwaylih Scale 1:50,000, Sheet No. 3154-II*; Natural Resources Authority: Amman, Jordan, 1993.

74. Powell, J.H. *Stratigraphy and Sedimentation of the Phanerozoic Rocks in Central and South Jordan. Part B: Kurnub, Ajlun, and Balqa Groups, Bull. 11, NRA*; Geological Directorate: Amman, Jordan, 1989; p. 130.

75. Malkawi, A.I.H.; Saleh, B.; Al-Sheriadeh, M.S.; Hamza, M.S. Mapping of landslide hazard zones in Jordan using remote sensing and GIS. *J. Urban Plan. Dev.* **2000**, *126*, 1–17. [CrossRef]

76. Saket, S.K.H. *Slope Instability on the Jordanian Highways*; Unpublished Report. Ministry of Public Works and Housing (MPWH): Amman, Jordan, 1974.

77. Al Rawashdeh, S.; Bassam, S. Studying land sliding and geospatial deformation based on conventional survey and GPS. In Proceedings of the Geospatial Scientific Summit, Dubai, United Arab Emirates, 12–13 November 2012.

78. Habib, A.; Jarvis, A.; Detchev, G.; Stensaas, D.; Moe, D.; Christopherson, J. Standards and specifications for the calibration and stability of amateur digital cameras for close-range mapping applications. *Int. Arch. Photogramm. Remote Sens. Spat. Inf. Sci.* **2008**, *37*, 1059–1064.

79. Habib, A.; Pullivelli, A.; Mitishita, E.; Ghanma, M.; Kim, E. Stability analysis of low-cost digital cameras for aerial mapping using different georeferencing techniques. *Photogramm. Rec.* **2006**, *21*, 29–43. [CrossRef]

80. Al-Rawabdeh, A.; He, F.; Habib, A. Comparative analysis using Multi-sensory data integration for extracting geotechnical parameters. In Proceedings of the ASPRS 2014 Annual Conference, Louisville, KY, USA, 23–28 March 2014.

81. He, F.; Habib, A.; Al-Rawabdeh, A. Planar constraints for an improved UAV-image-based dense point cloud generation. *Int. Arch. Photogramm. Remote Sens. Spat. Inf. Sci.* **2015**. [CrossRef]

82. Varnes, D. Slope movement types and processes. In *Landslides Analysis and Control*; Schuster, R.L., Krizek, R.J., Eds.; National Academy of Sciences: New York, NY, USA, 1978; Volume 176, pp. 12–33.

83. Van den Eeckhaut, M.; Poesen, J.; Govers, G.; Verstraeten, G.; Demoulin, A. Characteristics of the size distribution of recent and historical landslides in a populated hilly region. *Earth Planet. Sci. Lett.* **2007**, *256*, 588–603. [CrossRef]

84. He, F.; Habib, A. Linear approach for initial recovery of the exterior orientation parameters of randomly captured images by low-coast mobile mapping system. *Int. Arch. Photogramm. Remote Sens. Spat. Inf. Sci.* **2014**, *1*, 149–154. [CrossRef]

85. Lowe, D.G. Distinctive image features from scale-invariant key points. *Int. J. Comput. Vis.* **2004**, *60*, 91–110. [CrossRef]

86. Cho, W.; Schenk, T.; Madani, M. Resampling digital imagery to epipolar geometry. *Int. Arch. Photogramm. Remote Sens.* **1993**, *29*, 404–404.

87. Guzzetti, F.; Carrara, A.; Cardinali, M.; Reichenbach, P. Landslide hazard evaluation: A review of current techniques and their application in a multi-scale study, Central Italy. *Geomorphology* **1999**, *31*, 181–216. [CrossRef]

88. Kasai, M.; Ikeda, M.; Asahina, T.; Fujisawa, K. LiDAR-derived DEM evaluation of deep-seated landslides in a steep and rocky region of Japan. *Geomorphology* **2009**, *113*, 57–69. [CrossRef]

89. Woodcock, N.H. Specification of fabric shapes using an eigenvalue method. *Geol. Soc. Am. Bull.* **1977**, *88*, 1231–1236. [CrossRef]

90. Woodcock, N.H.; Naylor, M.A. Randomness testing in three-dimensional orientation data. *J. Struct. Geol.* **1983**, *5*, 539–548. [CrossRef]

91. Bentley, J.L. Multidimensional binary search trees used for associative searching. *Commun. ACM* **1975**, *18*, 509–517. [CrossRef]

92. Arya, S.; Mount, D.M.; Netanyahu, N.S.; Silverman, R.; Wu, A.Y. An optimal algorithm for approximate nearest neighbor searching fixed dimensions. *J. ACM* **1998**, *45*, 891–923. [CrossRef]

93. Pauly, M.; Gross, M.; Kobbelt, L.P. Efficient simplification of point-sampled surface. In Proceedings of the Conference on Visualization, Boston, MA, USA, 27 October–1 November 2002; IEEE Computer Society: Washington, DC, USA, 2002; pp. 163–170.

94. Lee, S.; Min, K. Statistical analysis of landslide susceptibility at Yongin, Korea. *Environ. Geol.* **2001**, *40*, 1095–1113. [CrossRef]

95. Coblentz, D.; Pabian, F.; Prasad, L. Quantitative geomorphometrics for terrain characterization. *Int. J. Geosci.* **2014**, *5*, 247–266. [CrossRef]

96. Çevik, E.; Topal, T. GIS-based landslide susceptibility mapping for a problematic segment of the natural gas pipeline, Hendek (Turkey). *Environ. Geol.* **2003**, *44*, 949–962. [CrossRef]

97. Lee, S.; Choi, J.; Min, K. Probabilistic landslide hazard mapping using GIS and remote sensing data at Boun, Korea. *Int. J. Remote Sens.* **2004**, *25*, 2037–2052. [CrossRef]

98. Lee, S. Application of logistic regression model and its validation for landslide susceptibility mapping using GIS and remote sensing data. *Int. J. Remote Sens.* **2005**, *26*, 1477–1491. [CrossRef]

99. Iwahashi, J.; Watanabe, S.; Furuya, T. Mean slope-angle frequency distribution and size frequency distribution of landslide masses in Higashikubiki area, Japan. *Geomorphology* **2003**, *50*, 349–364. [CrossRef]

100. Tarolli, P. High-resolution topography for understanding Earth surface processes: Opportunities and challenges. *Geomorphology* **2014**, *216*, 295–312. [CrossRef]

101. Hobson, R.D. Surface roughness in topography: A quantitative approach. In *Spatial Analysis in Geomorphology*; Chorley, R.J., Ed.; Methuen & Co.: London, UK, 1972; pp. 221–245.

102. Habib, A.F.; Shin, S.W.; Morgan, M.F. New approach for calibrating off-the-shelf digital cameras. *Int. Arch. Photogramm. Remote Sens. Spat. Inf. Sci.* **2002**, *34*, 144–149.

103. Habib, A.F.; Ghanma, M.S.; Al-Ruzouq, R.I.; Kim, E.M. 3-d Modelling of historical sites using low-cost digital cameras. In Proceedings of the XXth Congress of ISPRS, Istanbul, Turkey, 12–23 July 2004.

104. Habib, A.F.; Pullivelli, A.M. Camera stability analysis and geo-referencing. In Proceedings of the IEEE International Geoscience and Remote Sensing Symposium, Seoul, Korea, 25–29 July 2005; pp. 1169–1172.

105. Zhan, Q.; Molenaar, M.; Tempfli, K.; Shi, W. Quality assessment for geo-spatial objects derived from remotely sensed data. *Int. J. Remote Sens.* **2005**, *26*, 2953–2974. [CrossRef]

106. Tuermer, S.; Kurz, F.; Reinartz, P.; Stilla, U. Airborne vehicle detection in dense urban areas using HoG features and disparity maps. *IEEE J. Sel. Top. Appl. Earth Observ. Remote Sens.* **2013**, *6*, 2327–2337. [CrossRef]

107. Sim, J.; Wright, C.C. The Kappa statistics in reliability statistics: Use, interpretation, and sample size analysis. *J. Am. Phys. Ther. Assoc.* **2005**, *85*, 257–268.

108. Cohen, J. A coefficient of agreement for nominal scales. *Educ. Psychol. Meas.* **1960**, *20*, 37–46. [CrossRef]

109. Landis, J.R.; Koch, G.G. The measurement of observer agreement for categorical data. *Biometrics* **1977**, *33*, 159–174. [CrossRef] [PubMed]

110. Monserud, R.A.; Leemans, R. Comparing global vegetation maps with the Kappa statistic. *Ecol. Model.* **1992**, *62*, 275–293. [CrossRef]

remote sensing

MDPI

Article

Advanced Three-Dimensional Finite Element Modeling of a Slow Landslide through the Exploitation of DInSAR Measurements and in Situ Surveys

Vincenzo De Novellis [1], Raffaele Castaldo [1], Piernicola Lollino [2], Michele Manunta [1] and Pietro Tizzani [1,*]

1   Istituto per il Rilevamento Elettromagnetico dell'Ambiente, IREA-CNR, via Diocleziano 328,
    80124 Napoli, Italy; denovellis.v@irea.cnr.it (V.D.N.); castaldo.r@irea.cnr.it (R.C.);
    manunta.m@irea.cnr.it (M.M.)
2   Istituto di Ricerca per la Protezione Idrogeologica, IRPI-CNR, via Amendola 122 I, 70126 Bari, Italy;
    p.lollino@ba.irpi.cnr.it
*   Correspondence: tizzani.p@irea.cnr.it; Tel.: +39-081-7620635

Academic Editors: Roberto Tomas, Zhenhong Li, Richard Gloaguen and Prasad S. Thenkabail
Received: 10 June 2016; Accepted: 16 August 2016; Published: 19 August 2016

Abstract: In this paper, we propose an advanced methodology to perform three-dimensional (3D) Finite Element (FE) modeling to investigate the kinematical evolution of a slow landslide phenomenon. Our approach benefits from the effective integration of the available geological, geotechnical and satellite datasets to perform an accurate simulation of the landslide process. More specifically, we fully exploit the capability of the advanced Differential Synthetic Aperture Radar Interferometry (DInSAR) technique referred to as the Small BAseline Subset (SBAS) approach to provide spatially dense surface displacement information. Subsequently, we analyze the physical behavior characterizing the observed landslide phenomenon by means of an inverse analysis based on an optimization procedure. We focus on the Ivancich landslide phenomenon, which affects a residential area outside the historical center of the town of Assisi (Central Italy). Thanks to the large amount of available information, we have selected this area as a representative case study highlighting the capability of advanced 3D FE modeling to perform effective risk analyses of slow landslide processes and accurate urban development planning. In particular, the FE modeling is constrained by using the data from 7 litho-stratigraphic cross-sections and 62 stratigraphic boreholes; and the optimization procedure is carried out using the SBAS-DInSAR retrieved results by processing 39 SAR images collected by the Cosmo-SkyMed (CSK) constellation in the 2009–2012 time span. The achieved results allow us to explore the spatial and temporal evolution of the slow-moving phenomenon and via comparison with the geomorphological data, to derive a synoptic view of the kinematical activity of the urban area affected by the Ivancich landslide.

Keywords: 3D Finite Element model; landslides kinematics; Cosmo-SkyMed DInSAR measurements; Ivancich landslide (Assisi, Central Italy)

1. Introduction

The assessment of the kinematical evolution of the slow landslides is challenging for the analysis and zonation of the risk in urban areas. Indeed, the capability to detect the spatial kinematical variability of a slow landslide processes can represent a fundamental source of knowledge to support the land management decision for the development of infrastructures in urban areas. In this perspective, the evaluation of the landslide displacement field at ground level in a certain time span, can impart important information for the prediction of the expected damage to buildings and infrastructure. In addition, the understanding of the landslides kinematics is also crucial in defining efficient prevention and mitigation strategies and can be effectively pursued only if multidisciplinary data, both in terms of temporal and spatial coverage of the area is available.

The kinematics of the landslide phenomena has been analyzed in a large number of scientific studies; the approaches range from the analytical one-dimensional (1D) infinite slope models, suitable for landslides bodies with sliding surface depth about constant and significantly lower than the landslide length [1–5], to more sophisticated two- (2D) and three-dimensional (3D) Finite Element (FE) models aimed at detecting the different kinematical sectors along the slope area [6–9]. In particular, the recent development of three-dimensional numerical codes allows us to carry out simulations of the displacement rate field of a landslide process in a more realistic and accurate way. The simulations of the landslide kinematics are achieved through the implementation in a numerical domain, of the available a-priori information on the slope, as the topography, the landslide body geometry and the mechanical properties of the involved geomaterials. Such enhanced information is retrieved through a significant increase of the computational load of the 3D models with respect to 2D models. However, this computational complexity can be correctly reduced when the examined landslide shows the typical kinematical features of a landslide creep, i.e., landslide phenomena are characterized by very low pore water pressure variation rates and consequently, the displacement rates do not significantly change with time.

The reliability of the performed numerical models significantly improves when the same models are calibrated and validated by using the measurements collected by means of the available monitoring networks. To this purpose, the inverse analysis carried out via optimization procedures aimed at searching for the model best-fitting solution against the monitoring dataset, represents a very efficient tool in identifying the physical process that governs the observed phenomenon [8,9]. To accomplish an effective inverse analysis of a 3D landslide model, spatially distributed surface measurements are needed. In particular, a network of displacement measure points that covers a large portion of the landslide body is required to correctly constrain the three-dimensional FE model. However, surface deformation measurements based on in situ research (inclinometers, GPS, leveling) are often unavailable, or only partially available, due to the high costs related to these technologies, especially if needed for long-term research. In this context, satellite techniques offer an effective and reliable alternative or a complementary tool to traditional in situ research. In particular, the Differential Synthetic Aperture Radar (SAR) Interferometry (DInSAR) techniques are becoming a powerful tool for monitoring, with centimeter to millimeter accuracy, the spatial and temporal evolution of slow deformation phenomena [10]. In the last few years, the DInSAR technique has been used to detect, study and monitor the surface displacements related to mass movement and slope instabilities [11–13] through the availability of several advanced DInSAR algorithms able to retrieve deformation time series and velocity maps [14–17]. Among the DInSAR techniques currently available, the Small BAseline Subset (SBAS) approach [16,17] has well demonstrated its capability in monitoring the deformations related to mass movement phenomena with high spatial density of measure points [18]. The SBAS DInSAR measurements have already used the implement optimization procedure for the FE model of a landslide process, mainly in two-dimensional domains [8]. Although the proposed studies have allowed us to deeply explore the landslide kinematics along the considered 2D section, the achieved results do not permit us to fully clarify the role of the geometrical and geological three-dimensional features in the distribution of the deformation field at ground surface.

The present work (the Ivancich landslide, which affects a residential area outside the historical center of the Assisi town, Central Italy), has been selected as a representative case study to highlight the capability of advanced three-dimensional FE modeling as a complementary tool in performing an effective risk analyses of the slow landslide processes and accurate urban development planning (Figure 1).

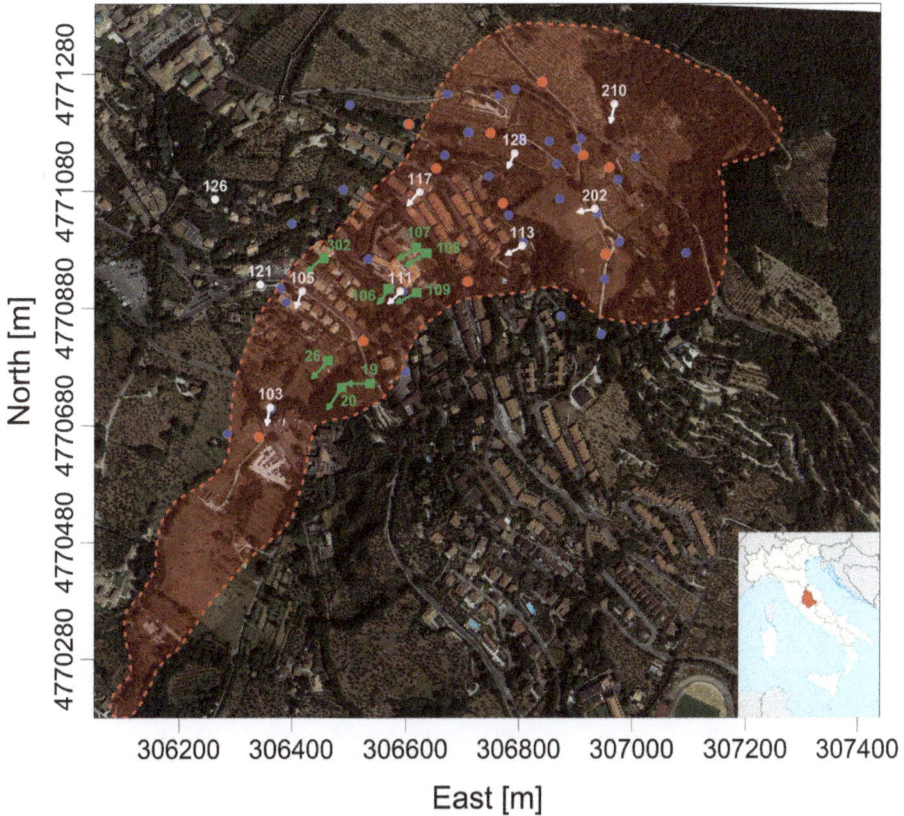

Figure 1. Study Area. Satellite optical image (source: Google Earth) of the Assisi surrounding area (Central Italy) affected by the Ivancich landslide. Insert map, lower right corner, reports the geographic location of the study area. Shaded area with red contour shows the zone involved by the recent slope instability phenomenon. All the circles represent borehole data: blue circles represent the 1998 survey; white circles are inclinometric boreholes (related to the same time period of the blue boreholes); red circles correspond to the 2006 survey. The green squares identify the previous inclinometric survey, acquired during 1982–1990 time interval. The arrows indicate the direction of the landslide movement.

The Ivancich landslide is an ancient landslide phenomenon and consists of a slow deep-seated mass movement, which can be classified as a typical earth-slide [19]. Although the rate of movement is very slow, the landslide has caused major damage to buildings, roads and infrastructures due to the effect of accumulation of ground differential displacements over time [20,21]. The first evidence of landslide activity were recorded in 1399 on the ancient Franciscan convent, which represented at that time the only anthropic structure of the area. In accordance with the demographic development that followed the end of World War II, the town of Assisi and its surrounding was intensely urbanized. Consequently around 1970, further evidence of landslide mobility started to appear in the form of growing damage to buildings, retaining walls, pipelines and road paving. The infrastructure damage led local authorities in carrying out several geological and geotechnical investigations aimed at studying the landslide process and designing the mitigation strategies [20,22]. Indeed, the area has been deeply investigated over the last 20 years and extensively monitored through in situ inclinometers, piezometers and GPS measurements [23–25]. At the same time, remote sensing techniques have been exploited in order to achieve dense spatial information of the unstable mass [18,26]. Such a large availability of measurements and in situ data makes the Ivancich landslide a real-scale laboratory in order to test and validate new approaches and techniques based on the integration of multi-source data. In this scenario, the FE modeling approach, based on the integration of available geological, geotechnical and satellite dataset, is aimed at investigating the 3D kinematical evolution of the landslide in the different sectors of the instability process. In particular, we exploit 39 SAR images collected by the Cosmo-SkyMed (CSK) constellation in the 2009–2012 time period, 7 litho-stratigraphic cross-sections and 62 stratigraphic boreholes.

Finally, the achieved outcomes allow us to investigate the spatial and temporal evolution of the slope affected by slow-moving phenomenon and via comparison with the geomorphological data, to derive insights into the kinematical activity map of the whole area affected by the Ivancich landslide.

2. Measurement Datasets

2.1. Geological and Geomorphological Data

From the 1970s to the present day, the Ivancich landslide has been deeply investigated from a geological and geotechnical point of view, so that the geometry of the active landslide has been quite accurately defined. The landslide morphology is characterized in the upper part by the presence of banks and detachment surfaces; secondary banks and smaller detachment niches are also present inside the landslide body; the lower zone of the landslide has a smaller slope inclination, showing an evident accumulation area with a tongue shape. The landslide has a total length of 1.4×10^3 m from the scarp of the source area to the nail of the toe area and the total surface area is estimated to be about $A = 4.1 \times 10^5$ m^2 (Figure 1); the slope presents an average inclination of about 21%. In addition, the geomorphological and topographic surveys revealed that the slope movement is an old translational slide, with a rotational component in the source area which moves along a well-defined shear band [18]. In particular the mass movement involves a debris deposit from 15 to 60 m in thickness, overlaying bedrock that consists of a pelitic sandstone unit and layered limestone. The performed in-situ investigations (also from the analysis of 62 stratigraphic boreholes) allow us to reconstruct with a high degree of accuracy, the depth of the bedrock and the geometry of the sliding surface (Figure 1). The unstable mass consists of a debris mass flowing over a stable bedrock composed of marl, sandstone and limestone (Figure 2a). Both transverse and longitudinal litho-stratigraphic sections highlighting the thickness of the landslide debris have been reconstructed (Figure 2b). The bedrock is characterized by a pelitic sandstone formation that consists mainly of marl, clay marl and calcareous marl and sandstone with variable thickness, from a few centimeters to several meters. The shallower part of the bedrock, in contact with the landslide detritus, is strongly weathered. The marl and marly limestone formation is characterized by calcareous marl stratification hardly distinguishable owing to the intense cleavage.

As explained before, the contact with the detritus is represented by a marked alteration band with a thickness not larger than 2 m, which can be recognized as the shear zone delimiting the landslide body. A fundamental constraint for the modeling approach, which is presented later on, derives from electric piezometer cell recordings, which do not measure relevant variations of the pore water pressures at the depth of the shear band. In particular, the ground water surface is measured to be only few meters above the shear band, resulting in very low piezometric heights, when compared to the total stress levels [18]. This further confirms the assumption of a landslide creep.

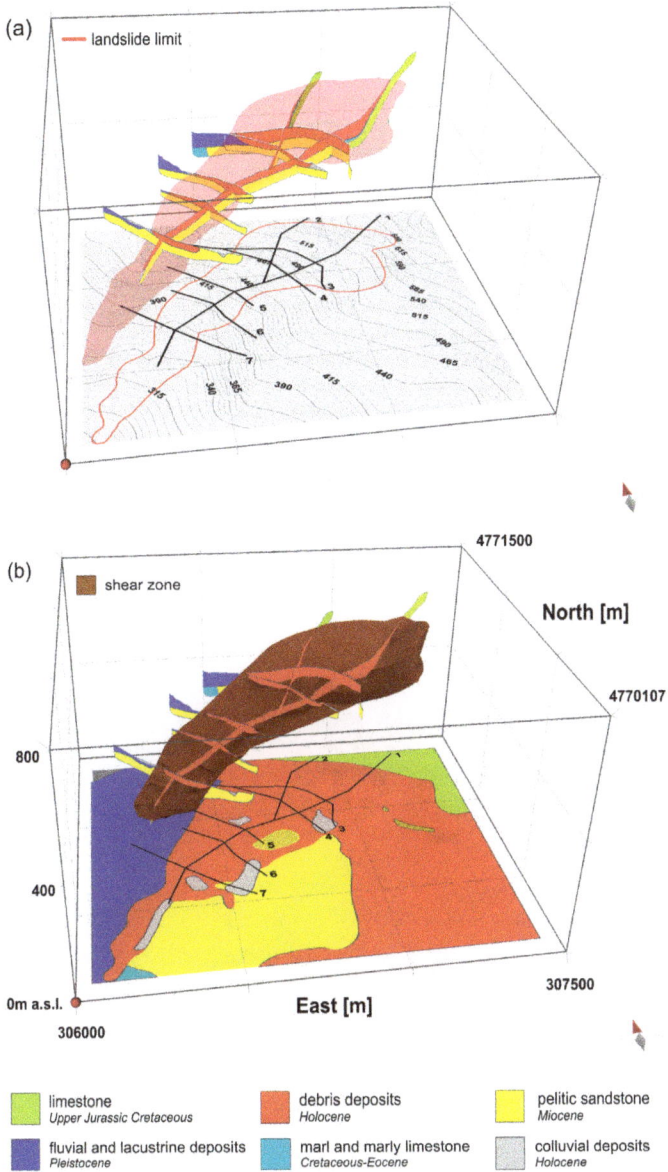

Figure 2. 3D view of geo-lithological data. Cross-sections of Ivancich area and the related traces are reported in the maps of panel (**a**) and (**b**) as black lines; (**a**) Contour of topographic map: the red line and the red shaded area indicate the boundaries and the area of the landslide, respectively; (**b**) Geological map: the brown surface represents the shear zone. In the legend we report the different lithologies of the cross section and geological map.

2.2. Inclinometric Measurements

From 1982 to 2008, numerous investigations of sub-surface landslide area, consisting of borehole inclinometer measurements [14,23], have been carried out in order to define the geometry of the sliding mass and to acquire data on the sub-surface displacement trend (Figure 1). All inclinometer data, except for n°121 and n°126 (outside the landslide area), indicate an abrupt increment of the cumulated displacements along the vertical profile, at depths ranging between 10 and 55 m, from the top to the toe of the landslide; the spatial correlation of this information has allowed us to identify the depth of the sliding surface throughout the landslide area. Therefore, the measurements acquired between 1998 and 2005, complementary to the satellite monitoring of the landslide kinematics, highlight the existence of a shear zone characterized by a thickness lower than 2 m, over which the landslide debris moves approximately as a rigid body [20].

The n°103 inclinometer shows a cumulated displacement at the ground surface of about 7 cm from January 1999 to October 2009 (displacement rate approximately equal to 0.6 cm/yr); for n°113 inclinometer, the measurements from December 1998 to December 2005 show a cumulated displacement at the ground surface of about 7.5 cm (approximately 1 cm/yr); for n°117 inclinometer, the measurements show a cumulated displacement at the ground surface of 6 cm, from December 1998 to July 2004 (approximately 1 cm/yr). Moreover, at the inclinometers located in the uppermost area of the landslide, slightly lower values of cumulated displacements were recorded: the inclinometers n°128 and n°202 respectively show about 3 cm between March 1999 and December 2008 (average displacement rate = 0.3 cm/yr) and about 4 cm between December 1998 and February 2008 (displacement rate = 0.4 cm/yr). Therefore, the inclinometer data indicate that the major landslide activity involves the central sector. It is worthwhile pointing out that the available inclinometers dataset [20,23,24] is utilized in order to spatially constrain the three-dimensional geometry of the shear-band, which delimits the landslide unstable mass at depth, as already demonstrated by Castaldo et al. [8].

2.3. Satellite Data

To rely on a large network of surface deformation measure points, we also benefit from remote sensing observations. In particular, we use the results achieved by applying the DInSAR technique for monitoring the surface deformations caused by the Ivancich landslide. DInSAR is an active microwave remote sensing technique that exploits the phase difference (i.e., the interferogram) between SAR image pairs acquired over the same area at different times, in order to extract information on the projection along the radar Line Of Sight (LOS) of the surface deformations that occurred between the two SAR acquisitions [27]. Among the several DInSAR techniques, in this work we benefit from the full resolution Small BAseline Subset (SBAS) approach [17,28] that has the capability to retrieve a very large number of measure points by processing only interferograms less affected by noise (small baseline interferograms) [16]. The SBAS DInSAR approach has been successfully applied in analyzing deformation phenomena caused by natural and human-induced processes, resulting a useful tool for the assessment and mitigation of the hydrogeological and urban risk [18,29–33]. Such a capability allows us to map the whole zone covered by the mass movement, by retrieving accurate and spatially dense information on the surface displacements affecting the investigated area.

The Ivancich landslide has been studied by using different SAR datasets to detect surface displacements and to carry out their long-term monitoring [18,26,30]. In this work, we benefit from the images acquired by the Cosmo-SkyMed (CSK) constellation. The CSK system is able to collect a sequence of SAR images of the Earth's surface with very short revisit time (4 days in the interferometric mode) and high spatial resolution (3 m in the stripmap mode and 1 m in the spotlight mode). These characteristics allow us to significantly increase, up to 15 times, the measure point density with respect to ERS-ENVISAT analyses, as demonstrated by Calò et al. [18]. This improvement is particularly significant in investigating the Ivancich landslide by integrating satellite measurements with in situ information since they allow a detailed analysis of the spatial distribution of the ground deformations. The CSK DInSAR measurements (Figure 3) already presented by Calò et al., [18], are retrieved by processing 39 images acquired from descending orbits between December 2009 and February 2012.

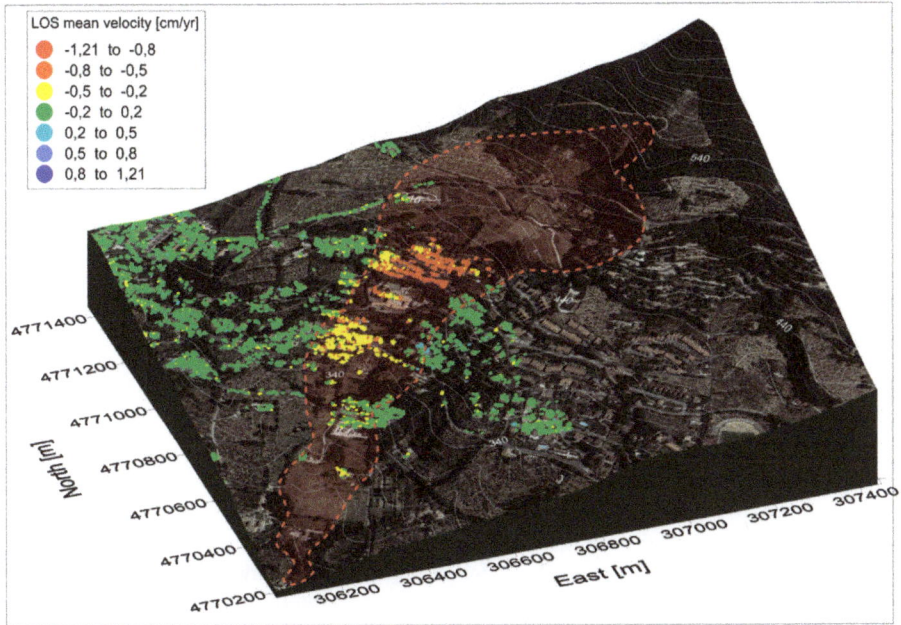

Figure 3. DInSAR measurements. CSK LOS mean deformation velocity map superimposed on the 3D view of satellite optical image with the DEM contour lines (grey lines). The red shade shows the zone involved by the recent instability phenomenon.

According to Bovenga et al., [26], Calò et al., [18] and Castaldo et al., [8], the DInSAR measurements show a quasi-linear LOS displacement trend (Supplementary Materials Figure S1), and a slight increase of the displacements over long time period. Similar outcomes have also been retrieved through the analysis of the measurements provided by a GPS network consisting of 16 stations [25]. Such a study shows that the maximum displacement values are observed in the central part of the landslide body and the vector direction follow the maximum slope line. The deformation trends have been evaluated on a long-term period and confirm a quasi-linear behavior with a value of squared correlation coefficient very close to the unity ($R^2 = 0.988$) [25].

The kinematic of Ivancich landslide is characterized, in the considered time interval (2009–2012), by a movement of the unstable mass directed downwards following the maximum slope line. The displacement rate is quite slow since the estimated maximum LOS mean velocities are of the order of 1 cm/yr as measured in the middle-upper area of the landslide body; lower displacement rates have been measured both in the lower and in the uppermost area of the landslide. The displacement rate values are in good agreement with those resulting from the inclinometer data (see Section 2.2), showing LOS velocities practically constant over time. As regards to the middle-upper sector of the landslide, Figure 3 also reveals a reduction of the LOS velocities measured for pixels located close to the right-hand side of the landslide body, in the range from −0.2 to −0.5 cm/yr, with respect to those observed in the central portion at corresponding elevations (−0.5 to −1.2 cm/yr). This result indicates a clear spatial variation of the landslide kinematical evolution, with higher velocities along the longitudinal axis and lower values along the lateral boundaries. In addition, the ground displacement velocity field, estimated via interferometric data processing, were in good agreement with the superficial boundaries of the landslide body as derived from the geomorphological analysis (Figure 3). However, this comparison is possible only for the coherent SAR pixels, which are located mainly in the central region of the unstable mass.

Finally, the SBAS-DInSAR displacement time-series also provide additional information regarding the effects of the rainfall regime on the landslide behavior. As a matter of fact, a comparative analysis of the landslide kinematics with the rainfall records for the same period [18] indicate the absence of a direct/immediate response of the surface deformation to the rainfall regime and furthermore, a complex temporal interaction between rainfall amount and ground movements. Accordingly in the advanced FE analysis described in the following section, we consider the possible effects due to rainfall as negligible and approximately assume the soil system as a single-phase material.

3. Three-Dimensional Finite Element Modeling

In this work, we exploit a numerical approach based on an optimized three-dimensional FE model that is constrained by DInSAR and field measurements. In particular, we combine the benefits of a deterministic numerical approach with statistical methods aimed at improving and optimizing the obtained numerical solution in order to analyze and interpret the ground deformations measured in the whole landslide area.

Based on the linearity of kinematics affecting the unstable mass, as observed according to the landslide monitoring dataset, we carry out the three-dimensional FE modeling of the active ground deformation field by assuming, as a first approximation, the soil material behavior as Newtonian fluid characterized by viscosity constant over time. We remark that this model can be applied mainly to landslide cases where the shear zone geometry is well formed and well known by means of accurate and detailed in-situ investigations. In this scenario, the distribution of the soil viscosity parameter within the whole numerical domain can be evaluated through an advanced procedure that implements a nonlinear optimization of the model parameters aimed at simulating the CSK DInSAR measurements. Accordingly, we can assume a steady-state viscous flow (Newtonian fluid) solved through the incompressible Navier–Stokes differential equations:

$$\begin{cases} -\nabla \cdot \mu \left(\nabla u + (\nabla u)^T \right) + \rho \left(u \cdot \nabla \right) u + \nabla p = \mathbf{F} \\ \nabla \cdot u = 0 \end{cases} \tag{1}$$

where u [m/s] is the deformation velocity vector; $\mathbf{F}$ [Pa/m] is the body force term; ρ [kg/m^3] is the density; p [Pa] is the pressure; and μ is the dynamic viscosity [Pa·s] (hereafter referred to as viscosity) [34].

The first step of the analysis deals with the setup of the a priori geo-lithological model. The interpretation of the available geological data provides a more accurate insight on the landslide geological conditions, as well as a more effective view capability of complex geological processes. In particular, in a geo-referenced system, the DEM (Digital Elevation Model), the geological map, the lithological cross-sections and the inclinometric borehole data are taken into account (Figure 4a,b) to define a realistic 3D geo-lithological model (1.5 × 1.5 × 0.1 km^3) of the Ivancich area.

Consequently, we have implemented the 3D geo-lithological model of the whole landslide divided into five geo-mechanical units within a numerical environment (see Table 1 for the corresponding physical properties): the limestone bedrock, in the upper part of the slope; the pelitic sandstone bedrock in the central and lower portions of the slope; the landslide deposit formed of unsorted debris; the colluvial deposit in the landslide toe area; the shear zone characterized by thickness of less than 2 m at depths ranging between 15 and 60 m from the ground surface.

The performed 3D geo-lithological model allows us to finely define the geometric features of the shear zone beneath the landslide deposit. Subsequently, we divided the shear zone into three sectors characterized by a homogenous viscosity value (Figure 4b), in accordance with both the geomorphological features along the slope and the kinematical behavior observed for the different portions of the landslide. The evidence of the existence of different landslide kinematical sectors is also supported by the analysis of the DInSAR measurements, which proves that the identified sectors are characterized by different kinematical rates.

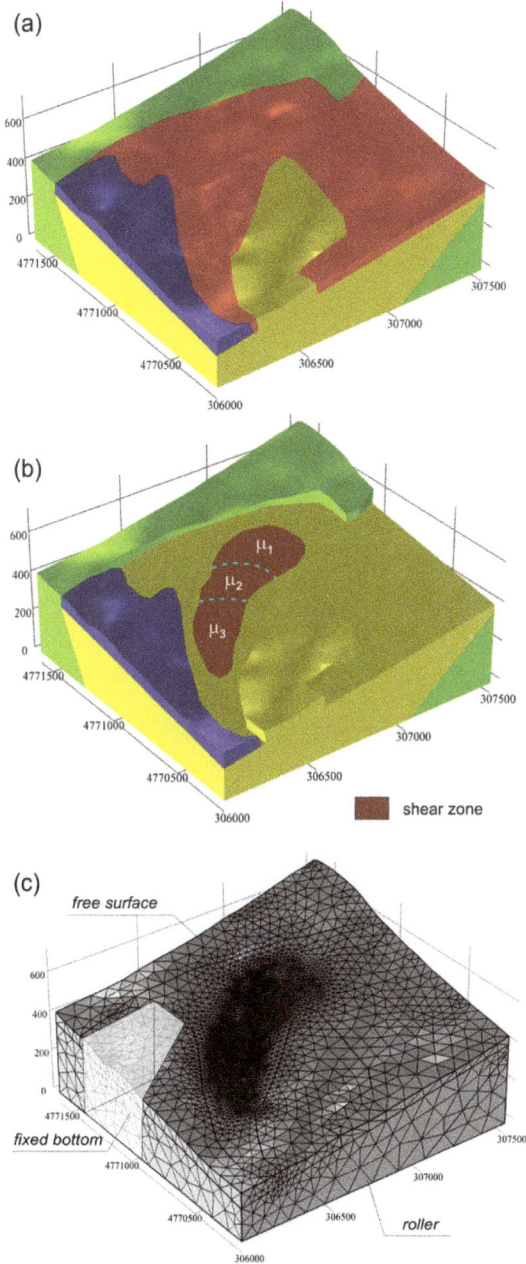

Figure 4. 3D FE Model: (**a**) Simplified geological model exploited in the FE environment; for each domain we report in Table 1 the physical and mechanical parameters relevant to the different lithologies, reported in Figure 2a; Panel (**b**) highlights the shear domain. The three sectors with different values of viscosity (μ) are also shown; the corresponding values are reported in Table 2; (**c**) Discretized numerical domain with the boundary conditions: fixed and roller ones.

Table 1. Input Parameters: Physical and Mechanical Properties of the Geo-Mechanical Units.

Rock Material	Density [kg/m³]	Young's Modulus [MPa]	Poisson Ratio [–]	Viscosity [Pa·s]
Limestone bedrock	2200	8×10^3	0.28	1×10^{22}
Pelitic-Sandstone bedrock	1850	7×10^3	0.26	1×10^{21}
Debris deposit	1600	1×10	0.24	1×10^{18}
Colluvial deposit	1700	6×10	0.24	1×10^{20}
Shear band	1600	1×10	0.23	–

Table 2. Optimized Parameters. Physical Parameters Bounds Used for the Shear Band Sectors in the Optimization and the Best Estimated Value.

Viscosity [Pa·s]	Lower Bound	Upper Bound	Estimated Value
μ_1	1×10^{14}	5×10^{16}	5.3×10^{15}
μ_2	1×10^{12}	5×10^{14}	2.5×10^{13}
μ_3	1×10^{11}	5×10^{13}	1×10^{12}

We remark that such an accurate definition of the shear zone domain is not possible in the 2D scenario [8], that represents, for the lack of geometrical constraints along the considered section, a rougher model. However, it is worth noting that the proposed subdivision has no significant impact on the applicability of the method and a finer discretization, although involving an increase of the computational load, can be easily applied in more complex scenarios.

The 3D numerical domain is discretized by creating a mesh that consists of about 620,000 tetrahedral elements. The element sizes range from 2 to 200 m; moreover, a mesh refinement was applied within the shear-zone domain where higher deformation rates were expected. The generated mesh is then validated through several resolution tests [35,36], which indicate a negligible mesh-dependency of the solution, since the use of a finer mesh would have affected the results by less than 2%.

Regarding the boundary conditions, the upper part of the geometry model which represents the ground surface of the slope, is considered as unconstrained (free surface); the bottom boundary is fixed. The four lateral side boundaries are instead characterized by null horizontal displacement. Finally, the inner domain is characterized by continuity between the different geological units.

According to Griffiths and Lane [37], we perform our modeling in two stages: gravity loading, aimed at defining a stress field representative of the current stress state of the slope; landslide process, i.e., the kinematics simulation. During the gravity loading, the stress state of the slope is defined by considering the slope as subjected to the soil gravitational loads and assuming elastic soil behavior. At this level, only the generated stress field is considered while the nodal displacements are kept equal to zero. In the landslide process, the previously calculated stresses are applied to the whole domain and the Newtonian approach is considered to simulate the kinematical trend of the landslide during the 2009–2012 time interval. The analysis implicitly assumes that the landslide body, delimited by the underlying shear zone, is unstable or marginally stable: this is specifically verified by means of a limit equilibrium calculation that provided a stability factor of the landslide body close to unity. It is worthwhile stressing that for the considered Newtonian viscous flow, soil is assumed to behave as a single-phase material. As described before, this assumption is supported by the available open-pipe piezometric measurements (Supplementary Materials Figure S2), locally indicating moderate variations of the water level within the landslide debris and by electric piezometer measurements showing very low variations of the pore water pressure at the depth of the shear band. Accordingly, we can assume that the pore water pressure regime plays a minor role in the landslide evolution process and separation between pore fluid and soil skeleton behavior is not needed for the present study. Note that, in order to simulate the landslide debris movement approximately as a rigid body, in the considered physical context the viscosity values of the layers overlaying the shear domain (surrounding rocks) are chosen with different order of magnitude (Table 1).

In order to integrate all the available data described in Section 2 and solve the field Equation (1), the COMSOL Multiphysics finite-element modeling code is used.

The modelled displacement rates are evaluated at the topographic surface, projected along the satellite line of sight (LOS) and compared with the DInSAR measurements. In particular, the optimization procedure is performed by exploiting the CSK SBAS-DInSAR results (Figure 3). To detect the model providing the best agreement between DInSAR measurements and FE model, we use the Monte Carlo statistical method, which is based on the exploration of the parameter values in a defined space of variables (coordinate search method) [38,39]. This procedure allows us to find, for the three different shear band sectors, the viscosity parameters that correspond to the model that best-fits the DInSAR measurements. To evaluate the similarity between the data vs. model, we use the Root Mean Square Error (RMSE) cost function; it is represented by a weighted average of the residuals, expressed as:

$$RMSE = \sqrt{\frac{1}{N} \sum_{i}^{N} \frac{(d_{i,obs} - d_{i,mod})^2}{\sigma_i}} \tag{2}$$

where $d_{i,obs}$ and $d_{i,mod}$ are the observed and modeled displacement of the i^{-th} point, σ_i is the standard deviation for the N points.

4. Results

A quantitative comparison between the CSK SBAS-DInSAR measurements (Figure 5a) and the achieved model results projected along LOS (Figure 5b), computed in correspondence of the SAR coherent pixels is performed; the Figure 5c represents the difference (residuals) between data and model results. The residuals map (Figure 5c) shows the good fit between data and model results, revealed also by the RMSE value that is equal to 0.15 cm/yr, smaller than the accuracy of the SBAS-DInSAR measurements [28]. In addition, the performed analysis also provides a good temporal fit between the DInSAR deformation time series and the modeled one for the considered time interval (Supplementary Materials Figure S1).

Starting from the analysis that best-fits the displacement monitoring data, we evaluate the related viscosity parameters of the three sector of shear zone (μ_1, μ_2, μ_3). The corresponding estimated values are reported in Table 2, along with the initial ranges adopted for the analysis.

The optimized FE model allows for the analysis of the overall ground velocities field. In Figure 6a we show the amplitude and direction of the achieved velocity field; in this map the amplitude of the velocity field reaches a maximum value of 1.5 cm/yr at altitudes between 320 and 400 m a.s.l., in the middle-upper part of the unstable area. In addition, the model results well represent the slight curvature in the horizontal plane of the landslide process along the South-East direction, as also highlighted by the modeled vectors (Figure 6a). A comparative analysis between the morphological units of the Ivancich slope (ancient and active edges landslide) as proposed by Pontoni [23,24] and the model results are also reported (Figure 6b); more than the 70% of the mapped geomorphological structures (units 6–11) are enclosed in the region where the 2009–2012 deformation process is active. The units 12 and 13 are marginally involved while the units 1–4 seem to be not active in the considered time interval (with mean velocity lower than 0.3 cm/yr). Our analysis also highlights the existence of a spatially correlated deformation pattern (0.5 cm/yr) superimposed on the cataloged ancient structures located in the North Western region of the landslide slope.

Finally, we show a map with the computed shear rate values (Figure 7a) that represents the ratio of the shear stress to the dynamic viscosity. It is interesting to note that the landslide portions where the structures are damaged, as many photographs recorded in the Ivancich surroundings clearly show (Figure 7b), are in accordance with the distribution of the anomaly of the shear rate. In fact, serious damage corresponding to the P2, P3 and P5 sites are localized along the landslide boundaries and in correspondence of highest values of shear rate.

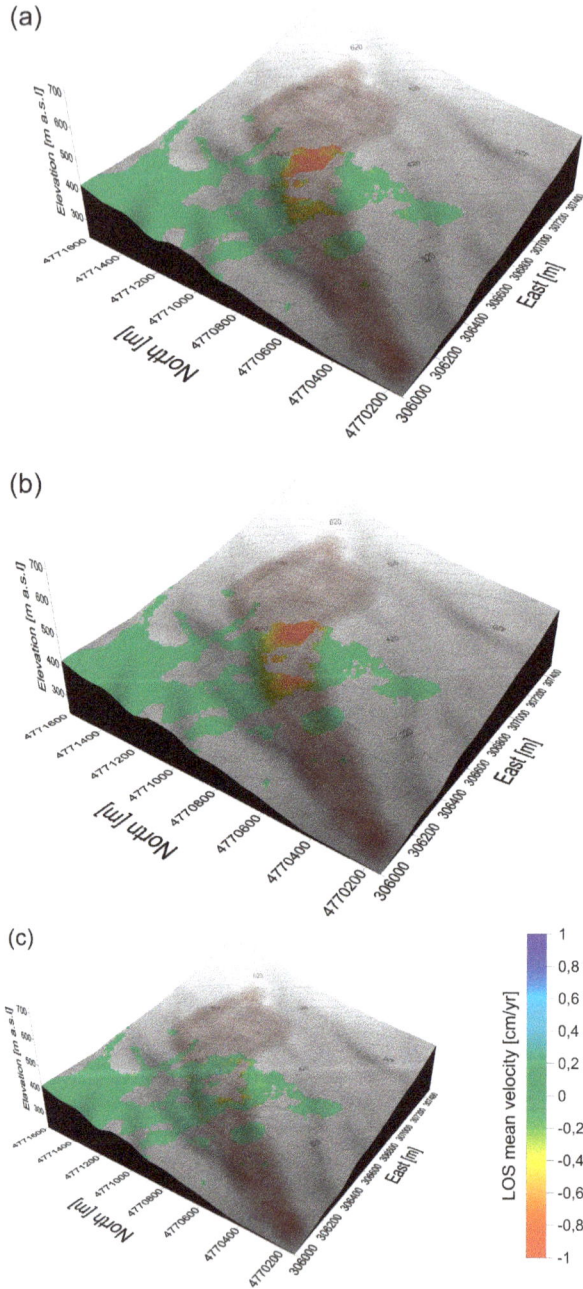

Figure 5. Geodetic inversion results: (**a**) Data; (**b**) Model and (**c**) Residuals relevant to the DInSAR velocity map superimposed on the filled contour DEM. The shading area indicates the unstable slope mass.

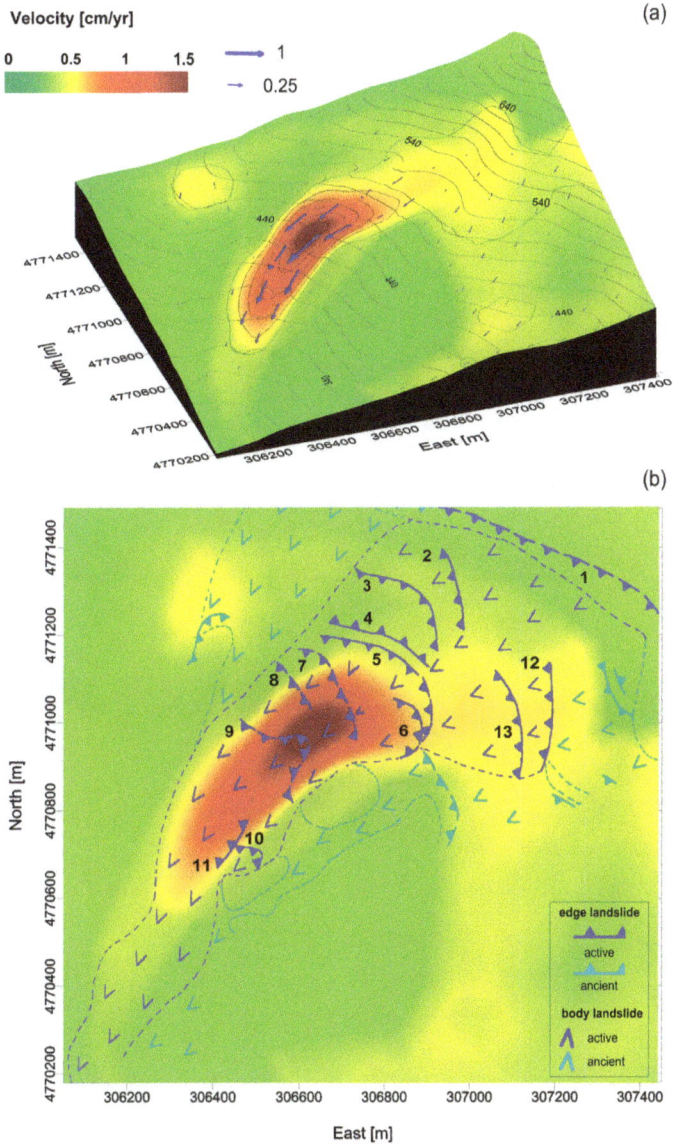

Figure 6. Analysis of FE Model results: (**a**) 3D view of modeled velocity vectors (blue arrows superimposed on the contour lines of the velocity field; (**b**) Modeled velocity map compared with the geomorphological evidences represented by the quiescent (cyan) and active (blue) elements as proposed by Pontoni [23,24].

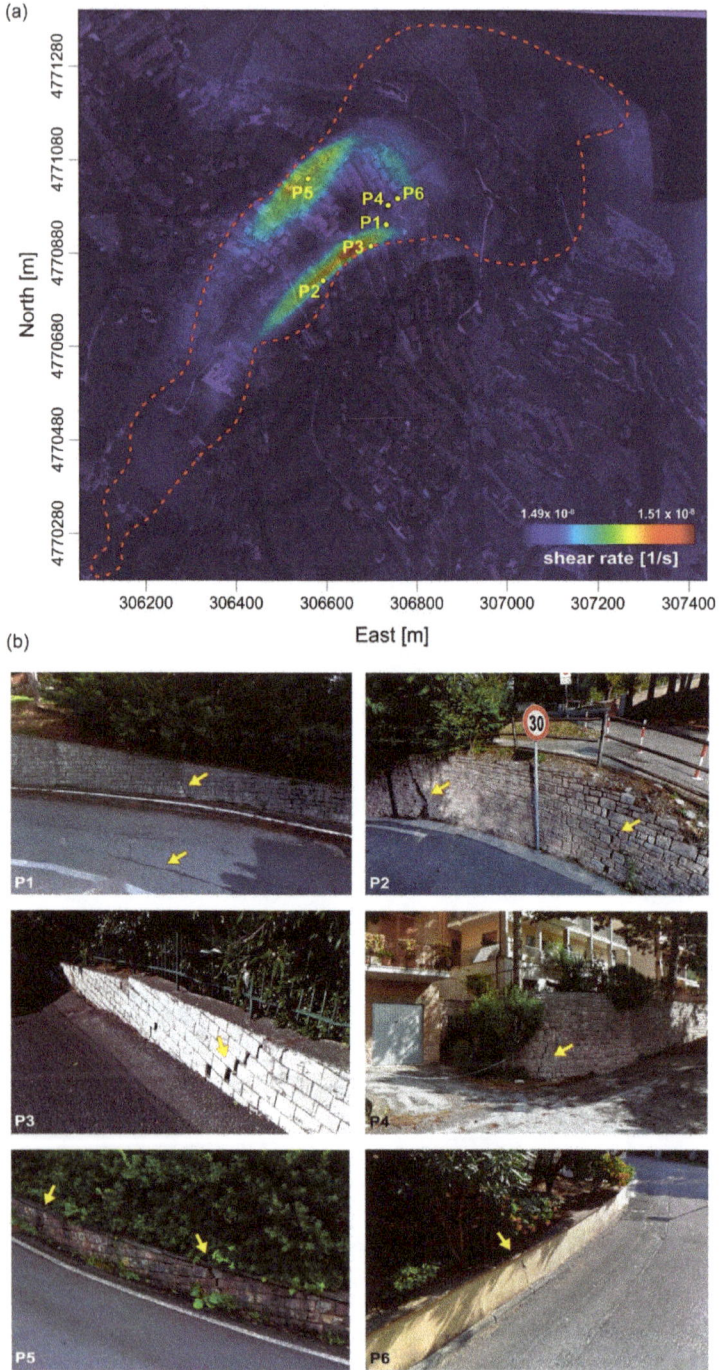

Figure 7. Shear rate map and structure damage: (**a**) Shear map superimposed on the satellite optical image of the Ivancich surrounding area (source Google Earth); (**b**) The pictures of the structure damage indicated with labels P1, P2, P3, P4, P5, P6 are shown.

5. Discussion

In this work, we perform a three-dimensional optimized model based on a fluid-dynamic physical scenario realizing an effective integration of satellite and terrestrial observations in the FE environment. In detail, based on the linearity of kinematics affecting the unstable mass as ensured by the DInSAR deformation time-series, we can assume the soil material behavior as Newtonian fluid, characterized by the viscosity parameter constant over time. Despite that, the Newtonian fluid assumption is not a rigorous constitutive model for the soil physical proprieties, it is useful to simulate the landslide kinematical behavior; we remark that this model can be applied only to a landslide-type where the shear zone geometry is well formed and well known by means of accurate and detailed in-situ investigations (Figure 2).

In this perspective, a three-dimensional kinematical model is performed, providing a full map of the current landslide velocities, both in terms of modules and corresponding directions. A spatial kinematic variability is also highlighted from the retrieved velocity map, since lower velocities are calculated along the lateral boundaries of the landslide with respect to the inner portions. The consistency of the model results with respect to the DInSAR velocity field (Figure 5) also represents a further back-validation of the 3D shear zone geometry. In addition, we compare the modeled velocity pattern with the main geomorphological features of the Ivancich slope (ancient and active edges landslide) highlighting a good match. In fact, the analysis reveals that more than the 70% of the mapped geomorphological structures are still active. The existence of a stable geological element made of in-place pelitic sandstone, along the left-hand side of the landslide at middle elevations, probably influences the direction of the phenomenon toward South-East, as supported by the geomorphological features of the study area (Figure 6). Moreover, we correlate the additional damage and the modeled shear rate. This study reveals that ruptures are most severe along the landslide boundaries between active and inactive sectors; this result is expected when the spatial gradient of the modeled velocities field in maximum (Figure 7). We argue that the information on the average landslide velocities and shear rate can provide an important contribution into predicting possible deformation scenarios for structures and infrastructures lying within the landslide area and for the land management of the whole area. Our results furnish a promising support for risk analysis of very slow landslides within a reasonable time span and for which the analysis can be considered as an expected representation of the coming years.

In order to show the advance of our 3D modeling respect to the 2D FE ones [8,18], we perform a comparative analysis between the retrieved velocity fields reported in Supplementary material (Figure S3). The achieved results point out a significant improvement of the residual values. It is worthwhile pointing out that the 3D optimized viscosity values are lower than those achieved via 2D modeling. These discrepancies could be related to: (i) the effect of the detailed shear zone geometry; (ii) a different confining effect and stress field of the three-dimensional model respect to the two-dimensional one. However, these results emphasize how the modeling approach is strongly influenced by the type of the performed dimensional analysis.

6. Conclusions

The achieved results clearly demonstrate how the integration of remote sensing, in-situ monitoring and FE modeling is strategic in performing detailed analyses of complex deformation scenarios. By benefiting from three-dimensional advanced modeling tools we can effectively investigate the overall mechanism of the evolution of the ground displacements related to slow landslide phenomena, as well as outline possible risk scenarios. As a matter of fact, if in-situ measurements or DInSAR-based analyses are capable of exploring the displacement trends of specific sectors of a landslide process, the three-dimensional model can provide a thorough physically-based conceptualization of the landslide kinematics, provided that the same model is calibrated and optimized against the aforementioned measurements.

The calibrated three-dimensional FE model provides more opportunities as a zonation tool capable of identifying and updating the status of the morphological structures originally mapped from in-situ investigations. The amplitude of the velocity field reaches a maximum value of 1.5 cm/yr at altitudes between 320 and 400 m a.s.l., in the middle-upper part of the unstable area; more than the 70% of the mapped geomorphological structures are enclosed in the region where the 2009–2012 deformation process is active. The model fully delineates the landslide portions that are subjected to different landslide velocities and this can be useful for land management purposes; therefore, the comparison between the 3D vs. 2D model results highlights that the sliding surface, calculated via the 3D geo-lithological model, is more realistic in respect to those realized in the 2D model.

Finally, we remark that the damage analysis detected in the Ivancich surroundings compared with the three-dimensional FE model results, reveals that ruptures are most severe along the boundary between active and inactive landslide sectors where the shear rate values are high. Hence, the proposed three-dimensional FE modeling tool represents a valuable support for landslide risk analyses and urban development planning within a specific territory area affected by complex slow-moving landslide processes.

Supplementary Materials: The following are available online at http://www.mdpi.com/2072-4292/8/8/670/s1. Figure S1: (**a**) CSK mean LOS velocity map superimposed on the 3D view of Satellite optical image with the DEM contour lines (grey lines). The shade of red shows the zone involved by the recent instability phenomenon. The blue triangles indicate the four SAR pixels of the CSK DInSAR measurements; (**b**) The plots allow comparing the time series of four SAR pixels (from upstream to downstream) and the fluid dynamic model projected in LOS. Figure S2: Limit Equilibrium (LE) analyses carried out in order to assess the stability conditions of the slope, based on the available geological, geomorphological and geotechnical dataset. In particular, a pore pressure distribution defined according to a finite element seepage analysis in agreement with the available piezometer measurements has been assigned in the LE calculation. According to the results of direct shear tests performed on samples taken at the depth of the sliding surface, a cohesion intercept equal to $c' = 0$ and a friction angle of $\varphi' = 15°$ have been considered as operative values of the shear strength parameters along the shear band and have been used for the LE analysis. The calculation results indicate that the stability factor of a landslide body corresponding to the central portion of the examined landslide, i.e. the most active zone of the landslide area, is FS = 0.99, whereas FS is equal to 1.02 for the whole landslide body. This confirms that the whole landslide body is at LE conditions. Figure S3: (**a**) Modelled LOS velocity map. The black line shows the SS' trace, already presented in [18]; (**b**) Extrapolated profile along SS' from 3D model vs DInSAR measurements; (**c**) model and DInSAR measurements from [18]; (**d**) residuals comparison.

Acknowledgments: This work has been supported by EC FP7 LAMPRE project (contract No. 312384) and the Italian Department of Civil Protection. Part of the presented research has been carried out through the I-AMICA (Infrastructure of High Technology for Environmental and Climate Monitoring-PONa3_00363) project of Structural improvement financed under the National Operational Programme (NOP) for "Research and Competitiveness 2007–2013", co-funded with European Regional Development Fund (ERDF) and National resources. The Cosmo-SkyMed satellite images were acquired within the EC FP7 DORIS project (contract No. 242212). The Digital Elevation Model of the investigated zone was acquired through the SRTM archive.

Author Contributions: V.D.N., R.C., P.L. and P.T. conceived and organized the research activity. M.M. processed the SAR images and supervised the integration of the satellite measurements within the modeling procedure. R.C. and V.D.N. performed the optimization modeling procedure. P.T. supervised the research activity. All authors co-wrote the manuscript.

Conflicts of Interest: The authors declare no conflict of interest.

References

1. Herrera, G.; Fernandez-Merodo, J.A.; Mulas, J.; Pastor, M.; Luzi, G.; Moserrat, O. A landslide forecasting model using ground based SAR data: The Portalet case study. *Eng. Geol.* **2009**, *105*, 220–230. [CrossRef]
2. Manconi, A.; Casu, F.; Ardizzone, F.; Bonano, M.; Cardinali, M.; De Luca, C.; Gueguen, E.; Marchesini, I.; Parise, M.; Vennari, C.; et al. Brief Communication: Rapid mapping of landslide events: The 3 December 2013 Montescaglioso landslide, Italy. *Nat. Hazards Earth Syst. Sci.* **2014**, *14*, 1835–1841. [CrossRef]
3. Ranalli, M.; Gottardi, G.; Medina-Cetina, Z.; Nadim, F. Uncertainty quantification in the calibration of a dynamic viscoplastic model of slow slope movements. *Landslides* **2010**, *7*, 31–41. [CrossRef]
4. Comegna, L.; Picarelli, L.; Bucchignani, E.; Mercogliano, P. Potential effects of incoming climate changes on the behaviour of slow active landslides in Italy. *Landslides* **2013**, *13*, 373–391. [CrossRef]

5. Bernardie, S.; Desramaut, N.; Malet, J.-P.; Gourlay, M.; Grandjean, G. Prediction of changes in landslide rates induced by rainfall. *Landslides* **2015**, *12*, 481–494. [CrossRef]

6. Conte, E.; Donato, A.; Troncone, A. A finite element approach for the analysis of active slow-movine landslides. *Landslides* **2014**, *11*, 723–731. [CrossRef]

7. Fernández-Merodo, J.A.; García-Davalillo, J.C.; Herrera, G.; Mira, P.; Pastor, M. 2D viscoplastic finite element modeling of slow landslides: The Portalet case study (Spain). *Landslides* **2012**, *11*, 29–42. [CrossRef]

8. Castaldo, R.; Tizzani, P.; Lollino, P.; Calò, F.; Ardizzone, F.; Lanari, R.; Manunta, M. Landslide kinematical analysis through inverse numerical modeling and differential SAR interferometry. *Pure Appl. Geophys.* **2014**, *172*, 1–14.

9. Schädler, W.; Borgatti, L.; Corsini, A.; Meier, J.; Ronchetti, F.; Schanz, T. Geomechanical assessment of the Corvara earthflow through numerical modeling and inverse analysis. *Landslides* **2015**, *12*, 495–510.

10. Sansosti, E.; Casu, F.; Manzo, M.; Lanari, R. Space-borne radar interferometry techniques for the generation of deformation time series: An advanced tool for Earth's surface displacement analysis. *Geophys. Res. Lett.* **2010**, *37*, 1–9. [CrossRef]

11. Bovenga, F.; Wasowski, J.; Nitti, D.O.; Nutricato, R.; Chiaradia, M.T. Using COSMO/SkyMed X-band and ENVISAT C-band SAR interferometry for landslides analysis. *Remote Sens. Environ.* **2012**, *119*, 272–285. [CrossRef]

12. Colesanti, C.; Wasowsky, J. Investigating landslides with space-borne Synthetic Aperture Radar (SAR) interferometry. *Eng. Geol.* **2006**, *88*, 173–199. [CrossRef]

13. Herrera, G.; Gutiérrez, F.; García-Davalillo, J.C.; Guerrero, J.; Notti, D.; Galve, J.P.; Cooksley, G. Multi-sensor advanced DInSAR monitoring of very slow landslides: the Tena valley case study (central Spanish Pyrenees). *Remote Sens. Environ.* **2013**, *128*, 31–43. [CrossRef]

14. Usai, S. A least-squares database approach for SAR interferometric data. *IEEE Trans. Geosci. Remote Sens.* **2003**, *41*, 753–760. [CrossRef]

15. Ferretti, A.; Prati, C.; Rocca, F. Permanent Scatterers in SAR interferometry. *IEEE Trans. Geosci. Remote Sens.* **2001**, *39*, 8–20. [CrossRef]

16. Berardino, P.; Fornaro, G.; Lanari, R.; Sansosti, E. A new algorithm for surface deformation monitoring based on small baseline differential SAR interferograms. *IEEE Trans. Geosci. Remote Sens.* **2002**, *40*, 2375–2383. [CrossRef]

17. Lanari, R.; Mora, O.; Manunta, M.; Mallorqui, J.; Berardino, P.; Sansosti, E. A small baseline approach for investigating deformations on full resolution differential SAR interferograms. *IEEE Trans. Geosci. Remote Sens.* **2004**, *42*, 1377–1386. [CrossRef]

18. Calò, F.; Ardizzone, F.; Castaldo, R.; Lollino, P.; Tizzani, P.; Guzzetti, F.M. Manunta, Enhanced landslide investigations through advanced DInSAR techniques: The Ivancich case study, Assisi, Italy. *Remote Sens. Environ.* **2014**, *142*, 69–82. [CrossRef]

19. Cruden, D.M.; Varnes, D.J. Landslide types and processes. In *Landslides, Investigation and Mitigation*; Transportation Research Board Special Report 247; Turner, A.K., Schuster, R.L., Eds.; National Academy Press: Washington, DC, USA, 1996; pp. 36–75.

20. Angeli, M.G.; Pontoni, F. The innovative use of a large diameter microtunneling technique for the deep drainage of a great landslide in an inhabited area: The case of Assisi (Italy). In *Landslides in Research, Theory and Practice*; Thomas Telford Publisher: London, UK, 2000; pp. 1666–1672.

21. Canuti, P.; Marcucci, E.; Trastulli, S.; Ventura, P.; Vincenti, G. Studi per la stabilizzazione della frana di Assisi. In Proceedings of the National Geotechnical Congress, Bologna, Italy, 14–16 May 1986; Volume 1, pp. 165–174.

22. Felicioni, G.; Martini, E.; Ribaldi, C. *Studio dei Centri Abitati Instabili in Umbria*; Pubbl. n° 979 del CNR-GNDCI; Rubettino Publisher: Rome, Italy, 1996. (In Italian)

23. Pontoni, F. Geoequipe Studio Tecnico Associato Geologia–Ingegneria, Italy. Unpublished Technical Report, 1999. (In Italian).

24. Pontoni, F. Geoequipe Studio Tecnico Associato Geologia–Ingegneria, Italy. Unpublished Technical Report, 2011. (In Italian).

25. Fastellini, G.; Radicioni, F.; Stoppini, A. The Assisi landslide monitoring: A multi-year activity based on geomatic techniques. *Appl. Geomat.* **2011**, *3*, 91–100. [CrossRef]

26. Bovenga, F.; Nitti, D.O.; Fornaro, G.; Radicioni, F.; Stoppini, A.; Brigante, R. Using C/X-band SAR interferometry and GNSS measurements for the Assisi landslide analysis. *Int. J. Remote Sens.* **2013**, *34*, 4083–4104. [CrossRef]

27. Franceschetti, G.; Lanari, R. *Synthetic Aperture Radar Processing*; CRC Press: Boca Raton, FL, USA, 1999.

28. Bonano, M.; Manunta, M.; Pepe, A.; Paglia, L.; Lanari, R. From previous C-band to new X-band SAR systems: Assessment of the DInSAR mapping improvement for deformation time-series retrieval in urban areas. *IEEE Trans. Geosci. Remote Sens.* **2013**, *51*, 1973–1984. [CrossRef]

29. Arangio, S.; Calò, F.; di Mauro, M.; Bonano, M.; Marsella, M.; Manunta, M. An application of the SBAS-DInSAR technique for the assessment of structural damage in the city of Rome. *Struct. Infrastruct. Eng.* **2014**, *10*, 1–15. [CrossRef]

30. Guzzetti, F.; Manunta, M.; Ardizzone, F.; Pepe, A.; Cardinali, M.; Zeni, G.; Reichenbach, P.; Lanari, R. Analysis of ground deformation detected using the SBAS-DInSAR technique in Umbria, Central Italy. *Pure Appl. Geophys.* **2009**, *166*, 1425–1459. [CrossRef]

31. Manunta, M.; Marsella, M.; Zeni, G.; Sciotti, M.; Atzori, S.; Lanari, R. Two-scale surface deformation analysis using the SBAS-DInSAR technique: A case study of the city of Rome, Italy. *Int. J. Remote Sens.* **2008**, *29*, 1665–1684. [CrossRef]

32. Scifoni, S.; Bonano, M.; Marsella, M.; Sonnessa, A.; Tagliafierro, V.; Manunta, M.; Lanari, R.; Ojha, C.; Sciotti, M. On the joint exploitation of long-term DInSAR time series and geological information for the investigation of ground settlements in the town of Roma (Italy). *Remote Sens. Environ.* **2016**, *182*, 113–127. [CrossRef]

33. Notti, D.; Calò, F.; Cigna, F.; Manunta, M.; Herrera, G.; Berti, M.; Meisina, C.; Tapete, D.; Zucca, F. A user-oriented methodology for DInSAR time series analysis and interpretation: Landslides and subsidence case studies. *Pure Appl. Geophys.* **2015**, *172*, 3081–3105. [CrossRef]

34. Ranalli, G. *Rheology of the Earth*; Chapmann & Hall: London, UK, 1995.

35. Zienkiewicz, O.C.; Taylor, R.L. *The Finite Element Method: Basic Formulation and Linear Problems*; McGraw-Hill: Burlington, MA, USA, 1988; Volume 1.

36. Zhang, Z.; Zhu, J.Z. Analysis of the superconvergent patch recovery technique and a posteriori error estimator in the finite element method (II). *Comput. Methods Appl. Mech. Engrgy.* **1998**, *163*, 159–170. [CrossRef]

37. Griffiths, D.V.; Lane, P.A. Slope stability analysis by finite elements. *Geotechnique* **1999**, *49*, 387–403. [CrossRef]

38. Tarantola, A. *Inverse Problem Theory and Methods for Model Parameter Estimation*; Society for Industrial and Applied Mathematics (SIAM): Philadelphia, PA, USA, 2005; pp. 1–343.

39. Sen, M.K.; Stoffa, P.L. *Global Optimization Methods in Geophysical Inversion*; Cambridge University Press: Cambridge, UK, 2013.

MDPI AG

St. Alban-Anlage 66

4052 Basel, Switzerland

Tel. +41 61 683 77 34

Fax +41 61 302 89 18

http://www.mdpi.com

Remote Sensing Editorial Office

E-mail: remotesensing@mdpi.com

http://www.mdpi.com/journal/remotesensing

www.ingramcontent.com/pod-product-compliance
Lightning Source LLC
Chambersburg PA
CBHW051708210326
41597CB00032B/5405